HYDRAULICS OF GROUNDWATER

HYDRAULICS OF GROUNDWATER

JACOB BEAR

Department of Civil and Environmental Engineering
Technion–Israel Institute of Technology
Haifa, Israel

DOVER PUBLICATIONS, INC.
Mineola, New York

Bibliographical Note

This Dover edition, published in 2007, is an unabridged, slightly corrected republication of the work published by McGraw-Hill, Inc., New York, 1979.

Library of Congress Cataloging-in-Publication Data

Bear, Jacob
    Hydraulics of groundwater / Jacob Bear.
        p. cm.
    Originally published: New York : McGraw-Hill, 1979. With corrections.
    Includes bibliographical references and index.
    ISBN 0-486-45355-3 (pbk.)
        1. Groundwater flow. 2. Groundwater. 3. Hydrodynamics. I. Title.

GB1197.7 .B4 2007
551.49—dc22

                                                                2006050786

Manufactured in the United States of America
Dover Publications, Inc., 31 East 2nd Street, Mineola, N.Y. 11501

To
Siona, Eitan, Alon, and Iris

HYDRAULICS OF GROUNDWATER

CONTENTS

Preface xi

1 Introduction 1

    1-1 Role of Groundwater in Water Resource Systems and Their Management 2
    1-2 Management of Groundwater Resources 10
    1-3 Scope and Organization of Book 13

2 Groundwater and Aquifers 19

    2-1 Definitions 19
    2-2 Moisture Distribution in a Vertical Profile 22
    2-3 Classification of Aquifers 24
    2-4 Hydraulic Approach to Flow in Aquifers 26
    2-5 Continuum Approach to Flow Through Porous Media 28
    2-6 Inhomogeneity and Anisotropy 31

3 Groundwater Balance 34

    3-1 Groundwater Flow and Leakage 35
    3-2 Natural Replenishment from Precipitation 37
    3-3 Return Flow from Irrigation and Sewage 41
    3-4 Artificial Recharge 42
    3-5 River–Aquifer Interrelationships 51
    3-6 Springs 53
    3-7 Evapotranspiration 57
    3-8 Pumpage and Drainage 58
    3-9 Change in Storage 59
    3-10 Regional Groundwater Balance 59

4   Groundwater Motion   60

    4-1   Darcy's Law   60
    4-2   Hydraulic Conductivity   66
    4-3   Aquifer Transmissivity   69
    4-4   Flow in Anisotropic Aquifers   71
    4-5   Dupuit Assumptions for a Phreatic Aquifer   74

5   Mathematical Statement of the Groundwater Forecasting Problem   83

    5-1   Aquifer Storativity   84
    5-2   Basic Continuity Equation   89
    5-3   Initial and Boundary Conditions   94
    5-4   Fundamental Equations for Flow in Aquifers   103
    5-5   Complete Mathematical Statement of a Groundwater Flow Problem   116
    5-6   Methods for Solving Groundwater Flow Problems   124
    5-7   Superposition   152
    5-8   Hydrologic Maps and Flow Nets   160
    5-9   Relationships between Flows in Isotropic and Anisotropic Aquifers   169
    5-10 Analytical Solutions of One-Dimensional Steady Flows in Aquifers   176
    5-11 Land Subsidence   184

6   Flow in the Unsaturated Zone   190

    6-1   Capillary Pressure and Retention Curves   191
    6-2   The Motion Equation   206
    6-3   Relative Permeability   209
    6-4   The Continuity Equations   213
    6-5   Mathematical Statement of Unsaturated Flow   218
    6-6   Methods of Solution   221

7   Groundwater Quality Problem (Hydrodynamic Dispersion)   225

    7-1   Dispersion Phenomena   227
    7-2   Occurrence of Dispersion Phenomena   231
    7-3   Coefficients of Dispersion   231
    7-4   The Equation of Hydrodynamic Dispersion   239
    7-5   Initial and Boundary Conditions   248
    7-6   Mathematical Statement of a Groundwater Pollution Problem   250
    7-7   Aquifer Dispersion Equation and Parameters   252
    7-8   Methods of Solution   259
    7-9   Some Simple Analytic Solutions   263
    7-10 Movement of Water Bodies Injected into Aquifers   276

8 Hydraulics of Pumping and Recharging Wells 300

8-1 Introduction 301
8-2 Steady Flow to a Well in a Confined Aquifer 304
8-3 Steady Flow to a Well in a Phreatic Aquifer 308
8-4 Steady Flow to a Well in a Leaky Confined Aquifer 312
8-5 Unsteady Flow to a Well in a Confined Aquifer 318
8-6 Unsteady Flow to a Well in a Phreatic Aquifer 331
8-7 Unsteady Flow to a Well in a Leaky Confined Aquifer 339
8-8 Partially Penetrating Wells 344
8-9 Multiple Well Systems 350
8-10 Wells Near Boundaries Treated by the Method of Images 357
8-11 Recharging and Pumping Wells in Uniform Flow 367
8-12 Well Losses and Specific Well Discharge 374
8-13 Hydraulics of Recharging Wells 377

9 Fresh Water–Salt Water Interface in Coastal Aquifers 379

9-1 Occurrence 379
9-2 Exact Mathematical Statement of the Problem 381
9-3 The Ghyben–Herzberg Approximation 384
9-4 Continuity Equation Based on the Dupuit Assumption 386
9-5 Stationary Interface 393
9-6 Approximate Solutions for a Moving Interface 406
9-7 Upconing Below a Well Pumping Above an Interface 414
9-8 Transition Zone 433

10 Modeling of Aquifer Systems 436

10-1 The Need for Aquifer Models 437
10-2 Model Calibration 439
10-3 Classification of Aquifer Models 440
10-4 Single Cell Models 442
10-5 Multiple Cell Models 447
10-6 Two Examples of Water Quality Models 455

11 Identification of Aquifer Parameters 463

11-1 Statement of the Inverse Problem 463
11-2 Pumping Tests 465
11-3 Inverse Methods for Determining Regional Aquifer
       Parameters 482

12 Use of Linear Programming in Aquifer Management 491

12-1 Brief Review of Linear Programming 492
12-2 Application of Linear Programming to Aquifer Management 495
12-3 Examples 501

Appendix A Derivation of the Basic Transport Equation by Averaging   513

Appendix B Averaging Along the Vertical   521

Problems (selected chapters)   523

Bibliography   541

Author Index   557

Subject Index   563

PREFACE

Groundwater plays an important role in the development and management of water resources. Consequently, there is an ever increasing demand for information on groundwater hydrology and on the hydraulics of the movement of water in aquifers.

In this book, an attempt is made to present the laws and equations which govern the flow and storage of groundwater, and of contaminants carried with the water, in aquifers. This is not a book on the theory of flow through porous media. For that, the reader is referred to appropriate texts, including my book, *Dynamics of Fluids in Porous Media*. My main objective in this book is to provide groundwater hydrologists, as well as engineers and planners who deal with the development and management of groundwater resources, with all the tools they need for forecasting the behavior of a regional aquifer system. The point of view is the practical one, bearing in mind the difficulties involved in defining the aquifer system and the conditions along its boundaries. Hence, the emphasis on the *hydraulic approach*, in which the flow in an aquifer is visualized essentially as a two-dimensional one in the horizontal plane.

This is also not a book on management in the sense of making decisions in order to achieve specified goals, subject to specified constraints. However, any procedure for making decisions, for example, with respect to the areal and temporal distributions of pumping from and artificial recharge of an aquifer, must be based on a knowledge of the aquifer's response (e.g., in the form of water levels, or concentration of contaminants in the aquifer) to the planned activities. The material presented here aims at providing the decision maker with the tools for determining this response. In this book, however, the emphasis is on ideas, principles, and the *accurate mathematical statement of the flow problem*, rather than on the actual solutions which depend in each case on the geometry of boundaries, on the specific values of the parameters, etc. Only seldom can such solutions for *regional* cases of practical interest be derived by analytic methods of solution.

Although the electric analog (RC-network) can provide solutions to the regional forecasting problem, most such problems are solved nowadays numerically by means of digital computers. An important advantage in solving the forecasting problem by the digital computer is that it enables the forecasting and the management problems to be solved simultaneously, as indeed they should. An introduction to numerical techniques is therefore presented. This is the background for the way the subject matter is presented in this book.

Although in writing this book and in selecting the material to be included in it, I had the practical aspects in mind, the various equations are rigorously developed from first principles. I believe that in doing so, a better insight is gained as to the physical phenomena taking place and the underlying assumptions.

Both the problem of water quantity (e.g., as expressed by water levels, or heads) and that of water quality (expressed, for example, in terms of concentration of dissolved solids) are treated in this book. Special attention is given to the problem of sea water intrusion encountered in coastal aquifers.

I believe that the subject matter I chose to include here should be sufficient for those who deal with the practical aspects of flow in aquifers. No attempt has been made to cover *everything*, as, obviously, such attempt would be hopeless to begin with. Instead, an attempt was made to present a complete general picture and especially to cover most principles, to present a methodology for treating groundwater problems and to include material which will be educational. On the basis of what is presented here, the reader should be able, when necessary, to pursue further studies and to understand advanced material continuously appearing in the professional literature.

The amount of mathematics required is minimal, as the emphasis is on the physical phenomena and the statement of the problems rather than on the solutions. However, the *mathematical language* is used. The reader is expected to know calculus, partial derivation, and the concepts of a vector, a gradient of a scalar, divergence of a vector, and what is a second rank tensor.

In writing this book, I had also in mind the problem of teaching this material, whether at a graduate or an undergraduate level, following an introductory course in hydrology. From my experience of teaching this subject as formal credit courses at the Technion—Israel Institute of Technology, Haifa, Israel, at M.I.T., at SUNY at Buffalo, at Princeton University, at the University of Hawaii in Honolulu, and at a large number of courses for engineers in the United States and around the world, I feel that a book devoted solely to the presentation of this subject in the way outlined above is missing. Many teachers use selected chapters in different books as their text or use no text at all, referring their students only to papers in professional journals. I believe that this book can serve as a single textbook for teaching groundwater flow. My experience shows that it can be taught as a regular 3-credit course during one semester of 12–14 weeks.

Finally, I should like to express my thanks to those who helped me to complete this book. Many thanks are due to my students at various places for their constructive remarks. Special thanks are due to the Departments of Civil Engineering at the State University of New York at Buffalo (especially to the Chairman of the department, Professor G. Lee and to Professor R. R. Rumer) and at

Princeton University (especially to the Chairman of the department, Professor A. S. Cakmak and to Professor G. F. Pinder) who made it possible for me to write most of this book while spending two fruitful semesters (in 1973 and 1976, respectively) as visiting professor in these departments.

1978   Jacob Bear
                                                                      Haifa, Israel

ONE

INTRODUCTION

In this book we intend to treat only groundwater, referring to both its quantity and quality. Obviously, dividing the water resources of a region into surface water and groundwater is often artificial, or questionable (and unwise from the management point of view). For example, where should spring water be placed? It is groundwater emerging at the ground surface and becoming surface runoff, yet by an appropriate groundwater management policy, water levels in the aquifer in the vicinity of a spring can be maintained below the spring's outlet such that the spring will dry up. The water previously emerging from the spring will now be stored in the aquifer. The same is true when an aquifer is in hydraulic contact with a river, or a lake. By manipulating water levels in the aquifer, we may affect the flow in the river and vice versa. In spite of these comments, we shall focus our attention on groundwater only, as our primary objective is to discuss the hydraulics of groundwater.

The main goal of the groundwater hydrologist, water resources engineer, or planner (or planning team), who deals with a groundwater system, or with a water resource system of which groundwater is a component, is the *management of the groundwater system*. Simply stated, and using the terminology of systems analysis only loosely, management of a system means making various *decisions* (that is, assigning numerical values to decision variables) aimed at modifying the *state* of a considered system. Location, rate, and time of pumping, or artificially recharging an aquifer, are examples of *decision variables*. Water levels, depth of land subsidence, and solute concentrations, as functions of location and time are examples of *state variables*. Our reason for modifying the state of a considered system, that is, to bring it from its existing state to another, more desirable one, is to achieve certain goals and objectives. We also wish to do so by the *best* set of decisions (= *policy*). This implies the existence of some criteria for comparing the outputs of the system (e.g., costs, pumping) and selecting the best policy. We may refer to this activity as solving the *management problem*.

However, in order to solve the management problem, we must be able to predict the *response* of the system to any proposed operation policy, and to obtain from it the new state of the system, given its initial one. Once the new state is known we can check whether it is feasible at all, that is, does not violate any of the constraints imposed on the system. Then we can compare outputs and responses in order to select, according to some criteria, the best policy. Referring now to a groundwater system, we have to be able to *forecast* water levels, salinities, spring discharges, land subsidence, etc., that is, the system's state variables resulting from any proposed operation policy, say, of pumping and artificial recharge. We may refer to this problem of forecasting the state of the groundwater system as the *forecasting problem*. No groundwater management problem can be solved without solving first the groundwater *forecasting problem* for feasible operation policies, in order to select the best solution by comparing the corresponding responses. In the more advanced management procedures, the two problems—the forecasting problem and the management—are solved simultaneously.

In this introductory chapter we shall identify the role played by groundwater in a water resource system and discuss what is involved in the management of such a system when it is based soley or partly on groundwater. The discussion itself will be rather superficial as its sole objective is to set the stage for the chapters which follow.

The last section defines the objectives and scope of this book and describes how we intend to achieve the stated objectives.

We shall assume here, as in the entire book, that the reader is familiar with general hydrology, of both surface and groundwater, at least at an introductory level, and that he knows the hydrologic cycle and the place of groundwater in it.

1-1 ROLE OF GROUNDWATER IN WATER RESOURCE SYSTEMS AND THEIR MANAGEMENT

In order to discuss the role that groundwater may play in the management of regional water resources, let us assume that both kinds of water—surface water and groundwater—are present in relatively significant quantities in a region.

Surface Water Versus Groundwater

Actually surface water (in lakes and streams) and groundwater (in aquifers) are not necessarily separate and independent water resources. Consider, for example, the interrelations between a river (or a lake) and an adjacent aquifer, or a river passing through a region under which a phreatic aquifer exists. If the river (or lake) bed is not completely clogged, water will flow through it from the river into the aquifer when water levels in the former are higher than in the latter, and vice versa. Base flow in streams is provided by groundwater. In this way, rivers and lakes in direct continuous hydraulic contact with adjacent, or underlying aquifers serve as boundaries to the flow domain in the latter. By controlling water levels in them, we can control the flow of water through them into or out of an aquifer.

Spring discharge is another example of groundwater emerging under certain conditions at the ground surface and becoming surface runoff. By controlling groundwater levels in the vicinity of a spring, its discharge is controlled, or even stopped completely.

The above considerations apply not only to water quantity, but also to water quality, defined, for example, by some chemical species or bacteria carried with the water. Polluted surface water may easily reach and pollute groundwater.

It is thus obvious that the management of regional water resources should always include both resources simultaneously, incorporating each of them in the overall system according to its individual features. In one way or another, any control of one resource will eventually, if not immediately, affect the other. The possible time lag may be due to storage and/or the relatively slow movement of groundwater and of pollutants carried by it. One should note, however, that the *water divide* delineating a groundwater basin and that delineating a surface one are not necessarily geographically identical and that depending on the geographical boundaries of any considered region, management may include transfer of water from one basin to another within the framework of regional conjunctive use.

Although it seems obvious that groundwater, when present in a region, should be used conjunctively with surface water within the framework of any development and management scheme, one finds in many parts of the world a certain degree of reluctance to include groundwater in the development and management of water resources.

Perhaps in part, this attitude stems from the fact that unlike water in streams and lakes, one cannot actually see groundwater in aquifers. However, in trying to rationalize this attitude, the following reasons are often given (Wiener, 1972):

(a) Exploitation of groundwater is energy consuming and expensive, especially when the water table, or the piezometric surface, is deep.
(b) Planning the development of groundwater resources requires long-term data, which usually are not available.
(c) Evaluation and planning groundwater resources requires highly trained personnel which are not available.
(d) It is difficult to predict the response of an aquifer (in terms of both water quantity and quality) to proposed activities.
(e) Groundwater projects are usually single purpose ones, namely, to supply water (whereas most surface water projects have multiple purposes).

Obviously in order to examine these arguments and to compare surface water with groundwater, one must know the local conditions. In general, however, it seems that at least in part, these arguments are based on lack of knowledge. For example:

(a) It is true that when pumping heads are large, energy costs may be significant (whereas energy may be produced from surface water). However, if one includes in the annual expenditures also the relatively high investments required for hydraulic structures, such as storage dams diversions, canals, and pipes, the overall economic picture may show a clear advantage of groundwater.

(b) Because of the large storage and slow motion involved, groundwater levels at any instant reflect the accumulated effect of a rather long period of time; changes are relatively small and slow, in comparsion with those of surface water. Hence, in general, shorter groundwater records give sufficient information for planning purposes, whereas much longer records are required in order to obtain a complete picture of the more frequent and rapid fluctuating behavior of surface water.

(c) It is true that a certain amount of knowledge is required which often is not part of the usual training of engineers. Most of this knowledge has been developed in the last two to three decades. Nowadays, however, this information is included more and more in the ordinary training of hydrologists and civil and agricultural engineers, or in special courses of continuing education. Most of the necessary theory is also included in the present text, as a contribution to the dissemination of information on groundwater. Consequently, the lack of skilled personnel can easily be overcome, even in regions where this subject has been neglected in the past.

(d) With modern hydrological tools, there is no difficulty in modeling the behavior of a groundwater system and forecasting its response (both quantity and quality wise). In general, the forecasts are reliable. Digital computers are often used when the complexity of the system warrants it.

(e) One can certainly not use groundwater for recreation, as one does in a large storage reservoir. Nevertheless, and perhaps to a more limited extent, groundwater projects may also serve multiple purposes. For example, in addition to water supply, drainage and reclamation of land may be achieved. We have already mentioned above the control of base flow in streams which may be achieved by controlling groundwater levels in adjacent aquifers. Artificial recharge can be used for disposal of reclaimed sewage water, using the purifying and mixing properties of the aquifer for augmenting the exploitable groundwater quantity.

Characteristics of Groundwater

Our main purpose in bringing these arguments is not to show that groundwater utilization is always superior and more advantageous, but to emphasize again that whenever the two resources are present, they should be used conjunctively according to their individual features. Following are some of the main characteristics of groundwater (Wiener, 1972).

Location Springs occur at *points*. Surface water flows along fixed curved *paths*. Their utilization usually requires the construction of regulative facilities which will make the water available only along certain portions of their path. Groundwater, on the other hand, underlies (when at all present) extended *areas*. If these coincide also with demand areas, there is no need for a surface distribution system, as the aquifer acts also as a conduit and each consumer can pump his share directly from the aquifer. This feature is of special interest in regions where development is gradual. More wells are sunk when an increase in pumpage is required. Often control structures for surface water (e.g., dams, or diversions) cannot be built in stages.

Flow and availability Fluctuations in surface flow may be significant. Minimum flows, including zero flow, occur often during the season of highest demand. On the other hand, climatic fluctuations in groundwater levels are usually small relative to the thickness of an aquifer, so that the large volume of water stored in the aquifer may serve as a buffer and also supply water in periods of drought. Whereas the regulation of surface flow requires hydraulic structures which are often rather costly, the regulation of groundwater flow is incorporated in the implementation of management schemes, namely through an appropriate areal and temporal distribution of pumpage and artificial recharge.

The regulation of groundwater flow is, therefore, in general, much less expensive. Base flow in streams and spring flow (including the drying up of streams, which means transforming surface flow into groundwater flow) can be regulated by controlling groundwater levels in their vicinity.

Annual and seasonal variability Annual and seasonal fluctuations are much more pronounced in surface than in groundwater flow. In surface flow, this means large losses of water by spillage in periods of excess water or the need for expensive regulatory structures (e.g., dams). In groundwater flow, storage is provided by the aquifer itself; spillage due to very high water levels near an outlet is relatively small and can easily be avoided by manipulating water levels through pumping.

Energy Energy must always be expended in order to lift groundwater to the ground surface. In general, capital investments in wells are low, but operating costs (i.e., cost of electricity or fuel) are relatively high.

Quality of water In many regions groundwater does not pose major biological or physical quality problems. Surface flow is much more susceptible to man made pollution, which usually requires costly treatment for its removal. This does not mean that groundwater cannot be affected by pollutants. For example, faulty sewage pipes or oil spillage may cause severe pollution of groundwater. In certain formations, pollutants may travel large distances in an aquifer without being modified. As for mineral quality, although the range of concentrations encountered is very large, in general one may observe that groundwater is more liable to pick up minerals in solution. The removal of such minerals is usually very expensive.

When groundwater does get polluted (e.g., by polluting solutes such as leachate from land fills carried down with the water from the ground surface, or by intrusion of groundwater of inferior quality into an aquifer), the restoration of quality and the removal of pollutants by mixing with and leaching by clean groundwater is a very slow hence lengthy, sometimes practically impossible, process (i.e., the process is practically irreversible). This is due to the very slow movement of groundwater especially in layers of very fine material, imbedded in formations of higher permeability, and to adsorption and ion-exchange phenomena on the surface of the solid matrix. These phenomena are especially significant when fine grained materials, such as clay are present in an aquifer. Adsorbed species continue to be fed into the groundwater flow for prolonged periods.

On the other hand, for certain polluting elements carried with the water, the above processes of adsorption and ion exchange are an advantage as they remove them from the water. The aquifer then plays the role of a filter and a purifier, taking advantage of the adsorptive capacity of the solid matrix.

In general, there is always the trend of salinization of groundwater by solutes brought down from the surface. Under natural conditions, an equilibrium is reached by the fact that water leaving the formation carries solutes with it.

However, when a management program calls for a reduction of outflow (i.e., increased pumping) and/or the introduction of more solutes (e.g., when the aquifer is artificially recharged with water of inferior quality, or when more soluble polluting sources are introduced on the ground surface), this equilibrium is destroyed and we observe an inevitable rise in the solute concentration of the groundwater, sometimes beyond permissible limits.

Impact on drainage problems The lowering of the phreatic surface by pumping may solve drainage problems in areas where the latter are produced by a high water table. The application of surface water in such cases may require a drainage system to maintain the water table at the desired depth.

In the case of marshes or of a water table which is close to the ground surface, the lowering of the water table will also reduce evapotranspiration, thus making more water available for beneficial use.

When artificial recharge (Sec. 3-4) is implemented, one should make sure that the rising water table of a phreatic aquifer will not create drainage problems.

Land subsidence When water is pumped out of a confined aquifer the intergranular stress in the solid matrix is increased even without changing the load at the ground surface. When relatively soft layers (e.g., clay, or silt) are present within the aquifer, they are compressed and we observe land subsidence. In certain areas this subsidence is very significant and limits, or even forces the stoppage, of pumping.

Data The main source for information about the movement and accumulation in an aquifer of water and solutes (or other pollutants) carried with the water, are measurements of water levels and solute concentrations in observation wells. Spring discharge and base flow are also sources of data. As already emphasized above, even with a relatively small amount of information, preliminary conclusions as to the feasibility of development can be drawn, with more information gathered as development proceeds. The construction of more refined models of aquifer behavior (and consequently of aquifer management) requires more data, well distributed over space and time.

Staged and gradual development The fact that groundwater is withdrawn through wells, with each well adding an increment of annual withdrawal, often located at the actual area of consumption, makes it easy to develop groundwater stage by stage as needs arise, or according to a development plan. Only a relatively small investment is required at each incremental development. Each facility is added when actually required. The economy of scale, so important in surface flow regulative structures, does not exist in this case. A large storage dam may not be fully utilized before a long period after its construction has elapsed, depending on the rate of growth of consumption.

Legal and institutional aspects Because large scale groundwater developments are relatively recent, an appropriate legal and institutional framework is in many instances rudimentary or nonexistent. This is so even in regions where this framework for the management of surface water is well established. Because of the interrelations between surface water and groundwater, and the recommended conjunctive use of both wherever called for by the local conditions, the two kinds of water should be incorporated within a single unified legal and administrative framework.

In establishing the legal and administrative framework for the exploitation of groundwater, it is important to consider some of its basic features. The entire aquifer may be regarded as a single basin from the point of view of its water balance. In the long run all consumers together cannot withdraw more than is made available by the water balance which takes into account all inputs (natural and artificial) and all outputs. Temporarily, excess of outflow over inflow is provided by reduction in storage. By pumping, each well produces a drawdown also in its vicinity and may affect the pumping of neighboring wells. The aquifer is also a basin with a certain internal flow pattern established by the pattern of pumping and recharge. Pollutants reaching the aquifer will be transported according to this flow pattern, and may reach areas at large distances from where they were introduced originally. In this way many wells downstream of a source of pollution may be affected.

Finally, the problems of groundwater quantity and quality cannot be handled separately, as they are interrelated. Sea water intrusion, for example, depends on the rate of fresh water flow from the aquifer to the sea. The movement of contaminants depends on the flow pattern, etc.

All these considerations lead to the conclusion that the management of an aquifer should be centralized and that it requires an appropriate legal and institutional framework. One cannot leave individual land owners to pump according to their needs, or to allow them to dump pollutants on their land.

Functions of Aquifers

It should be obvious from the discussion presented above that an aquifer is a system which is managed and operated as a unit to achieve various objectives, beyond serving as just a source for water. Let us briefly review some of the functions of aquifers.

Source for water This is the more obvious function. When an aquifer contains water stored in it in the far past, usually under different climatic conditions, its water should be considered a nonrenewable resource. However, in general, an aquifer is replenished annually from precipitation over the region overlying it or over its intake region (if it is a confined one). Thus, in general, groundwater is a *renewable* resource. Obviously only a certain part of the precipitation, depending on the distribution of storms, land topography and cover, permeability of soil, etc., infiltrates through the ground surface and replenishes the underlying phreatic aquifer. Aquifers can also be replenished from streamflow (with permeable beds) and floods. In many arid regions, aquifers in the low lands are replenished during a very short period once in several years, from flash floods originating in the mountains.

Under natural conditions, a quasi-equilibrium situation is maintained with inflow equal to outflow. Part of the replenishment can be intercepted by pumping thus reducing the outflow and establishing a new quasi-equilibrium. In this way the aquifer serves as a source for water.

Storage reservoir Every water resource system requires storage, especially when replenishment is intermittent and is subject to random fluctuations. A large volume of storage is available in the void space of phreatic aquifers. Just to give a rough idea, we can store 15×10^6 m^3 of water in a portion of an aquifer of 10 km $\times$ 10 km, with a storativity of 15 percent, raising the water table by 1 m. Using the technique of artificial recharge, large quantities of water can thus be stored in a phreatic aquifer. By doing so, water levels rise and outward (from the recharge area) gradients are established. These cause the stored water to spread over ever increasing areas and/or to leave the aquifer through its boundaries. Thus the stored water is gradually lost if not used. Nevertheless, due to the slow movement of groundwater, and with appropriate management schemes, these losses can be minimized. A possible management procedure is first to lower the water table by pumping in excess of natural replenishment, withdrawing the volume of water as a *one time reserve* stored in the aquifer between the initial and final water levels, and then use the dewatered zone for storage. In this way, storage is provided without raising the water table to too high elevations. Sometimes, we even start by producing a crater in the groundwater table and then filling it up for storage. Again, losses from storage are minimized (recalling always that losses exist also from surface storage by evaporation and seepage).

An aquifer can be used for long term storage, e.g., from a wet sequence of years to a dry one, for seasonal storage, from the rainy season to the dry one, or even for shorter periods. The selection of the type of storage, the right combination of surface and underground storage, etc., depends on local conditions, economics, etc.

The aquifer as a conduit Using the techniques of artificial recharge, water can be introduced into an aquifer at one point and be withdrawn by pumping at another point (or several other points). The injected water will flow through the aquifer from the high water levels of the region of recharge to the region of pumping, where water levels are lower, the rate of flow depending also on the aquifer's transmissivity. Large distribution systems, say to individual consumers spread out over large areas, may be avoided in this way. Obviously, there is a limit to the permissible rise in water levels, as well as to permissible drawdowns; these impose a limit to the use of an aquifer as a conduit.

The aquifer as a filter plant Using the techniques of artifical recharge, an aquifer may serve as a filter and purifier for water of inferior quality injected into it. This may take several forms:

1. By recharging an aquifer (through infiltration ponds) with surface water containing fine suspended load, we remove the latter by the time the water reaches the water table. The bottom of the pond and the soil column act as a filter.
2. Various chemicals may be removed by chemical reactions, by adsorption and by ion exchange phenomena on the solid surface of the porous matrix, especially when clay colloids are present. Of special interest is the reduction in organic matter content as well as the removal of taste, or bacteria and viruses, especially if the flow is of sufficient length and duration. Sometimes, minerals are added to the water by solution.
3. Mixing of injected water with indigenous water of an aquifer is achieved by their simultaneous movement in the aquifer due to both the mechanism of hydrodynamic dispersion and the geometry of the flow pattern.
4. Pumping near a river induces recharge from the latter into the aquifer. Filtering of the river water and purification are achieved by the flow through the aquifer material from the river to the wells.

In each case, the ability of an aquifer to upgrade the quality of water depends on the chemical and physical properties of the aquifer material and on the type of mineral and organic impurities contained in the water, taking into account that a significant part of the removal of impurities takes place at the phase of entry into the soil material (that is, bottom of an infiltration pond or vicinity of a recharge well).

Control of base flow This can be achieved in springs and streams by controlling water levels in the aquifer supplying water to them.

Aquifer as a water mine We have already mentioned above the possibility of mining a *one time reserve* stored in an aquifer between some initial phreatic surface and a planned ultimate one. The same is true (using some coefficient of efficiency of mining, due to hydrodynamic dispersion) in the case of the advancement of the interface in a coastal aquifer toward its planned position.

In general, the yield of an aquifer is a long term average of part of its replenishment (renewable resource). However, under certain conditions albeit very rarely, we may plan to completely mine an aquifer (like any other non-renewable resource), not worrying about what will happen once this source has been depleted. In this case mining is usually based only on economic considerations.

We have thus summarized the roles that groundwater can play in the management of regional water resources. We have also suggested that the aquifer be considered as a *system* which can perform different functions to achieve desired goals, and we have analyzed these functions.

1-2 MANAGEMENT OF GROUNDWATER RESOURCES

The management of a groundwater resource system, alone or conjunctively with a surface water one, aims at achieving certain goals through a set of decisions (= *policy*) related to the development and/or operation of the system. Certain aspects of the management of groundwater resources are discussed in Chap. 12 and need not be considered here. The brief introductory remarks presented here may serve as a background to the discussion of the physical behavior of water in aquifers presented in the following chapters.

Objectives and Constraints

Goals may be defined at different levels within a hierarchy of levels; national, regional (for example, provincial), at the community level and at the level of the individual consumer. As examples, we may mention improving living standards, increasing Gross National Product, improving life quality, redistribution of income, increasing net benefits, conservation of national resources, conservation and upgrading of environmental quality, etc. Often, more than a single goal is aimed at and we speak of multipurpose projects. Some of the goals are quantifiable, while others are intangible.

Here we consider only those goals which can be served directly or indirectly, at least in part, by the development and operation of a groundwater system.

Typically, the same goal or goals can be achieved by different policies. Management, therefore, includes the selection of the best policy which will lead to the achievement of a specified goal, or a number of goals simultaneously. This requires *criteria* on which to base the selection of the best policy, or some measure of the relative effectiveness with which the different alternative policies meet or approach the specified goals. The scalar function of the *decision variables* which measures the efficiency of the different alternative policies is called the *objective function*. Not all policies are feasible; some violate specified *constraints*, for example, social, economic, or technical, and should not be taken into account. We shall return in some detail to these concepts in Chap. 12.

More specifically, management of a specified aquifer usually means determining the numerial values of any or all of the following decision variables:

1. Areal and temporal distributions of pumpage,
2. Areal and temporal distributions of artificial recharge,
3. Water levels in streams and lakes in contact with an aquifer,

4. Quality of water to be used for artificial recharge,
5. Quality of pumped water,
6. Capacity of new installations for pumping and/or artificial recharge, their location, and time schedule of their construction,

such that any or all of the following hydrological *constraints* are satisfied:

1. Water levels everywhere, or at specified locations, should not rise above specified maximum elevations;
2. Water levels everywhere, or at specified locations, should not drop below specified minimum elevations;
3. The discharge of a spring should not drop below a specified minimum;
4. Base flow in a stream fed by groundwater emerging from an aquifer should not drop below a specified minimum;
5. The concentrations of certain species in solution in the water pumped at specified locations should not exceed specified threshold values;
6. Land subsidence should not exceed specified values;
7. Total pumpage should at least satisfy the demand for water in a given region;
8. Pumping (and/or artificial recharge) rates cannot exceed installed capacity of pumping (and/or artificial recharge);
9. The residence time for recharge water in an aquifer, before being pumped, should exceed a certain minimum period;
10. The length of an intruding sea water wedge should not exceed a specified value, etc.

Other, non-hydrological constraints are also possible (and, in fact, they always exist).

It is obviously possible that the water levels themselves, the concentrations of specified species in the water, as well as other state variables, may serve as decision variables and appear in the objective function in addition to their appearance in the constraints.

Examples of *objective functions* are:

1. Total net benefits (or present worth of total net benefits, if timing of costs and benefits is taken into account), from operating the system during a specified period of time, and we wish to maximize the value of this function;
2. Cost of a unit volume of water supplied to the consumer, and we wish to minimize the value of this function;
3. Total consumption of energy, and we wish to minimize the value of this function;
4. The sum of absolute values of the differences between certain desired water levels and actual ones (or sum of squares of differences), and we wish to minimize this sum, etc.

The time element is a very important feature of the considered system when the latter is not in a steady state, say during a period of development. In the decision making process, we have to take into account the fact that investments, costs, and benefits have different values when occurring at different times. Amounts of money at different times may be made equivalent by multiplying future amounts by a factor (*discount rate*) which becomes progressively smaller as we move into the more distant future.

Rates of pumping and/or artificial recharge may vary during a period of growth and produce an unsteady (transient) flow pattern in the aquifer. However, the time element is often introduced because of two additonal reasons. The first is the seasonal fluctuations of natural replenishment from precipitation (and/or streams) which serves as an uncontrolled input to the groundwater system. The importance of this factor depends on the time scale of the problem (e.g., months, seasons, or years). The second aspect stems from the nature of this input which fluctuates randomly from year to year. Various methods are available in order to take these random (or *stochastic*) fluctuations in input into account in solving the forecasting and the management problems.

In this case, it is not possible to determine a unique optimal management policy for the entire planning horizon, since such policy is influenced by information (including information on input) obtained up to the instant of time to which it relates. However, it is possible, for example, to pre-determine a *management strategy* which involves the determination of the optimal policy for each point in time in the form of a function of the system's states. When the state which the considered system has reached at any point in time becomes known, the optimal management policy is uniquely obtained from the specified strategy.

In addition to uncertainty introduced by the stochastic nature of the input, uncertainty is also introduced in two ways:
1. Uncertainty with respect to the model selected to represent the behavior of the groundwater system.
2. Uncertainty with respect to the various physical parameters of this model.

This observation is valid both for the physical model of the aquifer and for the management model which is based on it. In the latter model, we have additional parameters, e.g., economic and social, with various degrees of uncertainty. Various methods are available which enable the planner to derive a management policy in the face of these (and other) uncertainties.

We have tried to explain in a rather simple language, what we mean by the term *management of a groundwater system*. In one form or another, the ability to predict the response of the system is an intrinsic part of the procedure for determining any optimal management policy. For example, we must know the water levels which will occur in an aquifer as a result of the implementation of a proposed management policy (say, related to pumping) in order to examine whether they violate water level constraints. At the same time, if cost of energy is included in the objective function, it also depends on the depth to the resulting water levels.

Yield of an Aquifer

The *yield* of an aquifer is a term which requires some attention. Obviously, in the long run, unless we wish to mine all or part of the volume of water in storage, the volume of water withdrawn from an aquifer cannot exceed the aquifer's replenishment. However, this is only a limiting factor to the rate of withdrawal. To conserve an aquifer (especially from the quality point of view), we must maintain a certain rate of outflow of groundwater from it, which means maintaining certain minimal water levels, at least in the vicinity of outlets. Accordingly the withdrawal from an aquifer is the difference between all inflows and all outflows (see Chap. 3). Since inflows and outflows can be controlled by controlling water levels on which they depend (and artificial recharge is certainly a controlled input), the *rate of annual withdrawal is a decision variable*; it may vary within certain limits, according to our decision as to the rate of groundwater outflow we allow from the aquifer and to inflow into the aquifer from different sources.

Another important factor is the annual climatic fluctuations of natural replenishment. Because of the storage capacity of an aquifer, annual withdrawal is not directly related to the natural replenishment in the same year.

Altogether, the yield of an aquifer is a decision variable to be determined as part of its management. It is not necessarily a constant figure. It may vary from year to year, depending on the state of the aquifer (in terms of quantity and quality of water), the hydrologic constraints imposed on its operation (i.e., satisfying a water balance) and objectives and constraints included in the management model. We often use the term *optimal yield*, or *operational yield*. For general planning purposes, a single long-term average optimal yield can be calculated.

Until recently, another term—the *safe yield*—has often been used by hydrologists. Different definitions are available in the literature for this term, however, they are all based on more or less the same approach. Typically, a safe yield of an aquifer has been defined as the maximum annual withdrawal which will not produce undesired results. Among the latter are mentioned the impairment of water quality, increased energy costs by lowered water levels, and infringement of rights of other users in the same (or adjacent) basin (i.e., legal considerations) including the effect on base flow in streams. Terms such as *permissible yield*, *perennial yield*, *sustained yield*, with similar definitions have often been also used.

Once we understand the dynamic nature of a groundwater system, the large number of factors involved, the need to manage the system to achieve various goals, etc., it becomes obvious that the essentially hydrologic static concept of a safe yield is insufficient. The concept of *optimal yield*, or *operational yield*, incorporates in it both the hydrological features of the physical groundwater system and those of the management one.

1-3 SCOPE AND ORGANIZATION OF BOOK

In the introductory remarks to this chapter, we have introduced two *"problems"* which are of interest to the groundwater hydrologist or manager: the management problem and the forecasting one. We have emphasized that although the ultimate goal is the management of the aquifer system, no management is possible without solving first, or simultaneously, the forecasting problem.

The forecasting problem cannot be solved unless the considered groundwater system is *well defined*. We must know the geometry of the aquifer's boundaries, the conditions on the boundaries, uncontrolled inputs (e.g., from precipitation), and the various transport and storage coefficients. This information is usually obtained by solving the *identification* (or *calibration*, or *inverse*) *problem*. Whereas in the forecasting problem, we know all the aquifer parameters and seek the response of the aquifer to a proposed excitation, in the identification problem we seek the parameters, using known values of excitations and resulting responses as input information. These are obtained from records of past operations and resulting aquifer responses (e.g., water levels and solute concentrations *as observed in the field*).

The main objective of this book is to present and discuss the forecasting problem, where the forecasting is related to both groundwater quantity and groundwater quality. Chapter 11 introduces the identification problem, while Chap. 12 gives an introduction to groundwater management.

Accordingly, in the present chapter we have set the stage by discussing the role of groundwater in water resource systems and by introducing the reader to what is involved in the management of groundwater resources.

Chapter 2 gives the definition and classification of aquifers and introduces two concepts which serve as the basis for the entire presentation in this book. The first is the hydraulic approach in which we consider the flow through the entire thickness of an aquifer. The second is the continuum approach to the treatment of flow through porous media.

Chapter 3 reviews the regional groundwater balance and its components. Both natural and man-introduced components are considered. In each case, both the quantity and the quality of the water are discussed.

Chapter 4 presents the basic equations of groundwater motion, first the equation for three-dimensional flow, and then the integrated equations for flow in confined and phreatic aquifers. In doing so, the transmissivity is introduced as an aquifer parameter related to the hydraulic approach. The basic motion (or Darcy's) equation is extended to inhomogeneous fluids and to inhomogeneous and anisotropic aquifers.

The definitions of specific storativity and aquifer storativity are introduced in Chap. 5. With these definitions, the continuity equations are developed for the three types of aquifers—the confined, phreatic, and leaky—by integrating the continuity equation for three-dimensional flow over the vertical. It is shown how by doing so, the phreatic and other upper and lower boundary conditions are incorporated into the equations. Following a discussion of boundary and initial conditions, the structure of the complete mathematical statement of *any* groundwater flow problem is presented. Methods of solution, and especially numerical methods are reviewed.

Upon reaching this point, the reader should be able to state correctly and completely any problem of groundwater flow in an aquifer, in terms of a partial differential equation and appropriate boundary and initial conditions.

Chapter 6 considers flow in the unsaturated zone. This is important in view of the replenishment of the aquifer which usually has the form of unsaturated downward flow, and also because of the movement of pollutants with the unsaturated downward flow into aquifers (leachate from landfills, fertilizers spread over the ground surface and dissolved in water, etc.).

The problem of changes in groundwater quality, as affected by the various transport and accumulation phenomena, is discussed in Chap. 7. The main feature here is the introduction of hydrodynamic dispersion. The general equation of hydrodynamic dispersion (for both saturated and unsaturated flow) is developed. Following a discussion of the appropriate boundary and initial conditions, the complete, mathematical statement of the problem of movement and accumulation of a pollutant is presented. Special attention is given to the case of concentration dependent density. The integrated equation, following the hydraulic approach, is developed and discussed.

A review of hydraulics of groundwater flow to wells is presented in Chap. 8, both because of its practical interest to those dealing with groundwater extraction (and artificial recharge through wells), and as an example of the statement and solution of groundwater flow problems.

Chapter 9 deals with the important problem of sea water intrusion into coastal aquifers. Again the discussion leads to partial differential equations, based on the concept of essentially horizontal flow in an aquifer, the solution of which will yield the shape and position of an assumed sharp interface which bounds the fresh water flow domain.

Chapter 10 discusses the concepts and philosophy underlying the modeling of groundwater systems. Various kinds of models—for both water quantity and water quality—are presented.

Chapter 11 presents some of the methods used for solving the *inverse* (or the *identification*) problem. The objective of these methods is to identify the various coefficients which appear in aquifer models, including those appearing in the continuity and in the dispersion equations, the latter constituting possible models of aquifer behavior.

Finally, Chap. 12 serves as an introduction to the management of groundwater. It is presented in this book in order to relate the problem of forecasting the response of the aquifer, in the form of groundwater levels and concentrations of any considered solute to planned operations (e.g., pumping), considered in all the previous chapters, to that of making management decisions as to the desired levels to be assigned to these operations.

Classification of Groundwater Flow Problems

Attempts have often been made in the past to classify the various forecasting problems which may be encountered in the practice. Certainly, there exists no unique way for executing such classifications. To a large extent, the method of classification should depend on its objective. One possible objective is to facilitate access to a bank of models, computer programs or solutions for the various forecasting problems. Table 1-1 is an attempt of classification for the purpose

16

Table 1-1 Classification of groundwater problems

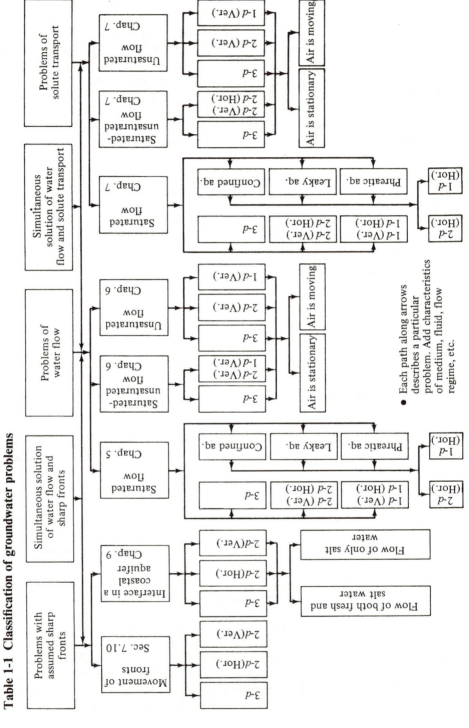

of summarizing the various forecasting problems, for the statement of which material is presented in this book. Only problems of practical interest are included. The identification problem and the management problem require separate classifications.

We first distinguish between flow of water only, transport of solutes and flow in the presence of (assumed) sharp fronts. The problems of water flow are essentially divided into two groups: saturated flow and unsaturated flow. A third group, which is a combination of these two, handles flows which are saturated in part of the domain and unsaturated in another part.

Saturated flow can be one of a general three-dimensional nature (with possible reductions to two- and one-dimensional flows). Or, under certain circumstances, it can be treated as one of essentially horizontal flow in an aquifer. In the latter case, we may further divide the problems according to the type of aquifer.

Unsaturated flow can be in three-, two- or one-dimensional spaces. The mixed saturated–unsaturated problems usually occur either as one of three-dimensional flow, or as one of two-dimensional flow in the vertical plane. A further possible classification of unsaturated flows is according to whether the air is considered a stationary, constant pressure phase, or a moving phase, such that the problem becomes one of multiphase flow.

In addition, for each of the above cases of flow, we may further classify problems according to:

1. Domain characteristics: homogeneous/inhomogeneous; isotropic/anisotropic; deformable/non-deformable.
2. Fluid characteristics: homogeneous/inhomogeneous; compressible/incompressible.
3. Flow regime: laminar/non-laminar; steady/unsteady.

We may also add such characteristics as:

4. Type of boundary conditions (first and/or second and/or third).
5. Isothermal/non-isothermal conditions.
6. Types of sources: point sources/distributed sources, etc.

The problems of solute transport (or groundwater quality problems) may also be classified in a similar way. First we distinguish between saturated flow, unsaturated flow, and mixed saturated–unsaturated flow. Then we divide the problems according to the dimensions of space involved. The integrated (or averaged), essentially horizontal flows in aquifers form, again, a separate class, further subdivided according to the type of aquifer.

In addition, for *each* of the above solute transport cases, we may further classify the problems according to domain characteristics, fluid characteristics, flow regime, type of boundary conditions, presence of temperature effects, and type of sources (see discussion above on flow problems). Here, however, we may add characteristics of the solutes involved and their chemical interactions amongst themselves (i.e., within the liquid phase), decay phenomena, and various surface activities (adsorption, ion exchange, solution, etc.) with the solid phase.

The solution of any solute transport problem depends on the knowledge of the velocity distribution in the flow domain. When water density remains constant (*ideal trace*), the two problems can be solved independently. First we solve the flow problem to obtain the velocity distribution and then use this information to solve the solute transport problem. When the water's density varies with the solute's concentration, the flow problem and the solute transport problem have to be solved simultaneously.

Of special interest is the class of saturated flow problems in which we assume (obviously as an approximation of reality) the presence of a sharp interface (= *front*) separating subdomains occupied by water of different quality. Two major classes of problems are of practical interest here: the fresh water–salt water interface in a coastal aquifer and the movement of fronts in general (especially in connection with artificial recharge or liquid waste disposal by injection into aquifers). The former problem is usually treated as one of flow in a vertical cross section (perpendicular to the coast) or as one of essentially horizontal flow in both the fresh water and the salt water zones. The latter problem is usually treated as one of essentially two-dimensional flow in an aquifer or as a problem in a three-dimensional space.

In each case we have, as above, to add specifications with respect to the flow domain, the fluids (in this case we have to characterize both fluids), the flow regime, etc.

All these problems are treated in this book. Using the material presented in various chapters, the reader should be able to recognize the various problems mentioned above and to state each one mathematically, exactly, or to a desired degree of approximation.

CHAPTER TWO

GROUNDWATER AND AQUIFERS

The purpose of this chapter is to introduce some of the basic definitions related to aquifers and to focus the attention on the flow of water in the saturated zone. Following a brief description of the moisture distribution along the vertical, the confined, leaky, and phreatic aquifers are defined. The concept of *essentially horizontal flow in aquifers*, or *the hydraulic approach*, is introduced as a powerful simplification justified in most situations of flow in aquifers (with a warning to avoid this approach whenever horizontal lengths of interest near partially penetrating sources and sinks, are smaller than the thickness of the considered aquifer). In general, the hydraulic approach is predominant in this book.

Actually the hydraulic approach constitutes already a second level of averaging, namely averaging (or integrating) over the thickness of the aquifer. The first averaging is that leading from the microscopic level of describing phenomena at points inside the void space to the macroscopic one, of viewing the porous medium as a continuum. This approach, based on the concept of a representative elementary volume of the porous medium, is discussed in Sec. 2-5.

2-1 DEFINITIONS

Groundwater, or *subsurface water*, is a term used to denote all the waters found beneath the surface of the ground. However, the groundwater hydrologist is primarily concerned with the water contained in the zone of saturation (Sec. 2-2), and uses the term *groundwater* to denote water in this zone. In drainage of agricultural lands, or agronomy, the term *groundwater* is sometimes used also to denote the water in the partially saturated layers above the water table. Practically all groundwater can be thought of as part of the *hydrologic cycle* (Fig. 2-1; see any textbook on Hydrology). Very small amounts, however, may enter the cycle from other sources (e.g., magmatic water).

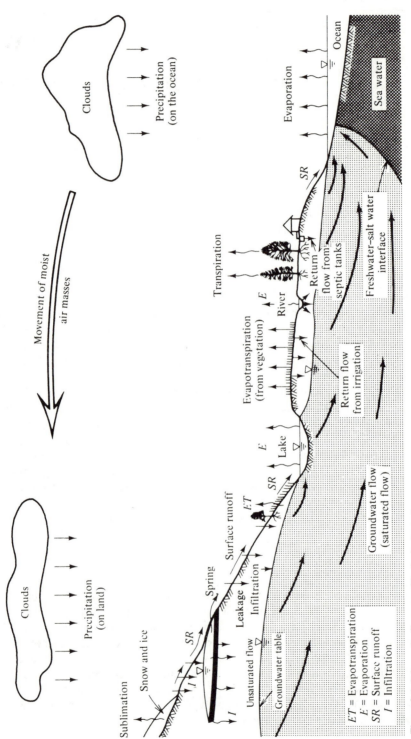

Figure 2-1 Schematic diagram of hydrologic cycle.

Clouds

Precipitation (on the ocean)

Ocean

Evaporation

Sea water

Clouds

Precipitation (on land)

Movement of moist air masses

Sublimation

Snow and ice

Transpiration

Return flow from septic tanks

Freshwater–salt water interface

Evapotranspiration (from vegetation)

River

E

Return flow from irrigation

Surface runoff

Spring

SR

ET

Lake

E

SR

Leakage

Infiltration

I

Unsaturated flow

Groundwater table

Groundwater flow (saturated flow)

ET = Evapotranspiration
E = Evaporation
SR = Surface runoff
I = Infiltration

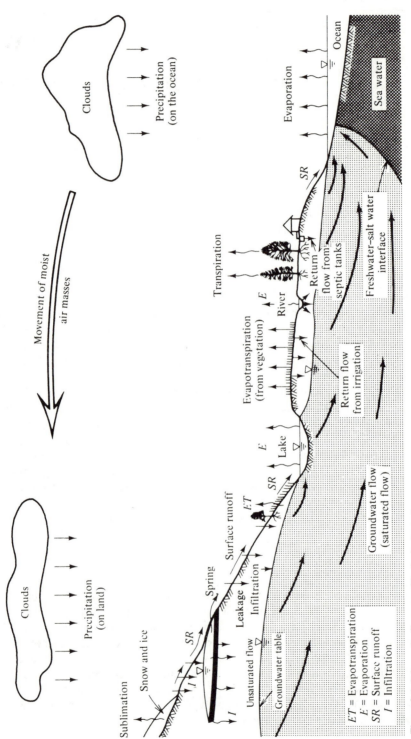

20

An *aquifer* is a geologic formation, or a group of formations, which (i) contains water and (ii) permits significant amounts of water to move through it under ordinary field conditions. Other terms often used are: *groundwater reservoir* (or *basin*) and *water bearing zone* (or *formation*). Todd (1959) traces the term aquifer to its latin orgin: *aqui* comes from *aqua*, meaning water, and *-fer*, from *ferre*, to bear.

In contradistinction, an *aquiclude* is a formation which may contain water (sometimes in appreciable quantities), but is incapable of transmitting significant quantities under ordinary field conditions. A clay layer is an example of an aquiclude. For all practical purposes, an *aquiclude* is considered an *impervious formation*.

An *aquitard* is a geologic formation which is of a semipervious nature; it transmits water at a very low rate compared to the aquifer. However, over a large (horizontal) area, it may permit the passage of large amounts of water between adjacent aquifers which it separates from each other. It is often referred to as a *semipervious formation* or a *leaky formation*.

An *aquifuge* is an impervious formation which neither contains nor transmits water.

That portion of the rock formation which is not occupied by solid matter is the *void space* (or *pore space*). In general, the void space may contain in part a liquid phase (water), and in part a gaseous phase (air). Only connected interstices can act as elementary conduits within the formation. Figure 2-2 (after Meinzer, 1942) shows several types of rock interstices. Interstices may range in size from huge limestone caverns to minute subcapillary openings in which water is held primarily by adhesive forces. The interstices of a rock formation can be grouped in two classes: original interstices (mainly in sedimentary and igneous rocks)

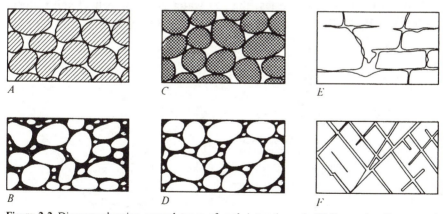

Figure 2-2 Diagram showing several types of rock interstices. A. Well-sorted sedimentary deposit having high porosity; B. Poorly sorted sedimentary deposit having low porosity; C. Well-sorted sedimentary deposit consisting of pebbles that are themselves porous, so that the deposit as a whole has a very high porosity; D. Well-sorted sedimentary deposit whose porosity has been diminished by the deposition of mineral matter in the interstices; E. Rock rendered porous by solution; F. Rock rendered porous by fracturing *(after Meinzer, 1942)*.

created by geologic processes at the time the rock was formed, and secondary interstices, mainly in the form of fissures, joints, and solution passages developed after the rock was formed.

2-2 MOISTURE DISTRIBUTION IN A VERTICAL PROFILE

Subsurface formations containing water may be divided vertically into several horizontal zones according to the relative proportion of the pore space which is occupied by water. Essentially, we have a *zone of saturation* in which all pores are completely filled with water, and an overlying *zone of aeration* in which the pores contain both gases (mainly air and water vapor) and water.

Figure 2-3 shows a schematic distribution of subsurface water in a homogeneous soil. Water (e.g., from precipitation and/or irrigation) infiltrates through the ground surface, moves downwards, primarily under the influence of gravity, and accumulates, filling all the interstices of the rock formation, above some impervious bedrock. The saturated zone in Fig. 2-3 is bounded from above by a *water table* (= *phreatic surface*). We shall see below that under different circumstances, the upper boundary can be an impervious one. The term groundwater defined in Sec. 2-1 is used by groundwater hydrologists to denote the water in the zone of saturation. Wells, springs, and effluent streams act as outlets of water from the zone of saturation. The phreatic surface is an imaginary surface at all points of which the pressure is atmospheric (conveniently taken as $p = 0$). In Fig. 2-3 it is revealed by the level at which water stands in a well just penetrating the aquifer. When the flow in the aquifer is essentially horizontal, the depth of penetration is immaterial. Actually saturation extends a certain distance above the water table, depending on the type of soil (Sec. 4-5).

The *zone of aeration* extends from the water table to the ground surface. It usually consists of three subzones: the soil water zone (or belt of soil water), the intermediate zone (or vadose water zone), and the capillary zone (or capillary fringe).

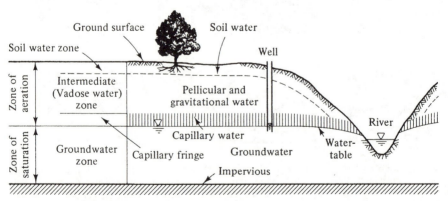

Figure 2-3 The distribution of subsurface water.

The *soil water zone* is adjacent to the ground surface and extends downward through the root zone. Vegetation depends on water in this zone, as the roots require aeration as well as water. The moisture distribution in this zone is affected by conditions at the ground surface (seasonal and diurnal fluctuations of precipitation, irrigation, air temperature, and humidity), and by the presence of a shallow water table. When the water table of the aquifer is deep, it does not affect the moisture distribution in this zone. Water in this zone moves downward during infiltration (e.g., from precipitation, flooding of the ground surface or irrigation), and upward by evaporation and plant transpiration. Temporarily, during a short period of excessive infiltration, the soil in this zone may be almost completely saturated.

After an extended period of gravity drainage without additional supply of water at the soil surface, the amount of moisture remaining in the soil is called *field capacity*. Below field capacity, the soil contains *capillary water* in the form of continuous films around the soil particles and meniscii between them, held by surface tension. Water in these films is moved by capillary action and is available to plants. Below some moisture content, called the *hygroscopic coefficient* (= maximum moisture which an initially dry soil will adsorb when brought in contact with an atmosphere of 50% relative humidity at 20°C), the water in the soil is called *hygroscopic water*. It also forms very thin films of moisture on the surface of soil particles, but the adhesive forces are very strong, so that this water is unavailable to plants.

The *intermediate zone* extends from the lower edge of the soil water zone to the upper limit of the capillary zone (Sec. 4-5). Its thickness depends on the depth of the water table below the ground surface; it does not exist when the water table is too high, in which case the capillary fringe may extend into the soil water zone, or even to the ground surface. Non-moving, or *pellicular*, water in the intermediate zone is held in place by hygroscopic and capillary forces. It is equivalent to the field capacity in the soil water zone. Temporarily, water moves downward through this zone as *gravitational water*.

The *capillary fringe* (Sec. 4-5) extends from the water table up to the limit of capillary rise of water. Its thickness depends on the soil properties and on the homogeneity of the soil, mainly on the pore size distribution. The capillary rise ranges from practically nothing in coarse material, to as much as 2 m to 3 m and more in fine materials (e.g., clay). Within the *capillary fringe* there is a gradual decrease in moisture content with height above the water table. Just above the water table, the pores are practically saturated. Moving higher, only the smaller connected pores contain water. Still higher, only the smallest connected pores are filled with water. Hence, the upper limit of the capillary fringe has an irregular shape. For practical purposes, some average smooth surface is taken as the upper limit of the capillary fringe, such that below it the soil is assumed practically saturated (say > 75%).

In the capillary fringe, the pressure is less than atmospheric and a vertical as well as horizontal flow of water may take place. When the saturated zone below the water table is much thicker than the capillary fringe, the flow in the latter is often neglected as far as groundwater flow is concerned. In drainage problems, the flow in the unsaturated zones may be of primary importance.

Obviously, numerous complications are introduced into the schematic moisture distribution described here by the great variability in pore sizes, the presence of layers of different permeability, and by the temporary movement of infiltrating water.

2-3 CLASSIFICATION OF AQUIFERS

Aquifers may be classed as confined or unconfined, depending upon the absence or presence of a water table.

A *confined aquifer* (Fig. 2-4), also known as a *pressure aquifer*, is one bounded from above and from below by impervious formations. In a well just penetrating such an aquifer, the water level will rise above the base of the upper confining formation; it may or may not reach the ground surface. A properly constructed *observation well* (or a *piezometer*) should have a relatively short screened section (yet not too short with respect to the size of the pores) such that it indicates the *piezometric head* (Sec. 2-5) at the point, say, the center of the screen. If the screen is too long it may connect zones of significantly different piezometric heads (unless the flow is horizontal) and disturb the flow pattern in the vicinity by providing a short-cut to the flow through it. The water level in it is then not necessarily the piezometric head at its center.

The water levels in a number of observation wells tapping a certain aquifer define an imaginery surface called the *piezometric surface* (or *isopiestic surface*). When the flow in an aquifer is essentially horizontal, such that equipotential surfaces are vertical, the depth of the piezometer opening is immaterial; otherwise, a different piezometric surface is obtained for piezometers which have openings at different elevations. Fortunately, except in the neighborhood of outlets such as partially penetrating wells or springs, the flow in aquifers is essentially horizontal.

An *artesian aquifer* is a confined aquifer (or a portion of it) where the elevations of the piezometric surface (say, corresponding to the base of the upper confining layer) are above ground surface. A well in such an aquifer will flow freely without pumping (*artesian well, flowing well*). Sometimes the term artesian is used to denote any confined aquifer.

Water enters a confined aquifer in a recharge area which is a phreatic aquifer formed where the confining strata terminate at, or close to the ground surface (Fig. 2-4).

A *phreatic aquifer* (also called *unconfined aquifer, water table aquifer*) is one in which a water table (= *phreatic surface*) serves as its upper boundary. Actually, above the phreatic surface, we have a capillary fringe which is often neglected in groundwater studies. A phreatic aquifer is directly recharged from the ground surface above it, except where impervious layers, sometimes of limited areal extent, exist between the phreatic surface and the ground surface.

Aquifers, whether confined or unconfined, that can lose or gain water through either or both of the formations bounding them from above or below, are called

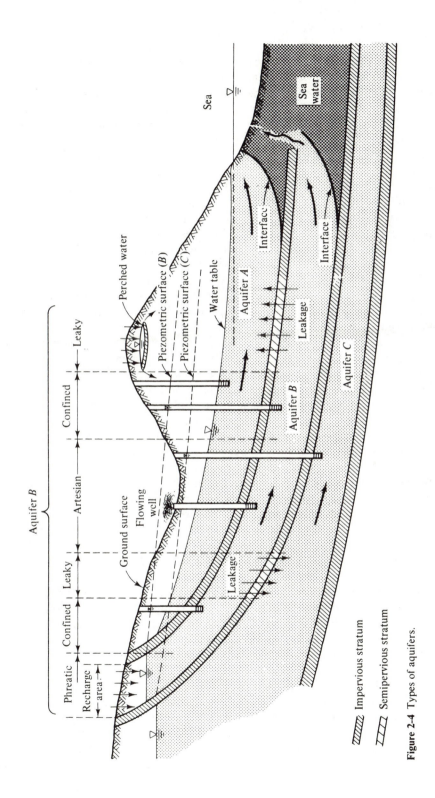

Figure 2-4 Types of aquifers.

Impervious stratum

Semipervious stratum

25

leaky aquifers. Although these bounding (semipervious) formations have a relatively high resistance to the flow of water through them, over the large horizontal areas of contact involved, significant quantities of water may leak through them into or out of an aquifer. The amount and direction of leakage is governed in each case by the difference in piezometric head which exists across the semipervious formation. Obviously, the decision in each particular case whether a certain stratum overlying or underlying an aquifer is an impervious formation, a semipervious one, or simply another pervious formation having a permeability which differs from that of the considered aquifer, is not a clear cut one. Usually, a layer which is thin relative to the thickness of the main aquifer and of a lower permeability, is considered a semipervious (or leaky) layer.

A phreatic aquifer (or part of it) which rests on a semipervious layer is a *leaky phreatic aquifer*. A confined aquifer (or part of it) which has at least one semipervious confining stratum is called a *leaky confined aquifer*.

Figure 2-4 shows several aquifers and observation wells. The upper phreatic aquifer (*A*) is underlain by two confined ones (*B* and *C*). In the recharge area, aquifer *B* becomes phreatic. Portions of aquifers *A*, *B*, and *C* are leaky, with the direction and rate of leakage determined by the elevations of the piezometric surface of each of these aquifers. The boundaries between the various confined and unconfined portions may vary with time as a result of changes in water table and piezometric surface elevations. A special case of phreatic aquifer is the *perched aquifer* (Fig. 2-4) which occurs wherever an impervious (or semipervious) layer of limited areal extent is located between the water table of a phreatic aquifer and the ground surface. Another groundwater body is then formed above this layer. Clay or loam lenses in sedimentary deposits often have shallow perched aquifers above them. Sometimes these aquifers exist only during a relatively short part of each year as they drain to the underlying phreatic aquifer.

Sometimes we refer to groundwater as confined, phreatic, or leaky, rather than to the geologic formation. Under certain conditions, a confined groundwater body at a given location may become a phreatic one and vice versa.

2-4 HYDRAULIC APPROACH TO FLOW IN AQUIFERS

In general, flow through a porous medium domain is three-dimensional. For example, the specific discharge vector **q** has the components q_x, q_y, and q_z which may be all different from zero. Also, the piezometric head, ϕ, usually varies in space, i.e., $\phi = \phi(x, y, z, t)$. However, since the geometry of most aquifers is such that they are thin relative to their horizontal dimensions (e.g., tens or hundreds of meters as compared to thousands of meters), a simpler approach can be introduced. According to this approach, we assume that the flow in the aquifer is everywhere essentially horizontal (*aquifer-type flow*), or that it may be approximated as such, neglecting vertical flow components. This is strictly true (not just an assumption) for flow in a horizontal, homogeneous, isotropic, confined aquifer, of constant thickness and with fully penetrating wells. Nevertheless, the approximation is still a good one when the thickness of the aquifer varies, but in such a

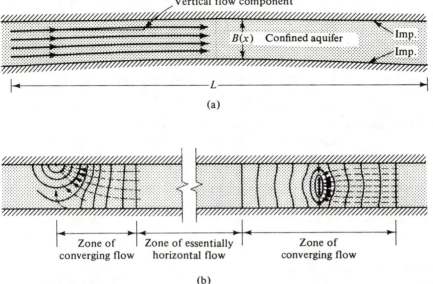

(a)

(b)

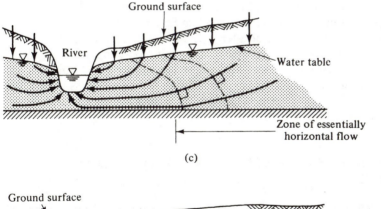

(c)

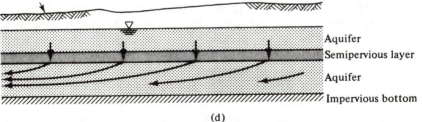

(d)

Figure 2-5 Examples of the hydraulic approach to flow in aquifers. (a) Flow in a confined aquifer with variable thickness: $B(x) \ll L$. (b) Flow in a confined aquifer with partially penetrating wells. (c) Flow in a phreatic aquifer with accretion. (d) Flow in a leaky confined aquifer.

way that the variations are much smaller than the average thickness (Fig. 2-5a). Whenever justified on the basis of the geometry (i.e., thickness versus horizontal length) and the flow pattern, the assumption of horizontal flow, which is equivalent to assuming vertical equipotentials, $\phi = \phi(x, y, t)$, greatly simplifies the mathematical analysis of the flow in the aquifer. The error introduced by this assumption is small in most cases of practical interest (see discussion in Sec. 4-3).

The assumption of essentially horizontal flow fails in regions where the flow has a large vertical flow component as, for example, in the vicinity of partially penetrating wells (Fig. 2-5b), or outlets in the form of springs, rivers, etc. However, even in these cases, at some distance from the source or the sink, the assumption of essentially horizontal flow is valid again. As a simple rule, one may assume aquifer-type flow at distances larger than 1.5 to 2 times the thickness of the aquifer at that vicinity. At smaller distances, equipotentials are no more vertical, the flow is three-dimensional and should be treated as such.

The assumption of essentially horizontal flow is applicable also to leaky aquifers (Fig. 2-5d). When the hydraulic conductivity of the aquifer is much larger than that of the semipermeable layer, and the thickness of the first is much larger than that of the latter, it follows from the law of refraction of streamlines (e.g., Bear, 1972, p. 26) that the flow in the aquifer is essentially horizontal, while it is essentially vertical in the semipermeable layer. These assumptions, which in cases of practical interest introduce very small errors, greatly simplify the analysis of flow in leaky aquifers.

The approximation that the flow is essentially horizontal in phreatic aquifers is the basis for the Dupuit assumption presented in Sec. 4-5 (Fig. 2-5c).

In this book, the concept of aquifer-type flow is widely employed in all types of aquifers, unless the flow is such that it must be considered as three-dimensional. In most cases, it will be shown how the aquifer flow equations are derived by averaging the basic, three-dimensional, flow equations along the thickness of the aquifer, using the assumption of vertical equipotential surfaces. This procedure is called the *hydraulic approach*.

2-5 CONTINUUM APPROACH TO FLOW THROUGH POROUS MEDIA

In an aquifer, flow takes place through a complex network of interconnected pores, or openings. However, when dealing with flow in an aquifer, we overlook the *microscopic* flow patterns *inside* individual pores and consider some fictitious average flow which takes place in the porous medium comprising the aquifer. By doing so, we are employing the concept of a *continuum*, which is common to most branches of physics. The obvious reason for employing the continuum approach in flow through a porous medium is that it is practically impossible to describe in any exact mathematical manner the complicated geometry of the solid surfaces that bound the flowing fluid. Consequently, although, in principle, we have at our disposal the basic equations governing the flow (say, the Navier–Stokes equations) and the boundary conditions, a solution at the microscopic level is precluded.

Soils and porous or fissured rocks are examples of porous media, as are ceramics, fibrous aggregates, filter paper, or sand filters. Somewhat less obvious, but still belonging, under certain conditions, to this group, are geological formations of karstic limestone, where the open passages (such as solution channels or caverns) may be of substantial size and far apart. Assembling the common features of these materials, we may define a porous medium as a portion of space occupied by *heterogeneous*, or multiphase material, at least one of the phases being a persistent, possibly deformable, solid phase. The solid phase is called the *solid matrix*. The domain which is not occupied by a solid matrix is referred to as the *void* (or *pore*) *space*. From the point of view of flow of fluids through the porous medium, only the interconnected pore space is of interest. Sometimes the interconnected pore space is called *effective pore space*. It is possible, however, that a certain portion of the interconnected void space may be practically ineffective as far as flow is concerned. This happens when the porous medium contains *dead-end pores*, i.e., pores or channels with only a narrow single connection to the interconnected pore space, so that almost no flow occurs through them.

In addition, we require that the solid matrix (and hence the pore space) be distributed throughout the domain occupied by a porous medium; solid must be present within each *representative elementary volume*.

We still need a definition for the *representative elementary volume* (REV) of a given porous medium. This is an essential step in passing from the *microscopic* level at which we consider what happens at each point within a phase inside each pore, to the *macroscopic* level of a continuum at which only averaged phenomena are considered.

Let $U(\mathbf{x})$ denote a volume of a spatial domain centered at a point whose position vector is $\mathbf{x}$, and let E denote the amount of some extensive property of the material system contained in U. The (average) density $\rho_{(E)}$ of E over U at time t is defined by

$$\rho_{(E)}[\mathbf{x}, t; U(\mathbf{x})] = \frac{E \text{ within } U(\mathbf{x})}{U(\mathbf{x})} \qquad (2\text{-}1)$$

In general, $\rho_{(E)}$ is a function of size, shape, and orientation of $U(\mathbf{x})$ at time t. In order to make $\rho_{(E)}$ depend on $\mathbf{x}$ only, we must select a volume $U = U_0$ bounded between two spheres: $U_{\min} = (\pi/6) L_{\min}^3$ and $U_{\max} = (\pi/6) L_{\max}^3$, such that for U_0:

$$\frac{\partial \rho_{(E)}[\mathbf{x}, t; U_0(\mathbf{x})]}{\partial U_0} = 0; \qquad \frac{\pi}{6} L_{\min}^3 < U_0 < \frac{\pi}{6} L_{\max}^3 \qquad (2\text{-}2)$$

If a range for U_0 can be found which is common to *all* points within a given spatial domain (R), one can define a field $\rho_{(E)}(\mathbf{x}, t)$ throughout (R), and treat (R) as a continuum for that E. The length $L(E)$ in the range $L_{\min} < L(E) < L_{\max}$, is called the scale of continuity of E in (R). If (R) is a multiphase system, it can be treated as a continuum in describing a process involving a set of E's in it, provided a common scale of continuity exists for all the E's. The volume U_0 is then the *representative elementary volume* (REV) of the material system within (R). Certainly $L_{\max}$ must be much smaller than the dimensions of (R).

To illustrate the definition of an REV, let P be a *mathematical* point inside a porous medium domain, it may fall inside the solid phase or the void space. Let $U^{(i)}$ be a spherical volume centered at P; $U^{(i)}$ should be much larger than a single pore or grain. We consider the property $E = U_v =$ volume of void space, and define for the volume $U^{(i)}$ the ratio $n^{(i)} = U_v^{(i)}/U^{(i)}$, where $U_v^{(i)}$ denotes the volume of the void space within $U^{(i)}$, $i = 1, 2, 3, \ldots$, gradually shrinking the size of $U^{(i)}$ around P: $U^{(1)} > U^{(2)} > U^{(3)} \ldots$ (Fig. 2-6).

For large values of $U^{(i)}$, the ratio $n^{(i)}$ may undergo gradual changes as $U^{(i)}$ is reduced, especially when the considered domain is inhomogeneous. Below a certain value of $U^{(i)}$, these changes, or fluctuations, tend to decay, leaving only small amplitude fluctuations that are due to the random distribution of pore sizes in the neighborhood of P. However, below a certain value $U^{(i)} = U_{\min}$, we suddenly observe large fluctuations in the ratio $n^{(i)}$. This happens as the dimensions of $U^{(i)}$ approach those of a single pore. Finally, as $U^{(i)} \to 0$, converging on the mathematical point P, $n^{(i)}$ will become either zero or one, depending on whether P is inside the solid matrix or inside the void space. Figure 2-6 shows the relationship between $n^{(i)}$ and $U^{(i)}$.

The representative elementary volume, U_0, which will make the ratio $n^{(i)}$ meaningful as a description of what happens at P, should be chosen such that $U_{\min} < U_0 < U_{\max}$ as indicated in Fig. 2-6. Then the ratio represents the medium's *(volumetric) porosity* $n(P)$ at point P:

$$n(P) = U_{0v}/U_0; \qquad \partial(U_{0v}/U_0)/\partial U_0 = 0; \qquad 0 < n < 1 \qquad (2-3)$$

or:

$$n(P) = \lim_{U^{(i)} \to U_0} n^{(i)}[U^{(i)}(P)] = \lim_{U^{(i)} \to U_0} U_v^{(i)}(P)/U^{(i)}(P) \qquad (2-4)$$

where U_{0v} is the volume of voids in U_0. The limit process in (2-4) is sometimes called the *extrapolated limit* (Hubbert, 1956). Obviously, the limit $U^{(i)} \to 0$ is meaningless. From the definition of the REV, it follows that its size is such that adding to it or subtracting from it one or several pores has no significant influence on the value of n.

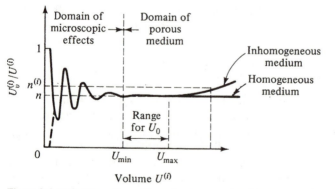

Figure 2-6 Definition of porosity and representative elementary volume.

The same procedure can be applied to any other property E.

Thus, by employing the definition of REV, we have replaced the actual medium by a fictitious continuum in which we may assign values of any property (whether of the solid matrix or of the fluid filling the void space) to *any* mathematical point.

The values assigned to a point $P(\mathbf{x})$ in the *continuum*, or macroscopic level of descriptions, are averaged ones, taken over the REV centered at P. Consider any tensorial field $G(\mathbf{x}', t)$, e.g., pressure, velocity, solute concentration, in a porous medium composed of several phases denoted by $\alpha = 1, 2, 3, \ldots$ (e.g., a solid, a gas, and a liquid). We have used the position vector $\mathbf{x}'$ to indicate points at the microscopic level, reserving the symbol $\mathbf{x}$ for the position vector of points at the macroscopic level. Each phase occupies at time t a certain volume $U_{0\alpha}$ of U_0

$$\sum_{(\alpha)} U_{0\alpha} = U_0; \qquad \theta_\alpha(\mathbf{x}, t) = U_{0\alpha}(\mathbf{x}, t)/U_0(\mathbf{x}) \tag{2-5}$$

where θ_α is the *volumetric fraction* of the α phase. Then, the average value $\bar{G}_\alpha(\mathbf{x}, t)$ of G at time t within the domain $U_{0\alpha}$ centered at $\mathbf{x}$ is defined by (App. A1)

$$\bar{G}_\alpha(\mathbf{x}, t) = \frac{1}{U_{0\alpha}} \int_{(U_{0\alpha})} G(\mathbf{x}', t; \mathbf{x})\, dU_\alpha(\mathbf{x}') = \frac{1}{U_{0\alpha}} \int_{(U_0)} G_\alpha(\mathbf{x}', t; \mathbf{x})\, dU \tag{2-6}$$

where G_α denotes the value of G in the α phase ($G_\alpha \equiv 0$ outside $U_{0\alpha}$). This average is also called the *intrinsic phase average of G in the α phase at time t*.

From the definition of θ_α above, it follows that another average may be defined

$$\theta_\alpha \bar{G}_\alpha(\mathbf{x}, t) = \frac{1}{U_0} \int_{(U_{0\alpha})} G(\mathbf{x}', t; \mathbf{x})\, dU_\alpha(\mathbf{x}') = \frac{1}{U_0} \int_{(U_0)} G_\alpha(\mathbf{x}', t; \mathbf{x})\, dU \tag{2-7}$$

It is called the *phase average of G* (or the *macroscopic value of G*) *in the α phase at time t*.

For field measurements, say of piezometric head in an observation well, to be of any use, they should be averaged values in the sense defined above. Hence, measuring devices should be designed such that they indeed read values averaged over an REV of the considered porous medium.

2-6 INHOMOGENEITY AND ANISOTROPY

In this book, unless otherwise specified, homogeneity and isotropy of a porous medium refer to its property permeability (k). In general, of course, these terms may be applied to other porous medium properties.

A porous medium domain is said to be homogeneous if its permeability is the same at all its points. Otherwise, the domain is said to be heterogeneous (or inhomogeneous). If, however, the permeability at a considered point is independent of direction, the medium is said to be *isotropic* at that point. Similar considerations apply to the hydraulic conductivity (K) and to transmissivity (T) of an aquifer, in the latter case, with the directions considered only in the xy plane (Sec. 4-3).

In general, we may distinguish two types of inhomogeneous aquifer domains:

Type 1, with a gradual change in transmissivity, and *Type 2*, with abrupt changes across well-defined surfaces of discontinuity.

In *Type 1* inhomogeneity, the variable transmissivity may be expressed as a function of the space coordinates in the form $T = T(x, y)$. In this form the transmissivity is represented in (4-33) which is applicable to flow in an inhomogeneous aquifer.

In *Type 2* inhomogeneity, each subdomain, enclosed by boundaries of discontinuity, is homogeneous in itself or is heterogeneous of *Type 1*, and should be treated as such (see Sec. 5-5). Across the boundary of discontinuity, there is a jump either in T or in its derivative normal to the boundary.

Under certain conditions, an inhomogeneous field of *Type 2* may be regarded as equivalent in its overall behavior to a (fictitious) homogeneous one. As an example, consider the case of a highly pervious sandy aquifer in which relatively impervious clay lenses are imbedded in a repetitive manner (Fig. 2-7). We may treat this aquifer as a homogeneous one as long as we are interested in phenomena, say head drop between two points, the length scale of which is much larger than the characteristic lengths (L_1 and L_2) of the repetitive unit indicated in Fig. 2-7. Homogeneity of a porous medium is thus judged by comparing the length scale of the phenomenon of interest with that of inhomogeneity of the porous medium (see also Sec. 2-5).

In a similar way, if an inhomogeneous medium of *Type 2* is non-repetitive, but the characteristic length scale of its inhomogeneity is much smaller than that of some phenomenon of interest, the medium may be considered as being equivalent in its behavior to a domain of *Type 1* inhomogeneity. Figure 2-8 shows how a layered aquifer (*Type 2* inhomogeneity) may be considered as an inhomogeneous aquifer with a gradual variation of hydraulic conductivity.

In many cases aquifers are *anisotropic*. This may happen, for example, when the sediments comprising the aquifer are such (e.g., flat shaped mica particles) that when deposited, the resulting porous medium has a higher permeability in one direction (usually the horizontal one, unless later tilting of the formation

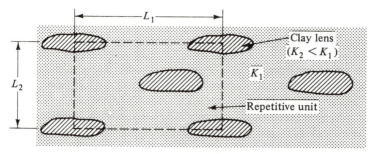

Figure 2-7 An inhomogeneous aquifer treated as an equivalent homogeneous one.

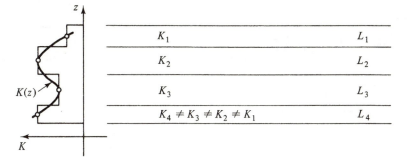

Figure 2-8 A layered aquifer.

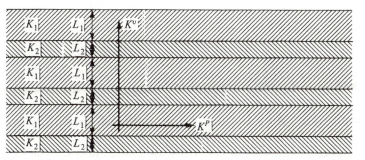

Figure 2-9 Aquifer composed of alternating layers exhibits anisotropy with $K^v \neq K^p$.

occurs) than in other directions. Both sedimentation and pressure of overlying material cause flat particles to be oriented with their longest dimension parallel to the plane on which they settle. Later this produces flow channels parallel to the bedding plane, which are different from those oriented normal to this plane, thus rendering the medium anistropic. In carbonate rocks, the flowing water dissolves the rock, producing solution channels that develop mainly in the direction of the flow. The rock becomes anisotropic, with a much higher permeability in the general direction of the channels. In some soils, structural fissures develop more readily in one direction than in others, and the soil will exhibit anisotropy. In certain rocks, fractures produce a very high permeability in the direction of the fractures.

An inhomogeneous material (*Type 2*) composed of alternating layers of different textures (Fig. 2-9) is equivalent in its behavior to an homogeneous anisotropic medium (Bear, 1972, p. 155). However, in order for a stratified formation of this kind to be considered as an equivalent homogeneous anisotropic aquifer, the thickness of the individual layers must be much smaller than lengths of interest. For example, it is meaningless to determine the equivalent permeability of such a formation from a core whose size is smaller than the thickness of a single layer.

THREE

GROUNDWATER BALANCE

The groundwater part of the hydrologic cycle is presented in Fig. 2-1 which schematically describes the entire hydrologic cycle. In the management of groundwater resources, man intervenes in the hydrologic cycle in order to achieve beneficial goals. This intervention takes the form of modifications imposed on the various components of the water balance.

Water and pollutants carried with it may enter an aquifer, or a considered portion of one, in the following ways:

1. Groundwater inflow through aquifer boundaries and leakage from overlying or underlying aquifers.
2. Natural replenishment (infiltration) from precipitation over the area.
3. Return flow from irrigation and septic tanks (or similar structures, including faulty water supply or sewage networks).
4. Artificial recharge.
5. Seepage from influent streams and lakes.

Water and pollutants carried with it may leave an aquifer in the following ways:

1. Groundwater outflow through boundaries and leakage out of the considered aquifer into underlying or overlying strata.
2. Pumping and drainage.
3. Seepage into effluent streams and lakes.
4. Spring discharge.
5. Evapotranspiration.

The difference between total inflow and total outflow of water and of pollutants during any period is stored in the aquifer, causing a rise in water levels and in the concentration of pollutants, respectively.

Let us review these components, which comprise the groundwater balance of a region, in order to facilitate the discussion in the following chapters which deal with them. The region's boundaries may be the natural boundaries of a groundwater basin (e.g., an impervious boundary, a water divide, or a river fully penetrating an aquifer), or any closed boundary drawn on a map (say, for administrative reasons).

In the present chapter, we shall use some of the aquifer concepts and definitions introduced in later chapters (especially in Chaps 4 and 5), assuming that the reader is familiar with them, at least in a general form, from previous studies and reading.

The balance, or budget, discussed below is only for groundwater in the saturated zone. It is obviously possible to discuss a water balance for the entire subsurface, including water in the unsaturated zone (see Sec. 10-6), or even a regional water balance which will also include surface water.

3-1 GROUNDWATER FLOW AND LEAKAGE

Inflow and Outflow through Aquifer Boundaries

When a boundary of an aquifer (or a portion of one) is pervious, groundwater may enter the aquifer through it from the outside (another aquifer or the remaining part of the aquifer). The flow is governed by the gradient of the water table, or the piezometric head, along the boundary.

Figure 3-1 shows a contour map and a portion of an aquifer, *ABCD*, for which a water balance is being established. Groundwater flows into the aquifer through the boundary *DAB* and out of the aquifer through *BCD*. Because the

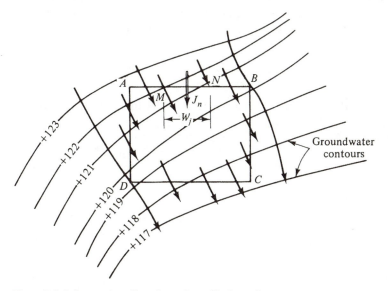

Figure 3-1 Inflow and outflow through aquifer boundary.

hydraulic gradient varies in magnitude and direction along the boundary and streamlines are in general not perpendicular to it, we divide the entire boundary into segments (like MN in Fig. 3-1) of length W_j each. Segments are chosen such that along each, the component of the hydraulic gradient normal to W_j can be satisfactorily represented by an average value J_{nj}, similarly, the transmissivity along the segment is represented by an average value T_j. The inflow through W_j is then given by the expression $W_j T_j J_{nj}$. We may then sum such expressions for the entire boundary with a positive value of J_{nj} for inflow and a negative one for outflow, and obtain $\sum_{(j)} W_j T_j J_{nj}$ for the instantaneous net inflow through aquifer boundaries (see also discussion in Sec. 10-4).

In general, the instantaneous net inflow may vary with time. Denoting the average net inflow during a water balance period Δt by $\overline{\sum_{(j)} W_j T_j J_{nj}}$ (e.g., by taking half the sum of net inflows computed at the beginning and at the end of Δt), the volume added to the aquifer by net inflow of groundwater is given by $\Delta t \overline{\sum_{(j)} W_j T_j J_{nj}}$.

Groundwater entering a balance area carries with it solutes (and other polluting elements) present in the formation (or adjacent aquifer) outside the boundary. As we shall see in Chap. 7, the modes of transport are convection and hydrodynamic dispersion. For a regional balance of the kind considered here, the transport by hydrodynamic dispersion is often neglected.

Groundwater leaving a balance area carries with it solutes (and other polluting elements) present in the aquifer on the inner side of the boundary. Often, especially if the balance area is not too large, it is assumed that complete mixing of groundwater takes place in the balance area so that the water leaving it carries with it the average concentration in the balance area (see Sec. 10-4).

Leakage

This subject is discussed in Sec. 5-4. The leakage q_v (volume of water per unit area and per unit time) through a semipermeable layer from an overlying (or underlying) aquifer with piezometric head ϕ_{ext} into an aquifer with piezometric head ϕ, is given by

$$q_v = K' \frac{\phi_{ext} - \phi}{B'} \tag{3-1}$$

where K' and B' are the hydraulic conductivity and the thickness, respectively, of the semipervious layer. If q_v as calculated by (3-1) is negative, we have leakage out of the aquifer. As leakage may vary from point to point, we may divide the area of the bottom and the top semipervious boundaries of an aquifer into elementary areas $(\Delta A)_i$ through each of which we calculate an average leakage $(\bar{q}_v)_i$ during Δt. Then $(\Delta t) \sum_{(i)} (\bar{q}_v)_i (\Delta A)_i$ gives the total net inflow into the aquifer by leakage during Δt.

As we lower the piezometric head in a pumped aquifer, the leakage may reverse its direction from outflow to inflow. In the case of artificial recharge, inflow may change into outflow.

The remarks given above with respect to inflow and outflow of solutes and other polluting elements carried with the water, are valid also with respect to leakage.

3-2 NATURAL REPLENISHMENT FROM PRECIPITATION

Phreatic aquifers can be replenished from above by precipitation falling directly over the ground surface overlying the aquifer, provided the ground surface is sufficiently pervious. Part of the area may be completely impervious (houses, streets, and highways, or an impervious rock or top soil which is practically impervious) and does not contribute to the natural replenishment of the aquifer beneath it. Confined aquifers are replenished by groundwater inflow from an adjacent phreatic aquifer which, in turn, is replenished from precipitation (e.g., aquifers B and C in Fig. 2-4).

The relationship between natural replenishment and total precipitation is governed, among others, by the following factors: type of precipitation, climatic conditions, soil moisture prior to storm, storm characteristics (duration, intensity, peak intensity), topography of ground surface, perviousness of ground surface and vegetation cover.

In principle, infiltration is unsaturated downward flow from the ground surface to the water table, and the theory presented in Chap. 6 is applicable. However, the use of this theory is not a practical way to determine the natural replenishment of an aquifer, as it requires detailed information on soil characteristics along the vertical column. Moreover, in general, for the purpose of management of a groundwater system, and in view of the buffer effect of the large volume of water in storage in the aquifer at any time, we are not interested in the variability in infiltration during any individual storm and not even that resulting from storms during the year, taking each storm as an instantaneous pulse. For most regional management purposes, we are interested in annual or seasonal replenishment. Within the framework of management models, we often assume that the natural replenishment is uniformly distributed throughout the year, or throughout the rainy season. In certain cases, where more details are required, monthly averages are used.

Several methods are available for estimating natural replenishment from annual or seasonal precipitation. For example, we may regard replenishment as an aquifer parameter, rather than relate it to precipitation. We can then employ any of the techniques for parameter identification described in Chap. 11.

Except for precipitation, which varies from one year to the next, all other factors affecting replenishment are constant in time, or vary only gradually (e.g., due to changes in land use). Hence, rather than refer to annual replenishment as an unknown variable, the natural replenishment is often first related to precipitation, for which a sufficient amount of data is usually available. One possible such relationship is

$$N = \alpha(P - P_0), \qquad P > P_0,$$
$$N = 0, \qquad\qquad P \le P_0, \tag{3-2}$$

where N is annual natural replenishment, α is a coefficient, P is annual precipitation, and P_0 is threshold precipitation. For example, for $\alpha = 0.9$ and $P_0 = 200$ mm/year, we obtain $N = 405$ mm/year for $P = 650$ mm/year. In this way, the number of unknown variables defining natural replenishment is reduced to only two: α and P_0. These are then regarded as parameters of the aquifer model. They may vary from one part of a considered aquifer to the next.

Another method often used for estimating natural replenishment, when detailed data on precipitation are available, is the use of any of the so called *Stanford Watershed Models* developed by Crawford and Linsley (1966). The Hydrocomp Simulation Program (Hydrocomp., 1968) is a more sophisticated version of this model. Many such models have been developed and published since the pioneering work of Linsley and Crawford (see a survey of such models given by Fleming, 1975, p. 190). Like the Stanford model, which since its first version has gone through a large number of development phases, most models of this kind simulate the hydrologic cycle, using a moisture accounting procedure of one form or another. A system of equations describes the interrelationships among the various elements of the model. During the simulation, a running record is maintained of all moisture entering the basin (or the considered part of it), stored in it, and leaving it as evapotranspiration, surface runoff, and groundwater. The latter is the natural replenishment considered in the present section.

As an example, consider the model used by Water Planning for Israel Ltd (Ben Zvi and Goldstoff, 1972; following Mero, 1969). Figure 3-2a shows the schematic structure of this model. The model, which uses daily rainfalls as input, is based on the assumption that the rainfall–surface runoff and rainfall–natural replenishment relationships are functions of soil moisture. Hence, the subsurface is divided into two moisture storage reservoirs. The upper one (U) simulates depression storage and interception storage (on vegetation); its maximum moisture capacity is U_{max}. The lower one (L) represents soil moisture in the vertical profile above the water table. The symbols L and U (in mm) denote the soil moisture in the reservoir (L) and (U), respectively. The rainfall over the watershed first fills up the moisture storage in (U) up to its maximum capacity U_{max} and only then starts contributing to the creation of surface runoff. For example, if on a certain day, the upper reservoir has still an available capacity of $U_{max} - U$, the excess rainfall which can contribute to surface runoff and/or to natural replenishment is $PM = P - \Delta U = P - (U_{max} - U)$. The surface runoff produced by PM depends on both PM and the soil's infiltration capacity F (in mm/day), where the latter depends on the moisture capacity L. In this model the empirical expression used for F is

$$F = F_{min} + (F_{max} - F_{min}) \exp(- aL/LN) \qquad (3\text{-}3)$$

where F_{max} and F_{min} (mm/day) are upper and lower bounds of F, respectively; L is moisture capacity (in mm) in the lower reservoir; LN (in mm) is a value of L corresponding to $F = 1.1 F_{min}$; and $a = \ln\left[(F_{max} - F_{min})/0.1 F_{min}\right]$ is a constant. Figure 3-2 shows how F varies with the soil moisture in (L). It is also possible to vary this relationship from one part of the watershed to the next. In the study reported by Ben Zvi and Goldstoff (1972), following Crawford and Linsley (1966), it is assumed that the infiltration capacity at any time varies linearly over the watershed from zero to F_{max} (Fig. 3-2c). From Fig. 3-2c it follows that the surface runoff, Q, produced by the net rainfall PM, is given by

$$Q = (PM)^2/2F \qquad \text{for } PM \leq F,$$
$$Q = PM - F/2 \qquad \text{for } PM \geq F \qquad (3\text{-}4)$$

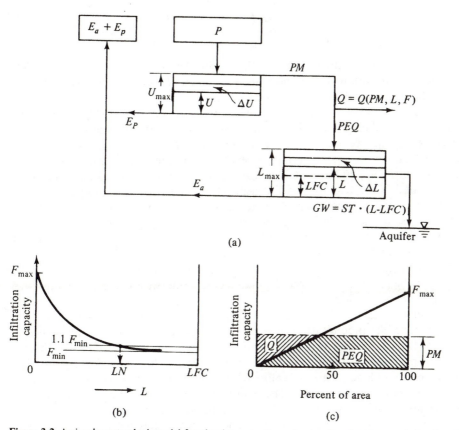

Figure 3-2 A simple watershed model for the determination of natural replenishment *(after Ben-Zvi and Goldstoff, 1972)*. (a) Schematic structure of model. (b) Variation of infiltration as a function of soil moisture. (c) Variation of infiltration over watershed.

Hence, the water balance indicates that the rate of downward percolation is

$$PEQ = PM - Q \tag{3-5}$$

where PEQ is in mm per day. This amount is now transferred into (L), where it is stored until the moisture storage in (L) reaches a constant value LFC, equal to field capacity (see Sec. 6-1). Any excess of moisture above this threshold continues to percolate to the underlying aquifer. Thus the daily contribution (in mm/day) to the natural replenishment of the aquifer is given by

$$GW = ST \cdot (L - LFC) \tag{3-6}$$

where ST is a coefficient characterizing the rate of drainage out of (L).

The model takes into account also evapotranspiration from the upper and lower moisture storages. First, *potential evapotranspiration*, E_p, is subtracted from the moisture in (U). Once this reservoir is completely dry, we have evaporation, E_a, from (L) as a function of the amount of moisture in it

$$E_a = E_p \cdot L/(LFC) \qquad L \leq LFC \tag{3-7}$$

$$E_a = E_p \qquad\qquad L > LFC \tag{3.-8}$$

Many more parameters are included in the more sophisticated models. Obviously, calibration (i.e., identification of all model parameters) is always required before a model can be used.

Thus depending on the degree of accuracy required and the amount and type of data available on precipitation, streamflow, etc., the natural replenishment of an aquifer can be determined. Although it is possible to couple the model described above, or a similar model leading to natural replenishment, directly with the multi-cell aquifer model for water level forecasting discussed in Sec. 10-5, the two models are usually handled separately. From the first model we obtain the natural replenishment which, in turn, is used as input into the second one. This is especially so when numerical models of forecasting are employed.

When we wish to predict water levels for a relatively short period of time, say two to three years, we introduce as input (assumed) values of future natural replenishment, say in the form of monthly averaged values. When, however, a forecasting problem is solved as part of a management one, we are in general interested in a much longer period, say a planning horizon of 15 or 20 years. We usually use then annual or seasonal (averaged) values. However, we have to take into account the fact that, like precipitation, natural replenishment is a *random phenomenon*. In order to take care of this feature, we then use one of the models which uses available past data to generate synthetic sequences of values, in order to generate a number of possible sequences of values of annual natural replenishment, each with some probability of occurrence. These, in turn, are introduced as time-dependent (deterministic) input in the forecasting and management models.

Most of the work to date on synthetic sequences has been carried out with respect to streamflow. However, the same methodology is also applicable to other random phenomena such as precipitation or natural replenishment. Summaries of this subject are given, among many others, by Chow (1964), Fiering and Jackson (1971), and Matalas and Wallis (1976).

A relatively simple example of an equation used for generating a synthetic sequence of monthly flow is (Thomas and Fiering, 1962)

$$Q_{i+1} = \bar{Q}_{j+1} + b_j(Q_i - \bar{Q}_j) + s_{j+1}(1 - r_j^2)^{1/2} t_i \tag{3-9}$$

where Q_i and Q_{i+1} are discharges during the ith and $(i + 1)$st months, respectively, counted from the start of the generated sequence. $\bar{Q}_j$ and $\bar{Q}_{j+1}$ are the mean monthly discharges during the jth and $(j + 1)$st months, respectively, within a repetitive annual cycle of 12 months, b_j is the regression coefficient for estimating flow in the $(j + 1)$st month from the flow in the jth month, s_{j+1} is the standard deviation of the flows in the $(j + 1)$st month, r_j is the correlation coefficient between the flows in the jth and $(j + 1)$st months, and t_i is a random normal and independent variate, with zero mean and unit variance. Usually it is assumed that annual events are not correlated.

The main advantages of the procedure for generating synthetic sequences of events (natural replenishment in the case considered here), are (1) the possibility of creating records which are longer than the historical ones and (2) that different sequences with different probability of occurrence can be generated and used as inputs in various models. This is especially important for management purposes.

Precipitation water which becomes natural replenishment does not reach the aquifer immediately. The lag of time depends on the depth of the water table and the hydraulic properties of the soil. However, the mechanism can be visualized approximately as one of displacement (at some more or less constant degree of saturation) whereby water is continuously added at the top of its column (i.e., at ground surface) and removed from the column at the bottom (i.e., at the water table). In this way, if we think of water as labeled, say by some contaminant, we may have a considerable lag of time (even tens of years) before the latter reaches the water table. However, from the point of view of water quantity, the actual replenishment practically does not lag behind the precipitation producing it.

This phenomenon is of special importance when water quality is being considered. In addition, as water passes through the soil column from the ground surface to the water table, changes may take place in the quality of the water. Precipitation water is not distilled water. Depending on the area, the air pollution conditions, and the distance from the sea, precipitation water may already contain dissolved matter. Observations close to the coast of Israel showed up to 25 p.p.m. Cl^- in rainwater (with a fast reduction farther inland). Rainwater will further pick up salinity at the ground surface and upon passing through the top soil layer. Some of the dissolved species (e.g. Cl^-) undergo no changes as the water percolates downward. Others may undergo changes as a result of interaction (e.g., adsorption, ion exchange) with the solid matrix. For example, due to adsorption, the downward movement of some heavy metals will be slowed down many times with respect to the movement of the Cl^-. Their arrival at the water table may be delayed for many years. The occurrence of clay layers may appreciably affect adsorption and ion exchange phenomena along the column. This subject is of special importance in artificial recharge through infiltration ponds with reclaimed sewage, or when this kind of water is used for irrigation.

3-3 RETURN FLOW FROM IRRIGATION AND SEWAGE

Even in efficient irrigation practices, a certain portion of the water applied to an area is not used up as *consumptive use*, but infiltrates, eventually reaching the water table. We shall refer to this contribution to an aquifer's replenishment as *return flow* from irrigation. It may amount to as much as 20–40 percent of the volume of water used for irrigation. It includes also seepage from open channels and leakage from faulty pipes. As the irrigation becomes more efficient, this percentage is reduced. The water used for irrigation may be that pumped from an underlying aquifer (hence the term *return flow*), surface water or water imported from other regions. Obviously, return flow carries with it salts and other impurities: those contained in the original irrigation water, actually augmented by evaporation, and those picked up upon passage through the ground surface and the root zone.

In fact, sometimes, return flow is created on purpose in order to leach salts from the root zone, often overlooking the fact that when an aquifer is present under the area and in the absence of adsorption and other modifying phenomena (e.g., in the case of Cl^-), the leached salts eventually reach the underlying aquifer. For the sake of simplicity we are using here the term "salts", but this should be understood to include also dissolved fertilizers, pesticides, and many other kinds of (potential) groundwater pollutants present in the root zone and on the ground surface.

If we denote the concentration of salts in the irrigation water by C_I and the maximum permissible salt concentration in the soil solution (without causing undesirable losses in production) by $C_{max}(>C_I)$, then out of the volume of irrigation water, V_I, the minimum volume of water, V_L, required for leaching the soil, such that an equilibrium will be maintained at C_{max} is

$$V_L = V_I C_I / C_{max} \tag{3-10}$$

Hence, the amount of salt continuously added to the aquifer is $V_L C_{max} (= V_I C_I)$. Actually, a volume larger than V_L as defined by (3-10) is required because of the inefficiency of the leaching process. If we increase V_I, consumptive use will practically remain unchanged, V_L will increase, causing a reduction in C_{max}. Obviously, we have to take into account that part of the leaching is also performed during rainy seasons by the natural replenishment from precipitation that in general has a relatively low salt content.

The quality problem associated with return flow and leaching should be more carefully studied when reclaimed sewage (or other kinds of contaminated water) is used for irrigation.

3-4 ARTIFICIAL RECHARGE

Artificial recharge may be defined as man's planned operations of transferring water from the ground surface into aquifers. This is in contradistinction to *natural replenishment* (or *natural recharge*), considered in Sec. 3-2 above, whereby water from precipitation and surface runoff reaches the aquifer without man's intervention. Whereas natural replenishment is an *uncontrolled* (by man) *input* to the groundwater system, artificial recharge is a *controlled input*. The quantity, quality, location, and time of artificial recharge are decision variables, the values of which are determined as part of the management policy of a considered groundwater system.

Objectives

Artificial recharge may be practiced in order to achieve various objectives. Among them, we may list the following.

1. *Control of regional hydrological regime.* By artificially recharging an aquifer, water levels, or piezometric heads, are raised. By manipulating these levels

(obviously, taking also the effect of pumping into account), we can control the rate and direction of flow in an aquifer, control the movement of water bodies of inferior quality, e.g., sea water intrusion, control spring discharge, and control seepage to or out of adjacent water bodies (rivers and lakes), etc. Because of the very low value of storativity of a confined aquifer, relatively small volumes of recharge are needed in order to produce a large rise in piezometric head elevations in such an aquifer. This fact is used, for example, to control sea water intrusion in coastal confined aquifers (e.g., Laverty and van der Goot, 1955).

2. *Storage of water.* Water can be stored in an aquifer, to be pumped at a later time. Phreatic aquifers may serve as very large storage reservoirs. Water is stored in the void space of such aquifers (see Sec. 5-1), and can be put to use at a later time by pumping (Fig. 3-3). For example, a portion of a phreatic aquifer of (horizontal) area of 100 km² and storativity of 15% can store as much as $150 \times 10^6 \, m^3$ of water if water levels are raised by 10 m. As groundwater is never at rest, if no use is made of the stored water, it will gradually flow out of the area as groundwater outflow. However, due to the relatively slow movement of water, appropriate management will result in recovering most of the stored water.

Both long-term and short-term shortage may be practiced. In years with excess surface runoff, water may be diverted from streams and lakes to be stored in aquifers for use in dryer years. Short-term storage may be practiced in order to make a more efficient use of the water supply lines. Water may be delivered to a demand area at a constant rate throughout the year, to be stored in the aquifer when supply exceeds demand, and pumped by local wells to supplement demand in excess of direct supply.

Obviously, storage of water in any quantity of economical and hydrological significance is possible only in phreatic aquifers, where the storativity is related to the porosity (or at least to the effective one) and not, as in a confined aquifer, to the elastic properties of the water and the solid skeleton of the aquifer (see Sec. 5-1).

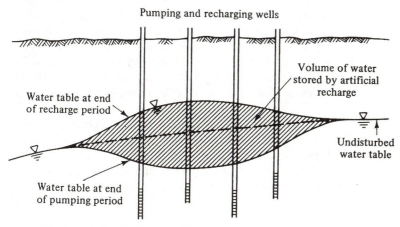

Pumping and recharging wells

Volume of water stored by artificial recharge

Water table at end of recharge period

Undisturbed water table

Water table at end of pumping period

Figure 3-3 Storage of water in an aquifer.

In each particular case a comparison should be made between storage in aquifers and storage in surface reservoirs. Among the points to be considered in such comparison, we could mention

Storage in Aquifers

Cost of recharge wells; cost (and availability) of land for infiltration basins.

Cost of pumping wells (in excess of ordinary pumping).

Cost of energy.

Loss of water by evapotranspiration (only if water levels are very close to ground surface).

Loss of water by seepage (can be avoided by proper management).

Storage in Surface Reservoirs

Cost (and availability) of land.

Cost of dams and other regulating installations.

Recovery of energy by producing hydroelectric power.

Benefits from recreation.

Loss of water by evaporation.

Loss of water by infiltration.

Availability of good sites (geologically).

Possibility of contamination.

Distance from area of demand may require costly transportation.

In many instances, especially in arid and semiarid regions, storage in (phreatic) aquifers has been proven to be more economical than surface storage.

3. *Control of water quality.* As the water introduced into an aquifer and the indigenous water of the aquifer move, they mix as a result of hydrodynamic dispersion (Chap. 7). Mixing is also achieved by wells which pump simultaneously from the two kinds of water (Sec. 7-10). We can control the quality (in terms of dissolved matter) of pumped water by manipulating pumping and artificial recharge, thus controlling the movement of the water bodies introduced into the aquifer and the mixing that takes place in the aquifer and in the pumping wells.

The water used for artificial recharge may be either water of a quality higher than that of the indigenous water of the aquifer, or of an inferior quality. In the former case we improve the quality of the pumped water. In the latter case, we lower the quality. Yet if the resulting quality is still within the permissible range, we may upgrade the efficiency of the entire water resource system by increasing the total quantity of water thus made available and, perhaps, making use of water which may otherwise be unacceptable from the quality point of view.

Due to the very slow movement of water in the aquifer, a period of years, sometimes many years, may elapse between the time water is introduced into an aquifer and the time it is pumped. During that time phenomena such as chemical reactions among constituents present in the water, interaction with the solid skeleton (adsorption and ion exchange), decay (e.g., radioactive), and filtering may take place. Thus the aquifer acts to improve the quality of the injected water.

It is for that last reason that very often reuse of reclaimed sewage water is implemented in conjunction with artificial recharge.

Suspended fine material in surface water used for artificial recharge can be removed by the filtering that takes place as the water percolates through the bottom of a spreading basin and the soil underlying it on its downward way to the aquifer.

Of special interest is the improvement of water quality (e.g., removal and destruction of microorganisms) as the recharge water percolates through the unsaturated zone.

In addition to these major objectives, we may also mention the following additional ones.

1. Supplementing the difference between the demand for groundwater and the natural replenishment of an aquifer.
2. Disposal of liquid waste into deep formations where it will stay, or move very slowly (sometimes for thousands of years) towards outlets. It is always important to verify by thorough geohydrological investigations that indeed there exists no possibility of contact between the injected waste and groundwater which will eventually be used.
3. Using the aquifer as a conduit or a water distribution system. By recharging and pumping, water levels are raised and lowered, respectively. It is therefore possible to create a flow pattern within the aquifer from the area of recharge to that of withdrawal by pumping, with the aquifer serving as a conduit. Wells distributed over an area may withdraw water for local use, thus avoiding the need for a distribution system.
4. Maintenance of high water levels (or heads) to prevent land subsidence or other undesirable phenomena which result from lowered water levels (e.g., damage to foundations).
5. Conservation of water. For example, water used only for cooling can be re-circulated by injecting the warm water back into the aquifer from which it is pumped.

In most cases, artificial recharge is implemented to achieve a number of goals and in conjunction with the utilization of surface water. However, in spite of many advantages, one should carefully examine in each case the danger of permanently (or at least for a long time) damaging an aquifer's water quality by recharging it with toxic or nondegradable pollutants.

Methods

Artificial recharge can be implemented by several methods. The choice of method for each particular case depends on the source of water, the quality of the water, the type of aquifer, the topographical and geological conditions, type of soil, economic conditions, etc.

(a) Methods for enhancing infiltration In these methods, the objective is to increase infiltration by various agrotechniques which affect ground surface roughness, slope, vegetation cover, etc. The purpose is to extend the time and area over which infiltration from surface runoff takes place. Both the slopes of the watershed and the drainage channel network can be treated to achieve this purpose. For example, small (rock and wire) check dams in the natural channels will cause the water to spread over a larger area.

(b) Surface spreading methods Here water is diverted to specially constructed ponds or basins, and allowed to infiltrate through their pervious bottom. Sometimes ditches, dug along ground surface contours, are used instead of basins. Figure 3-4 shows a typical scheme of a project in which water is diverted from an intermittent stream (usually the peak of large floods, carrying large quantities of silt and debris, are not diverted) to a settling basin (where most of the fine material is removed) and then to infiltration basins. Wells are located at some distance from the infiltration basins to allow for a certain minimum retention time (say, one year) before pumping.

Two objectives are achieved by the project shown in Fig. 3-4. Storage (say, if water in the river is available in winter and is needed for irrigation in summer) and purification. The latter is related to the filtering of fine materials, mainly in the

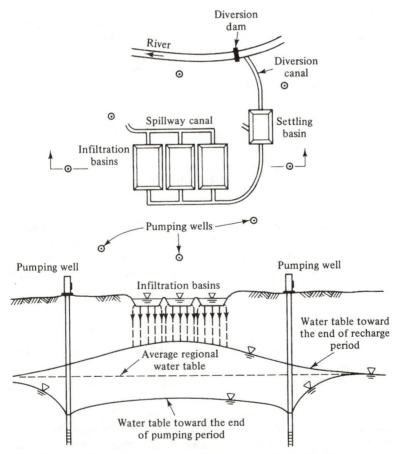

Figure 3-4 Schematic project of recharge by means of infiltration basins.

settling basin, but also through the soil layer just beneath the infiltration basins.

Ditches and furrows are also sometimes used instead of basins. Excess irrigation, especially during nonirrigation seasons, can also be used as a method for artificially recharging the underlying aquifer.

Obviously surface spreading techniques should be implemented only when the recharged aquifer is a phreatic one, and when no impervious layer of significant areal extent is present between the bottom of the spreading basins and the water table.

The economy of artificial recharge by surface spreading techniques depends to a large extent upon the maintenance of high infiltration rates. Depending on the type of soil, rates of 3–15 m/day (that is, 15 m^3/m^2/day) have been observed in gravel, up to 3 m/day in gravel and sand, up to 2 m/day in fine sand and sandstone, and up to 0.5 m/day in sand and silt. Values lower than 0.5 m/day have also been reported. All these figures represent initial values, because as infiltration continues through an artificial recharge basin, its bottom becomes gradually clogged. A typical curve showing the reduction in infiltration rate is presented in Fig. 3-5. The initial reduction in infiltration rate is caused by dispersion and swelling of soil particles after wetting. The subsequent increase results from elimination of entrapped air by solution in the water. The following reduction in infiltration rate, which has, more or less, an exponential form, is due to the clogging of the soil pores at the bottom of the basin and just beneath it. This clogging is due to the retention of suspended solids (when present in the recharge water, as, for example, when water is diverted from flash floods), growth of algae and bacteria (when nutrients are present in the water), entrained or dissolved gases released from the water, and precipitation of dissolved solids and chemical reactions between dissolved solids and the soil particles and/or the native water present in the void space.

When the infiltration rate in a basin drops below some design value, its use as a recharge basin is discontinued. By drying it, cleaning, and sometimes scraping the top 2–5 cm of bottom material, etc., the infiltration rate is brought back almost to its initial value and the basin is put back to use. The frequency of cleaning depends, of course, on the local conditions (type of water and soil) and may be as often as every

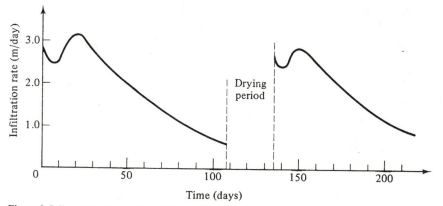

Figure 3-5 Example of variation of infiltration rate with time.

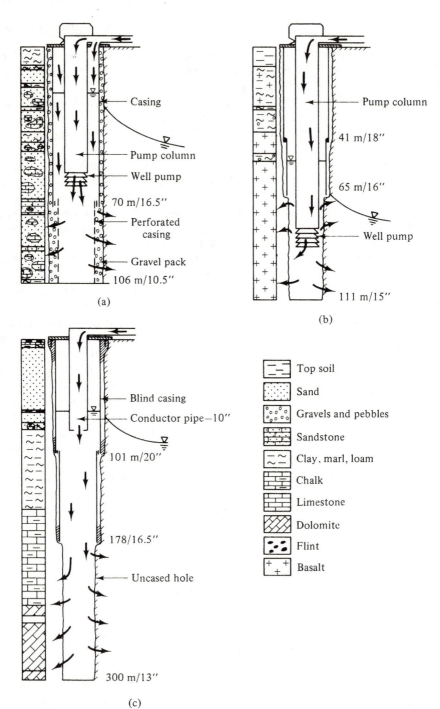

(a)

(b)

(c)

Casing

Pump column

Well pump

70 m/16.5″

Perforated
casing

Gravel pack

106 m/10.5″

Pump column

41 m/18″

65 m/16″

Well pump

111 m/15″

Blind casing

Conductor pipe—10″

101 m/20″

178/16.5″

Uncased hole

300 m/13″

Top soil

Sand

Gravels and pebbles

Sandstone

Clay, marl, loam

Chalk

Limestone

Dolomite

Flint

Basalt

Figure 3-6 Typical recharging wells. (a) In sandstone. (b) In basalt. (c) In limestone *(Harpaz, 1971)*.

three months or once in 2–3 years. Sometimes it is found economical to pretreat the water (chlorination, polyelectrolytes, etc.) to increase the period of effective operation.

Sometimes, pits and shafts are used instead of basins.

(c) Artificial recharge through wells Artificial recharge can be carried out through ordinary pumping wells, or through specially constructed recharging wells. It is also possible to design a dual purpose well. Figure 3-6 shows some typical recharging wells in use in Israel (Harpaz, 1971), where artificial recharge is an intrinsic part of the operation of the national water resources system.

The phenomenon of clogging, accompanied by a reduction in injection rate, occurs also in wells, and for reasons similar to those mentioned above in connection with clogging of infiltration basins. However, in wells clogging is usually under anaerobic conditions. Because the velocity of the injected water decreases as water travels away from a well, the deposition of fines will occur at some distance from a recharging well, making cleaning more difficult.

Artificial recharge through wells is practiced (a) for recharging confined aquifers, (b) when extended impervious layers are present between the ground surface and an underlying phreatic aquifer, (c) when land is expensive or unavailable, and (d) when existing pumping wells can be used for recharge, thus eliminating the need for costly artificial recharge installations. Because clogging is more severe in wells, artificial recharge through wells is mostly implemented with higher quality water (often drinking water quality). Pretreatment (chlorination, sedimentation, filtration, etc.) is the rule rather than the exception. In spite of all these precautions, clogging does take place. Figure 3-7 shows some examples of reduction of injection rate in wells in a sandstone aquifer in Israel.

When the injection rate drops to below some design value, renovation is possible by various chemical treatments (e.g., acidation, oxygen supply, enzymes), by pumping at a high rate (backwashing), or some combination of the various methods.

(d) Induced recharge This term is used for the case when withdrawal installations, in the form of a gallery or an array of shallow wells are located at a relatively small distance from a river, or a lake, and parallel to it. By withdrawing water through these installations, the groundwater table is lowered, thus inducing the movement of water from the river, or the lake, into the aquifer, provided, of course, that the river bed is not completely clogged. Figure 3-8 shows a typical cross section with induced recharge. The gallery intercepts water originally drained to the river, but also water from the river.

By induced recharge we can achieve two goals: recharge the aquifer by river water, to be pumped for beneficial uses, without constructing any recharge installations (the aquifer itself is used as a conduit), and filtration and purification of the river water as it travels through the aquifer towards the abstraction installations. The aquifer acts as a large slow sand filter. Detention time of 2–3 months may serve as a typical example.

The brief review of artificial recharge presented above is by no means a full

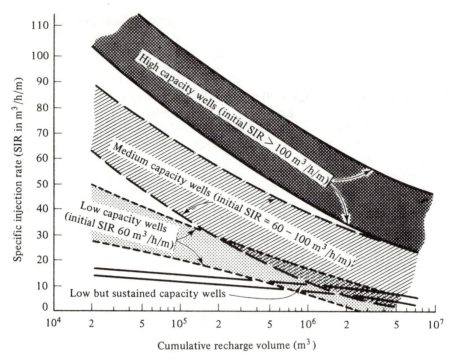

Figure 3-7 Observed specific injection rates (SIR) towards the end of recharge period in recharging-pumping wells in a sandstone aquifer *(Harpaz, 1971)*.

coverage of this important subject for those who manage groundwater resources. It is presented here both as an element in the water balance, and also as an example of a technique which requires the knowledge of most of the material presented in this book for the design of its implementation. For example:

Determining travel time of water injected into aquifers (Chap. 4).
Forecasting water level changes caused by artificial recharge in different aquifers. (Chap. 5).
Studying the movement of the water infiltrating from spreading basins through the unsaturated zone. (Chap. 6).
Determining the changes of quality that take place both in the aquifer and in the unsaturated zone (Chap. 7).
Determining the movement of an (assumed) abrupt front between injected and indigenous water (Sec. 7.10).
Determining the cone of build-up around recharging wells (Chap. 8).
Preventing sea water intrusion by maintaining a pressure barrier (Chap. 9).

A vast amount of literature is available on artificial recharge. Among summaries available on this subject we may mention the works of Schiff (1955, 1964), Baumann (1965), Schwarz and Rebhun (1968), Harpaz (1971), Marvin *et al.* (1971), Fetter and Holzmacher (1974), and Huisman (1964, 1975).

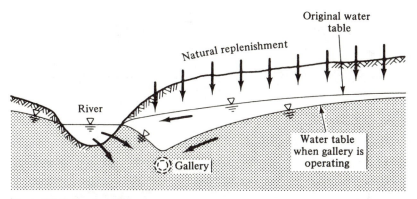

Figure 3-8 Induced recharge.

We may also mention the Symposium on Artificial Recharge and Management of Aquifers, held at Haifa, Israel, by the International Association of Scientific Hydrology in 1967, a Symposium on Artificial Ground Water Recharge, held in 1970 at Reading, England, by the Water Research Association, and the Second International Symposium on Underground Waste Management and Artificial Recharge, by the American Association of Petroleum Geologists, held in New Orleans in 1973.

Annotated bibliographies were published by Todd (1959a), International Association of Scientific Hydrology (1970), Signor *et al.* (1970), Boen *et al.* (1971), and Bize *et al.* (1972).

3-5 RIVER – AQUIFER INTERRELATIONSHIPS

Rivers passing through a region underlain by a phreatic aquifer (and in special cases even by a confined aquifer) may either contribute water to the aquifer or serve as its drain. Much of the low water flow in streams (base flow) is derived from groundwater whose water table elevations in the vicinity of a stream are higher than the stream. Such streams are called *effluent streams* (Fig. 3-9a). On the other hand, when the water level in a stream is higher than the water level in an adjacent (or underlying) aquifer, water will flow from the river to the aquifer. The river is then called an *influent river* (Figs 3-9b and c). When a stream cuts through an impervious layer, establishing a direct contact with an underlying confined aquifer, the stream may be either an influent one or an effluent one, depending on whether the piezometric heads in the aquifer are above or below the water level in the stream (Fig. 3-9d). The same stream can be an influent one along one stretch and an effluent one along another. Or, it can be both influent and effluent at the same point, as shown in Fig. 3-9e.

Obviously, the entire discussion presented above is based on the assumption that the river bed is not completely clogged and that water can flow freely through the river bed. Otherwise, there is no hydraulic contact between the water in the

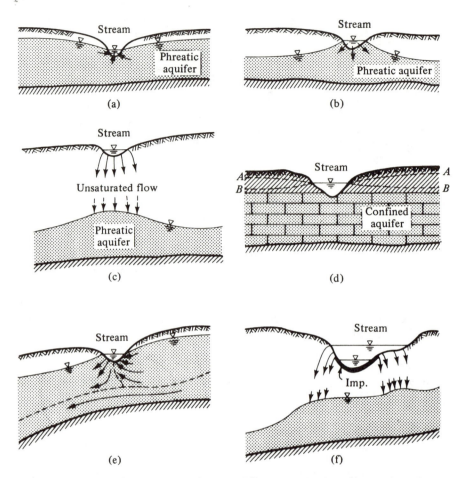

Figure 3-9 River–aquifer interrelationships. (a) Effluent stream. Groundwater drains into stream. (b) Influent stream (shallow water table). River contributes to groundwater flow. (c) Influent stream (deep water table). River contributes to groundwater flow. (d) Influent stream (piezometric surface B), or effluent one (piezometric surface A) intersecting a confined aquifer. (e) A stream which is both influent and effluent. (f) A partly clogged influent stream.

river and in the aquifer and no relationship exists between the two. It is possible that the profile of a stream is such that its deeper part, accommodating for low flows, is completely clogged, while above a certain level, the river bed is pervious (Fig. 3-9f).

When the water table under an influent stream is sufficiently deep, a mound is formed in the former by the percolating water (Figs. 3-9c and f).

The volume of water contributed to an aquifer by streamflow (or drained into a stream from an aquifer), is part of the regional water balance. The rate of flow in either direction, also when the stream bed is partly clogged, can be calculated.

In view of the different possible situations shown in Fig. 3-9, the river may play several roles when solving a groundwater forecasting problem.

(a) The river may act as a *boundary of specified head* to the flow domain in the adjacent aquifer. This situation is shown in Figs 3-9a, b, d, and e. We have in mind an aquifer-type flow, which is based on the assumption of essentially horizontal flow in the aquifer, so that the shape of this boundary is a curve in the xy plane. This is a good approximation, overlooking the details of the flow net under the stream. A somewhat different boundary condition should be employed when the river bed is semipervious (Sec. 5-5).

(b) The river serves as a *source*, contributing water to the aquifer. The rate of leakage depends on the depth of water in the river and on the permeability of the riverbed. This can be a line (actually curve) source, or a strip of some width, when the river is sufficiently wide. The main point is that the rate of seepage is independent of the water levels in the aquifer (Figs 3-9c and f).

Thus, depending on the elevation of the water table, both situations (a) and (b) above are possible at different periods of time for the same stretch of river (and certainly for different stretches of the same river).

Finally, we note that when a river is sufficiently large in terms of the rate of flow, the exchange of water between it and an aquifer practically does not affect the flow, and hence the depth of flow, in it. However, in small streams, the leakage itself may lower the water level in the stream and even completely dry up the stream. We may encounter a passage from condition (a) to (b) as defined above.

Often, the type of situation to be realized is not known a priori, when solving a forecasting problem and some trial and error, or iteration technique is required.

3-6 SPRINGS

A spring is a point (sometimes a small area) through which groundwater emerges from an aquifer to the ground surface. The discharge of some springs is small and of no significance in the groundwater balance; however, some are very large and dominate the flow pattern in their vicinity. For example, probably the only outlets of the large limestone aquifer along the coast of Israel,—one of the two major aquifers of the country—are two springs: the Yarkon Spring and the Taninim Spring.

Figure 3-10 shows several types of springs. A *depression spring* (Fig. 3-10a) occurs when a high water table intersects the ground surface. A *perched spring* (Fig. 3-10b) occurs when an impervious layer which underlies a phreatic aquifer intersects the ground surface. A confined aquifer can be drained in the form of a spring either through a pervious fault or fissure reaching the ground surface, or where it becomes exposed at the ground surface.

The instantaneous rate of discharge of a spring depends on the difference between the elevations of the water table (or piezometric head) in the aquifer in the vicinity of the spring, and the elevation of the spring's threshold (point A in Fig. 3-10). During the dry season, the spring's discharge is derived from water

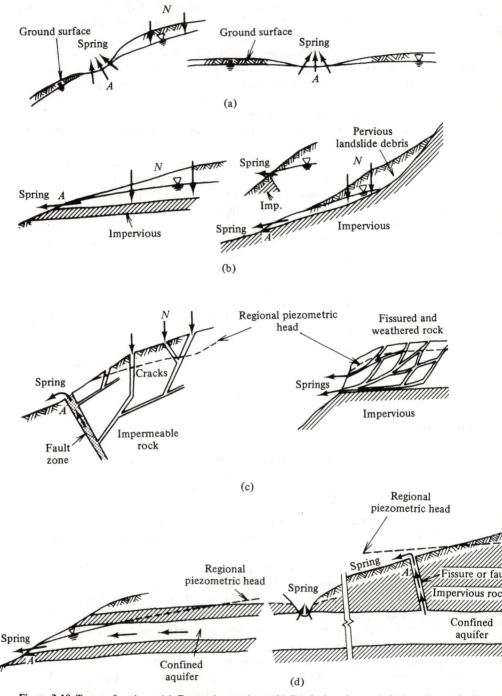

Figure 3-10 Types of springs. (a) Depression springs. (b) Perched springs. (c) Springs in cracked, impermeable rock. (d) Springs in a confined aquifer.

stored in the aquifer. Hence, water levels in the aquifer will gradually decline and with it the spring's discharge.

The spring may completely dry up when water levels fall below its threshold. The relationship between the rate of decline of discharge thus depends on the storage characteristics of the aquifer (storativity and geometry of aquifer, e.g., areal extent).

Figure 3-11 shows a typical portion of a spring's hydrogram, the *recession* (or *depletion*) portions of this hydrogram, corresponding to the dry seasons. On a semilog paper (with time on the linear scale) the recession curve usually plots as a straight line.

Using the simple model of a spring draining an aquifer, shown in Fig. 3-12, with $Q = \alpha_1 h$, $\alpha_1 =$ a constant, we obtain during the recession period

$$Q\, dt \equiv \alpha_1 h\, dt = -\, SA\, dh \tag{3-11}$$

By solving this equation, with $t = t_0$, $h = h_0$, $Q = Q_0 = \alpha_1 h_0$, we obtain

$$t - t_0 = (SA/\alpha_1) \ln (h_0/h) = (SA/\alpha_1) \ln (Q_0/Q)$$

$$Q(t) = Q_0 \exp\left[-\frac{\alpha_1}{SA}(t - t_0) \right] \tag{3-12}$$

which plots as a straight line on a semilog paper (Q on the logarithmic scale).

To obtain an interpretation of α_1 above, consider another simple model of steady flow to a spring, shown in Fig. 3-13. As we shall learn below (Sec. 4-5), the steady rate of flow, Q, in this model is given by

$$Q = WK \frac{h_L^2 - h_0^2}{2L} = WK \frac{h_L + h_0}{2} \frac{h_L - h_0}{L} \tag{3-13}$$

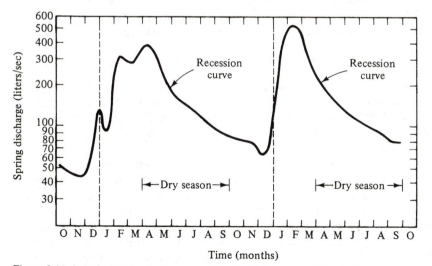

Figure 3-11 A typical spring hydrogram with significant seasonal fluctuations.

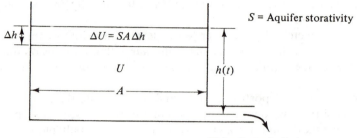

Figure 3-12 A simple model of a spring draining an aquifer.

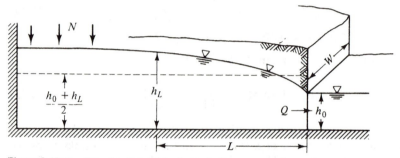

Figure 3-13 Another simple model of steady flow to a spring.

The product $K(h_L + h_0)/2 = T$ represents an average transmissivity of the aquifer, $h_L - h_0$ represents a difference of head above the spring's threshold. Thus, with $Q = WT(h_L - h_0)/L$, we may express α_1 by $\alpha_1 = (W/L)\,T$, where L is a characteristic length of the aquifer, or we may express the coefficient α_1/SA by $\alpha_2 T/S$, where $\alpha_2 = (W/LA)$ is a coefficient (dims. L^{-2}) representing the geometry of the aquifer.

We may conclude by expressing the spring's recession curve by the formula

$$Q = Q_0 \exp\left[-\beta(t - t_0)\right] \tag{3-14}$$

where we have some indication as to the nature of the coefficient $\beta = \alpha_2 T/S$.

Often, the aquifer contributing to the spring is made of several separate subregions, each with its own characteristic coefficient β. Formulas other than (3-14) are also presented in the literature. Mero (1963) presents a detailed analysis of spring hydrographs.

Since, in one form or another, the coefficient, or coefficients, appearing in the expression describing a spring's recession curve are related to the aquifer's geometry, transmissivity, and storativity, it is possible to investigate these properties of an aquifer by analyzing the hydrograph of the spring's discharge. We should note that in the above models, no pumpage or recharge takes place during the analyzed depletion period. It is obvious that when wells operate in the region they affect the water levels, which, in turn, affect the spring's discharge. In other words, spring discharge can be controlled (and that includes drying up a spring if it is indicated by the optimal management scheme) by controlling water levels in its vicinity.

In a regional study, a spring serves as a boundary condition. Usually it is considered a fixed head (= elevation of physical threshold) boundary condition. However, we have to watch for the possibility that at some point in time, which is a priori unknown, water levels may drop to below the threshold; the spring then dries up and ceases to act as a boundary of the flow domain. As water levels rise, it may return to its role as boundary of the flow domain. Another point to watch for is the possibility that as the rate of flow increases, a layer of water of a certain thickness covers the spring, thus making the water level at the spring a function of the discharge rate.

In a management problem, a constraint of minimum spring discharge (say to supply downstream consumers or to maintain wildlife) is sometimes imposed.

3-7 EVAPOTRANSPIRATION

This is another mechanism by means of which groundwater may leave an aquifer. *Evaporation* is the net transfer of water from the liquid phase to the vapor one. *Transpiration* is the process by means of which plants remove moisture from the soil and release it to the atmosphere as vapor. *Evapotranspiration*, a combination of the above two processes, is the term used to describe the total water removal from an area partly covered by vegetation, by transpiration, evaporation from soil (actually from the water present in the void space of unsaturated soil), from snow, and from open water surfaces (lakes, streams, and reservoirs).

The amount of energy required to evaporate 1 cm^3 of water is 597 calories. The source of energy for the process of evapotranspiration in the hydrological cycle is solar energy. However, the actual amount of energy available for evapotranspiration depends on the type of surface and the degree of cloudiness.

Given this amount of energy, the actual evapotranspiration also depends on temperature, air pressure, wind, salinity of water, and the curvature of the air–water interface through which evaporation takes place. Obviously, evapotranspiration requires the availability of water. The term *potential evapotranspiration* is often used to define the evapotranspiration that would occur were there an adequate supply of soil moisture at all times. Actual evapotranspiration is less than, or at most equal to, potential evapotranspiration. The latter is affected mainly by meteorological factors, whereas the former depends on plant and soil conditions.

Various methods are available for determining actual and potential evapotranspiration. Among the better known methods for determining potential evapotranspiration are the Thornthwaite formula (Thornthwaite, 1954; Thornthwaite and Hare, 1965), the Blaney and Criddle formula for estimating potential evapotranspiration when water supply to vegetation cover is not a limiting factor (Blaney and Criddle, 1950), and the Penman method (Penman, 1948). For determining actual evapotranspiration we have the water balance method (based on lysimeter studies, water level fluctuations, or soil moisture balances), and the moisture flux method (e.g., Rider, 1957).

Unless the groundwater table is within 1–1.5 m from ground surface, evaporation from groundwater is practically zero. White (1932) reports 10 percent of pan evaporation at a depth of 1 m. We do have some evaporation from water in the unsaturated zone, but this loss does not enter a water balance which is formulated here for the saturated zone only. When the water table is near ground surface, evapotranspiration may constitute a significant factor in the water balance. Certain plants have very deep roots and take up water by transpiration even from a rather deep water table (e.g., 15–20 m).

3-8 PUMPAGE AND DRAINAGE

Water can be withdrawn from an aquifer for beneficial usages by means of shallow dug wells, tubular deep wells, horizontal wells (also known as radial collector wells) and galleries. The reader is referred to the literature (e.g., American Water Works Assoc., 1958; Campbell and Lehr, 1973) for details on well design and construction.

Often wells penetrate an aquifer only partially. In three-dimensional flows they are line sinks of finite length, producing in their vicinity a converging flow pattern toward them. Under the assumption of essentially horizontal flow in an aquifer, they are point sinks, producing a radially converging flow in their vicinity (to be superimposed on whatever other flow pattern exists in the aquifer). Very seldom should the fact that the well has actually a finite diameter be taken into account.

In consolidated materials, wells are often completed as uncased holes. In unconsolidated materials a gravel pack filter is placed around the well's screen to prevent sand from entering the well. The permeability of the gravel pack is higher than that of the formation.

While pumping wells are point sinks in the flow domain in an aquifer, galleries are line sinks of finite length. Actually, in the vicinity of a gallery (also surrounded by a gravel pack), the flow is two-dimensional in the vertical plane. However, in view of the assumption of essentially horizontal flow in an aquifer, in regional studies we regard it as a line sink in the xy plane. A well can pump water as long as the water table at its location is higher than the elevation of the pump installed in it. For water to enter a gallery, the water table should be above its bottom.

In a regional water balance, we are often interested only in the total withdrawal by pumpage during the balance period. In a detailed forecasting problem, the areal distribution of pumpage is important.

A drainage system (open channels, or buried drains) is usually installed in order to control the elevation of the (ground) water table (say, to maintain water levels below the root zone). Groundwater will then leave the aquifer through this system (say, to a nearby stream) whenever the water table is higher than the drains. The overall behavior is similar to that of a depression spring (Sec. 3-6). The volume of water drained out of an aquifer in this way should not be left out of the water balance.

Usually a second important objective of a drainage system is the removal (with the drainage water) of salts flushed down to the water table. In fact, every advanced irrigation system is always supplemented by a drainage system. Salts drained out of an aquifer in this way should be taken into account in a regional salt balance.

3-9 CHANGE IN STORAGE

The difference between all inflows and outflows during a balance period accumulates in the considered aquifer region. In a phreatic aquifer, water is stored in the void space (i.e., that portion of the void space not occupied already by water and from which air can readily be displaced by water). In a confined aquifer, water is stored on account of water and solid matrix compressibility. In the first case, increased storage is followed by a rise of the phreatic surface. In the second case, by a rise in the piezometric head. The details of the mechanisms are explained in Sec. 5-1. At this point we shall introduce (with no further explanations) the definition of *aquifer storativity*, S, as the volume of water added to a unit horizontal area of aquifer per unit rise in the water table elevation. Over an area A, a volume of water U_w stored in an aquifer causes the water table to rise by Δh

$$U_w = S \times A \times \Delta h \tag{3-15}$$

If the rise is not uniform and the storativity varies from point to point, we can always divide the balance area into N subareas such that

$$U_w = \sum_{j=1}^{N} S_j \times A_j \times (\Delta h)_j \tag{3-16}$$

Obviously, excess of outflow over inflow produces a decline in the water table or in the piezometric surface.

3-10 REGIONAL GROUNDWATER BALANCE

We can now summarize the regional groundwater balance by the following equation

$$
\begin{Bmatrix} \text{Groundwater} \\ \text{inflow} \end{Bmatrix} - \begin{Bmatrix} \text{Groundwater} \\ \text{outflow} \end{Bmatrix} + \begin{Bmatrix} \text{Natural} \\ \text{replenishment} \end{Bmatrix} + \begin{Bmatrix} \text{Return} \\ \text{flow} \end{Bmatrix}
$$

$$
+ \begin{Bmatrix} \text{Artificial} \\ \text{recharge} \end{Bmatrix} + \begin{Bmatrix} \text{Inflow from} \\ \text{streams and lakes} \end{Bmatrix} - \begin{Bmatrix} \text{Spring} \\ \text{discharge} \end{Bmatrix} - \begin{Bmatrix} \text{Evapo-} \\ \text{transpiration} \end{Bmatrix}
$$

$$
- \begin{Bmatrix} \text{Pumpage} \\ \text{and drainage} \end{Bmatrix} = \begin{Bmatrix} \text{Increase in volume of} \\ \text{water stored in aquifer} \end{Bmatrix}
$$

where all terms are expressed as volume of water during the balance period.

FOUR

GROUNDWATER MOTION

As part of the *hydrologic cycle*, groundwater is always in motion from regions of natural and artificial replenishment to those of natural and artificial discharge. Bodies of stagnant, usually saline, water trapped in various porous geological formation do exist, but as long as they do not participate in the hydrologic cycle, they are of little interest to the groundwater hydrologist.

One of the main characteristics of groundwater motion is that it occurs at very, sometimes extremely, low velocities. However, because of the large cross-sectional areas through which this motion takes place, large quantities of water are transported.

This section deals with the basic laws governing the motion of groundwater in aquifers, and with the porous matrix and aquifer properties appearing in these laws. Only saturated flow is considered here. Unsaturated flow is discussed in Chap. 6. In saturated flow, water completely fills the *void space* of the considered porous medium domain. The *continuum approach*, introduced in Sec. 2-5 is employed, and all variables and parameters have already their average meaning in a porous medium regarded as a continuum.

4-1 DARCY'S LAW

Empirical One-Dimensional Form

In 1856, Henry Darcy investigated the flow of water in vertical homogeneous sand filters in connection with the fountains of the city of Dijon (France). Figure 4-1 shows the experimental set-up he employed (Darcy, 1856). From his experiments, Darcy concluded that the rate of flow (i.e., volume of water per unit time), Q, is (a) proportional to the cross-sectional area A, (b) proportional to $(h_1 - h_2)$, and (c) inversely proportional to the length L, where the symbols are defined in Fig. 4-1. When combined, these conclusions give the famous *Darcy formula* (or *law*)

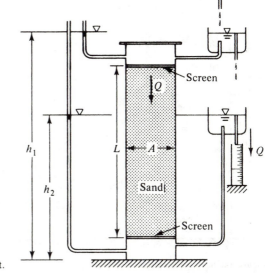

Figure 4-1 Darcy's experiment.

$$Q = KA(h_1 - h_2)/L \qquad (4\text{-}1)$$

where K is a coefficient of proportionality to be discussed in Sec. 4-2 below. The lengths h_1 and h_2 are measured with respect to some arbitrary (horizontal) datum level.

One can easily recognize that here h is the piezometric head and $h_1 - h_2$ is the difference in piezometric head across the filter of length L. As the piezometric head describes (in terms of head of water) the sum of pressure and potential energies of the fluid per unit weight, $(h_1 - h_2)/L$ is to be interpreted as *hydraulic gradient*. Denoting this gradient by $J(=(h_1 - h_2)/L)$ and defining the *specific discharge*, q, as the volume of water flowing per unit time through a unit cross-sectional area normal to the direction of flow, we obtain

$$q = KJ \qquad (4\text{-}2)$$

where $q = Q/A$ and $J = (h_1 - h_2)/L$, as another form of Darcy's formula.

Figure 4-2 shows how Darcy's law (4-1) may be extended to flow through an inclined homogeneous porous medium column

$$Q = KA(\phi_1 - \phi_2)/L; \qquad q = K(\phi_1 - \phi_2)/L = KJ; \qquad \phi = z + p/\gamma \quad (4\text{-}3)$$

where p is pressure and γ is specific weight of water.

The *energy loss* $\Delta\phi = \phi_1 - \phi_2$ is due to friction in the flow through the narrow tortuous paths of the porous medium. Actually, in Darcy's law, the kinetic energy of the water has been neglected, as, in general, changes in the piezometric head along the flow path are much larger than changes in the kinetic energy.

The quotient p/γ appearing in (4-3) and in Fig. 4-2, is called the *pressure*

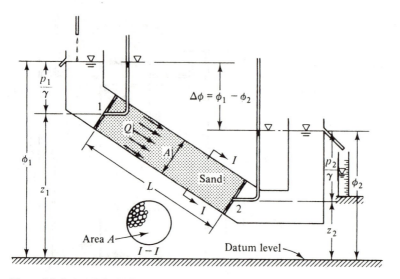

Figure 4-2 Seepage through an inclined sand filter.

head. It represents the *pressure energy* per unit weight of water of specific weight γ at that point. For a compressible fluid under isothermal conditions, $\gamma = \gamma(p)$, the pressure head is defined by

$$\int_{p_0}^{p} \frac{dp}{\gamma(p)}$$

so that the piezometric head becomes

$$\phi^* = z + \int_{p_0}^{p} \frac{dp}{\gamma(p)} \tag{4-4}$$

This piezometric head was introduced by Hubbert (1940). The length z appearing in Fig. 4-2 (z_1 for point 1 and z_2 for point 2) represents the *elevation head*, that is, potential energy per unit weight of water. The sum of the pressure head and the elevation head is the *piezometric head* ϕ.

It is important to note that (4-3) states that the flow takes place from a higher piezometric head to a lower one and not from a higher to a lower pressure. In the case shown in Fig. 4-2, $p_1/\gamma < p_2/\gamma$, that is, the flow is in the direction of *increasing* pressure; however, it is in the direction of *decreasing* head. It is *only* in the special case of horizontal flow, that is, $z_1 = z_2$, that we may write

$$Q = KA(p_1 - p_2)/\gamma L \tag{4-5}$$

Actually, flow takes place only through part of the cross-sectional area A of the column of porous medium shown in Fig. 4-2, the remaining part being occupied by the solid matrix of the porous medium. Since it can be shown that the average areal porosity is equal to the volumetric porosity n, the portion of the area A available to flow is nA. Accordingly, the *average velocity V* of the flow through the column is (Bear, 1972, p. 23)

$$V = Q/nA = q/n \tag{4-6}$$

Sometimes, even in the flow of a single homogeneous fluid, part of the fluid in the pore space is immobile (or practically so). This may occur when the flow takes place in a fine textured medium where adhesion (i.e., the attraction to the solid surface of the porous matrix of the molecules of fluid adjacent to it) is important, or when the porous matrix includes a large portion of dead-end pores. In this case, one may define an *effective porosity with respect to the flow through the medium*, n_{ef} ($<n$), such that

$$V = q/n_{ef} \tag{4-7}$$

One should distinguish clearly between the specific discharge q (to be used, for example, for determining the volume of fluid passing through a given cross-sectional area), and the average velocity, or simply "velocity," (to be used, for example, for front, or particle, movements). Both concepts should not be confused with the (actual or microscopic) local velocity of the fluid at (microscopic) points inside the pore space. Considering dimensions, one should note that q, V, and K have the same dimensions

$$[q] = L/T; \quad [V] = L/T; \quad [K] = L/T \tag{4-8}$$

The coefficient of proportionality, K, appearing in Darcy's law (4-3) is called *hydraulic conductivity* of the porous medium. Some authors call it *coefficient of permeability*. It depends on properties of both the porous matrix and the fluid. A discussion of K is given in the following section.

Extensions of Darcy's Law

The experimentally derived equation of motion in the form of Darcy's law (4-3) is limited to one-dimensional flow of a homogeneous incompressible fluid. When the flow is three-dimensional, the generalization of (4-3) and (4-6) is

$$\mathbf{q} = K\mathbf{J} = -K \operatorname{grad} \phi; \quad \mathbf{V} = \mathbf{q}/n \tag{4-9}$$

where $\mathbf{V}$ is the velocity vector, with components V_x, V_y, and V_z, $\mathbf{q}$ is the specific discharge vector with components q_x, q_y, q_z in the directions of the cartesian, xyz, coordinates, respectively, and $\mathbf{J} = -\operatorname{grad} \phi \equiv -\nabla\phi$ is the hydraulic gradient, with components $J_x = -\partial\phi/\partial x$, $J_y = -\partial\phi/\partial y$, $J_z = -\partial\phi/\partial z$, in the xyz directions, respectively. When the flow takes place through a homogeneous isotropic medium, the coefficient K is a constant scalar, and (4-9) may be written as three equations

$$\left. q_x = KJ_x = -K\partial\phi/\partial x = nV_x; \quad q_y = KJ_y = -K\partial\phi/\partial y = nV_y; \atop q_z = KJ_z = -K\partial\phi/\partial z = nV_z \right\} \tag{4-10}$$

or, for flow in any direction indicated by the unit vector **1s**

$$q_s = \mathbf{q} \cdot \mathbf{1s} = -K\partial\phi/\partial s = nV_s \tag{4-11}$$

It should be recognized that **V** in (4-6) through (4-11), as elsewhere in this book, unless specified otherwise, has the meaning of an average velocity, $\overline{\mathbf{V}}$, although the bar symbol is omitted. Using the concepts of Sec. 2-5, we have from (2-6) and (2-7) for any velocity component in the direction j

$$\overline{V}_j(\mathbf{x}, t) = \frac{1}{U_{0v}} \int_{(U_{0v})} V_j \, dU_v \quad \text{and} \quad n\overline{V}_j \equiv q_j = \frac{1}{U_0} \int_{(U_{0v})} V_j \, dU_v \quad (4\text{-}12)$$

where V_j is the jth component of the local (microscopic) velocity at a point inside the fluid. Equation (4-12) also relates the average velocity to the specific discharge. A more detailed analysis is given in Appendix A2.

Equations (4-9) through (4-11) remain valid also for three-dimensional flow through inhomogeneous media, where $K = K(x, y, z)$, as long as the medium is also isotropic. A discussion on inhomogeneity is given in Sec. 2-6.

It should be emphasized that the velocity **V** as defined by Darcy's law is that of the fluid relative to the solid. This comment may be of little practical importance in most groundwater flows; however, under certain conditions, the movement of the solid grains may play an important role (e.g., in the case of subsidence, Sec. 5-11).

Although (4-9) is presented here just as a formal generalization of the one-dimensional experimental equation of motion to three-dimensional flow, there exist a large number of theories which support this extension (see texts on flow through porous media, e.g., Bear, 1972, p. 104; Gray and O'Neill, 1976).

In an isotropic medium, K is a scalar and the vectors **q** and grad ϕ are co-linear. From (4-9) and the definition of grad ϕ, it also follows that the vector **q** is everywhere normal to the equipotential surface $\phi = \text{constant}$.

When flow takes place in a homogeneous isotropic medium, it is sometimes more convenient to write (4-9) in the form

$$\mathbf{q} = -\text{grad}(K\phi) = -\text{grad } \Phi, \quad \Phi = K\phi, \quad [\Phi] = L^2/T \quad (4\text{-}13)$$

where Φ is called the *specific discharge potential*. It is incorrect to write (4-13) for any medium which is not homogeneous and isotropic. Other forms of (4-9) are

$$\left. \begin{aligned} \mathbf{q} &= -\frac{k}{\mu}(\text{grad } p + \rho g \mathbf{1z}) = -\frac{k}{\mu}\left[\frac{\partial p}{\partial x}\mathbf{1x} + \frac{\partial p}{\partial y}\mathbf{1y} + \left(\frac{\partial p}{\partial z} + \rho g \right)\mathbf{1z} \right], \\ \mathbf{q} &= -\frac{k}{\mu}(\text{grad } p - \rho \mathbf{g}) \end{aligned} \right\} \quad (4\text{-}14)$$

where $K = k\rho g/\mu$, μ is the dynamic viscosity of the fluid, $\rho = \gamma/g$ is the density of the fluid, k is the medium's *permeability* (Sec. 4-2) and $\mathbf{g} = -g \mathbf{1z}$ is the vector of gravity acceleration directed downward. These forms are applicable when $\rho \neq \text{const.}$

For a compressible fluid, where $\rho = \rho(p)$, the motion equation (4-9) may be written as

$$\mathbf{q} = -K \text{ grad } \phi^* \quad (4\text{-}15)$$

where ϕ^* is *Hubbert's potential* defined by (4-4).

Range of Validity

As the specific discharge **q** increases, the relationship between the specific discharge **q** and the hydraulic gradient **J** gradually deviates from the linear relationship expressed by Darcy's law (say in the form of (4-2) or (4-9)). Figure 4-3 shows this deviation. Therefore, it seems reasonable to define a range of validity for Darcy's linear law.

In flow through conduits, the *Reynolds number, Re*, which is a dimensionless number expressing the ratio of inertial to viscous forces acting on the fluid, is used as a criterion to distinguish between laminar flow occurring at low velocities and turbulent flow occurring at higher velocities (see any text of fluid mechanics). The critical value of *Re* between laminar and turbulent flow in pipes is around 2000. By analogy, a Reynolds number is defined also for flow through porous media

$$Re = qd/v \tag{4-16}$$

where d is some representative length of the porous matrix, and v is the kinematic viscosity of the fluid. Although by analogy to the Reynolds number for pipes, d should be a length representing the cross section of an elementary channel of the porous medium, it is customary (probably because of the relative ease of determining it) to employ for d some representative length of the grains (in an unconsolidated porous medium). Often the mean grain diameter is taken as the length dimension d in (4-16). Sometimes d_{10}, that is, the diameter such that 10 percent by weight of the grains are smaller than that diameter, is mentioned in the literature as a representative grain diameter. Collins (1961) suggested $d = (k/n)^{1/2}$, where k is permeability (Sec. 4-2) and n is porosity, as the representative length d. Ward (1964) used $k^{1/2}$ for d.

In spite of the various definitions for d in (4-16), practically all evidence indicates that *Darcy's law is valid as long as the Reynolds number does not exceed some value between 1 and 10.*

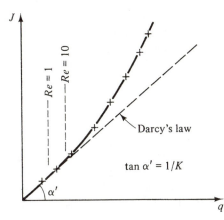

Figure 4-3 Schematic curve relating J to q.

Flow at Large Re

In the range of validity of Darcy's law, i.e., $Re < 1$–10, the viscous forces are predominant. As the velocity of the flow increases, a region of a gradual transition is observed, from laminar flow, where viscous forces are predominant, to still laminar flow, but with inertial forces governing the flow. Often, the value of $Re = 100$ is mentioned as the upper limit of this transition region in which Darcy's law is no more valid (Fig. 4-3). Some authors explain the deviation from the linear law by the separation of the flow from the solid walls of the solid matrix caused at large Re by the inertial forces at (microscopic) points of the pore space where the flow diverges or is curved. At still higher values of Re (say, $Re > 150$–300), the flow becomes really turbulent.

There is no universally accepted non-linear motion equation (that is, relationship between $\mathbf{J}$ and $\mathbf{q}$) which is valid for $Re > 1$–10. Many such relationships appear in the literature. Most of them have the general form (Forchheimer, 1901)

$$J = Wq + bq^2, \qquad \text{or:} \qquad J = Wq + bq^m, \qquad 1.6 \le m \le 2 \qquad (4\text{-}17)$$

where W and b are constants. For example, Kozeny and Carman (see Scheidegger, 1960) suggested

$$J = 180\alpha \frac{(1 - n)^2 v}{gn^3 d^2} q + \frac{3\beta(1 - n)}{4gn^3 d} q^2 \qquad (4\text{-}18)$$

where α and β are shape factors, v is the fluid's kinematic viscosity and d is the grains' diameter. Ergun (1952) suggested a similar equation, but with 150 replacing 180α and 1.75 replacing $3\beta/4$. Ward (1964) proposed

$$J = \frac{v}{gk} q + \frac{0.55}{g\sqrt{k}} q^2; \qquad k = d^2/360 \qquad (4\text{-}19)$$

Most groundwater flows occur at Re well within the laminar flow range, where the linear Darcy law is applicable. Flow at large Re may sometimes occur in Karstic formations or in aquifers in the vicinity of outlets, i.e., very close to wells, springs, etc. (see example in Sec. 8-2).

There are indications that there also exists a lower limit to the validity of Darcy's law, with some threshold gradient necessary to initiate flow. This is true especially for very small pores (see, for example, Kutilek, 1969). However, this phenomenon is of no significance in aquifer flows of practical interest.

4-2 HYDRAULIC CONDUCTIVITY

The coefficient of proportionality K appearing in the various forms of Darcy's law discussed above is called hydraulic conductivity. In an isotropic medium it may be defined, using (4-2), as the specific discharge per unit hydraulic gradient. It is a scalar (dimensions L/T) that expresses the ease with which a fluid is transported through a porous matrix. It is, therefore, a coefficient which depends on both matrix and fluid properties. The relevant fluid properties are density ρ and viscosity μ (or in the combined form of kinematic viscosity ν). The relevant solid matrix properties are mainly grain- (or pore-) size distribution, shape of grains (or pores), tortuosity, specific surface, and porosity. The hydraulic conductivity K may be expressed as (Nutting, 1930)

$$K = k\rho g/\mu = kg/\nu \qquad (4\text{-}20)$$

where k (dimensions L^2)—called the *permeability*, or the *intrinsic permeability*, of the porous matrix—depends solely on properties of the solid matrix.

With (4-20), Darcy's law (4-9) may be written as

$$\mathbf{q} = -(k\rho g/\mu)\,\text{grad}\,\phi \qquad (4\text{-}21)$$

Various formulas relating k to the various properties of the solid matrix are presented in the literature. Some of these formulas are purely empirical, as, for example

$$k = cd^2 \qquad (k \text{ in cm}^2, \ d \text{ in cm}) \qquad (4\text{-}22)$$

where c is a coefficient with cg/ν in the range between 45 for clayey sand, and 140 for pure sand (often the value of 100 is used as an average), and d is the effective grain diameter, d_{10}, defined in Sec. 4-1.

Another example is the Fair and Hatch (1933) formula developed from dimensional considerations and verified experimentally

$$k = \frac{1}{\beta}\left[\frac{(1-n)^2}{n^3}\left(\frac{\alpha}{100}\sum_{(m)}\frac{P_m}{d_m}\right)^2\right]^{-1} \qquad (4\text{-}23)$$

where β is a packing factor, found experimentally to be about 5, α is a sand shape factor, varying from 6.0 for spherical grains to 7.7 for angular ones, P_m is the percentage of sand held between adjacent sieves, and d_m is the geometric mean diameter of the adjacent sieves.

Purely theoretical formulas are obtained from theoretical derivations of Darcy's law. Often, such formulas include numerical coefficients which have to be determined empirically. An example is the Kozeny–Carman equation

$$k = C_0\frac{n^3}{(1-n)^2\,M_{\mathrm{s}}^2} \qquad (4\text{-}24)$$

where M_{s} is the specific surface area of the porous matrix (defined per unit volume of solid) and C_0 is a coefficient for which Carman (1937) suggested the value of 1/5.

Under certain conditions, the permeability, k, may vary with time. This may be caused by external loads which change the structure and texture of the porous matrix by subsidence and consolidation, by the solution of the solid matrix (which over prolonged times may produce large channels and cavities), and by the swelling of clay, if present within the void space. When a soil contains argillaceous material, drying of the soil may shrink the clay, especially bentonite, causing the permeability to air of the dried soil to be higher than for water. Fresh water in a soil sample may cause the clay to swell as compared with salt water, thereby reducing k. Biological activity in the medium may produce a growth

Table 4-1 Typical values of hydraulic conductivity and permeability†

$-\log_{10} K(cm/sec)$	-2	-1	0	1	2	3	4	5	6	7	8	9	10	11
Permeability		Pervious				Semipervious				Impervious				
Aquifer		Good				Poor				None				
Soils		Clean gravel	Clean sand or sand and gravel		Very fine sand, silt, loess, loam, solonetz									
				Peat		Stratified clay			Unweathered clay				
Rocks					Oil rocks			Sandstone	Good limestone, dolomite	Breccia, granite			
$-\log_{10} k(cm^2)$	3	4	5	6	7	8	9	10	11	12	13	14	15	16
$\log_{10} k(md)$	8	7	6	5	4	3	2	1	0	-1	-2	-3	-4	-5

† From Bear, Zaslavsky, and Irmay, 1968.

which tends to clog the matrix, thus reducing k with time. Clogging may also be caused by fines carried by the water (e.g., in artificial recharge).

Various units are used in the practice for the hydraulic conductivity K (dimensions L/T). Hydrologists prefer the unit m/day (meters per day). Soil scientists often use cm/sec. In the USA, as in many countries using the English system of units, two other units are commonly employed by hydrologists. One is a *laboratory*, or *standard*, *hydraulic conductivity* defined as the total discharge (Q) of water at 60°F, expressed in gallons per day, through a porous medium cross-sectional area (A) expressed in ft^2 under a hydraulic gradient $\{(\phi_1 - \phi_2)/L\}$ of 1 ft/ft. With this definition, the units of K are gal/day ft^2. In a similar way, a *field*, or *aquifer*, *hydraulic conductivity* is defined as the discharge of water at field temperature, through a cross-sectional area of an aquifer one foot thick and one mile wide under a hydraulic gradient of 1 ft/mile. The unit is the same as for the laboratory K. Following are some conversions among these units.

$$1 \text{ US gal/day ft}^2 = 4.72 \times 10^{-5} \text{ cm/sec} = 4.08 \times 10^{-2} \text{ m/d}$$

Permeability k (dims. L^2) is measured in the metric system in cm^2 or in m^2. In the English system, the unit is ft^2. For water at 20°C, we have the conversion

$$1 \text{ cm/sec} = 1.02 \times 10^{-5} \text{ cm}^2$$

Reservoir engineers use the unit *darcy* defined by

$$1 \text{ darcy} = \frac{1 \text{ cm}^3/\text{sec/cm}^2 \times 1 \text{ centipoise}}{1 \text{ atmosphere/cm}}$$

with

$$1 \text{ darcy} = 9.8697 \times 10^{-9} \text{ cm}^2 = 1.062 \times 10^{-11} \text{ ft}^2$$

$$= 9.613 \times 10^{-4} \text{ cm/sec (for water at } 20°C)$$

$$= 1.4156 \times 10^{-2} \text{ US gal/min ft}^2 \text{ (for water at } 20°C)$$

Table 4.1 gives a summary of some values of hydraulic conductivity and permeability (Irmay in Bear, Zaslavsky and Irmay, 1968). In this table, following the US Bureau of Reclamation, K is expressed in units of hydraulic conductivity class: $K_c = -\log_{10} K \, (\text{cm/sec})$.

4.3 AQUIFER TRANSMISSIVITY

Consider the flow through the confined aquifer of thickness B shown in Fig. 4-4. If the aquifer is homogeneous and isotropic, with hydraulic conductivity K, the total discharge in the $+x$ direction, Q_x, through the area WB normal to flow is given by Darcy's law

$$Q_x = -KBW\partial\phi/\partial x \equiv KBWJ_x; \quad \mathbf{J}' \equiv -\text{grad}'\phi; \quad J_x = -\partial\phi/\partial x \quad (4\text{-}25)$$

The discharge per unit width of aquifer, Q'_x, normal to the direction of the flow, is

$$Q'_x \equiv Q_x/W = KBJ_x \equiv TJ_x; \quad T = Q'_x/J_x \quad (4\text{-}26)$$

A similar expression can be written for flow in the y direction. In vector form, we may write

$$\mathbf{Q}' = -T\nabla'\phi; \quad \nabla'(\ ) = \frac{\partial(\ )}{\partial x}\mathbf{1x} + \frac{\partial(\ )}{\partial y}\mathbf{1y} \equiv \text{grad}'\phi \quad (4\text{-}27)$$

where the prime symbol indicates that the operation is in the xy plane only.

The product KB, denoted by T, which appears whenever the flow through the entire thickness of the aquifer is being considered, is called *transmissivity*. It is an aquifer characteristic which is defined by the rate of flow per unit width through the entire thickness of an aquifer per unit hydraulic gradient. The concept is valid only in two-dimensional, or aquifer-type flow. In three-dimensional flow through porous media, the concept of transmissivity is meaningless.

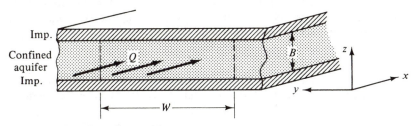

Figure 4-4 Flow through a confined aquifer.

Let an aquifer's thickness, B, vary such that $B(x, y) = b_2(x, y) - b_1(x, y)$, where $b_1(x, y)$ and $b_2(x, y)$ are the elevations of the fixed bottom and ceiling of the confined aquifer. The total discharge through the aquifer can then be expressed by

$$
\left.\begin{aligned}
Q'_x &= \int_{b_1(x,y)}^{b_2(x,y)} q_x \, dz = -\int_{b_1(x,y)}^{b_2(x,y)} K(\partial\phi/\partial x) \, dz \\
Q'_y &= \int_{b_1(x,y)}^{b_2(x,y)} q_y \, dz = -\int_{b_1(x,y)}^{b_2(x,y)} K(\partial\phi/\partial y) \, dz
\end{aligned}\right\}
\tag{4-28}
$$

In the general case, $\mathbf{Q'} = \mathbf{Q'}(x, y)$, $\phi = \phi(x, y, z)$ and $K = K(x, y, z)$.

Whenever we have to integrate a derivative, or to differentiate an integral, we make use of *Leibnitz' rule* (see any text on advanced calculus)

$$
\frac{\partial}{\partial x}\int_{b(x)}^{a(x)} f(x, t)\, dt = \int_{b(x)}^{a(x)} \frac{\partial f(x, t)}{\partial x}\, dt + f(x, a)\frac{\partial a}{\partial x} - f(x, b)\frac{\partial b}{\partial x}
\tag{4-29}
$$

Applying it to (4-28), we obtain for the special case of $K = K(x, y)$

$$
\left.\begin{aligned}
Q'_x &= -KB\frac{\partial\tilde\phi}{\partial x} - K\left[\tilde\phi\frac{\partial B}{\partial x} - \phi(x, y, b_2)\frac{\partial b_2}{\partial x} + \phi(x, y, b_1)\frac{\partial b_1}{\partial x}\right] \\
Q'_y &= -KB\frac{\partial\tilde\phi}{\partial y} - K\left[\tilde\phi\frac{\partial B}{\partial y} - \phi(x, y, b_2)\frac{\partial b_2}{\partial y} + \phi(x, y, b_1)\frac{\partial b_1}{\partial y}\right] \\
\mathbf{Q'} &= -KB\nabla'\tilde\phi - K[\tilde\phi\nabla'B - \phi(x, y, b_2)\nabla'b_2 + \phi(x, y, b_1)\nabla'b_1]
\end{aligned}\right\}
\tag{4-30}
$$

where

$$
\tilde\phi(x, y) = \frac{1}{B}\int_{b_1(x,y)}^{b_2(x,y)} \phi(x, y, z)\, dz
$$

is the average piezometric head along a vertical line at point (x, y). Appendix B gives some more details including the case of a moving boundary (consolidation).

If we now assume essentially horizontal flow, that is, vertical equipotentials, $\tilde\phi \simeq \phi(x, y, b_2) \simeq \phi(x, y, b_1)$, (4-30) may be approximated by

$$
\mathbf{Q'} = -T(x, y)\nabla'\tilde\phi; \qquad T(x, y) = K(x, y)B(x, y)
\tag{4-31}
$$

The error resulting from employing (4-31), based on the assumption of essentially horizontal flow, is given by the second term on the right-hand side of (4-30). We increase the error when expressing $\mathbf{Q'}$ by

$$
\mathbf{Q'} = -K\nabla'(B\tilde\phi)
\tag{4-32}
$$

Another case of interest is $K = K(x, y, z)$, but we *start with the assumption of essentially horizontal flow*, that is, $\phi = \phi(x, y)$ in (4-28). Then (4-28) leads to

$$
\mathbf{Q'} = -T(x, y)\nabla'\phi; \qquad T(x, y) = \int_{b_1}^{b_2} K(x, y, z)\, dz \equiv \bar{K}B
\tag{4-33}
$$

In general, we have

$$\nabla'\tilde{\phi} = \widetilde{\nabla'\phi} - (\tilde{\phi}/B)\,\nabla'B + [\phi(x, y, b_2)\,\nabla'b_2 - \phi(x, y, b_1)\,\nabla'b_1]/B \quad (4\text{-}34)$$

When $\phi(x, y, b_1) \cong \phi(x, y, b_2) \cong \tilde{\phi}(x, y)$, then $\nabla'\tilde{\phi} = \widetilde{\nabla'\phi}$ (that is, the gradient of average head is equal to the average of the head gradient). Equations (4-27), (4-31), and (4-33) are identical if ϕ is understood to mean $\tilde{\phi}$.

The above discussion serves as a justification for employing the concepts of aquifer flow (Sec. 2-4) and aquifer transmissivity also in the cases of inhomogeneous hydraulic conductivity and variable thickness.

For a layered aquifer with horizontal stratification, $K = K(z)$ and (4-33) becomes

$$\mathbf{Q}' = -T\nabla'\phi, \qquad T = \int_0^B K(z)\,dz \quad (4\text{-}35)$$

where we have assumed essentially horizontal flow in the aquifer. When the aquifer is composed of N distinct homogeneous layers, each with thickness B_i and hydraulic conductivity K_i, (4-35) reduces to

$$\mathbf{Q}' = -T\nabla'\phi, \qquad T = \sum_{i=1}^{N} B_i K_i \quad (4\text{-}36)$$

Following the discussion above, (4-36) is valid as an approximation also when $B_i = B_i(x, y)$ and $K_i = K_i(x, y)$.

As indicated in Sec. 2-4, the assumption of essentially horizontal flow may be extended also to leaky aquifers (Fig. 2-5d). Accordingly, the concept of transmissivity may also be extended to such aquifers, with T defined by (4-31). This point is further discussed in Sec. 5-4.

4.4 FLOW IN ANISOTROPIC AQUIFERS

Anisotropy of aquifers was introduced in Sec. 2-6.

In Secs. 4-1 through 4-3, we have been considering aquifers composed only of isotropic materials. Then, the hydraulic conductivity, K, and hence also the transmissivity, T, is a scalar, and the vectors $\mathbf{q}$ and $\mathbf{J}$ (which is everywhere perpendicular to the equipotentials $\phi = \mathrm{const}$) are collinear.

Darcy's law, which in the form of (4-10) is written for a homogeneous isotropic medium, when rewritten for a homogeneous anisotropic medium takes the form (Bear, 1972, p. 137)

$$\left.\begin{aligned} q_x &= K_{xx}J_x + K_{xy}J_y + K_{xz}J_z \\ q_y &= K_{yx}J_x + K_{yy}J_y + K_{yz}J_z \\ q_z &= K_{zx}J_x + K_{zy}J_y + K_{zz}J_z \end{aligned}\right\} \quad (4\text{-}37)$$

In (4-37), q_x, q_y, and q_z are the components in the x, y, and z directions, respectively, of the specific discharge vector $\mathbf{q}$; J_x, J_y, J_z are the components of the hydraulic gradient vector $\mathbf{J}$; $K_{xx}, K_{xy}, \dots, K_{zz}$ are nine constant coefficients. In an inhomogeneous medium, each of these coefficients may vary in space.

The nine coefficients appearing in (4-37) are components of the *second rank tensor of hydraulic conductivity* of an anisotropic medium (Bear, 1972, p. 137). A detailed discussion on the nature of second rank tensors and on methods for treating them is beyond the scope of this book. The reader is referred to texts on tensor analysis (e.g., Morse and Feshbach, 1953; Spiegel, 1959; Aris, 1962). Symbolically, we write

$$[K] \equiv \begin{bmatrix} K_{xx} & K_{xy} & K_{xz} \\ K_{yx} & K_{yy} & K_{yz} \\ K_{zx} & K_{zy} & K_{zz} \end{bmatrix} ; \qquad [K] = \begin{bmatrix} K_{xx} & K_{xy} \\ K_{yx} & K_{yy} \end{bmatrix} \qquad (4\text{-}38)$$

in three- and two-dimensional spaces, respectively.

The hydraulic conductivity tensor is symmetric, that is, $K_{xy} = K_{yx}, K_{xz} = K_{zx}$, and $K_{yz} = K_{zy}$. This means that actually only six distinct components in three-dimensional flow, and three such components in two-dimensional flow, are needed for fully defining the hydraulic conductivity.

Equations (4-37) may be written in several compact forms, for example

$$q_i = K_{ij}J_j; \qquad \mathbf{q} = \mathbf{K} \cdot \mathbf{J} \qquad (4\text{-}39)$$

where subscripts i and j stand for x_i, x_j, respectively, with $x_1 \equiv x$, $x_2 \equiv y$, and $x_3 \equiv z$. In the first form of (4-39), we have employed *Einstein's summation convention* (or the *double index summation convention*) according to which in any product of terms, a suffix (here subscript) repeated twice and only twice is held to be summed over the entire range of values (1, 2, 3 and 1, 2 for three- and two-dimensional spaces, respectively). The component $K_{x_i x_j} (\equiv K_{ij})$ may be interpreted as the coefficient which, when multiplied by the component J_{x_j} of the hydraulic gradient $\mathbf{J}$, gives the contribution of the latter to the specific discharge component q_{x_i} in the x_i direction. The total specific discharge q_{x_i} is the sum of partial specific discharges caused by J_{x_1}, J_{x_2}, and J_{x_3}.

While the hydraulic conductivity of a porous medium (expressed symbolically by $\mathbf{K}$) is independent of the coordinate system which we happen to use, the magnitude of the components K_{ij} depends on the chosen coordinate system. Texts on tensor analysis give the rules of transformation of these components from one coordinate system to another, (see, e.g., Bear, 1972). It is also shown in these texts that it is always possible to find three mutually orthogonal directions in space such that when these directions are chosen as the coordinate system for expressing the components K_{ij}, we find that $K_{ij} = 0$ for all $i \neq j$ and $K_{ij} \neq 0$ for $i = j$. These directions in space are called the *principal directions of the anisotropic medium* (actually of the permeability of the medium).

When the principal directions are used as the coordinate system, (4-38) becomes

$$[K] \equiv \begin{bmatrix} K_{xx} & 0 & 0 \\ 0 & K_{yy} & 0 \\ 0 & 0 & K_{zz} \end{bmatrix} ; \qquad [K] \equiv \begin{bmatrix} K_{xx} & 0 \\ 0 & K_{yy} \end{bmatrix} \qquad (4\text{-}40)$$

and (4-37) reduces to

$$q_x = K_x J_x, \qquad q_y = K_y J_y, \qquad q_z = K_z J_z \tag{4-41}$$

where $K_{xx} \equiv K_x, \; K_{yy} = K_y, \; K_{zz} \equiv K_z$.

An aquifer composed of alternating horizontal layers exhibits anisotropy with a hydraulic conductivity in a direction normal to the layers (K^v) smaller than that in a direction which is parallel to them (K^p) (see Bear, 1972, p. 154).

When the above considerations are applied to flow in an anisotropic aquifer, where flow is in the horizontal xy plane, we obtain

(a) The equation of flow in an anisotropic aquifer:

$$Q'_x = T_{xx} J_x + T_{xy} J_y; \qquad Q'_y = T_{yx} J_x + T_{yy} J_y; \qquad T_{xy} = T_{yx} \tag{4-42}$$

or, in the compact forms

$$\mathbf{Q}' = \mathbf{T} \cdot \mathbf{J}'; \qquad Q'_i = T_{ij} J_j; \qquad i, j = 1, 2 \tag{4-43}$$

where the summation convention is applied to the last form.

(b) If x' and y' are principal directions, then

$$[\mathbf{T}] \equiv \begin{bmatrix} T_{x'x'} & 0 \\ 0 & T_{y'y'} \end{bmatrix} \tag{4-44}$$

and (4-42) becomes

$$Q'_{x'} = T_{x'x'} J_{x'}; \qquad Q_{y'} = T_{y'y'} J_{y'} \tag{4-45}$$

(c) The transformation from any xy coordinates to the principal $x'y'$ coordinates is obtained by using the following relationships

$$\frac{T_{x'x'}}{T_{y'y'}} = \frac{T_{xx} + T_{yy}}{2} \pm \left[\left(\frac{T_{xx} - T_{yy}}{2} \right)^2 + T_{xy}^2 \right]^{1/2} \tag{4-46}$$

When $T_{x'x'}$ and $T_{y'y'}$ are given, and xy are cartesian coordinates rotated clockwise by an angle θ, with respect to $x'y'$, we obtain

$$\left. \begin{aligned} \frac{T_{xx}}{T_{yy}} &= \frac{T_{x'x'} + T_{y'y'}}{2} \pm \frac{T_{x'x'} - T_{y'y'}}{2} \cos 2\theta, \\[2mm] T_{xy} &= -\frac{T_{x'x'} - T_{y'y'}}{2} \sin 2\theta \end{aligned} \right\} \tag{4-47}$$

Flow at Large *Re* in an Anisotropic Medium

The effect of medium anisotropy on flow at large *Re* is more complicated. Barak and Bear (1973), who investigated this problem, suggested the following equation of motion

$$J_i = (v/g) w_{ij} q_j + \beta'_{ijkl} q_j q_k q_l / g q + \beta''_{ijk} q_j q_k / g \tag{4-48}$$

as a good approximation for a Newtonian fluid. In (4-48), w_{ij}, β''_{ijk}, and β'_{ijkl} are tensors of the second, third, and fourth orders, respectively, which represent matrix properties only. At low Re, the last two terms on the right-hand side of (4-48) vanish. The last term describes the effect of matrix non-symmetry.

4-5 DUPUIT ASSUMPTIONS FOR A PHREATIC AQUIFER

A phreatic aquifer is defined in Sec. 2-3 as one in which a water table (phreatic surface) serves as its upper boundary. In Sec. 2-2, we have introduced the fact that actually above a phreatic surface, which is an imaginary surface, at all points of which the pressure is atmospheric, moisture does occupy at least part of the pore space (Fig. 2-3). The capillary fringe was introduced as an approximation of the actual distribution of moisture in the soil above a phreatic surface.

Figure 4-5a shows how the actual moisture distribution is approximated by a step distribution, assuming that no moisture is present in the soil above a certain level. This step defines the height, h_c, of the capillary fringe. Obviously, this approximation is justified only when the thickness of the capillary fringe thus defined is much smaller than the distance from the phreatic surface to the ground surface. Note also that the capillary fringe better approximates reality in the case of a poorly graded soil than in a well-graded soil (Fig. 4-5b). In the capillary fringe (as in the entire aerated zone above the phreatic surface), pressures are negative; therefore, they cannot be monitored by observation wells which serve as piezometers. A special device, called *tensiometer*, is needed in order to measure the negative pressures in the aerated zone (Fig. 6-6b and further details in Sec. 6-1). Water levels in observation wells that terminate below the phreatic surface give elevations of points on the phreatic surface. Using a sufficient number of such points, we can draw contours of this surface.

Thus, the capillary fringe approximation means that we assume a saturated zone up to an elevation h_c above the phreatic surface, and no moisture at all above it. In this case, the upper end of the capillary fringe may be taken as the *groundwater table*, as the soil is assumed saturated below it. However, when h_c is much smaller than the thickness of an aquifer below the phreatic surface, and this is indeed the situation encountered in most aquifers, the hydrologist often neglects the capillary fringe. He then assumes that the (phreatic) aquifer is bounded from above by a phreatic surface. This is also the assumption underlying the presentation in this book.

An estimate of h_c can be obtained, for example, from (Mavis and Tsui, 1939)

$$h_c = \frac{2.2}{d_H} \left(\frac{1-n}{n} \right)^{3/2} \tag{4-49}$$

where h_c is in inches, and d_H is the mean grain diameter, also in inches and n is porosity. Another expression is (Polubarinova-Kochina, 1952, 1962)

$$h_c = \frac{0.45}{d_{10}} \frac{1-n}{n} \tag{4-50}$$

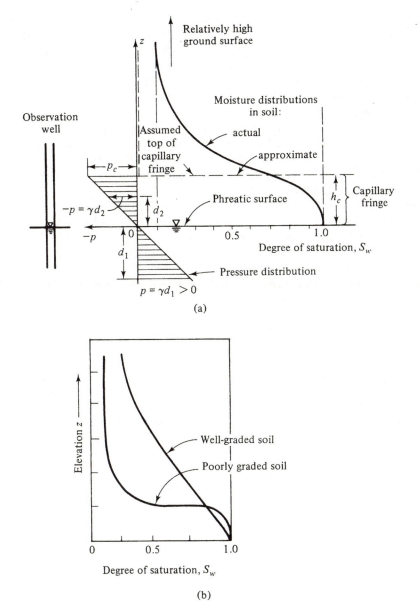

Figure 4-5 (a) Approximation of phreatic surface and capillary fringe. (b) Retention curves in soil.

where both h_c and the effective particle diameter are in centimeters. Silin Bekchurin (1958) suggested a capillary rise of 2–5 cm in coarse sand, 12–35 cm in sand, 35–70 cm in fine sand, 70–150 cm in silt, and 2–4 m and more in clay. Equations (4-49) and (4-50) can be compared with the rise of water in a capillary tube of radius r: $h_c = 2\sigma/r$, where σ is the surface tension of the water.

Both ϕ and $\mathbf{q}$ vary from point to point within a phreatic aquifer. In order to obtain the specific discharge $\mathbf{q} = \mathbf{q}(x, y, z, t)$ at every point, we have to know the piezometric head $\phi = \phi(x, y, z, t)$. In Sec. 5-5 below, we shall show how to derive $\phi = \phi(x, y, t)$ by solving certain partial differential equations. However, because the phreatic surface is never horizontal (except for special cases like water at rest), even in the case of an aquifer with a horizontal impervious bottom, equipotentials are never vertical (Fig. 4-6a) and we have to derive $\phi = (x, y, z, t)$ by solving a partial differential equation in the three-dimensional xyz space. Adding also the fact that on the phreatic surface we have to specify a non-linear boundary condition, and that the location of this surface is unknown before the problem is solved, we immediately realize that this is not a practical way of solving common problems of flow in phreatic aquifers.

The Dupuit assumptions discussed below are probably the most powerful tool for treating unconfined flows. In fact, it is the only simple tool available to most engineers and hydrologists for solving such problems.

Dupuit (1863) based his assumptions on the observation that in most ground-water flows, the slope of the phreatic surface is very small. Slopes of 1/1000 and 10/1000 are commonly encountered. In steady flow without accretion in the vertical two-dimensional xz plane (Fig. 4-6a), the phreatic surface is a streamline. At every point P along this streamline, the specific discharge is in a direction tangent to the streamline and is given by Darcy's law

$$q_s = -K d\phi/ds = -K \, dz/ds = -K \sin \theta \qquad (4\text{-}51)$$

since along the phreatic surface $p = 0$ and $\phi = z$. As θ is very small, Dupuit suggested that $\sin \theta$ be replaced by the slope $\tan \theta = dh/dx$. The assumption of small θ is *equivalent* to assuming that equipotential surfaces are vertical (that is, $\phi = \phi(x)$ rather than $\phi = \phi(x, z)$) and the flow is essentially horizontal. Thus, the Dupuit assumptions lead to the specific discharge expressed by

$$q_x = -K dh/dx, \qquad h = h(x) \qquad (4\text{-}52)$$

In general, $h = h(x, y)$ and we have

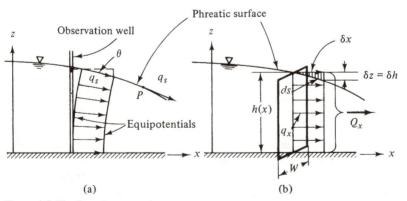

Figure 4-6 The Dupuit assumptions.

$$q_x = -K\partial h/\partial x, \qquad q_y = -K\partial h/\partial y; \qquad h = h(x, y) \qquad (4\text{-}53)$$

or

$$\mathbf{q} = -K\nabla'h; \qquad \nabla'(\ ) \equiv \{\partial(\ )/\partial x\}\, \mathbf{1x} + \{\partial(\ )/\partial y\}\, \mathbf{1y} \qquad (4\text{-}54)$$

Since **q** is thus independent of elevation, the corresponding total discharge through a vertical surface of width W (normal to the direction of flow; Fig. 4-6b) is

$$Q_x = -KWh\partial h/\partial x, \quad Q_y = -KWh\partial h/\partial y; \qquad h = h(x, y) \qquad (4\text{-}55)$$

or, in the compact vector form

$$\mathbf{Q} = -KWh\nabla'h = -KW\nabla'(h^2/2) \qquad (4\text{-}56)$$

Per unit width, we obtain

$$\mathbf{Q'} \equiv \mathbf{Q}/W = -Kh\nabla'h = -K\nabla'(h^2/2) \qquad (4\text{-}57)$$

In (4-55) through (4-57), the aquifer's bottom is horizontal. It should be emphasized that the Dupuit assumptions may be considered as a good approximation in regions where θ is indeed small and/or the flow is essentially horizontal.

The important advantage gained by employing the Dupuit assumptions is that $\phi = \phi(x, y, z)$ has been replaced by $h = h(x, y)$, that is, z does not appear as an independent variable. Also, since at a point on the free surface, $p = 0$ and $\phi = h$, we assume that the vertical line through the point is also an equipotential line on which $\phi = h = \text{const}$. In general h varies also with time so that $h = h(x, y, t)$.

In order to obtain a better understanding of what is involved in the Dupuit assumptions, let us attempt to determine the exact expression for flow in a phreatic aquifer where $\phi = \phi(x, y, z, t)$ and $h = h(x, y, t)$. We obtain it by integrating along the vertical from the bottom of the aquifer, $\eta = \eta(x, y)$, which need not be horizontal, to the phreatic surface at elevation $h = h(x, y, t)$. For flow in the $+x$ direction, assuming $K = \text{const.}$ or $K = K(x, y)$, we obtain

$$Q'_x = \int_{\eta(x,y)}^{h(x,y,t)} q_x\, dz = -K \int_{\eta(x,y)}^{h(x,y,t)} (\partial\phi/\partial x)\, dz$$

$$= -K\left\{ \frac{\partial}{\partial x}\int_\eta^h \phi\, dz - \phi\bigg|_h \frac{\partial h}{\partial x} + \phi\bigg|_\eta \frac{\partial \eta}{\partial x} \right\}$$

$$= -K\left\{ \frac{\partial}{\partial x}[(h - \eta)\tilde\phi] - \phi\bigg|_h \frac{\partial h}{\partial x} + \phi\bigg|_\eta \frac{\partial \eta}{\partial x} \right\} \qquad (4\text{-}58)$$

where the average head is

$$\tilde\phi = \frac{1}{h - \eta}\int_\eta^h \phi\, dz$$

and $\phi|_h \equiv h$ on the phreatic surface.

Equation (4-58) involves no approximation. If we now *assume* a vertical equipotential, i.e.

$$\phi\bigg|_h \, (= h) \cong \phi\bigg|_\eta \cong \tilde\phi,$$

(4-58) reduces to

$$Q'_x = -K(h - \eta)\frac{\partial h}{\partial x}, \qquad \text{or } \mathbf{Q}' = -K(h - \eta)\nabla'h \qquad (4\text{-}59)$$

which is the same as (4-57), written for a non-horizontal bottom.

For a horizontal bottom, $\eta = 0$, (4-58) can be written as

$$Q'_x = -K\frac{\partial}{\partial x}(h\tilde{\phi} - h^2/2), \qquad \text{or } \mathbf{Q}' = -K\nabla'(h\tilde{\phi} - h^2/2) \qquad (4\text{-}60)$$

By comparing (4-60) with (4-57), we see that we have replaced $h\tilde{\phi} - h^2/2$ by $h^2/2$ in the Dupuit assumptions. The error reduces to zero as $\tilde{\phi} \to h$. Bear (1972, p. 363) gives an estimate of the error involved in replacing $\phi'' \equiv h\tilde{\phi} - h^2/2$ by $h^2/2$ in (4-57)

$$0 < \frac{h^2/2 - \phi''}{h^2/2} < \frac{i^2}{1 + i^2}; \qquad i \equiv dh/dx \qquad (4\text{-}61)$$

so that the error is small as long as $i^2 \ll 1$, where i is the slope of the phreatic surface. When the medium is anisotropic, with $K_x \neq K_z$ (x, z principal directions), i^2 in (4-61) should be replaced by $(K_x/K_z)i^2$.

Dupuit–Forchheimer Discharge Formula

As a simple example of the application of (4-52), consider the case of steady unconfined flow through a homogeneous formation between two reservoirs with vertical faces (Fig. 4-7). Following the Dupuit assumptions, the total discharge in the x direction per unit width, through a vertical cross section of height $h(x)$ is given by (4-57)

$$Q'_x \equiv Q' = qh(x) = -Kh(x)\,dh/dx = \text{const.}; \qquad Q'\,dx = -Kh(x)\,dh \qquad (4\text{-}62)$$

By integrating this expression between the boundary at $x = 0$, where $h = h_0$, and any distance x, where $h = h(x)$, we obtain

$$Q'\int_{x^*=0}^{x} dx^* = -K\int_{h^*=h_0}^{h(x)} h^*(x^*)\,dh^*: \qquad Q'x = K\frac{h_0^2 - h^2(x)}{2} \qquad (4\text{-}63)$$

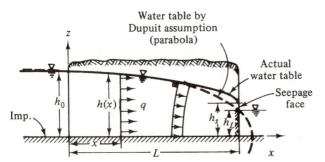

Figure 4-7 Steady unconfined flow between two reservoirs.

Equation (4-63) describes a water table, $h = h(x)$, which has the shape of a parabola passing through $x = 0$, $h = h_0$. If we know $h(x)$ at some distance x, we can use (4-63) to derive Q' (obviously, if K is known). The boundary condition at the other end, $x = L$, however, is somewhat more complicated.

Whenever a phreatic surface approaches the downstream external boundary of a flow domain, it always terminates on it at a point that is above the water table of the body of open water present outside the flow domain. Points A in Figs. 4-8a, b, and c are such points. The segment AB of the boundary above the water table and below the phreatic surface is called the *seepage face*. Along the seepage face, water emerges from the porous medium into the external space, trickling down along the seepage face. In Figs. 4-8c and b, the phreatic surface at A is tangent to the external boundary; in Fig. 4-8a, it is tangent to a vertical line at A.

Back to our problem, because of the presence of a seepage face which terminates at a point on the (*unknown*) phreatic surface, h_s in Fig. 4-7 is unknown. Instead, when the Dupuit assumptions are employed, we approximate the situation by overlooking the presence of the seepage face and assuming that the water table at $x = L$ passes through $h = h_L$. Using this as the downstream boundary condition, we obtain from (4-63)

$$Q' = K \frac{h_0^2 - h_L^2}{2L} \tag{4-64}$$

known as the *Dupuit–Forchheimer discharge formula*.

The parabolic water table is shown in broken line in Fig. 4-7. A discrepancy exists mainly at the boundaries at $x = 0$, where the water table should be tangent to the horizontal line, whereas the parabola has a slope of

$$\left. dh/dx \right|_{x=0} = -Q'/Kh_0$$

and at $x = L$, where the seepage face is neglected. Otherwise, the discrepancy between the curves derived by the exact theory of the phreatic surface boundary, and by the Dupuit approximation is negligible. A simple rule is that at distances

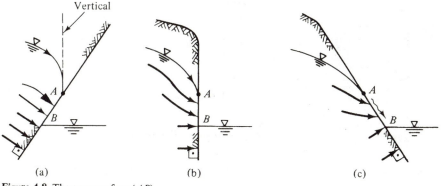

(a)   (b)   (c)

Figure 4-8 The seepage face (AB).

from the downstream end, larger than 1.5–2 times the height of the flow domain, the solution based on the Dupuit assumption is sufficiently accurate for all practical purposes.

Moreover, it can be shown (Bear, 1972; p. 367) that (4-64) *is accurate* as far as the rate of discharge is concerned, although (4-63) does not give the accurate water table elevations $h = h(x)$.

Flow Through a Stratified Phreatic Aquifer

As a second example, consider the case of phreatic flow through the stratified aquifer shown in Fig. 4-9. The flow is made up of two parts, each corresponding to one of the layers. Using the Dupuit assumption of vertical equipotentials, we obtain for the total discharge, $\mathbf{Q}'$, per unit width through the system

$$Q' = -K'a\,dh/dx - K''(h-a)\,dh/dx = \text{const.} \tag{4-65}$$

Upon integration from $x = 0, h = h_0$ to $x = L, h = h_L$ (i.e., neglecting the presence of a seepage face), we obtain

$$Q' = \frac{K''}{2L}(h_0 - h_L)\left[h_0 + h_L - 2a + 2a(K'/K'')\right] \tag{4-66}$$

We could obtain the approximate shape of the phreatic surface by integrating from $x = 0$ to any x. Bear (1972, p. 370) shows that (4-66) is again an accurate expression for Q', although we have made use of the Dupuit assumptions and neglected the seepage face.

Equation (4-66) may be rewritten as

$$Q' = K'a\frac{h_0 - h_L}{L} + \frac{K''}{2L}\left[(h_0 - a)^2 - (h_L - a)^2\right] \tag{4-67}$$

which shows that when the Dupuit assumptions are employed, the total flow may be obtained as the sum of a confined-type flow in the lower layer, and a phreatic-type flow in the upper one.

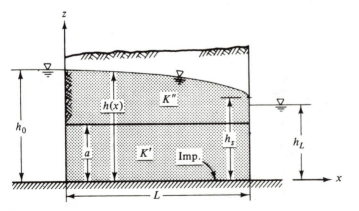

Figure 4-9 Unconfined flow in a horizontally stratified aquifer.

Finally consider the case of unconfined flow in an aquifer with vertical strata shown in Fig. 4-10. Using the Dupuit assumptions, we obtain from (4-62) and (4-63) for the region $0 \leq x \leq L'$

$$h^2 = h_0^2 - 2Q'x/K'; \qquad h_{L'}^2 = h_0^2 - 2Q'L'/K' \qquad (4\text{-}68)$$

Similarly, we obtain for the region $L' \leq x \leq L$

$$h^2 = h_{L'}^2 - 2Q'(x - L')/K''; \qquad h_L^2 = h_{L'}^2 + 2Q'(L - L')/K'' \qquad (4\text{-}69)$$

Hence

$$Q' = \frac{h_0^2 - h_L^2}{2[L''/K'' + L'/K']} \qquad (4\text{-}70)$$

$$h^2 = h_0^2 - \frac{h_0^2 - h_L^2}{K'[L''/K'' + L'/K']}x, \qquad 0 \leq x \leq L'$$

$$h^2 = h_L^2 + \frac{h_0^2 - h_L^2}{K''[L''/K'' + L'/K']}(L - x), \qquad L' \leq x \leq L \qquad (4\text{-}71)$$

Again, the expression for Q' in (4-70) can be shown to be an accurate one (Bear, 1972, p. 373). Equation (4-70) can easily be extended to N strata

$$Q' = (h_0^2 - h_L^2)/2 \sum_{i=1}^{N} (L_i/K_i) \qquad (4\text{-}72)$$

The Dupuit assumptions cannot be applied in regions where the vertical flow component is not negligible. Such flow conditions occur as a seepage face is approached (Fig. 4-11a) or at a crest (*water divide*) in a phreatic aquifer with accretion (Fig. 4-11b). Another example is the region close to the impervious vertical boundary of Fig. 4-11a. It is obvious that the assumption of vertical equipotentials fails at, and in the vicinity of such a boundary. Only at distances $x > \sim 2h_0$ have we equipotentials that may be approximated as vertical lines or surfaces. It is important to note here that in cases with accretion, a horizontal (or almost so) water table is not sufficient to justify the application of the Dupuit assumptions. One must verify that vertical flow components may indeed be

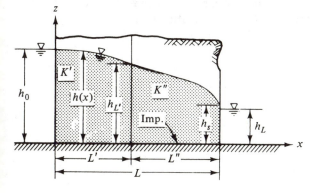

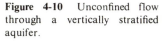

Figure 4-10 Unconfined flow through a vertically stratified aquifer.

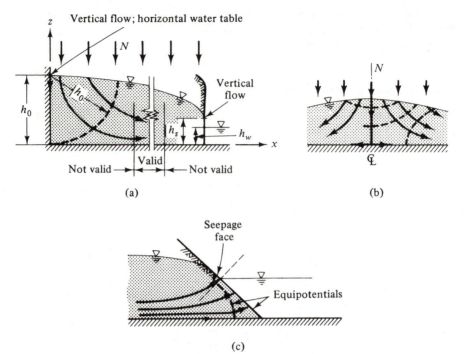

Figure 4-11 Regions where Dupuit assumptions are not valid.

neglected, before applying the Dupuit assumptions. Another case to which the Dupuit assumptions should be applied with care is that of unsteady flow in a decaying phreatic surface mound. Although no accretion takes place, yet at, and in the vicinity of, the crest the flow is vertically downward. At a distance of say 1.5 − 2 times the thickness of the flow, the approximation of vertical equipotentials is again valid.

In spite of what was said above, in regional studies, the Dupuit assumptions, because of their simplicity and the relatively small error involved, are usually applied also to those (relatively small) parts of an investigated region where they are not strictly applicable. One should, however, be careful in making use of results (say, water levels) derived for these parts of an investigated region.

FIVE

MATHEMATICAL STATEMENT OF THE GROUNDWATER FORECASTING PROBLEM

The basic laws governing the flow of water in confined and phreatic aquifers are presented in the previous section. However, if we observe (4-9) closely, we notice that we actually have here one equation with two dependent variables: $\mathbf{q}(x. y, z, t)$ and $\phi(x, y, z, t)$, or three equations in four unknowns ϕ, q_x, q_y, q_z, if we refer to components of $\mathbf{q}$. This means that one additional equation is required in order to obtain a complete description of the flow regime in an aquifer. Similarly, we have $\mathbf{Q}'(x, y, t)$ and $\phi(x, y, t)$ in the single equation (4-27) and $\mathbf{Q}'(x, y, t)$ and $h(x, y, t)$ in the single equation (4-57). The additional basic law that we have to invoke is that of *conservation of matter, or mass,* which here takes the form of a *continuity equation.*

One should not be surprised that in Sec. 4-5 we did succeed in solving (4-57) for some simple cases. We actually used there the equation of continuity, which in that case took the simple form of $Q' = \text{const.}$

Our objective in what follows is to develop the continuity equations for different types of aquifers. The distribution of $\phi = \phi(\mathbf{x}, t)$ in an aquifer is obtained by solving these equations, subject to appropriate boundary and initial conditions.

We shall first consider the basic equations and boundary conditions for three-dimensional flows. Then, the equations for flow in aquifers will be developed for confined, leaky, and phreatic aquifers. We shall derive these (integrated, or averaged) aquifer equations in two ways. First, by merely assuming that the flow in an aquifer is essentially horizontal (see Secs 4-3 and 4-5), and writing a balance for a control volume which has the height of the saturated flow domain in the aquifer, and secondly, by integrating the point continuity equation over the vertical height of the aquifer. In this way, the conditions on the confined, leaky, or phreatic, upper and lower boundaries of the aquifers will be incorporated in the resulting integrated equations. It will be of interest to note that although the latter method is more rigorous, the results are identical.

With the material presented in this section, one should be able to state mathematically *any* groundwater flow problem. The problem of the movement of pollutants dissolved in the water is discussed in Chap. 7.

The determination of the future distribution of piezometric heads $\phi = \phi(\mathbf{x}, t)$, is, in fact, the solution to the groundwater forecasting problem referred to in Chap. 1. We are looking for future piezometric heads, or water levels, produced in a given (by geometry and properties) aquifer by any planned schedule of future pumping and artificial recharge activities and anticipated natural replenishment.

5-1 AQUIFER STORATIVITY

Specific Storativity

Let us start by introducing the concept of *effective stress* (or *intergranular stress*), first introduced by Terzaghi (1925).

Figure 5-1a shows a cross section through a confined aquifer. To simplify the discussion, we shall consider a granular non-cohesive matrix with grain sizes such that molecular and interparticle forces are negligible. Figure 5-1b shows the details at any internal horizontal elemental plane AB in an aquifer, whether confined or phreatic; it is also valid for an elemental surface of the impervious ceiling of an aquifer ($A'B'$).

The total load of soil and water (and actually also everything that adds load at the ground surface, including atmospheric pressure) above the considered plane is balanced by a stress (force per unit area) σ' in the solid matrix and by a pressure p in the water (Fig. 5-1b)

$$\sigma = \sigma' + p \tag{5-1}$$

where σ is the *total stress* resulting from the overburden load. Each of the three terms appearing in (5-1) is a force divided by the total area, A, of the considered plane. Strictly speaking, we should have taken into account the fact that whereas

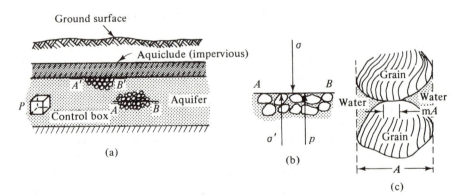

Figure 5-1 Pressure and intergranular stress in an aquifer.

σ acts over the entire area under consideration, the water pressure acts only over part of A and so does the force carried by the solid matrix. Nevertheless, it can be shown (e.g., Lambe and Whitman, 1969; Bear, 1972, p. 54) that the *effective stress* σ', as defined above, is a good approximation of the stress transmitted through the skeleton (and hence it is also called *intergranular stress*), and that we may assume that indeed p acts over the entire area A. In (5-1), positive σ and σ' denote compression.

Equation (5-1) is derived by considering vertical forces only. However, the discussion can be extended to the general case of three-dimensional space (e.g., Bear, 1972, p. 55; Verruijt, 1965, 1969).

When changes in the overburden load take place, changes will be produced also in σ' and p

$$d\sigma = d\sigma' + dp \tag{5-2}$$

If we keep $\sigma = $ const., but change the pressure, for example by pumping from the aquifer, or by artificially recharging it, we have

$$d\sigma = 0 = d\sigma' + dp, \qquad d\sigma' = -dp \tag{5-3}$$

which means that a corresponding change is produced in the intergranular stress. Thus, *a reduction of water pressure by pumping from a well results in an increase in the load borne by the solid skeleton of the aquifer.*

Now, the water in the aquifer is compressible. Although this compressibility is small, it plays an important role mainly in confined aquifers. We define a coefficient of compressibility of water, β, by

$$\beta = -\frac{1}{U_w}\frac{\partial U_w}{\partial p} = \frac{1}{\rho}\frac{\partial \rho}{\partial p} \tag{5-4}$$

where U_w and ρ are volume and density, respectively, of a given mass of water subjected to pressure changes. The minus sign indicates a decrease in volume as pressure increases. For β independent of pressure, we obtain from (5-4)

$$U_w = U_{w0}\exp[-\beta(p - p_0)]; \qquad \rho = \rho_0\exp[\beta(p - p_0)] \tag{5-5}$$

where U_{w0} and ρ_0 correspond to the reference pressure p_0.

The solid matrix of the aquifer is elastic and not rigid. By subjecting it to a change in the intergranular stress, it will undergo deformation. This deformation involves a movement of the solid, or the solid particles and their rearrangement, such that the porosity of the porous medium is changed. We assume that the elasticity of the solid or the solid particles is much smaller (relative to the solid matrix as a whole), so that their volume remains unchanged.

The elastic property of the solid matrix is expressed by its coefficient of compressibility, α, defined by

$$\alpha = -\frac{1}{U_b}\frac{\partial U_b}{\partial \sigma'}, \tag{5-6}$$

where U_b is the bulk volume of a porous medium. As emphasized above, we are considering here only vertical compressibility, the lateral deformation in the aquifer is assumed negligible.

Since the volume of solids U_s in U_b remains constant, we have

$$U_s \equiv (1 - n) U_b = \text{const.}; \qquad \frac{\partial U_s}{\partial \sigma'} = 0; \qquad \frac{1}{U_b} \frac{\partial U_b}{\partial \sigma'} = \frac{1}{(1 - n)} \frac{\partial n}{\partial \sigma'};$$

$$\alpha = -\frac{1}{1 - n} \frac{\partial n}{\partial \sigma'} = \frac{1}{1 - n} \frac{\partial n}{\partial p} \qquad (5\text{-}7)$$

which relates α to the changes in porosity, n, resulting from changes in water pressure.

Consider now the vicinity of a point in an aquifer, where water pressure is reduced by pumping. As indicated by (5-3), this results in an increase in the inter-granular stress transmitted by the solid skeleton of the aquifer. This, in turn, causes the aquifer to be compacted, reducing its porosity. At the same time, as a result of pressure reduction, the water will expand according to (5-4). Together, the two effects—the slight expansion of water and the small reduction in poros-ity—cause a certain amount of water to be released from storage in an aquifer. Thus, by releasing water from storage in an aquifer, we produce in it a reduction in water pressure. Conversely, in response to adding water to a unit volume of aquifer, the pressure in it will rise, accompanied by a reduction in the inter-granular stress, which, in turn, increases the porosity. If we assume both water and solid matrix to be perfectly elastic, within the range of the considered changes, the two processes are reversible. In reality, however, changes in a granular matrix are irreversible (see Sec. 5-11).

Based on the above considerations, we can now define a *specific storativity*, S_{0p}, of the porous medium of an aquifer as the volume of water released from storage (or added to it) in a unit volume of aquifer per unit decline (or rise) in pressure

$$S_{0p} = \Delta U_w / U_b \Delta p \qquad (5\text{-}8)$$

or, per unit change in the piezometric head ϕ^* (defined by (4-4))

$$S_0 \equiv S_{0\phi^*} = \Delta U_w / U_b \Delta \phi^* \qquad (5\text{-}9)$$

S_0 has the dimensions of L^{-1}. From (5-9) it follows that by adding a volume ΔU_w to a volume U_b of aquifer, the piezometric head there will rise by $\Delta \phi^* = \Delta U_w / U_b S_0$.

One should note that S_{0p} and $S_{0\phi^*}$ are actually *defined* by (5-8) and (5-9), respectively, without analyzing their internal relationship to the compressibilities of water and solid matrix.

Aquifer Storativity

In a similar way we can define a *storativity for a confined aquifer*, S, as the volume of water released from storage (or added to it) per unit horizontal area of aquifer and per unit decline (or rise) of piezometric head, ϕ

$$S = \Delta U_w / A \Delta \phi \qquad (5\text{-}10)$$

S is dimensionless. The reasons for the relationship between the amount of water released and the change of head are now obvious. The volume of aquifer from which water is released is $A \times B$, where A is horizontal area and B is the thickness of the confined aquifer (Fig. 5-2).

It is important to understand that like T, S is an aquifer property. If we work under the assumption of essentially horizontal flow in an aquifer (Sec. 2-4), we should use the parameters T and S. If, however, we wish to consider three-dimensional flow in an aquifer, we should use the parameters K and S_0. Although it is possible to relate K to T by $T = KB$ and S_0 to S by $S = S_0 B$, one should avoid mixing the two concepts, as in principle they are related to two different flow models.

We can also define a storage coefficient for a phreatic aquifer. Consider a unit (horizontal) area of a phreatic aquifer. The volume of water stored in a phreatic aquifer is indicated by the water table (see Sec. 5-4). If, as a result of the flow in the aquifer, a volume of water will leave this area in excess of the volume of water entering it, the water table will drop. We may define the *storativity of a phreatic aquifer* in the same way as we defined above the storativity of a confined aquifer, except that here the drop, Δh, is of the water table (Fig. 5-3)

$$S = \Delta U_w / A \Delta h \tag{5-11}$$

In spite of the similarity in the definition, the storativity in the two types of aquifer is due to different reasons. In a confined aquifer, it is the outcome of water and matrix compressibility. In a phreatic aquifer, water is mostly drained

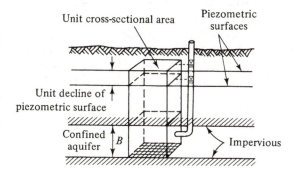

Figure 5-2 Definition sketch for storativity in a confined aquifer.

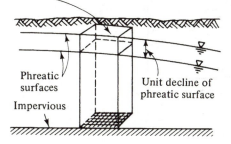

Figure 5-3 Definition sketch for storativity in a phreatic aquifer.

from the volume of pore space between the two positions of the phreatic surface. The storativity of a phreatic aquifer is, therefore, sometimes referred to as *specific yield*, S_y; it gives the yield of an aquifer per unit area and unit drop of the water table (see further discussion in Sec. 6-1).

Recalling that actually the water table is an approximate concept, we understand that water is actually being drained from the entire column of soil up to the ground surface. Bear (1972, p. 485) shows that when the soil is homogeneous and the fluctuating water table is sufficiently deep, the above definition for specific yield still holds (see Sec. 6-1).

One should be careful not to identify the specific yield with the porosity of a phreatic aquifer. As water is being drained from the interstices of the soil, the drainage is never a complete one. A certain amount of water is retained in the soil against gravity by capillary forces. After drainage has stopped, the volume of water retained in an aquifer per unit (horizontal) area and unit drop of the water table is called *specific retention*, S_r. Thus

$$S_y + S_r = n \tag{5-12}$$

For this reason S_y ($<n$) is sometimes called *effective porosity*. Here, again, one should note that we have been referring to the approximate concept of a water table. However, for a homogeneous soil and a sufficiently deep water table, the above definition for S_r holds (see Sec. 6-1).

Figure 5-4 shows the relationships between S_y, S_r, and particle size.

When drainage occurs, it takes time for the water to flow, partly under unsaturated flow conditions, out of the soil volume between two positions of a water table, at t and at $t + \Delta t$. This is especially true if the lowering of the water table is rapid. Under such conditions, the specific yield becomes time dependent, gradually approaching its ultimate value (Fig. 5-5). When the water level is rising or falling slowly, the changes in moisture distribution have time to adjust continuously and the time lag vanishes. This phenomenon of time dependency of the

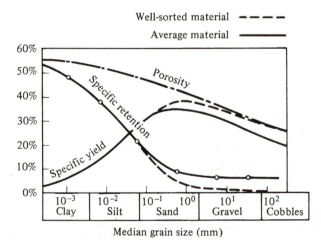

Figure 5-4 Relationship between specific yield and grain size *(from Conkling et. al., 1934, as modified by Davis and DeWiest, 1966).*

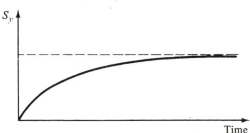

Time   **Figure 5-5** Time-dependent specific yield.

specific yield should not be overlooked in the analysis of pumping tests. water balances for short periods, etc. (see Neuman, 1972).

When the water table is lowered, the pressure drops throughout the aquifer below it. In principle, this pressure drop should cause water to be released from storage in the aquifer, due to the elastic properties of the aquifer and the water. However, when we calculate the total volume of water released from storage in the aquifer per unit area and unit decline of head: $(\Delta U_w)_1 = S_0 h$ due to the elastic storage and $(\Delta U_w)_2 = S_y$ due to the actual drainage of water from the pore space, we have $S_0 h \ll S_y$ so that $(\Delta U_w)_1$ can be neglected (see further discussion on this point in Sec. 5-2).

Typical values of S in a confined aquifer are of the order of 10^{-4}–10^{-6}, roughly 40 percent of which result from the expansion of the water and 60 percent from the compression of the medium. In a sandy phreatic aquifer, we may have S_0 of the order 10^{-7} cm^{-1}, whereas S_y may be 20–30 percent (see Fig. 5-4).

We shall return to the definition of aquifer storativity, both for a confined aquifer and for a phreatic one, in Sec. 5-2 where the aquifer equations will be derived by averaging the three-dimensional flow equations along the vertical.

5-2 BASIC CONTINUITY EQUATION

Mass Conservation Equation

In this section, we shall develop the basic equation describing three-dimension flow in a porous medium. One way of deriving the basic mass balance equation is given in App. A-6, leading to (A-24)

$$\frac{\partial (n\rho)}{\partial t} = -\operatorname{div} \rho \mathbf{q} \tag{5-13}$$

Nevertheless, let us derive this equation by using more elementary considerations.

Consider a *control volume* (or *control box*) having the shape of a rectangular parallel-piped box of dimensions δx, δy, δz centered at some point $P(x, y, z)$ inside the flow domain in an aquifer (Figs. 5-1 and 5-6). A control box may have any arbitrary shape, but once its shape and position in space have been fixed, they remain unchanged during the flow, although the amount and identity of the material in it may change with time. In the present analysis, water and solids

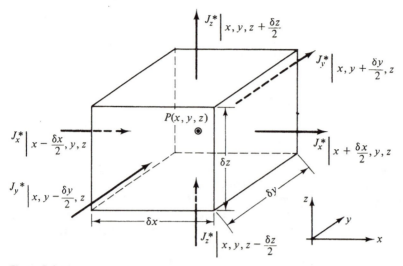

Figure 5-6 Mass conservation for a control volume.

enter and leave the box through its surfaces, and our objective is to write a balance, or a statement of conservation, for the mass of water entering, leaving, and being stored in the box. In hydrodynamics, this is called the *Eulerian approach*, contrary to the *Lagrangian approach* in which a fixed mass of water is followed as it moves through a porous medium.

Let the vector **J*** denote the mass flux (i.e., mass per unit area per unit time) of water of density ρ at point $P(x, y, z)$

$$\mathbf{J}^* = \rho\mathbf{q} \tag{5-14}$$

Referring to Fig. 5-6, the excess of inflow over outflow of mass during a short time interval δt, through the surfaces which are perpendicular to the x direction, may be expressed by the difference

$$\left[J_x^* \Big|_{x-\delta x/2, y, z} - J_x^* \Big|_{x+\delta x/2, y, z} \right] \delta y\, \delta z\, \delta t$$

In the y and z directions we have, in a similar way

$$\left[J_y^* \Big|_{x, y-\delta y/2, z} - J_y^* \Big|_{x, y+\delta y/2, z} \right] \delta x\, \delta z\, \delta t$$

and

$$\left[J_z^* \Big|_{x, y, z-\delta z/2} - J_z^* \Big|_{x, y, z+\delta z/2} \right] \delta x\, \delta y\, dt$$

respectively.

By adding the last three expressions, we obtain an expression for the total excess of mass inflow over outflow during δt

$$\left[\frac{J_x^*|_{x-\delta x/2,y,z} - J_x^*|_{x+\delta x/2,y,z}}{\delta x} + \frac{J_y^*|_{x,y-\delta y/2,z} - J_y^*|_{x,y+\delta y/2,z}}{\delta y} \right.$$

$$\left. + \frac{J_z^*|_{x,y,z-\delta z/2} - J_z^*|_{x,y,z+\delta z/2}}{\delta z} \right] \delta x \, \delta y \, \delta z \, \delta t \quad (5\text{-}15)$$

where $\delta x \, \delta y \, \delta z = \delta U$ is the volume of the box. By dividing the expression in (5-15) by δU and δt, we obtain the mass inflow over outflow per unit volume of porous medium and per unit time. Then, by letting the box converge on the point P, that is, by letting $\delta x, \delta y, \delta z \to 0$, the excess of inflow over outflow per unit volume of medium (around P) and per unit time becomes

$$-(\partial J_x^*/\partial x + \partial J_y^*/\partial y + \partial J_z^*/\partial z), \quad \text{or:} \quad -\text{div }\mathbf{J}^*; \quad \mathbf{J}^* = \rho \mathbf{q} \quad (5\text{-}16)$$

By the principle of mass conservation, in the absence of sources and/or sinks of mass, the excess of mass as expressed by (5-16) must be equal to the change of mass, m, during δt within the box. Since $m = \rho n \, \delta U$, this mass accumulation in the box during δt can be expressed as

$$m|_{t+\Delta t} - m|_t \equiv \left[(\rho n)|_{t+\Delta t} - (\rho n)|_t \right] \delta U \quad (5\text{-}17)$$

Again, by dividing the expression on the right-hand side of (5-17) by δU and δt, and letting $\delta t \to 0$, we obtain $\partial(\rho n)/\partial t$ as the mass of water accumulated per unit volume and unit time at point P. Hence, the mass balance at $P(x, y, z)$ can now be stated as

or
$$\left. \begin{array}{l} -(\partial J_x^*/\partial x + J_y^*/\partial y + J_z^*/\partial z) = \partial \rho n/\partial t \\[2mm] -\text{div}(\rho \mathbf{q}) = \partial \rho n/\partial t \end{array} \right\} \quad (5\text{-}18)$$

with each side of the equation expressing added mass of water per unit time and unit volume of porous medium around P.

Now, for the conditions of (5-3), and for $\rho = \rho(p)$ only

$$\frac{\partial(n\rho)}{\partial t} = \rho \frac{\partial n}{\partial t} + n \frac{\partial \rho}{\partial t} = \left(\rho \frac{\partial n}{\partial p} + n \frac{\partial \rho}{\partial p} \right) \frac{\partial p}{\partial t} \quad (5\text{-}19)$$

By making use of (5-3), (5-4), (5-6), (5-7), and (4-4), we obtain from (5-19)

$$\frac{\partial(\rho n)}{\partial t} = \rho[\alpha(1-n) + \beta n] \frac{\partial p}{\partial t} = g\rho^2[\alpha(1-n) + n\beta] \frac{\partial \phi^*}{\partial t};$$

$$\frac{\partial \phi^*}{\partial t} = \frac{1}{\rho g} \frac{\partial p}{\partial t} \quad (5\text{-}20)$$

Hence, the expression $g\rho^2[\alpha(1-n) + \beta n] \equiv S_{0\phi^*}^*$ may be interpreted as specific mass storativity related to potential change (Bear, 1972, p. 204). It gives the mass of water added to storage (or released from it) in a unit volume of porous medium per unit rise (or decline) of potential ϕ^*. By comparing this definition with that of specific (volume) storativity as expressed by (5-9), we obtain

$$S_{0\phi*}^* = \rho S_{0\phi*}; \qquad S_{0\phi*} = \rho g[\alpha(1 - n) + n\beta] = gS_{0p}^* = \rho gS_{0p} \qquad (5\text{-}21)$$

where the subscript $\phi*$ denotes the measured quantity.

We have thus arrived at an expression which relates the specific volume storativity to the elastic properties of the medium and the water. In practice, however, it is difficult to determine α (and n) so that rather than determine them experimentally and then use (5-21) to determine $S_{0\phi*}$, we prefer to determine $S_{0\phi*}$ directly from field observations of $\phi*$.

Henceforth, we shall simplify the notation, using S_0 for $S_{0\phi*}$.

For $\rho = \rho(p, c)$, where c is some solute concentration

$$\frac{\partial(n\rho)}{\partial t} = \left(\rho \frac{\partial n}{\partial p} + n \frac{\partial \rho}{\partial p} \right) \frac{\partial p}{\partial t} + n \frac{\partial \rho}{\partial c} \frac{\partial c}{\partial t} \qquad (5\text{-}22)$$

Returning now to (5-18) and inserting in it $\partial \rho n / \partial t$ from (5-20) and (5-21), we obtain the continuity equation

$$-\operatorname{div}(\rho \mathbf{q}) = \rho S_0 \partial \phi* / \partial t; \qquad S_0 \equiv S_{0\phi*} = \rho g[\alpha(1 - n) + \beta n] \qquad (5\text{-}23)$$

If $\rho = \rho(p, c)$, in view of (5-22), we cannot use the potential $\phi*$ or ϕ. The mass conservation equation is then

$$\frac{\partial(\rho n)}{\partial t} + \operatorname{div} \rho \mathbf{q} = \left\{ \rho S_{0p} \frac{\partial p}{\partial t} + n \frac{\partial \rho}{\partial c} \frac{\partial c}{\partial t} \right\} + \operatorname{div} \rho \mathbf{q} = 0 \qquad (5\text{-}24)$$

where $S_{0p} = \alpha(1 - n) + \beta n$, and $\mathbf{q}$ is expressed by (4-14). This is the second equation that we mentioned at the beginning of this section.

Continuity Equation

When the flow is steady (that is, $\partial \phi* / \partial t = 0$) and/or when both fluid and solid matrix are incompressible (that is, $S_0 = 0$ and $\rho = $ const.), or assumed so (as in an unconfined aquifer), (5-23) reduces to

$$\operatorname{div} \mathbf{q} = 0 \qquad (5\text{-}25)$$

which actually states volume continuity. Recall that $\operatorname{div} \mathbf{A} = \partial A_x / \partial x + \partial A_y / \partial y + \partial A_z / \partial z$ for any vector $\mathbf{A}$, where A_x, A_y, and A_z are components of $\mathbf{A}$.

Our next step now is to introduce an equation of motion (i.e., an expression for $\mathbf{q}$) into the continuity equation (5-23). At this point, we should note that Darcy's law, as presented in Chap. 4, gives the motion of groundwater with respect to the solid matrix, while in (5-23) $\mathbf{q}$ is with respect to the fixed coordinate system (the Eulerian approach). Thus, to be accurate, we should take account of the fact that since here we are considering a deformable, or consolidating, medium, we should also account in some way for the movement of the solid matrix with respect to the fixed coordinate system (see Sec. 5-11).

Bear (1972, p. 205) presents a detailed analysis of this topic. Here we shall assume that for all practical purposes:

(a) The velocity of the solids is so small (with respect to $\mathbf{q}/n$) that $\mathbf{q}$ in (5-23) and (5-25) may still be expressed by Darcy's law (4-15).

(b) K is a constant, although $\rho = \rho(p)$, or, if the medium is inhomogeneous, K varies in space independent of the variability of ρ.

(c) S_0 and K are unaffected by variations in n due to matrix deformability. It is assumed that these variations are small relative to the initial n. The same is true for ρ.

(d) $\mathbf{q} \cdot \text{grad } \rho \ll n \, \partial \rho / \partial t$ (so that (5-18) actually reduces to $\rho \, \text{div} \mathbf{q} + \partial(\rho n)/\partial t = 0$), that is, we assume that spatial variations in ρ are much smaller than the local, temporal ones.

Under these assumptions, (5-23) can be written in terms of the single variable ϕ^* (still using the symbol ϕ^* to indicate that we are dealing with a compressible fluid)

$$-\text{div} \mathbf{q} \equiv \text{div}(K \cdot \text{grad } \phi^*) = S_0 \, \partial \phi^* / \partial t \qquad (5\text{-}26)$$

Taking into account the velocity of solids (Sec. 5-11), we obtain

$$\text{div}(K \cdot \text{grad } \phi^*) = (\alpha + n\beta) \, \partial \phi^* / \partial t \qquad (5\text{-}26a)$$

For a homogeneous isotropic medium, (5-26) reduces to

$$K\nabla^2\phi^* \equiv K \, \text{div}(\text{grad } \phi^*) \equiv K\left(\frac{\partial^2\phi^*}{\partial x^2} + \frac{\partial^2\phi^*}{\partial y^2} + \frac{\partial^2\phi^*}{\partial z^2} \right) = S_0 \frac{\partial \phi^*}{\partial t} \qquad (5\text{-}27)$$

For an isotropic, but inhomogeneous medium, (5-26) becomes

$$\frac{\partial}{\partial x}\left(K \frac{\partial \phi^*}{\partial x} \right) + \frac{\partial}{\partial y}\left(K \frac{\partial \phi^*}{\partial y} \right) + \frac{\partial}{\partial z}\left(K \frac{\partial \phi^*}{\partial z} \right) = S_0 \frac{\partial \phi^*}{\partial t} \qquad (5\text{-}28)$$

For a nonhomogeneous, anisotropic medium, where the principal axes are in the x, y, and z directions, (5-26) becomes

$$\frac{\partial}{\partial x}\left(K_x \frac{\partial \phi^*}{\partial x} \right) + \frac{\partial}{\partial y}\left(K_y \frac{\partial \phi^*}{\partial y} \right) + \frac{\partial}{\partial z}\left(K_z \frac{\partial \phi^*}{\partial z} \right) = S_0 \frac{\partial \phi^*}{\partial t} \qquad (5\text{-}29)$$

Obviously, in (5-28) and (5-29) $K = K(x, y, z)$ must be continuous and have a continuous first derivative everywhere in the considered flow domain (see discussion in Sec. 5-3). Finally, if the flow is steady and/or when both water and solid matrix are assumed to be incompressible, the right-hand side of (5-26) through (5-29) vanishes. For example, (5-27) reduces to the well-known Laplace equation

$$\nabla^2\phi^* \equiv \frac{\partial^2\phi^*}{\partial x^2} + \frac{\partial^2\phi^*}{\partial y^2} + \frac{\partial^2\phi^*}{\partial z^2} = 0 \qquad (5\text{-}30)$$

When using any of these equations, one should always bear in mind the various assumptions made along their development. For problems of practical interest in groundwater flow (and this is the point of view of this book), (5-26) through (5-30) should be sufficient. However, it is always possible, whenever the situation calls for it, to remove any of the assumptions made here, and arrive at different, usually more complicated, equations (Bear, 1972, Sec. 6-3).

From the definitions of Hubbert's potential ϕ^* in (4-4), it follows that for $\rho = \rho(p)$ only

$$\frac{\partial \phi^*}{\partial t} = \frac{1}{\rho g} \frac{\partial p}{\partial t}; \qquad \nabla \phi^* = \frac{1}{\rho g} \nabla p + \nabla z \qquad (5\text{-}31)$$

As a practical approximation, we often replace in all the continuity equations above, as elsewhere in this book, $\nabla \phi^*$ by $\nabla \phi$ and $\partial \phi^*/\partial t$ by $\partial \phi/\partial t$. It is also usually assumed that $K = k\rho g/\mu \simeq k\rho_0 g/\mu_0$, independent of the effect of pressure on ρ and μ.

In Sec. 5-9 a method is presented which can be used for transforming any problem given in a homogeneous anisotropic domain into one in an equivalent isotropic domain.

One should remember that (5-18), (5-23), and (5-24) have been developed for a domain with no sources and/or sinks. If sources and/or sinks are present, they should be represented by an additional term on the left-hand side of (5-18) expressing the rate at which mass of water is added per unit time and unit volume of porous medium around P. Equations (5-23) and (5-24) should be modified in a similar way.

Coordinate Systems

We have chosen here to develop the various equations using cartesian coordinates, as they are the ones more often employed in hydrological studies. There is no difficulty in converting these equations into any other coordinate system, if the situation arises. One may be aided by the following relationships for any vector $\mathbf{A}$:

In polar coordinates, (r, θ)

$$\nabla \cdot \mathbf{A} \equiv \text{div } \mathbf{A} = \frac{1}{r} \left[\frac{\partial}{\partial r}(rA_r) + \frac{\partial A_\theta}{\partial \theta} \right] \qquad (5\text{-}32a)$$

In cylindrical coordinates (r, θ, z)

$$\nabla \cdot \mathbf{A} \equiv \text{div } \mathbf{A} = \frac{1}{r} \left[\frac{\partial}{\partial r}(rA_r) + \frac{\partial A_\theta}{\partial \theta} + \frac{\partial}{\partial z}(rA_z) \right] \qquad (5\text{-}32b)$$

5-3 INITIAL AND BOUNDARY CONDITIONS

Each of the equations (5-26) through (5-30) presented in the previous section is a partial differential equation which describes a *class of phenomena*; the equations themselves contain no information (e.g., the shape of the flow domain) related to any specific case of flow through a porous medium. Therefore, each equation has an infinite number of possible solutions, each of which corresponds to a particular case of flow through a porous medium domain.

To obtain from this multitude of possible solutions one particular solution corresponding to a certain specific problem of interest, it is necessary to provide supplementary information that is not contained in the equations. The supplementary information that, together with the partial differential equation, defines an individual problem, should include specifications of:

(a) The geometry of the domain in which the considered flow takes place, with possibly parts of the boundary being at infinity; here boundaries have the idealized shape discussed in App. A-5.
(b) Values of all relevant physical coefficients (for example, K, S_0).
(c) *Initial conditions* which describe the initial state of the fluid in the considered flow domain.
(d) Statements on how the fluid in the considered domain interacts with its surroundings, i.e., conditions on the boundaries of the considered flow domain.

Initial and boundary conditions for flow through porous media are discussed in the present section. More details can be found in various texts on flow through porous media (e.g., Bear, 1972, Chap. 7).

Obviously, boundary and initial conditions are motivated by the physical reality of the flow. Hence, they are first determined (or assumed, on the basis of available information and past experience) in the field and only then expressed in mathematical forms. Different boundary conditions lead to different solutions, hence the importance of correctly determining the conditions which exist along the real boundaries.

A mathematical problem that corresponds to a physical reality should satisfy the following basic requirements:

(a) The solution must *exist* (existence).
(b) The solution must be *uniquely determined* (uniqueness).
(c) The solution should *depend continuously on the data* (stability).

The first requirement states simply that a solution does, in fact, exist. The second requirement stipulates completeness of the problem — leeway or ambiguity should be excluded unless inherent in the physical situation. The third requirement means that a variation of the given data (e.g., boundary and initial conditions) in a sufficiently small range leads to an arbitrary small change in the solution. This requirement is also valid for approximate (e.g., numerical) solutions. We require that a small error in satisfying the equation be reflected by only a small deviation of the approximate solution from the true one. If small errors in the data do not result in correspondingly small errors in the solution, we should be inclined to think that our mathematical model of the physical phenomenon is badly formulated.

Any problem that satisfies these three requirements is called a *well-* (or *properly*) *posed problem*.

In mathematical texts on the solution of partial differential equations (e.g., Garabedian, 1964), before an actual solution is undertaken, a rigorous analysis is usually carried out to prove *existence, uniqueness*, and *stability* of the solution to various boundary value problems. In the present text we shall assume implicitly, without proof, that all mathematically stated problems are well posed. This follows from the fact that our problems are not mere mathematical exercises. where it is not always obvious that they are well posed, but are attempts to describe actual physical phenomena.

Let D denote a considered flow domain. In a three-dimensional space it is bounded by a surface S; in two dimensions by a curve C. In certain cases, a flow domain may be imagined to extend to infinity (unbounded domain). A fixed boundary S may be described mathematically (in a cartesian xyz coordinate system) by an equation whose general form is

$$F(x, y, z) = 0$$

Similarly, C in the two dimensions x and y can be described by

$$F(x, y) = 0$$

The unit vector $\mathbf{1n}$ in the direction of the normal to the surface S is given by

$$\mathbf{1n} = [(\partial F/\partial x)\, \mathbf{1x} + (\partial F/\partial y)\, \mathbf{1y} + (\partial F/\partial z)\, \mathbf{1z}]/|\text{grad } F| \equiv \nabla F/|\nabla F|$$

where $\mathbf{1x}$, $\mathbf{1y}$, and $\mathbf{1z}$ are unit vectors in the x, y, and z directions, respectively; $|\nabla F|^2 \equiv |\text{grad } F|^2 = (\partial F/\partial x)^2 + (\partial F/\partial y)^2 + (\partial F/\partial z)^2$. Similar expressions can be written for the unit vector normal to the curve C.

In (5-26) through (5-30), the dependent variable is ϕ^* for which we seek a solution in the form of $\phi^* = \phi^*(x, y, z, t)$. Henceforth, we shall assume $\phi^* \approx \phi$ (see comment following (5-31)), so that the dependent variable will be $\phi = \phi(x, y, z, t)$ and therefore, we have to specify our initial and boundary conditions also in terms of ϕ.

Initial Conditions

Initial conditions include the specification of ϕ at all points within the domain D at some initial time, usually denoted as $t = 0$. This can be written schematically as specifying

$$\phi = f(x, y, z, 0) \tag{5-33}$$

for all points x, y, z inside D; f is a known function.

Boundary Conditions

To facilitate the discussion on boundary conditions, consider the case of flow described in Fig. 5-7. Although the figure represents a two-dimensional flow field, we shall continue to consider boundary conditions for a three-dimensional one; the passage to two dimensions is straightforward and requires no further explanations. Whenever a domain will be considered to be anisotropic, we shall assume that x, y, z are principal directions.

The various types of boundary conditions encountered in flow through porous media are the following (Fig. 5-7):

(a) Boundary of prescribed potential The potential, ϕ, is prescribed for all points of this boundary

$$\phi = f_1(x, y, z), \quad \text{or} \quad \phi = f_2(x, y, z, t) \text{ on } S \tag{5-34}$$

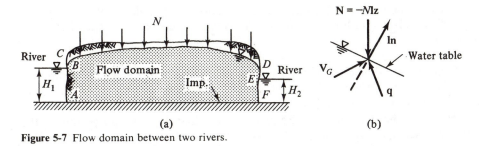

(a)   (b)

Figure 5-7 Flow domain between two rivers.

where f_1 and f_2 are known functions. A boundary of this kind occurs whenever the flow domain is adjacent to a body of open water. Segments AB and FE in Fig. 5-7 are examples of a boundary of prescribed potential. On AB, the potential is $\phi = H_1$. If H_1 varies with time, it may take the form of $\phi = H_1(t)$. On FE the potential is $\phi = H_2$.

A special case of (5-34) is

$$\phi = \phi_0 \text{ on } S, \qquad \phi_0 = \text{const.} \tag{5-35}$$

that is, the boundary is an equipotential surface. Actually AB and EF of Fig. 5-7 are equipotential boundaries. Since the piezometric head is the same at all points of a body of stationary water, this piezometric head is also the boundary condition at points on the interface between a porous medium and this body of water.

In the theory of partial differential equations, a problem in which only this type of boundary condition is encountered is called a *Dirichlet, or a first type, boundary value problem.*

(b) Boundary of prescribed flux On a boundary of this type, the flux normal to the boundary surface is prescribed for all points. We may express this flux as

$$q_n = f(x, y, z, t) \text{ on } S \tag{5-36}$$

where f is a known function, and $q_n = \mathbf{q} \cdot \mathbf{1n}$ denotes the component of the specific discharge $\mathbf{q}$ normal to S. For an isotropic medium, (5-36) can be expressed in terms of ϕ as

$$q_n = -K \operatorname{grad} \phi \cdot \mathbf{1n} = f(x, y, z, t);$$

$$\operatorname{grad} \phi \cdot \mathbf{1n} \equiv \frac{1}{|\operatorname{grad} F|} \left(\frac{\partial \phi}{\partial x} \frac{\partial F}{\partial x} + \frac{\partial \phi}{\partial y} \frac{\partial F}{\partial y} + \frac{\partial \phi}{\partial z} \frac{\partial F}{\partial z} \right) \tag{5-37}$$

where $F(x, y, z) = 0$ is the equation describing S.

A special case of this type of boundary is the impervious boundary (AF in Fig. 5-7), where the flux normal to the boundary vanishes everywhere. This condition can be expressed as

$$q_n \equiv \mathbf{q} \cdot \mathbf{1n} = \begin{cases} \dfrac{\partial \phi}{\partial x} \dfrac{\partial F}{\partial x} + \dfrac{\partial \phi}{\partial y} \dfrac{\partial F}{\partial y} + \dfrac{\partial \phi}{\partial z} \dfrac{\partial F}{\partial z} = 0, & \text{for isotropic media} \\[2ex] K_x \dfrac{\partial \phi}{\partial x} \dfrac{\partial F}{\partial x} + K_y \dfrac{\partial \phi}{\partial y} \dfrac{\partial F}{\partial y} + K_z \dfrac{\partial \phi}{\partial z} \dfrac{\partial F}{\partial z} = 0, & \text{for anistropic media} \end{cases} \tag{5-38}$$

In the theory of partial differential equations, a problem having only this type of boundary condition is called a *Neumann, or second type, boundary value problem.*

(c) Semipervious Boundary This type of boundary occurs when the porous medium domain is in contact with a body of water continuum (or another porous medium domain), however, a relatively thin semipervious layer separates the two domains. Let ϕ denote the piezometric head in the considered domain and ϕ_0 that in the external one. If we approximate by assuming no storage in the semipermeable thin layer (see Neuman, 1972), then q_n through that layer (positive if inflow) is

$$q_n = (\phi_0 - \phi)/\sigma'; \qquad \sigma' = B'/K' \qquad (5\text{-}39)$$

where σ' is the resistance of the semipermeable layer, equal to the ratio of its thickness, B', to its hydraulic conductivity, K'. With (5-39) expressing the inflow through S in (5-36) and (5-37), we obtain for an anisotropic medium

$$-\boldsymbol{K} \cdot \operatorname{grad} \phi \cdot \mathbf{1n} = (\phi_0 - \phi)/\sigma' \qquad (5\text{-}40)$$

or, for x, y, z principal directions

$$\frac{1}{|\operatorname{grad} F|}\left(-K_x \frac{\partial \phi}{\partial x}\frac{\partial F}{\partial x} - K_y \frac{\partial \phi}{\partial y}\frac{\partial F}{\partial y} - K_z \frac{\partial \phi}{\partial z}\frac{\partial F}{\partial z}\right) + \frac{\phi}{\sigma'} = \frac{\phi_0}{\sigma'} \qquad (5\text{-}41)$$

This is a *mixed boundary condition (boundary condition of third type; Cauchy boundary condition).*

(d) Unsteady phreatic surface with accretion The phreatic (or free) surface was already discussed in Sec. 2-2, where it was defined as the surface on which $p = 0$. We shall neglect here the capillary fringe above this surface.

The location and shape of the free surface are unknown. In fact, their determination constitutes part of the required solution. Following the usual procedure, however, we must specify for the boundary being considered (a) its geometry and (b) the condition to be satisfied at all points along it.

Since the pressure at all points of the free surface, S, is taken as $p = 0$, we have from $\phi(x, y, z, t) = z + p(x, y, z, t)/\gamma$

$$\phi(x, y, z, t) = z, \quad \text{or:} \quad \phi(x, y, z, t) - z = 0 \text{ on } S \qquad (5\text{-}42)$$

S is thus a boundary of prescribed potential. Equation (5-42) gives at any time t a relationship between the coordinates of points of the free surface. It may, therefore, be considered the equivalent to $F(x, y, z, t) = 0$, describing the geometry of this surface

$$F(x, y, z, t) \equiv \phi(x, y, z, t) - z = 0 \text{ on } S \qquad (5\text{-}43)$$

The difficulty, however, stems from the fact that the distribution $\phi(x, y, z, t)$, and hence $F(x, y, z, t)$, is unknown before the problem is solved. We have here a vicious circle: in order to determine ϕ we must know the boundary's location F, and in order to know where this boundary is, we have to know ϕ. Iterative techniques are sometimes used (especially in numerical solutions) to overcome this difficulty. In addition, the boundary condition itself, as developed below, is very complicated. Analytical techniques, except for some particular classes of problems, are, therefore, seldom used.

The unsteady phreatic surface with accretion is a surface on which a certain property, G, is maintained constant (here $F = \text{const.} = 0$). For such a surface, we have the relationship (Bear, 1972, p. 72)

$$\frac{DG}{Dt} \equiv \frac{\partial G}{\partial t} + \mathbf{V}_G \cdot \text{grad } G = 0 \qquad (5\text{-}44)$$

where $D(\ )/Dt$, the substantial derivative, is made equal to zero because no change in G takes place as particles carrying the property G move at a velocity $\mathbf{V}_G$, which is the velocity of propagation of the free surface.

Continuity requires that at the moving free surface (CD in Fig. 5-7)

$$(\mathbf{q} - \mathbf{N}) \cdot \mathbf{1n} = n_e \mathbf{V}_G \cdot \mathbf{1n} \qquad (5\text{-}45)$$

where n_e is *effective porosity* and N is the rate of accretion (positive downward; $\mathbf{N} = -N\mathbf{1z}$). Equation (5-45) states that the velocity of the surface depends on the velocities of the water on both sides of the surface; the effective porosity n_e is employed here because as the surface moves, only part of the water is removed from the void space, while the remaining part is retained. When the pores are sufficiently large n_e is close to the porosity n (see Fig. 5-7, where $n_e \equiv S_y$).

By combining (5-44) with (5-45), into which we insert $G \equiv F$, we obtain

$$\frac{\partial F}{\partial t} + \frac{1}{n_e}(\mathbf{q} - \mathbf{N}) \cdot \text{grad } F = 0 \qquad (5\text{-}46)$$

Inserting F from (5-43), we obtain for an anisotropic medium

$$\frac{\partial \phi}{\partial t} - \frac{1}{n_e}\left[K_x\left(\frac{\partial \phi}{\partial x}\right)^2 + K_y\left(\frac{\partial \phi}{\partial y}\right)^2 + K_z\left(\frac{\partial \phi}{\partial z}\right)^2 - \frac{\partial \phi}{\partial z}(K_z + N) + N \right] = 0$$

$$(5\text{-}47)$$

This (*nonlinear equation!*) then is the boundary condition to be satisfied on an unsteady free surface with accretion.

For an isotropic medium (5-47) reduces to

$$\frac{\partial \phi}{\partial t} - \frac{K}{n_e}\left[\left(\frac{\partial \phi}{\partial x}\right)^2 + \left(\frac{\partial \phi}{\partial y}\right)^2 + \left(\frac{\partial \phi}{\partial z}\right)^2 - \frac{\partial \phi}{\partial z}\left(1 + \frac{N}{K}\right) + \frac{N}{K} \right] = 0 \qquad (5\text{-}48)$$

In the absence of accretion, we set $N = 0$. When the flow is steady, the phreatic surface is stationary and we set $\partial \phi / \partial t = 0$.

(e) Seepage face Finally, as explained in Sec. 4-5, we have to consider the seepage face (or surface) which is always present when a phreatic surface ends at the downstream external boundary of a flow domain. Segments CB and DE in Fig. 5-7 are seepage surfaces.

The phreatic surface is tangent to the boundary of the porous medium at points C and D (Fig. 5-7; see also Fig. 4-11). Along a seepage surface, water emerges from the flow domain, trickling downward to the adjacent body of water.

Being exposed to the atmosphere, the pressure along a seepage face is atmospheric pressure ($p = 0$), and hence the boundary conditions along such a surface is

$$\phi = z \qquad (5\text{-}49)$$

The geometry of the seepage face is known (as it coincides with the boundary of the porous medium), except for its upper limit (points C and D in Fig. 5-7) which is also lying on the (a priori) unknown phreatic surface. The location of this point is, therefore, part of the required solution. In unsteady flow, the location of the upper limit of the seepage face varies with time.

Bear (1972, pars. 7.1.7 and 7.1.9) mentions two additional types of boundary conditions which should be used when the presence of a capillary fringe is assumed.

With the boundary conditions described in the present section and the partial differential equations described in the previous one, *any* problem of flow can now be stated mathematically by

(a) defining the flow domain by an equation (or equations) of the surface bounding it; part of the boundary may be at infinity; sometimes the boundary is unknown a priori;

(b) specifying the partial differential equation to be satisfied by ϕ (which is commonly used as the dependent variable, although other variables, for example, p or ρ, could also be used) at all points within the flow domain, and $t > 0$;

(c) specifying the boundary conditions that must be satisfied by ϕ at all points of the boundary and at all times;

(d) prescribing initial conditions at all points of the flow domain, when the flow is unsteady.

Before leaving this subject of boundary conditions, we should mention one more type of boundary and that is the *surface of discontinuity in permeability*.

Abrupt Change in Permeability

Equations (5-28) and (5-29) describe flow in inhomogeneous domains. However, the distributions $K = K(x, y, z)$, (or $K_x = K_x(x, y, z)$, etc. in an anisotropic domain) must be continuous up to and including the first derivative in order for these equations to have solutions. If a discontinuity in K or in ∇K exists along certain surfaces (or lines in a two-dimensional flow), the only way to solve the problem is to divide the flow domain by these surfaces (or lines) into subdomains such that within each of them $K = K(x, y, z)$ will be continuous and so will its first derivative. This means that we have to specify the conditions that ϕ has to satisfy along each such boundary in order to have a well-posed problem for each subdomain.

As an example, consider the inhomogeneous, two-dimensional flow domain shown in Fig. 5-8. The domain D is composed of two homogeneous subdomains,

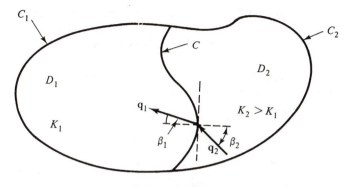

Figure 5-8 Boundary between regions of different permeability.

D_1 with $K_1 = $ const. and D_2 with $K_2 = $ const. In Sec. 2-6 this type of inhomo-geneity was referred to as inhomogeneity of *Type 2*. Further, let the problem to be solved in D be such that the partial differential equation to be satisfied is (5-28). However, since K is discontinuous along C, we cannot solve for ϕ, satisfying (5-28) in D. We therefore decompose the problem into two subproblems, denoting the potential in D_1 by ϕ_1 and in D_2 by ϕ_2. Simultaneous solutions are then sought for ϕ_1 and ϕ_2, satisfying the considered partial differential equation, here (5-27), in their respective subdomains, and specified boundary conditions on the external boundaries C_1 (for ϕ_1) and C_2 (for ϕ_2). If each of the two subdomains was in-homogeneous (that is, $K_1 = K_1(x, y, z)$ in D_1 and $K_2 = K_2(x, y, z)$ in D_2), the equation to be solved would have been (5-28). Since C constitutes part of the boundary of both D_1 and D_2, we must also specify the boundary conditions along it. We need two conditions, one for D_1 and one for D_2.

The two conditions are obtained as follows.

(a) Since the elevation z and pressure p are the same when a point on C is ap-proached from both sides, we have

$$\phi_1 = \phi_2 \text{ on } C \tag{5-50}$$

(b) The second condition on C is obtained from the requirement of continuity of flux across the boundary. This is expressed through the components of the specific discharge normal to the boundary

$$(q_n)_1 = (q_n)_2; \quad \mathbf{q}_1 \cdot \mathbf{1n} = \mathbf{q}_2 \cdot \mathbf{1n}; \quad (\mathbf{q}_1 - \mathbf{q}_2) \cdot \mathbf{1n} = 0 \text{ on } C \tag{5-51}$$

In this general form, (5-51) is valid also on a boundary S in three-dimensional flow. The exact expression of (5-51) in terms of ϕ_1 and ϕ_2 depends on the nature of the material occupying D_1 and D_2 (with respect to isotropy). For example, if K_1 and K_2 are both isotropic, (5-51) in two-dimensional flow reduces to

$$K_1 \frac{\partial \phi_1}{\partial n} = K_2 \frac{\partial \phi_2}{\partial n} \text{ on } C \tag{5-52}$$

where n is distance measured along the normal.

Thus, for the case shown in Fig. 5-8, the two boundary conditions to be satisfied on C are (5-50) and (5-51). Unfortunately, each of these equations includes both ϕ_1 and ϕ_2 and therefore the two problems (for D_1 and D_2) must be solved simultaneously.

From (5-50), it follows that $\partial\phi_1/\partial s = \partial\phi_2/\partial s$, where s is distance measured along the tangent to C (in Fig. 5-8). This can also be expressed as

$$(q_s)_1/K_1 = (q_s)_2/K_2 \tag{5-53}$$

where both K_1 and K_2 are isotropic. By combining (5-51) with (5-53), we obtain

$$\frac{K_1}{(q_s)_1/(q_n)_1} = \frac{K_2}{(q_s)_2/(q_n)_2} \; ; \qquad \frac{K_1}{\tan\beta_1} = \frac{K_2}{\tan\beta_2} \tag{5-54}$$

where $\tan\beta_1 = (q_s)_1/(q_n)_1$, $\tan\beta_2 = (q_s)_2/(q_n)_2$, and β_1 and β_2 are the angles which $\mathbf{q}_1$ and $\mathbf{q}_2$ make with the normal to C. This means that along such a boundary, the incident streamline is refracted. Equation (5-54) is the *law of refraction of streamlines* for two-dimensional flow when both subdomains are occupied by isotropic media.

Bear (1972, par. 7.1.10) discusses the laws of refraction of streamlines and of equipotentials also for three-dimensional flows and for cases where D_1 and D_2 are occupied by anisotropic media.

From (5-54) it follows that when $K' \gg K''$, $\beta' \gg \beta''$ and the refracted streamline approaches the normal to the common boundary on passing from one medium to another, less pervious than the first. When $K'' \gg K'$, $\beta'' \gg \beta'$ and the refracted streamline tends to become almost parallel to the common boundary on passing from a less pervious (e.g., semipervious) to a more pervious medium. This justifies the assumption of "practically horizontal flow" in a leaky aquifer (Sec. 2-4). The two cases mentioned here are shown in Fig. 5-9.

Another type of boundary, which is similar in its nature to the free surface one, is the interface between two regions occupied by different fluids (where like in the case of free surface, the concept of an abrupt interface is an approximate one). This boundary, and the conditions along it, are discussed in Sec. 9-2.

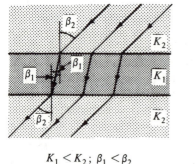

$$K_1 < K_2 ; \beta_1 < \beta_2 \qquad\qquad K_1 > K_2 ; \beta_1 > \beta_2$$

Figure 5-9 Refraction of streamlines at an interface between two permeabilities.

5-4 FUNDAMENTAL EQUATIONS FOR FLOW IN AQUIFERS

In Sec. 5-2, we have developed partial differential equations for the general case
of flow through porous media. In investigating groundwater flow in aquifers,
however, these equations are applicable whenever the flow is three dimensional
or two dimensional in the vertical plane. When the flow in the aquifer is treated
as *essentially two-dimensional flow in the horizontal plane* (see discussion on this
concept in Sec. 2-4), the governing equations take on modified forms of the
general equations presented in Sec. 5-2. These modified forms, for confined,
phreatic, and leaky aquifers, are discussed in the present section.

Whenever the concept of essentially horizontal flow is justified, in a leaky
or a confined aquifer, and this is the case in most regional groundwater problems,
the problem is treated with the averaged head

$$\tilde{\phi} = \tilde{\phi}(x, y, t) = \frac{1}{B} \int_{(B)} \phi(x, y, z, t)\, dz$$

as the dependent variable, rather than as a problem in $\phi(x, y, z, t)$. Obviously,
the reduction by one space dimension greatly simplifies the problem (and requires
much less information on the aquifer and its properties). In fluid mechanics, this
simplification is known as the *hydraulic approach* (see comment following (5-30)).
To obtain $\tilde{\phi}$, the average is taken along a vertical line extending from the bottom
to the top of the confined or leaky aquifer.

With these assumptions, we may apply the statement of balance to the volume
of water transported through an aquifer.

The aquifer properties which we need in order to state the balance of volume
are the *transmissivity, T,* as defined by (4-31), (4-33), or (4-35) and the *aquifer
storativity, S,* as defined by (5-10).

Flow in a Confined Aquifer

Consider the flow in the confined aquifer shown in Fig. 5-10. Writing the continuity
of flow (balance of volume) for the control box of volume $B\delta x\, \delta y$, we obtain

$$\delta t \left\{ \delta y \left[Q'_x\left(x - \frac{\delta x}{2}, y \right) - Q'_x\left(x + \frac{\delta x}{2}, y \right) \right] \right.$$

$$\left. + \delta x \left[Q'_y\left(x, y - \frac{\delta y}{2} \right) - Q'_y\left(x, y + \frac{\delta y}{2} \right) \right] \right\} = S(\delta x\, \delta y)[\phi(t + \delta t) - \phi(t)]$$

(5-55)

where $\mathbf{Q}' = -\mathbf{T}\cdot \text{grad}\, \phi$ is the discharge per unit width in the aquifer. Each side
of (5-55) expresses the excess (in terms of volume) of inflow over outflow of water
in the control box during the balance period δt. Dividing both sides of (5-55)
by $\delta x\, \delta y\, \delta t$ and passing to the limit of $\delta x, \delta y, \delta t \to 0$, we obtain

$$-\nabla' \cdot \mathbf{Q}' \equiv -\text{div}\, \mathbf{Q}' \equiv -(\partial Q'_x/\partial x + \partial Q'_y/\partial y) = S\, \partial \phi/\partial t;$$

$$\nabla' \cdot (\mathbf{T}\cdot \nabla' \phi) \equiv \text{div}'(\mathbf{T}\cdot \text{grad}'\, \phi) = S\, \partial \phi/\partial t$$

(5-56)

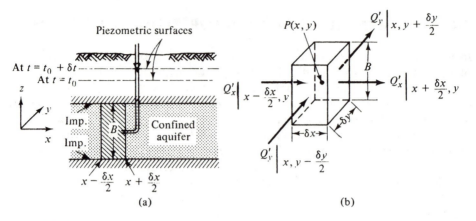

Figure 5-10 Flow in a confined aquifer.

where $\nabla'\cdot$, ∇', div$'$ and grad$'$ are symbols for the operations in the xy plane only (for example, $\nabla'\phi = (\partial\phi/\partial x)\,\mathbf{1x} + (\partial\phi/\partial y)\,\mathbf{1y}$).

If we also have distributed sources and sinks, $N(x, y, t)$, in the aquifer, expressing excess of inflow (say, by artificial recharge) over outflow (say, by pumping), per unit area and unit time, then (5-56) becomes

$$\text{div}'\,(\mathbf{T} \cdot \text{grad}'\phi) + N(x, y, t) = S\,\partial\phi/\partial t \tag{5-57}$$

For point sources and sinks (e.g., recharging and pumping wells, respectively), we may express the net inflow in (5-57) as $\sum_{(i)} N(x_i, y_i, t)\,\delta(x - x_i, y - y_i)$, where N is the rate of net inflow at a point (x_i, y_i) and time t and δ is the *Dirac delta function*. Note that the dimensions of $N(x, y, t)$ in (5-57) are L/T, while those of $N(x_i, y_i, t)$ are L^3/T.

Equation (5-57) is the fundamental continuity equation for flow in a confined aquifer. The two terms on the left-hand side express excess inflow over outflow per unit area and unit time by groundwater flow and by net replenishment, respectively. The term on the right-hand side expresses the resulting change in storage (also as volume per unit area and per unit time).

Several cases are of interest:

(*a*) Inhomogeneous anisotropic aquifer, $T_x(x, y) \neq T_y(x, y)$. Equation (5-57) becomes

$$\frac{\partial}{\partial x}\left(T_x \frac{\partial\phi}{\partial x} \right) + \frac{\partial}{\partial y}\left(T_y \frac{\partial\phi}{\partial y} \right) + N(x, y, t) = S\frac{\partial\phi}{\partial t} \tag{5-58}$$

x and y are principal directions; $T_x = T_x(x, y)$, $T_y = T_y(x, y)$.

(*b*) Inhomogeneous isotropic aquifer; $T = T(x, y)$. Equation (5-58) reduces to

$$\frac{\partial}{\partial x}\left(T \frac{\partial\phi}{\partial x} \right) + \frac{\partial}{\partial y}\left(T \frac{\partial\phi}{\partial y} \right) + N(x, y, t) = S\frac{\partial\phi}{\partial t} \tag{5-59}$$

(*c*) Homogeneous isotropic aquifer, $T = $ const. Equation (5-59) reduces to

$$T\left(\frac{\partial^2 \phi}{\partial x^2} + \frac{\partial^2 \phi}{\partial y^2}\right) + N(x, y, t) \equiv TV^2\phi + N(x, y, t) = S\frac{\partial \phi}{\partial t} \qquad (5\text{-}60)$$

(d) Steady flow, or negligible S, and $N = 0$ in a homogeneous isotropic aquifer. Equation (5-60) reduces to the well-known Laplace equation

$$\nabla^2\phi \equiv \frac{\partial^2 \phi}{\partial x^2} + \frac{\partial^2 \phi}{\partial y^2} = 0 \qquad (5\text{-}61)$$

The various aquifer equations developed above could have been derived also by integrating any of the three-dimensional flow equations of Sec. 5-2 over a vertical line. As an example, consider the integration of (5-27) written for a homogeneous isotropic aquifer of constant thickness B. By integrating, using Leibnitz' rule (4-29), and replacing ϕ^* by ϕ, we obtain

$$K\left(\int_0^B \frac{\partial^2 \phi}{\partial x^2}\,dz + \int_0^B \frac{\partial^2 \phi}{\partial y^2}\,dz + \int_0^B \frac{\partial^2 \phi}{\partial z^2}\,dz\right) = S_0\int_0^B \frac{\partial \phi}{\partial t}\,dz$$

Since $B = \text{const}$.

$$K\int_0^B \frac{\partial^2 \phi}{\partial x^2}\,dz = K\frac{\partial^2}{\partial x^2}\int_0^B \phi\,dz = KB\frac{\partial^2 \tilde{\phi}}{\partial x^2}; \qquad \tilde{\phi}(x, y, t) = \frac{1}{B}\int_0^B \phi(x, y, z, t)\,dz;$$

$$K\int_0^B \frac{\partial^2 \phi}{\partial y^2}\,dz = KB\frac{\partial^2 \tilde{\phi}}{\partial y^2}; \qquad \int_0^B \frac{\partial \phi}{\partial t}\,dz = B\frac{\partial \tilde{\phi}}{\partial t} \qquad (5\text{-}62)$$

$$K\int_0^B \frac{\partial^2 \phi}{\partial z^2}\,dz = K\int_0^B \frac{\partial}{\partial z}\left(\frac{\partial \phi}{\partial z}\right)dz = -q_z\big|_B + q_z\big|_0$$

where $q_z\big|_B = q_z\big|_0 = 0$ for completely impervious horizontal top and bottom boundaries. In what follows, we shall present a more general condition for an impervious boundary.

Thus, by integrating (5-27), we obtain

$$T\left(\frac{\partial^2 \tilde{\phi}}{\partial x^2} + \frac{\partial^2 \tilde{\phi}}{\partial y^2}\right) = S\frac{\partial \tilde{\phi}}{\partial t}; \qquad T = KB, \quad S = S_0 B \qquad (5\text{-}63)$$

which is identical to (5-60) with $N = 0$ and $\phi = \tilde{\phi}$.

When the impervious top and bottom of the aquifer are fixed, but not horizontal planes, that is, $B = b_2(x, y) - b_1(x, y)$, a similar integration can also be performed (see Sec. 4-3). This time let us integrate (5-26). We obtain

$$-\int_{b_1(x,y)}^{b_2(x,y)}\left(\frac{\partial q_x}{\partial x} + \frac{\partial q_y}{\partial y} + \frac{\partial q_z}{\partial z}\right)dz = S_0\int_{b_1(x,y)}^{b_2(x,y)} \frac{\partial \phi}{\partial t}\,dz \qquad (5\text{-}64)$$

Using Leibnitz' rule (4-29) and App. B, with the compact notation

$$\nabla' \cdot \mathbf{A}' = \frac{\partial A_x}{\partial x} + \frac{\partial A_y}{\partial y}; \qquad \mathbf{A}' = A_x\mathbf{1x} + A_y\mathbf{1y}$$

we obtain for the left-hand side of (5-64)

$$\int_{b_1}^{b_2} (\nabla \cdot \mathbf{q})\, dz = \int_{b_1}^{b_2} \left(\nabla' \cdot \mathbf{q}' + \frac{\partial q_z}{\partial z} \right) dz = \nabla' \cdot \int_{b_1}^{b_2} \mathbf{q}'\, dz - \mathbf{q}'|_{b_2} \cdot \nabla' b_2 + \mathbf{q}'|_{b_1} \cdot \nabla' b_1$$

$$+ \; q_z|_{b_2} - q_z|_{b_1} = \nabla' \cdot \int_{b_1}^{b_2} \mathbf{q}'\, dz + \mathbf{q}|_{b_2} \cdot \nabla(z - b_2) - \mathbf{q}|_{b_1} \cdot \nabla(z - b_1) \quad (5\text{-}65)$$

The equation describing a fixed impervious bottom, or top, at elevations $z = b(x, y)$ is

$$F = z - b(x, y) = 0$$

with a normal $\mathbf{1n}$ defined by $\mathbf{1n} = \nabla F/|\nabla F|$. For such a surface, we have

$$q_n \equiv \mathbf{q} \cdot \mathbf{1n} = 0; \qquad \mathbf{q} \cdot \nabla F \equiv \mathbf{q} \cdot \nabla(z - b) = 0 \quad (5\text{-}66)$$

Hence, (5-65) reduces for a confined aquifer to

$$\int_{b_1}^{b_2} \nabla \cdot \mathbf{q}\, dz = \nabla' \cdot \int_{b_1}^{b_2} \mathbf{q}'\, dz = \nabla' \cdot B\tilde{\mathbf{q}}'; \qquad \tilde{\mathbf{q}}' = \frac{1}{B} \int_{b_1}^{b_2} \mathbf{q}'\, dz$$

For the right-hand side of (5-64), we obtain $BS_0(\partial \tilde{\phi}/\partial t)$. Had we assumed that because of the elasticity of solid skeleton, both b_2 and b_1 may also be functions of time (and not fixed), we would have also obtained then (App. B)

$$\int_{b_1(x,y,t)}^{b_2(x,y,t)} \frac{\partial \phi}{\partial t}\, dz = \frac{\partial}{\partial t}(B\tilde{\phi}) - \phi\bigg|_{b_2} \frac{\partial b_2}{\partial t} + \phi\bigg|_{b_1} \frac{\partial b_1}{\partial t}$$

$$= B\frac{\partial \tilde{\phi}}{\partial t} + \left(\tilde{\phi}\frac{\partial B}{\partial t} - \phi\bigg|_{b_2} \frac{\partial b_2}{\partial t} + \phi\bigg|_{b_1} \frac{\partial b_1}{\partial t} \right) \approx B\frac{\partial \tilde{\phi}}{\partial t}$$

where the approximation results from the assumption of vertical equipotentials, that is, $\tilde{\phi} \approx \phi|_{b_2} \approx \phi|_{b_1}$.

The left-hand side of (5-64) yields for $b_1 = b_1(x, y, t)$ and $b_2 = b_2(x, y, t)$

$$\int_{b_1(x,y,t)}^{b_2(x,y,t)} (\nabla \cdot \mathbf{q})\, dz = \nabla' \cdot \int_{b_1}^{b_2} \mathbf{q}' dz + \mathbf{q}\bigg|_{b_2} \cdot \nabla F_2 - \mathbf{q}\bigg|_{b_1} \cdot \nabla F_1 = \nabla' \cdot \int_{b_1}^{b_2} \mathbf{q}\, dz$$

$$- \; n\frac{\partial F_2}{\partial t} + n\frac{\partial F_1}{\partial t} = \nabla' \cdot \int_{b_1}^{b_2} \mathbf{q}\, dz + n\frac{\partial B}{\partial t}; \qquad n\bigg|_{b_2} \approx n\bigg|_{b_1}$$

This result follows from the observation that for a moving impervious boundary defined by $F = 0$, we have $dF/dt \equiv \partial F/\partial t + (\mathbf{q}/n) \cdot \nabla F = 0$. Then (5-64) leads to

$$- \nabla' \cdot B\tilde{\mathbf{q}}' = BS_0 \frac{\partial \tilde{\phi}}{\partial t} + n\frac{\partial B}{\partial t}$$

(see App. B for more details).

Finally, assuming, in analogy to (5-6), that

$$\alpha = -\frac{1}{B}\frac{\partial B}{\partial \bar{\sigma}} = -\frac{1}{B}\frac{\partial B}{\partial \bar{p}}$$

and $n\,\partial B/\partial t \cong n\alpha B\rho g\,\partial\tilde{\phi}/\partial t$, we obtain

$$-\nabla\cdot B\tilde{\mathbf{q}}' = B(S_0 + n\alpha\rho g)\frac{\partial\tilde{\phi}}{\partial t} = S\frac{\partial\tilde{\phi}}{\partial t}\;;$$

$$S = \rho g(\alpha + n\beta)\,B$$

or

$$-\nabla\cdot\mathbf{Q}' = S\,\partial\tilde{\phi}/\partial t, \qquad \tilde{\phi} = \tilde{\phi}(x,y,t), \qquad \mathbf{Q}' = B\tilde{\mathbf{q}}' \qquad (5\text{-}67)$$

where the aquifer storativity, S, is defined as follows.

For fixed upper and lower bounds: $S = BS_0 = B\rho g[\alpha(1-n) + \beta n]$

For a consolidating medium, $B = B(t)$: $S = B[S_0 + \rho g\alpha n] = B\rho g(\alpha + \beta n)$

$\mathbf{Q}'$ in (5-67) can be expressed by $\mathbf{Q}' = -\mathbf{T}\cdot\nabla'\tilde{\phi}$, following (4-43). It is of interest to note that (5-67) obtained as an exact expression by averaging is identical to (5-56) obtained by the single assumption of horizontal flow.

Flow in a Leaky Confined Aquifer

The leaky aquifer is defined in Sec. 2-3. A confined leaky aquifer is essentially a confined aquifer except that its ceiling and/or bottom are semipervious so that water may leak into or out of the aquifer. To be qualified as a leaky aquifer, the leakage into or out of the aquifer should take place through top and/or bottom layers which, in general, are thin (with respect to the aquifer's thickness) and the hydraulic conductivity of which is much smaller than that of the aquifer.

Again, we assume that the flow in the main aquifer is essentially horizontal and therefore, in view of the law of refraction discussed in the previous section, the flow in the (top and/or bottom) semipervious layers is essentially vertical. As an approximation, we ignore storage in the semipervious layer. In general a leaky confined aquifer may be overlain or underlain by other leaky phreatic or confined aquifers, so that the whole system of aquifers has to be solved simultaneously (see below).

Consider the case of the inhomogeneous isotropic leaky confined aquifer shown in Fig. 5-11a. Figure 5-11b shows the control box for which the following (volumetric) water balance is written

$$\delta t\left\{\delta y\left[Q'_x\left(x - \frac{\delta x}{2},y\right) - Q'_x\left(x + \frac{\delta x}{2},y\right)\right]\right.$$

$$\left. + \delta x\left[Q'_y\left(x,y - \frac{\delta y}{2}\right) - Q'_y\left(x,y + \frac{\delta y}{2}\right)\right] + \delta x\,\delta y(q_{v2} - q_{v1})\right\}$$

$$= S\,\delta x\,\delta y[\phi(t + \delta t) - \phi(t)] \qquad (5\text{-}68)$$

where $\phi(x,y,t)$ is the piezometric head in the considered aquifer and q_{v1} and q_{v2}

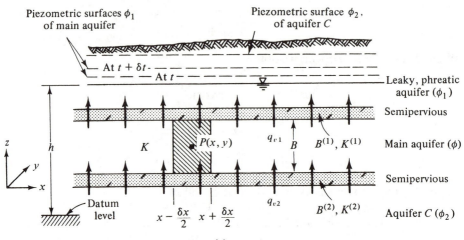

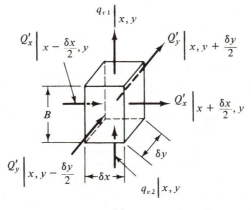

Figure 5-11 Flow in a leaky confined aquifer.

are vertical leakage rates [dims $L^3/L^2/T$] through the top and bottom semipervious layers, respectively. With $\phi_1(x, y, t)$ and $\phi_2(x, y, t)$ denoting the piezometric heads above the upper semipervious layer (in this case in the leaky phreatic aquifer overlying the main aquifer) and below the lower semipervious layer (i.e., in aquifer C), respectively, we have

$$q_{v1} = K^{(1)} \frac{\phi - \phi_1}{B^{(1)}} = \frac{\phi - \phi_1}{\sigma^{(1)}}; \qquad q_{v2} = K^{(2)} \frac{\phi_2 - \phi}{B^{(2)}} = \frac{\phi_2 - \phi}{\sigma^{(2)}}$$

where $\sigma^{(1)} = B^{(1)}/K^{(1)}$ and $\sigma^{(2)} = B^{(2)}/K^{(2)}$ are the resistances (dim T) of the semipervious layers. Hantush (1949, 1964) calls $K^{(i)}/B^{(i)} = 1/\sigma^{(i)}$ the *coefficient of leakage*. It is defined as the rate of flow across a unit (horizontal) area of a semipervious layer into (or out of) an aquifer under one unit of head difference across this layer.

Inserting the expressions for q_{v1} and q_{v2} into (5-68), dividing both sides by $\delta x\, \delta y\, \delta t$ and letting $\delta x, \delta y, \delta t \to 0$, leads to

$$-\nabla' \cdot \mathbf{Q}' + q_{v2} - q_{v1} = S\frac{\partial \phi}{\partial t}$$

For an isotropic medium

$$\frac{\partial}{\partial x}\left(T\frac{\partial \phi}{\partial x}\right) + \frac{\partial}{\partial y}\left(T\frac{\partial \phi}{\partial y}\right) + \frac{\phi_1 - \phi}{\sigma^{(1)}} + \frac{\phi_2 - \phi}{\sigma^{(2)}} = S\frac{\partial \phi}{\partial t};$$

$$\mathbf{Q}' = -T\nabla'\phi.$$

$$\left.\right\} \quad (5\text{-}69)$$

As in (5-57) through (5-60), we may also consider distributed sources and sinks and add N on the left-hand side of (5-69).

For a homogeneous isotropic aquifer, (5-69) becomes

$$\frac{\partial^2 \phi}{\partial x^2} + \frac{\partial^2 \phi}{\partial y^2} + \frac{\phi_1 - \phi}{\lambda^{(1)2}} + \frac{\phi_2 - \phi}{\lambda^{(2)2}} = \frac{S}{T}\frac{\partial \phi}{\partial t} \qquad (5\text{-}70)$$

where $\lambda^{(i)} = (T\sigma^{(i)})^{1/2}$, $i = 1, 2$, is another leaky aquifer parameter, called *leakage factor*, that determines the areal distribution of the leakage. Equation (5-69), or (5-70), is the basic continuity equation describing groundwater flow in a leaky confined aquifer, in the absence of sources and/or sinks.

As in the case of the confined aquifer, (5-69) or (5-70) for a confined-leaky aquifer can also be obtained by integrating (5-26), or (5-27) over the vertical.

For this case, let the leakage flux be denoted by $\mathbf{q}_l$ (not necessarily in the vertical direction). Instead of the condition (5-66), we now have (for a fixed bound)

$$\mathbf{q} \cdot \mathbf{1n} = \mathbf{q}_l \cdot \mathbf{1n}, \qquad \text{or:} \quad \mathbf{q} \cdot \nabla F = \mathbf{q}_l \cdot \nabla F = \mathbf{q}_l \cdot \nabla (z - b)$$

Hence, (5-65) for this case reduces to

$$\int_{b_1(x,y)}^{b_2(x,y)} (\nabla \cdot \mathbf{q})\, dz = \nabla' \cdot \int_{b_1}^{b_2} \mathbf{q}'\, dz + \mathbf{q}_l\Big|_{b_2} \cdot \nabla (z - b_2) - \mathbf{q}_l\Big|_{b_1} \cdot \nabla (z - b_1) \qquad (5\text{-}71)$$

For a horizontal fixed semipervious boundary, the last two terms on the right-hand side of (5-71) reduce to the vertical leakage $q_z|_{b_2} - q_z|_{b_1} = q_{v2} - q_{v1}$.

Again the final equation derived in this way is identical to (5-69) obtained by the assumption of horizontal flow and vertical leakage.

The material presented in App. B can be used to derive an averaged equation for the case of moving upper and/or lower boundaries, $b_1 = b_1(x, y, t)$, $b_2 = b_2(x, y, t)$, for example, due to consolidation.

Effect of Storage in a Semipervious Layer

Sometimes we wish to take into account the fact that changes in piezometric head, and hence in storage, take place within the semipervious layer so that the rates of leakage q_{v1} and q_{v2} are *time dependent*. We then denote the piezometric heads in the upper and lower semipervious layers by ϕ' and ϕ'', respectively. The rates of leakage to be inserted in (5-69) are given by

$$q_{v1} = -K^{(1)} \partial\phi'/\partial z, \qquad q_{v2} = -K^{(2)} \partial\phi''/\partial z$$

where gradients are taken at the interfaces between the main aquifer and the top and bottom semipervious layers. We then have two more unknown dependent variables, $\phi'(z, t)$ and $\phi''(z, t)$, which require two additional partial differential equations. These are

$$\frac{\partial}{\partial z}\left(K^{(1)} \frac{\partial\phi'}{\partial z} \right) = S_s^{(1)} \frac{\partial\phi'}{\partial t} \quad \text{and} \quad \frac{\partial}{\partial z}\left(K^{(2)} \frac{\partial\phi''}{\partial z} \right) = S_s^{(2)} \frac{\partial\phi''}{\partial t}$$

where $S_s^{(1)}$ and $S_s^{(2)}$ are specific storativities of the top and bottom semipervious layers, respectively, and we assume only vertical flow in these layers. Obviously, at the interfaces between the main aquifer and the top and bottom semipervious layers, $\phi' = \phi$ and $\phi'' = \phi$, respectively.

Bredehoeft and Hanshaw (1968), Wolff (1970), and Bredehoeft and Pinder (1970) discuss this problem of the effect of release (or absorption) of water from (or into) storage in a semipervious layer, when the head in the latter varies during leakage.

To demonstrate the effect of delayed storage, consider the upper semipermeable layer shown in Fig. 5-11a. Assume that the permeability contrast is such that the flow in the semipermeable layer is essentially vertical. Let a steady flow be established through the layer, then assume that a stepwise reduction of head is produced in the main aquifer by pumping. After a sufficiently long time, a new steady state will be established (linear distribution of head). However, during this period, the reduction of head in the semipermeable layer will lag behind that corresponding to the new steady state. Let $s'(z', t) = \phi'|_{t=0} - \phi'|_t$ denote the difference between the initial head and the changing head along the semipermeable layer. The problem of determining $s'(z, t)$ is stated by the following partial differential equation and initial and boundary conditions for $K^{(1)} = $ constant

$$K^{(1)} \frac{\partial^2 s'}{\partial z'^2} = S_s^{(1)} \frac{\partial s'}{\partial t} \qquad \begin{aligned} & s'(z', 0) = 0, \quad t \le 0 \\ & \\ ; & \quad s'(0, t) = 0, \qquad t > 0 \\ 0 < z' < B^{(1)} & \\ & s'(B^{(1)}, t) = \begin{cases} 0, & t \le 0 \\ H_0, & t > 0 \end{cases} \end{aligned}$$

The solution of this problem is given by (Carslaw and Jaeger, 1959, p. 310)

$$\frac{s'}{H_0} = \sum_{n=0}^{\infty}\left[\mathrm{erfc}\,\frac{(2n + 1) B^{(1)} - z'}{2(K^{(1)}t/S_s^{(1)})^{1/2}} - \mathrm{erfc}\,\frac{(2n + 1) B^{(1)} + z'}{2(K^{(1)}t/S_s^{(1)})^{1/2}} \right] \qquad (5\text{-}72)$$

Figure 5-12 (Bredehoeft and Pinder, 1970) shows this solution graphically.

From s', one can determine the flow into the pumped aquifer produced by the stepwise head change

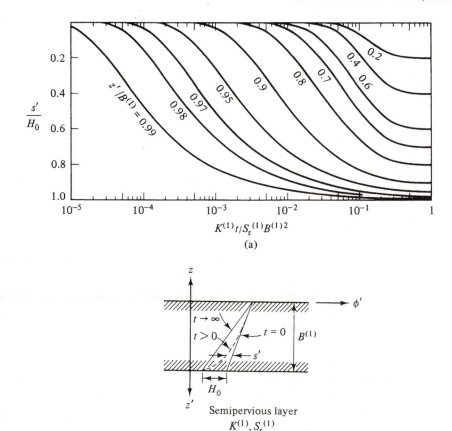

(a)

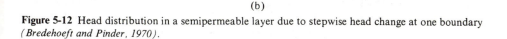

(b)

Figure 5-12 Head distribution in a semipermeable layer due to stepwise head change at one boundary (*Bredehoeft and Pinder, 1970*).

$$\Delta q = -K^{(1)} \frac{\partial s'}{\partial z}\bigg|_{z=B^{(1)}}$$

$$= \frac{K^{(1)} H_0}{B^{(1)}(\pi K^{(1)} t / B^{(1)2} S_s^{(1)})} \cdot \left\{ 1 + 2 \sum_{n=1}^{\infty} \exp\left[-n^2 / (K^{(1)} t / B^{(1)2} S_s^{(1)}) \right] \right\} \quad (5\text{-}73)$$

This flow is plotted in Fig. 5-13. From the figure, it follows that a long time may elapse before steady flow is re-established in the semipervious layer.

Flow in Phreatic Aquifer

Again, we shall derive the necessary equation by considering a control box in a phreatic aquifer (Fig. 5-14). We shall add a rate of accretion $N = N(x, y, t)$, positive when vertically downward. This accretion may be the net effect of natural replenishment, artificial recharge, and pumping. All these inputs and outputs

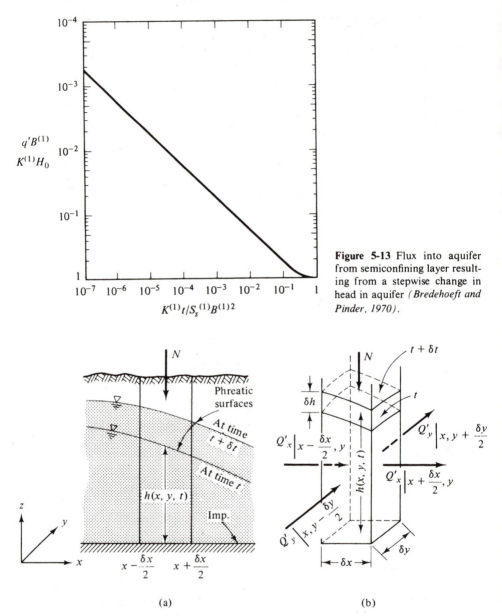

Figure 5-13 Flux into aquifer from semiconfining layer resulting from a stepwise change in head in aquifer (*Bredehoeft and Pinder, 1970*).

Figure 5-14 Flow in a phreatic aquifer.

can be introduced as distributed sources and sinks or as point ones. In the latter case, the Dirac delta function is used to describe them

$$N(x, y, t) = \sum_{(i)} N(x_i, y_i, t)\, \delta(x - x_i, y - y_i)$$

(see comment following 5-57). It is also possible to distinguish (using separate terms) between sources and sinks in the form of accretion applied to the water table and those located inside the flow domain.

The balance equation based on the Dupuit assumption of horizontal flow (Sec. 4-5), and with no internal sources and/or sinks, is

$$\delta t \left\{ \delta y \left[Q'_x \left(x - \frac{\delta x}{2}, y \right) - Q'_x \left(x + \frac{\delta x}{2}, y \right) \right] \right.$$

$$+ \delta x \left[Q'_y \left(x, y - \frac{\delta y}{2} \right) - Q'_y \left(x, y + \frac{\delta y}{2} \right) \right] + N \delta x \delta y \right\}$$

$$= S_y (\delta x \delta y) \left[h(t + \delta t) - h(t) \right] \tag{5-74}$$

where we have neglected the elastic storativity, as the storativity due to drainage from the pore space is much larger than that resulting from the elasticity of the water and the solid matrix: $S_y \gg S_0 h$ (Bear, 1972, p. 376). In (5-74), S is the storativity or specific yield, S_y, of the phreatic aquifer. Expressing $\mathbf{Q}'$ by (4-57), and following the usual procedure of dividing both sides of (5-74) by $\delta x\, \delta y\, \delta t$ and letting $\delta x, \delta y, \delta t \to 0$, we obtain for an inhomogeneous isotropic aquifer, $K = K(x, y)$

$$\frac{\partial}{\partial x} \left(Kh \frac{\partial h}{\partial x} \right) + \frac{\partial}{\partial y} \left(Kh \frac{\partial h}{\partial y} \right) + N = S \frac{\partial h}{\partial t}, \qquad S \equiv S_y \tag{5-75}$$

For a homogeneous aquifer, $K = $ const., we obtain

$$\frac{\partial}{\partial x} \left(h \frac{\partial h}{\partial x} \right) + \frac{\partial}{\partial y} \left(h \frac{\partial h}{\partial y} \right) + \frac{N}{K} = \frac{S}{K} \frac{\partial h}{\partial t} \tag{5-76}$$

This is the basic continuity equation for groundwater flow in a phreatic aquifer with a horizontal impervious base. It is called the *Boussinesq equation*.

As in the previous cases, (5-76) can also be derived by integrating the basic equation of three-dimensional flow through porous media. In this case, we shall denote the elevations of the fixed impervious, but not necessarily horizontal, bottom by $b_1 = b_1(x, y)$, and of the phreatic surface by $h = h(x, y, t)$, so that the thickness of the flow domain will be $B = h - b_1 = B(x, y, t)$. Then we obtain from (5-26)

$$\int_{b_1(x,y)}^{h(x,y,t)} \left[\frac{\partial q_x}{\partial x} + \frac{\partial q_y}{\partial y} + \frac{\partial q_z}{\partial z} \right] dz + S_0 \int_{b_1(x,y)}^{h(x,y,t)} \frac{\partial \phi}{\partial t} dz = 0,$$

$$\frac{\partial}{\partial x}(Q'_x) + \frac{\partial}{\partial y}(Q'_y) - q_x \bigg|_h \frac{\partial h}{\partial x} + q_x \bigg|_{b_1} \frac{\partial b_1}{\partial x} - q_y \bigg|_h \frac{\partial h}{\partial y} + q_y \bigg|_{b_1} \frac{\partial b_1}{\partial y} + q_z \bigg|_h - q_z \bigg|_{b_1}$$

$$+ S_0 B \frac{\partial \tilde{\phi}}{\partial t} + S_0 \left[\tilde{\phi} \frac{\partial B}{\partial t} - \phi \bigg|_h \frac{\partial h}{\partial t} + \phi \bigg|_{b_1} \frac{\partial b_1}{\partial t} \right] = 0 \tag{5-77}$$

where we have introduced

$$\mathbf{Q}' = \int_{b_1}^{h} \mathbf{q}' \, dz \quad \text{and} \quad \phi\bigg|_{h} = h.$$

With the approximation $\tilde{\phi} = \phi|_{b_1} \approx \phi|_{h} = h$, the last term on the left-hand side of (5-77) vanishes. Let us rewrite (5-77) for this case in the more compact form

$$\nabla' \cdot \mathbf{Q}' - \mathbf{q}'\bigg|_{h} \cdot \nabla' h + \mathbf{q}'\bigg|_{b_1} \cdot \nabla' b_1 + q_z\bigg|_{h} - q_z\bigg|_{b_1} + S_0 B \frac{\partial h}{\partial t} = 0 \qquad (5\text{-}78)$$

Referring now to the boundary condition, (5-46), on the phreatic surface, where $z = h(x, y, t)$, $F \equiv z - h(x, y, t) = 0$, we obtain

$$n_e \frac{\partial F}{\partial t} + (\mathbf{q} - \mathbf{N}) \cdot \nabla F = n_e \frac{\partial(z - h)}{\partial t} + (\mathbf{q} - \mathbf{N}) \cdot \nabla(z - h) = 0$$

or

$$-\mathbf{q}'\bigg|_{h} \cdot \nabla' h + q_z\bigg|_{h} = n_e \frac{\partial h}{\partial t} - N$$

Substituting this relationship in (5-78) yields

$$\nabla' \cdot \mathbf{Q}' + \mathbf{q}'\bigg|_{b_1} \cdot \nabla' b_1 - q_z\bigg|_{b_1} + (n_e + S_0 B) \frac{\partial h}{\partial t} - N = 0 \qquad (5\text{-}79)$$

Finally, since $n_e \gg S_0 B$ and $b_1(x, y)$ is an impervious bottom for which $\mathbf{q}'|_{b_1} \cdot \nabla' b_1 - q_z|_{b_1} = 0$ (see (5-66)), we obtain

$$-\nabla' \cdot \mathbf{Q}' + N = n_e \frac{\partial h}{\partial t}; \qquad \mathbf{Q}' = B\tilde{\mathbf{q}}'; \qquad n_e \equiv S_y \equiv S \qquad (5\text{-}80)$$

which is the same as (5-75) when $\mathbf{Q}'$ is expressed by the Dupuit assumptions.

We use the symbol S ($\equiv n_e$) to indicate the storativity of a phreatic aquifer, as the definition of storativity remains the same for a phreatic and a confined aquifer.

We have thus incorporated the phreatic boundary condition (as we did above for the impervious and leaky ones) in the equation of flow in the aquifer. This is referred to as the hydraulic approach. In the hydraulic flows, also called one-dimensional flows, we neglect the variations in velocity, pressure, etc., transverse to the main flow direction and in every cross section perpendicular to the flow, express the flow conditions in terms of average values over the cross section. In this way, the number of space dimensions is reduced by one.

Here also, by the integration, the problem in $\phi(x, y, z, t)$ has been reduced to one in $h(x, y, t)$. At the same time, the problem no more involves the non-linear condition (5-48) on the boundary. An essential simplification is thus achieved.

Linearization

Equation (5-75), or (5-76), is a non-linear one (because of the product $h \, \partial h/\partial x$). We may regard the product Kh in (5-75) and in (5-76) as the transmissivity, T, of the phreatic aquifer. However, unlike the transmissivity in a confined aquifer, here it may vary both in space and in time, as $h = h(x, y, t)$.

Two methods of linearization are often applied to (5-75) in order to facilitate a solution.

(i) Assume that $T = \overline{T} + \mathring{T}$; $\overline{T}$ $(\gg \mathring{T})$ is the average constant transmissivity of the phreatic flow and $\mathring{T}$ is a deviation from the average. Then (5-75) reduces to the linear equation in h

$$\overline{T}\left(\frac{\partial^2 h}{\partial x^2} + \frac{\partial^2 h}{\partial y^2}\right) + N = S\,\partial h/\partial t; \qquad \overline{T} = K\overline{h} \qquad (5\text{-}81)$$

to be compared with (5-60).

(ii) We rewrite the right-hand side of (5-76) as $(S/h)\,\partial(h^2/2)/\partial t$ and assume that S/h may be considered as a constant $S/\overline{h}$, where $T = K\overline{h}$. Then (5-76) reduces to

$$\left(\frac{\partial^2 h^2}{\partial x^2} + \frac{\partial^2 h^2}{\partial y^2}\right) + \frac{2N}{K} = \frac{S}{T}\frac{\partial h^2}{\partial t} \qquad (5\text{-}82)$$

which is a linear equation in h^2.

Equation (5-81) is the one commonly used to describe unsteady groundwater flow in phreatic aquifers. The approximation involved in the linearization (further to that introduced by the Dupuit assumptions) is justified in view of the relatively small changes in h (with respect to the total thickness h) in most phreatic aquifers. Whenever the situation is different, (5-75) or (5-76) should be used.

By replacing h in (5-81) by ϕ (measured from the same datum level as h), (5-60) and (5-81) become identical. We may, therefore, regard (5-81) with h replaced by ϕ, as the general continuity equation describing flow in both phreatic and confined aquifers. For a phreatic aquifer this is true whenever linearization is justified.

Flow in a Leaky Phreatic Aquifer

In this case, the phreatic aquifer is located above a semipermeable layer, which, in turn, overlies a leaky confined aquifer. Figure 5-11 shows such a case. The continuity equation can be easily derived by considering a control box in the phreatic aquifer, taking into account a leakage (q_{v1}) between the leaky confined aquifer and the overlying leaky phreatic one. Obviously, the direction of q_{v1} depends on whether $h > \phi$, or $\phi > h$. We would then obtain

$$\frac{\partial}{\partial x}\left(Kh\frac{\partial h}{\partial x}\right) + \frac{\partial}{\partial y}\left(Kh\frac{\partial h}{\partial y}\right) + N - \frac{h - \phi}{\sigma^{(1)}} = S\frac{\partial h}{\partial t} \qquad (5\text{-}83)$$

where the piezometric head in the leaky confined aquifer, ϕ, is measured from the same datum level as h. Here S $(\equiv S_y)$ stands for the storativity of the phreatic aquifer. This is the basic continuity equation describing groundwater flow in a leaky phreatic aquifer. It can be obtained by integration. We start from (5-79), noting that n_e $(\equiv S) \gg S_0 B$ and that $\mathbf{q}'|_{b_1} \cdot \nabla' b_1 - q_z|_{b_1} \equiv \mathbf{q} \cdot \nabla(z - b_1) =$

$\mathbf{q}_l \cdot \nabla(z - b_1)$, where $\mathbf{q}_l$ denotes the leakage through b_1. For a horizontal semi-pervious layer, $\nabla' b_1 = 0$, $\mathbf{q}_l \cdot \nabla z \equiv q_z|_{b_1} \equiv q_{v1} = (\phi - h)/\sigma^{(1)}$.

As was already emphasized above, when we have a system of leaky aquifers, each equation will also include the piezometric head in the underlying and/or overlying aquifer. This means that a continuity equation must be written for each of the aquifers and the system of equations must be solved simultaneously. Sometimes, delayed storage in a semipervious layer is taken into account by writing also a continuity equation for that layer as shown above.

Whenever we consider an inhomogeneous aquifer, with $T = T(x, y)$, the distribution $T(x, y)$ must be continuous up to and including the first derivative. If surfaces of discontinuity in T or in ∇T exist within the considered flow domain, we have to divide the aquifer into subdomains along the lines of discontinuity and solve simultaneously for all subdomains.

It may be of interest to note that when the aquifer is anisotropic, that is $T_x \neq T_y$, a procedure presented in Sec. 5-9 can be employed in order to transform the problem into one dealing with an equivalent isotropic aquifer (Bear, 1972, Sec. 7.4).

Mathematically, (5-58), (5-59), (5-60), (5-81), and (5-82) are second order linear partial differential equations of the parabolic type. They are often called heat conduction equations, or diffusion equations, as they are encountered in these fields. Equation (5-61) is also a second order linear partial differential equation, but of the elliptic type; it is known as the Laplace equation.

When necessary, they can easily be written in any other coordinate system by expressing $\nabla \cdot (T \nabla \phi)$ or $\nabla^2 \phi$ properly in that coordinate system. For example, in radial coordinates

$$\nabla^2 \phi \equiv \frac{1}{r} \frac{\partial}{\partial r} \left(r \frac{\partial \phi}{\partial r} \right) + \frac{1}{r^2} \frac{\partial^2 \phi}{\partial \theta^2} = \frac{\partial^2 \phi}{\partial r^2} + \frac{1}{r} \frac{\partial \phi}{\partial r} + \frac{1}{r^2} \frac{\partial^2 \phi}{\partial \theta^2}$$

5-5 COMPLETE MATHEMATICAL STATEMENT OF A GROUNDWATER FLOW PROBLEM

As was already explained in Sec. 5-3, a complete mathematical statement of a groundwater flow problem (and a correct mathematical statement is always the first step of solving a problem, no matter which method of solution is to be applied) consists of five parts.

(a) *Specifying the geometry* of the (two-dimensional) flow-domain in the aquifer.
(b) *Determining which dependent variable (or variables)* is to be used. Usually we use $\phi(x, y, t)$ for flow in confined and in leaky confined aquifers, and $h(x, y, t)$ for flow in phreatic and in leaky phreatic aquifers. When the linearized equation (5-81) is used, we often replace $h(x, y, t)$ by $\phi(x, y, t)$.
(c) *Stating the continuity equation* describing the flow in the aquifer (depending on the type of aquifer and on its properties).
(d) *Specifying the initial conditions* $\phi = \phi(x, y, 0)$, or $h = h(x, y, 0)$ at some initial time referred to as $t = 0$.

(e) *Specifying the conditions* which ϕ or its derivatives have to satisfy *on the boundaries* specified in (a). Regions under investigation need not extend to the *natural* external boundaries of a considered aquifer, such as rivers or lakes in contact with the aquifer, impervious faults, etc. Many reasons (sometimes merely economical or political ones) may dictate boundaries other than natural ones. Since boundary conditions actually introduce the effect of the environment on the considered groundwater system, the isolation of any portion of an aquifer is permitted, provided we specify the appropriate conditions to be satisfied along its boundaries.

When a natural boundary is used (e.g., an impervious fault or a river), the boundary conditions are obvious. For any other arbitrarily drawn boundary, separating a portion of the aquifer from the rest of it, the boundary conditions are a priori unknown. In fact, in order to derive ϕ or $\partial\phi/\partial n$ along such a boundary, one has to solve simultaneously for the aquifer domains on both sides of the boundary and thus learn what happens on the boundary itself.

Boundary Conditions for Flow in Aquifers

The detailed description of boundary conditions encountered in flow through porous media is presented in Sec. 5-3. Since in aquifers we have essentially two-dimensional flow in the horizontal plane, with $\phi = \phi(x, y, t)$, only three types of boundary conditions are encountered along the external boundary C of a considered flow domain.

1. Boundary of prescribed head In this case

$$\phi = f_1(x, y, t) \quad \text{on} \quad C \tag{5-84}$$

with the *equipotential boundary* $\phi = $ const., or $\phi = f_2(t)$, as a special case; f_1 and f_2 are known functions. A known potential boundary is encountered whenever the aquifer is in direct hydraulic contact with a river or a lake in which the water level is known.

A special case of this kind of boundary is a spring. The outlet threshold is at a fixed elevation. Therefore, water emerges from the aquifer into the atmosphere ($p = 0$) at that fixed elevation and hence this is a boundary of fixed piezometric head. Sometimes there exists above the threshold a water layer which may vary with the rate of flow. However, if the piezometric heads in the aquifer in the vicinity of the spring are lower than this threshold, the spring dries up and ceases to serve as a boundary to the flow domain. It is thus a boundary of fixed potential only as long as water heads in the vicinity are above the spring's outlet; they drop toward the spring (= loss of head in the converging flow in the aquifer) and in the spring itself.

2. Boundary of prescribed flux Along such boundary

$$Q'_n = f_3(x, y, t) \tag{5-85}$$

where f_3 is a known function. For an isotropic medium, this condition may be expressed as

$$\partial \phi / \partial n = f_4(x, y, t), \quad \text{on} \quad C \tag{5-86}$$

where n is distance measured normal to the boundary; f_4 is a known function. In an anisotropic aquifer, we have to express Q'_n in terms of the various components of the tensor T and the hydraulic gradient $J \ (\equiv -\nabla'\phi)$.

A special case of this boundary is the *impervious boundary*. For this case $Q'_n = 0$ and (5-86) becomes

$$\partial \phi / \partial n = 0 \quad \text{on} \quad C \tag{5-87}$$

In addition to the obvious case of an impervious boundary, condition (5-87) is also encountered in the following two cases.

(*a*) Along a *streamline* which is used as a boundary of a flow domain. This stems from the definition of a streamline as a curve which is everywhere tangent to the flow velocity, with no flow component existing normal to it.

(*b*) Along a *water divide* in the aquifer. By definition, a water divide is a curve with no flow crossing it. Unlike a water divide in surface runoff, a water divide in an aquifer may continuously change its shape and position. Nevertheless, it is sometimes used as boundary to the flow domain, approximating its changing shape and position by average fixed ones (see comment at the end of Sec. 4-5 in connection with the water divide in a phreatic aquifer).

3. Semipervious boundary This boundary condition is encountered when a clogged (e.g., by a thin layer of silt or clay) river bed serves as a boundary of the flow domain. The river may be either an effluent or an influent one (see Sec. 3-5). Because of the resistance to the flow offered by the semipervious layer, the water level (or piezometric head) in the river (point A of Fig. 5-15) differs from that at a point P in the aquifer, on the other side of the semipervious boundary. This situation is shown schematically for a confined aquifer in Fig. 5-15. Assuming horizontal

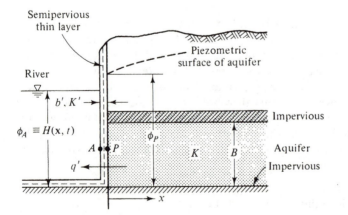

Figure 5-15 A clogged river bed serving as a semipervious boundary.

flow, we have

$$q' = K\left.\frac{\partial \phi}{\partial x}\right|_{x=0} = \frac{\phi|_{x=0} - H}{\sigma'}$$

or:

$$T\frac{\partial \phi}{\partial x} - B\frac{\phi - H}{\sigma'} = 0 \quad \text{on} \quad C \qquad (5\text{-}88)$$

where $H = H(\mathbf{x}, t)$ is known on C: $T = KB$ and $\sigma' = b'/K'$.

In general, using (5-40) we shall have the condition

$$(-T \cdot \nabla'\phi) \cdot \mathbf{1n} - \frac{H(x, y, t) - \phi}{\sigma'/B} = 0 \quad \text{on} \quad C \qquad (5\text{-}89)$$

We have here a mixed boundary condition, or a boundary condition of the third type.

The same situation may occur in a phreatic aquifer in contact with a river with a clogged bed. However, if the river is only slightly penetrating the aquifer, it does not necessarily act as a boundary. Instead it may be regarded as a source (or sink) and therefore be incorporated in the source (or sink) terms appearing in the aquifer equation, e.g., (5-80) (see discussion in Sec. 3-5). When flow in a phreatic aquifer is being considered, the boundary conditions are the same as those listed previously, except that we usually express them in terms of $h(x, y, t)$ rather than in terms of ϕ: $h = h(x, y, t)$, $h = $ const., $\partial h/\partial n = f(x, y, t)$, $\partial h/\partial n = 0$, etc., all on C.

In reality, aquifer boundaries of the types discussed here are seldom vertical surfaces; usually they deviate from the vertical, sometimes to a rather large extent. Nevertheless, if the assumption underlying our investigation is that of aquifer-type flow, so that the various equations derived in Sec. 5-4 are valid, we must approximate the geometry of all aquifer boundaries as vertical surfaces. The vertical cross section in Fig. 5-16 shows how aquifer boundaries are approximated (broken lines). Boundary I is an equipotential boundary; boundary II is an impervious boundary, while III represents a boundary between the confined and the phreatic portions of the aquifer. One should note that fluctuations in the water table in the phreatic portion of the aquifer will cause boundary III to shift. Ob-

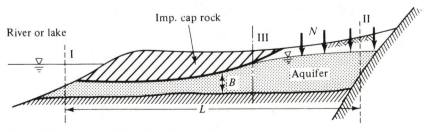

Figure 5-16 Approximation of aquifer boundaries.

viously these approximations are valid only when the concept of essentially horizontal flow in an aquifer is applicable, i.e., when $L \gg B$ (Fig. 5-16). Note that in Fig. 5-16, as in all other hydrological cross sections in this book, the scales are very much distorted.

Example

Let us summarize the discussion on how a groundwater flow problem should be stated mathematically by the following hypothetical example.

Figure 5-17 shows a map and a cross section of an area in which investigations on the development of groundwater resources take place. The aquifer is in the alluvium filling an intermontane valley with an impervious bedrock. The shaded area on the map indicates where existing or planned intensive pumping

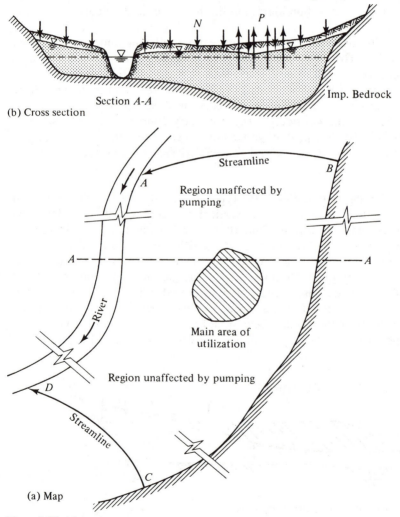

Figure 5-17 A hypothetical investigated area.

takes place. The aquifer is replenished from precipitation directly above it. Let $N(x, y, t)$ express the replenishment which may be a function of both location and time. We may also include artificial replenishment in N. The investigations are aimed at determining the response of the aquifer, say in the form of changes in water level elevations, to various planned pumping policies (or changes in existing ones). A pumping policy here is the specification of pumping rates at different wells as a function of time. Let the planned pumping be expressed symbolically by $P(x, y, t)$.

We have summarized above what should be included in a well-posed mathematical statement of a groundwater flow problem. Following this summary, the mathematical statement of the regional study considered here is as follows.

(a) The external boundary of the flow domain is $ABCDA$ (Fig. 5-17). There should be no question with respect to the segments CB and AD, as they are natural boundaries of the flow domain of interest. Because the river fully penetrates the aquifer down to the impervious bottom. the region on each side of the river (assumed, of course, to be perennial) behaves independently. This assumption of no interference could have remained valid also as a good approximation for a partially penetrating river, except when the aquifer is deep (relative to horizontal distances of interest), provided an hydraulic contact is maintained between the water in the river and in the aquifer. The boundaries in the other two directions pose a more difficult problem. Whenever natural boundaries do not exist, and the aquifer extends far beyond the area of planned operations, we must locate the boundary in a region sufficiently removed from that area, so that the flow regime there is practically unaffected. We then choose in this remote area either an equipotential line, or a line of specified potentials (using observed water levels), or a streamline (which serves as an impervious boundary) as a boundary. In Fig. 5-17, we have chosen two streamlines $(BA$ and $CD)$. At the end of the analysis, we should always verify if our assumption of practically unaffected regions is valid. We do so by repeating the analysis with a boundary located at a larger distance from the region of pumping. If the drawdowns in the region of pumping are practically unaffected, our original boundary was a good one. Otherwise, we have to remove it farther away from the area of pumping until any change in boundary location does not affect water levels in the region of interest.

(b) We have now to choose the unknown variable, the behavior of which in the xy plane we wish to forecast by solving the problem, and the partial differential equation which this unknown has to satisfy at all points inside the investigated domain $ABCDA$. Since the aquifer here is phreatic with a non-horizontal bottom, we have several possibilities:

1. Let $h(x, y, t)$ and $\eta(x, y)$ denote the elevations of the water table and of the impervious bottom, respectively, above some common datum level. Then, the dependent variable $h(x, y, t)$ must satisfy the following partial differential equation based on (5-80), within the investigated region:

$$\frac{\partial}{\partial x}\left[K(h - \eta)\frac{\partial h}{\partial x} \right] + \frac{\partial}{\partial y}\left[K(h - \eta)\frac{\partial h}{\partial y} \right] + N(x, y, t) - P(x, y, t) = S\frac{\partial h}{\partial t} \quad (5\text{-}90)$$

where $S \equiv S_y$ is the specific yield (or effective porosity) of the aquifer. For an aquifer with a horizontal bottom (or practically so) we set $\eta = 0$ in (5-90). Equation (5-90) is non-linear in h.

2. Although for a numerical solution by means of digital computer (see Sec. 5-6), the non-linearity in h does not constitute a major problem (see for example, Prickett and Lonnquist, 1971), we can linearize (5-90) by assuming that the variations in the saturated thickness $B(x, y, t) = h(x, y, t) - \eta(x, y)$ are much smaller than $B(x, y, t)$ itself. Then, with $B(x, y, t) = \bar{B}(x, y) + b'(x, y, t), b' \ll \bar{B}$, and $T(x, y) = K\bar{B}$, we obtain the linearized equation

$$\frac{\partial}{\partial x}\left(T \frac{\partial h}{\partial x} \right) + \frac{\partial}{\partial y}\left(T \frac{\partial h}{\partial y} \right) + N(x, y, t) - P(x, y, t) = S \frac{\partial h}{\partial t} \qquad (5\text{-}91)$$

Or, replacing the symbol $h(x, y, t)$ by $\phi(x, y, t)$, where the piezometric head $\phi(x, y, t)$ is measured from the same datum level as h, we obtain

$$\frac{\partial}{\partial x}\left(T \frac{\partial \phi}{\partial x} \right) + \frac{\partial}{\partial y}\left(T \frac{\partial \phi}{\partial y} \right) + N(x, y, t) - P(x, y, t) = S \frac{\partial \phi}{\partial t} \qquad (5\text{-}92)$$

as the equation describing the flow in the investigated domain.

(c) The boundary conditions in the present case are

(i) $\phi = $ const. along AD.

Note that if the river is rather wide, the boundary does not necessarily coincide with the shore line (see boundary along the lake in Fig. 5-16). If the water surface in the river has an appreciable slope along AD, we may take this into account by specifying the boundary conditions as

$$\phi = f(x, y) \quad \text{or:} \quad h = f(x, y) \text{ along } AD; \qquad f = \text{a known function.}$$

Streamlines act as impervious boundaries. Hence

(ii) $\partial\phi/\partial n = 0$, or $\partial h/\partial n = 0$ on AB, BC, and CD

where n is distance measured normal to the boundary at each point.

(d) Initial conditions in the form

$$h = g(x, y) \quad \text{or} \quad \phi = g(x, y); \qquad g = \text{a known function}$$

have to be specified at some initial instant of time.

This completes the mathematical statement of the groundwater flow problem in the region shown in Fig. 5-17.

Obviously, the input and output functions $N(x, y, t)$ and $P(x, y, t)$ have to be specified. We must also have complete information on $T(x, y)$ and $S(x, y)$. In order to arrive at an analytic solution, certain constraints are imposed regarding the continuity of the functions $K(x, y)$ and $T(x, y)$ appearing in (5-90) through (5-92) and on their first derivatives. The functions themselves have to be specified in analytic forms, and this is practically impossible. However, when numerical methods are employed (Sec. 5-6) and these are the main methods for solving regional groundwater flow problems of practical interest, no such constraints exist. Analog methods (Sec. 5-6) are also free of these constraints.

For example, let Fig. 5-16 represent a cross section of the investigated region shown in Fig. 5-17a. In this case, we have along the boundary III an abrupt change from a confined aquifer, the flow in which is described by (5-60), possibly with a term $-P(x, y, t)$ added to the left-hand side to represent pumpage, to a phreatic one, the flow in which is described by either (5-90) or (5-91). The value of S in the two equations is, obviously, also completely different. In addition, the boundary III itself shifts as the water table fluctuates. When the problem is solved by numerical techniques, using digital computers, no special difficulties arise. The computer program is made to find out whether at a given point (x, y) the water level is above or below the ceiling of the confined aquifer and then assigns to that point the appropriate values of S, T, etc.

In an analytic solution, whenever abrupt changes in the various physical parameters occur, we have to subdivide the investigated flow domain into subdomains in each of which these parameters vary in a continuous fashion, applying the appropriate boundary conditions (Sec. 5-3) to the boundaries between adjacent subdomains. The flow in each subdomain may then be described by a separate partial differential equation. Again a numerical solution is usually not constrained by requirements of continuity (and continuity of first derivatives) of the various parameters appearing in the equations.

Other examples of mathematical statements of regional groundwater flow problems can easily be constructed, using the procedure outlined above. Equation (5-92) is the partial differential equation for confined aquifers and most often also for phreatic ones if $h = \bar{h} + h'$, $h' \ll \bar{h}$, $T = K\bar{h}$. Equation (5-69) is used for leaky confined aquifers, where we may add $R(x, y, t) - P(x, y, t)$ to the left-hand side to describe recharge and pumping. We use (5-83) for a leaky phreatic aquifer, or its linearized form, adding $R(x, y, t) - P(x, y, t)$ on the left-hand side. However, in the cases of leaky aquifers, or in cases where aquifers are interconnected (e.g., by wells tapping several aquifers simultaneously), the mathematical statement will include a set of partial differential equations (and accompanying initial and boundary conditions), each for a single aquifer. The solution has to be derived simultaneously for the entire set.

The problem as stated in the present section, of seeking the response $\phi = \phi(x, y, t)$ of an aquifer to planned operations (pumping and/or artificial recharge), as well as to anticipated natural replenishment, may be called the forecasting problem. Given the information on the aquifer (T, S, and boundaries), we wish to forecast the response to specified pumping schedules, artificial recharge programs, specified (i.e., deterministic) sequences of natural replenishment from precipitation, changes in boundary conditions (e.g., lowering the water level in a lake), etc. The problem of determining the aquifer's parameters themselves (and these may also include the characteristics of the aquifer's natural replenishment) is often referred to as the identification (or the inverse) problem (Chap. 11).

Methods of solution of the forecasting problem are discussed in the following section.

5-6 METHODS FOR SOLVING GROUNDWATER FLOW PROBLEMS

In the previous section, we have shown how to state any problem of forecasting water levels, $h = h(x, y, t)$, or piezometric heads, $\phi = \phi(x, y, t)$, in a given aquifer, as a boundary value problem. To obtain $h(x, y, t)$, or $\phi(x, y, t)$, we have to solve a partial differential equation, subject to specified initial and boundary conditions. The geometry of the flow domain as well as its various transport and storage coefficients are known. Each of the partial differential equations is based on a set of assumptions which simplify the complex real world.

In principle, three classes of methods exist for solving such problems.

(*a*) Analytical methods.
(*b*) Methods based on the use of models and analogs.
(*c*) Numerical methods.

When more than one method can be applied to a given problem, the selection of the most suitable one depends on the availability of skilled manpower and on time and costs required for reaching a solution. One should also take into account the objective of the investigation and how the results are to be used. For example, the fact that the solution of a forecasting problem serves as an input to a management one should affect the selection of a solution method.

The purpose of the following paragraphs is to serve as a background for the selection of a solution method, given a forecasting problem. The reader is referred to specialized texts for additional information on specific methods.

Analytical Methods

Analytical methods are superior to any of the other ones and should be used whenever possible. One advantage, for example, is that the solution can be applied to different values of the parameters and inputs involved and that it clearly shows the influence of each parameter. Some examples of simple, one-dimensional cases for which analytical solutions are derived, are presented in Sec. 5-10. Well hydraulics (Chap. 8) is another class of problems for which analytical solutions have been derived, although in most cases the solution takes the form of an infinite series which has to be integrated numerically. In some of the more complex problems, when a lag of time in the release of water from storage is being considered, an analytical solution method cannot be applied.

Analytical methods are also used for certain two-dimensional flows in the vertical plane. It is even possible to solve analytically certain steady state problems in the vertical plane involving a phreatic surface or a sharp interface between two fluids (see Bear, 1972, Secs 8.3 and 9.6).

For more information on analytical solutions, the reader is referred to any text on partial differential equations, especially to those parts related to the elliptic Laplace equation and to the parabolic heat conduction, or diffusion, equation. Of special interest should be the book on *Conduction of Heat in Solids* (Carslaw and Jaeger, 1946) which contains many solutions of the heat conduction equation which is actually the same as (5-56) and (5-61) describing flows in aquifers. Bear (1972, Chap. 7) reviews some analytical methods of solution.

Unfortunately, in most regional studies of practical interest an analytical solution is not possible, mainly because of the irregularity of the shape of aquifer boundaries. The types of boundary conditions may vary along these boundaries. Furthermore, in most cases, the considered aquifer flow domain is inhomogeneous with respect to its storativity and transmissivity and the spatial distributions of these parameters cannot be presented in the form of analytical expressions. The same is true for initial conditions and for the various inputs and outputs (natural replenishment, artificial recharge, and pumping). Finally, the partial differential equation for a phreatic aquifer is nonlinear (unless we linearize it as a justified approximation).

As a consequence, analytical methods are seldom applied in the practice to the solution of regional forecasting problems.

Analog Methods

Analogs (and models) may be regarded as *special purpose computers*, when compared with digital computers, the use of which is discussed later in this section. An analog is usually constructed to solve a particular flow problem, although as long as the geometry and parameters of an aquifer remain unchanged, once constructed, an analog can be used for solving a number of cases which differ from each other only in the rates of input and output.

Sometimes an analog is regarded as an apparatus used for *simulating* a regime in an aquifer. Every aspect of aquifer behavior, which is relevant to the solution of the forecasting problem, is reproduced in an analog. This includes the interaction among various parts of the aquifer system. Then, we excite the analog by planned future activities (e.g., pumping) and observe and record the analog's response, which is analogous (except for a scale) to that of the real aquifer system.

We have thus two systems: the investigated system, referred to as the *prototype system*, and the *analog system*. These two systems are said to be analogous if the characteristic equations governing their dynamic and kinematic behavior are similar in form. This is possible only if there is a one-to-one correspondence between elements belonging to the two systems. For every element in the prototype system, there must be present in the analog system an element having a similar excitation-response relationship.

This is obviously the case when the two systems involve the same physical phenomena, as in the case of a sand box model simulating flow in an aquifer; both systems involve flow through a porous material. Generally, however, the physical dimensions of the two systems need not be the same. Often the term "model" is used to denote an analog having the same dimensions as those of a prototype. Analogs, on the other hand, are based on the *analogy* between systems belonging to entirely different physical categories. In an analogy, similarity is recognized by the following: (a) to each dependent variable and all its derivatives in the equations describing a prototype system there corresponds a variable with corresponding derivatives in the equations of the analog system, and (b) the dependent variables and their derivatives are related to each other in the same manner in the two sets of equations. Obviously the reason underlying the analogy is not just the similarity of the equations—this is merely a clever device for recognizing it. Actually the analogy (e.g., between Darcy's law and Ohm's law, or the corresponding continuity equations for water flowing in a porous medium and electricity flowing through a conductor) stems from the fact that the equations in both systems describe the same basic principles of conservation and transport that govern physical phenomena.

Flow through porous media, described by any of the equations developed in Secs. 5-2 and 5-4 (e.g., (5-29), (5-30) or (5-56)), can be studied by the following models and analogs.

(a) Sand box model.
(b) Vertical and horizontal Hele–Shaw analogs.
(c) Electric analogs of the electrolytic tank type, of the conducting paper type and of the RC-network type.
(d) Ion motion analog.
(e) Membrane analog.

The application of all these models and analogs to the solution of problems, such as three-dimensional flow in a porous medium, two-dimensional flow in a vertical plane and flow in an aquifer, is described in numerous books and articles. For example, Bear *et al.* (1968) and Bear (1972) present detailed reviews of these models and analogs, including their scaling, structure, and applications. Table 5-1 (Bear, 1972) summarizes the main features of the various models and analogs. As for their applicability to the solution of aquifer forecasting problems, we may add the following comments.

(i) The sand box model, the ion motion analog and the membrane analog are of very little practical use to the solution of aquifer forecasting problems. Their use also involves many technical difficulties.
(ii) The vertical Hele–Shaw analog is a relatively simple, yet a very useful tool for the solution of sea water intrusion in coastal aquifers (also in the presence of multiple aquifers). In fact, one can easily simulate in it the entire water balance of a coastal strip. It is restricted, however, to two-dimensional flows in the vertical plane. Another advantage (e.g., in studies of pollution in inhomogeneous aquifers) is that streamlines and path-lines can be made visible. The analog solves *exactly* problems with a phreatic surface and/or an interface. It does not simulate dispersion.
(iii) The horizontal Hele–Shaw analog can be used for studying any problem of essentially two-dimensional flow in a confined aquifer (and approximately in a phreatic one, assuming that the transmissivity is independent of changes in water table elevations). One can observe streamlines, path-lines and fronts between water bodies of different quality. It does not simulate dispersion.
(iv) The electrolytic tank and conducting paper analogs are usually used for steady flow only. In the former one can study both two- and three-dimensional flows; in the latter only two-dimensional flows. The presence of a phreatic surface requires special techniques.

Table 5-1 Applicability of models and analogs*

Feature	Sand box model	Hele–Shaw analog		Electrolytic	Electric analogs		Membrane analog
		Vertical	Horizontal		RC-network	Ion motion	
Dimensions of field	two or three	two	two	two or three	two or three	two (horizontal)	two (horizontal)
Steady or unsteady flow	both	both	both	steady	both	steady	steady
Simulation of elastic storage	yes, for two dimensions	yes	yes	yes, for two dimensions	yes	no	no
Simulation of capillary fringe and capillary pressure	yes	yes	no	no	no	no	no
Simulation of phreatic surface	yes[1]	yes[1]	no	yes[2]	no[3]	no	no
Simulation of anisotropic media[4]	yes	yes $k_x \neq k_z$ [5]	yes $k_x \neq k_v$ [5]	yes	yes	yes $k_x \neq k_y$	yes $k_x = k_z$
Simulation of medium inhomogeneity	yes	yes[5]	yes[5]	yes	yes	yes	no
Simulation of leaky formation	yes	yes	yes	yes[5]	yes	no	no
Simulation of accretion	yes	yes	yes	yes, for two dimensions	yes	no	yes
Flow of two liquids with an abrupt interface	approximately	yes	yes (no gravity)	no[6]	no[6]	yes (no gravity)	no
Hydrodynamic dispersion	yes	no	no	no	no	no	no
Simultaneous flow of two immiscible fluids	yes	no	no	no	no	no	no
Observation of streamlines and pathlines	yes, for two dimensions, near transparent walls for three dimensions	yes	yes	no	no	no	no

(1) Subject to restrictions because of the presence of a capillary fringe. (2) By trial and error for steady flow.

(3) By trial and error for steady flow, or, as an approximation, for relatively small phreatic surface fluctuations.

(4) By scale distortion in all cases, except for the RC-network and sometimes the Hele–Shaw analog where the hydraulic conductivity of the analog can be made anisotropic.

(5) With certain constraints. (6) For a stationary interface by trial and error.

* After Bear, 1972.

(v) The RC-network is a most suitable tool for the solution of regional fore-casting problems. It simulates the flow as described by (5-58), but can also be applied to a sequence of aquifers separated by semipervious layers. When necessary, it can also simulate three-dimensional flows. Water level fluctua-tions, $\phi(t)$ or $h(t)$, for any point in the flow domain, can be observed on an oscilloscope. Only flows described by linear equations and boundary con-ditions can be simulated. A disadvantage of this analog is its cost, especially that of peripheral equipment for input, output, and boundary conditions.

In the 50's and 60's the RC-network (Bear and Schwartz, 1966) was ex-tensively used as it was at that time practically the only tool for solving com-plex regional problems. However, in recent years, with the large strides in numerical solution techniques, using digital computers, there is no advantage in using them anymore except under special circumstances. Occasionally, local conditions, like lack of computers, programs, and skilled personnel, may make the use of the above analogs economically more attractive.

Since management models are run on digital computers, and the results of forecasting, as we shall see in Chap. 12, serve as input to such models, an important disadvantage of the RC-network (and of course of the other analogs) is that it cannot easily be directly connected to a management model run on a digital computer, feeding information on water levels directly into it. In recent years attempts have been made to construct *hybrid models*, i.e., to connect an RC-net-work directly to a digital computer. The main idea was to take advantage of the fact that the RC-network solves the partial differential equations, even in three dimensions, very fast, and let the computer solve the management problem. How-ever, it seems that while the interface equipment (analog to digital and digital to analog converters) is very complicated and expensive, no real advantage (e.g., economic), is achieved (Hefez, 1972; Hefez *et al.*, 1975a).

Like in any other method of solution, in order to design and construct an analog, we need complete information on the flow domain, its boundaries, parameters, etc. When these are not completely known, it is possible to base the initial construction of an analog on whatever data are available and on a rough estimate of missing information, and to calibrate the analog by reproducing in it the known past history of the aquifer. The various analog parameters are then adjusted until a satisfactory fit is obtained between the analog's response (e.g., water levels, $h = h(x, y, t)$) and that actually observed in the field. A calibration procedure is recommended also when we think that we know all the parameters.

A further discussion on calibration, the need for it and its methodology, is presented in Chap. 10 in connection with conceptual aquifer models.

In conclusion, except for special circumstances, the use of analogs for fore-casting flow regimes in aquifers is not recommended.

Numerical Methods

Computer based numerical methods are, nowadays, practically the major tool for solving large scale groundwater forecasting problems as encountered in prac-tice. In recent years, parallel to the advance in computer technology, much effort has been devoted, in many parts of the world, to the development of the methodology and techniques for numerical solution (sometimes referred to as *numerical simulation*) of the partial differential equations that govern the flow of water in aquifers of various types (Sec. 5-4). Numerous publications have been issued on numerical solutions of groundwater flow problems. In most cases, the end product of research has been a computer program which has been made readily available to any user. Many of the programs have been published in the open literature (e.g., Pinder, 1970; Prickett and Lonnquist, 1971, 1973; Trescott *et al.*, 1976) so that any hydrologist can apply them to his specific groundwater problems without having to develop a program of his own. Certain modifications are sometimes necessary in order to make an available program applicable to a specific problem on hand.

While thoughtless application of numerical solutions using available programs is not recommended, most of these programs have been documented in such a way that a hydrologist with a minimal background in numerical methods and programming can safely use them.

The use of ready programs for the solution of problems involving the equations of unsaturated flow (Sec. 6-6) and those of hydrodynamic dispersion (Sec. 7-8) is much more difficult and should not be considered a matter of routine at this stage.

Among specialized books published on this subject, we may mention those of Remson *et al.* (1971), who deal mainly with finite difference techniques, and Pinder and Gray (1977), who deal with finite element methods. A larger number of books is available on the solution of partial differential equations by numerical methods (e.g. Richtmeyer (1957), Rosenberg (1969), Forsythe and Wasow (1960), McCracken and Dorn (1964), Smith (1965), Mitchell (1971), Brebbia and Connor (1974)).

The objective of the following paragraphs is to introduce the reader to some of the numerical methods of solution which are commonly used for solving groundwater flow problems. No attempt has been made to present a thorough or a comprehensive review, but rather to show the main features of these methods and to indicate how they can be applied to solve forecasting problems. The actual computer programs are not considered.

Only saturated groundwater flow is considered in the present section. Comments on numerical solutions of unsaturated flow and hydrodynamic dispersion are given in Secs 6-6 and 7-8, respectively.

Some of the ideas presented here are discussed again, from a different point of view in Sec. 10-5, in connection with multiple-cell models.

Finite Difference Methods

In most numerical methods of solving a partial differential equation, the first step is to replace the latter by a set of *algebraic difference equations*. These are relationships among values of the dependent variables, say ϕ, at neighboring points in the considered $\mathbf{x}t$ space, where $\mathbf{x}$ denotes the position vector in the employed coordinate system, and t denotes time. The continuous variable $\phi =$

$\phi(\mathbf{x}, t)$, appearing in a partial differential equation, is thus replaced by a discrete variable. The set of finite difference equations is then solved numerically on a digital computer, to yield values of the dependent variable at a predetermined number of discrete points (grid points) in the investigated $\mathbf{x}t$ domain.

In what follows, we shall limit the discussion only to essentially horizontal flows in aquifers, governed by equations such as (5-57) through (5-61) and (5-75).

In order to solve one of the above equations in the xy plane, the flow domain, R, enclosed by a boundary B is divided into cells by a mesh of grid lines as shown in Fig. 5-18. The distance between grid lines need not be constant throughout the flow domain (see Fig. 10-4). Also, the distance between grid lines parallel to the x axis and those parallel to the y one need not be the same. Yet, unless a specific need for a variable grid spacing arises, $\Delta x_i = \Delta y_i = a$ is maintained constant. The grid lines then form a network of squares. The time interval of interest is divided into time increments Δt_k, also not necessarily equal to each other.

Trescott *et al.* (1976) suggest that to avoid large truncation errors and possible convergence problems, one should use $\Delta x_i/\Delta x_{i-1} \leq 1.5$ whenever a variable grid is employed. Also they suggest that the grid should be oriented such that a minimum number of nodes are outside the considered aquifer domain. In an anisotropic aquifer it is preferable to orient the grid with its axes parallel to the principal directions.

As stated above, the partial differential equation is replaced by a set of (algebraic) finite difference equations written in terms of values of the dependent variable, say ϕ, at the grid points. Denoting the exact solution of the partial differential equation by ϕ_E, the exact solution of the difference equation by ϕ_D and the numerical solution of the difference equations by ϕ_N, we call the difference $|\phi_E - \phi_D|$ the *truncation error* and $|\phi_D - \phi_N|$ the *numerical, round-off, error*. The condition for the *convergence* of the solution is that $|\phi_E - \phi_D| \to 0$ everywhere in the solution domain. The condition for stability is that everywhere in the solution domain, $|\phi_D - \phi_N| \to 0$. The problem is to find ϕ_N such that over the whole

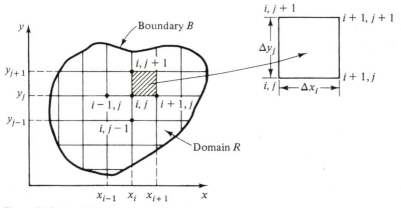

Figure 5-18 A grid for a numerical solution.

domain of interest $|\phi_E - \phi_N|$ is smaller than some a priori specified error criterion. As $(\phi_E - \phi_N) = (\phi_E - \phi_D) + (\phi_D - \phi_N)$, the total error is made up of the sum of the truncation error and the round-off error. The former is due to the form selected for the finite difference equation, and is often the larger part of the total error.

The finite difference equations can be obtained in two ways. A first, mathematical, approach is to approximate the derivatives appearing in the partial differential equations. Here we shall introduce this approach. The second, physical, approach is to consider the water balance of an element of area $\Delta x_i \, \Delta y_j$. This is actually the first step in developing any continuity equation (e.g., (5-55)), using the control box approach, without proceeding to the next step of letting Δt and the dimensions of the control box go to zero. This approach is also discussed in Sec. 10-5 in connection with a multiple-cell aquifer model. We shall explain the mathematical approach first.

Basic Finite Difference Schemes

Let $\phi = \phi(x)$ be a sufficiently smooth function so that it can be expanded in the following Taylor series about x in the positive direction (Fig. 5-19)

$$\phi(x + \Delta x) = \phi(x) + \Delta x \frac{d\phi}{dx}\bigg|_x + \frac{(\Delta x)^2}{2} \frac{d^2\phi}{dx^2}\bigg|_x + \frac{(\Delta x)^3}{3!} \frac{d^3\phi}{dx^3}\bigg|_x + \cdots \quad (5\text{-}93)$$

where we have used the notation $f(x) \equiv f|_x$. From (5-93), by dividing by Δx, we obtain

$$\frac{d\phi}{dx}\bigg|_x = \frac{\phi(x + \Delta x) - \phi(x)}{\Delta x} + 0(\Delta x) \quad (5\text{-}94)$$

where the term $0(\Delta x)$ represents the remaining terms of the series.

A term A is said to be of order $(\Delta x)^n$ written $0\{(\Delta x)^n\}$, if a positive constant C, independent of Δx, can be found such that $|A| < C|(\Delta x)^n|$. Thus, as $|(\Delta x)^n| \to 0$, $|A| \to 0$ at least as rapidly. In (5-94), the first term on the right-hand side is a finite difference analog to the first derivative at point x

$$\frac{d\phi}{dx}\bigg|_x \cong \frac{\phi(x + \Delta x) - \phi(x)}{\Delta x} \quad (5\text{-}95)$$

It contains values of the dependent variable at two adjacent grid points and the increment, Δx, of the independent variable. The error in using this analog is of the order of the first term which is truncated

$$\frac{\Delta x}{2} \frac{d^2\phi}{dx^2}$$

This is a *forward difference* approximation of the derivative $d\phi/dx$.

In a similar way, $\phi(x)$ may be expanded into a Taylor series about x in the

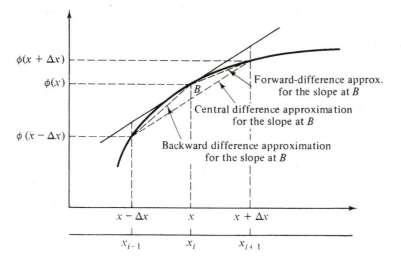

Figure 5-19 Geometric interpretation of the forward, backward, and central difference approximations of the derivative of a function.

negative direction (Fig. 5-19)

$$\phi(x - \Delta x) = \phi(x) - \Delta x \frac{d\phi}{dx}\bigg|_x + \frac{(\Delta x)^2}{2} \frac{d^2\phi}{dx^2}\bigg|_x - \frac{(\Delta x)^3}{3!} \frac{d^3\phi}{dx^3}\bigg|_x + \cdots \quad (5\text{-}96)$$

from which we obtain the *backward difference* approximation

$$\frac{d\phi}{dx}\bigg|_x \cong \frac{\phi(x) - \phi(x - \Delta x)}{\Delta x} \quad (5\text{-}97)$$

Again, the *truncation error* is $0(\Delta x)$.

By subtracting (5-96) from (5-93), we obtain the *central difference* analog

$$\frac{d\phi}{dx}\bigg|_x \cong \frac{\phi(x + \Delta x) - \phi(x - \Delta x)}{2\Delta x} \quad (5\text{-}98)$$

where all terms containing expressions of $(\Delta x)^2$ and higher powers have been dropped.

Figure 5-19 gives a geometrical interpretation of the above approximations for the slope $d\phi/dx$ at a point. The difference between the different approximations decreases with Δx.

To obtain an approximation to the second derivative, we add (5-93) and (5-96)

$$\phi(x + \Delta x) + \phi(x - \Delta x) = 2\phi(x) + \frac{2(\Delta x)^2}{2} \frac{d^2\phi}{dx^2} + 0\{(\Delta x)^4\}$$

hence

$$\frac{d^2\phi}{dx^2}\bigg|_x \cong \frac{\phi(x + \Delta x) - 2\phi(x) + \phi(x - \Delta x)}{(\Delta x)^2} \tag{5-99}$$

with a truncation error $0\{(\Delta x)^2\}$.

A similar analysis may be carried out with respect to time as the dependent variable. In this case, we are only interested in the first derivative.

Let us now apply the above approximations to the Laplace equation

$$\frac{\partial^2\phi}{\partial x^2} + \frac{\partial^2\phi}{\partial y^2} = 0$$

which describes steady flow in a homogeneous isotropic aquifer.

The numerical analog, using a forward difference simulation, with the notation of Fig. 5-18 and $\Delta x_i = \Delta x$ for all i and $\Delta y_j = \Delta y$ for all j, is

$$\frac{\phi_{i+1,j} - 2\phi_{i,j} + \phi_{i-1,j}}{(\Delta x)^2} + \frac{\phi_{i,j+1} - 2\phi_{i,j} + \phi_{i,j-1}}{(\Delta y)^2} = 0 \tag{5-100}$$

For $\Delta x = \Delta y$, we obtain

$$\phi_{i,j} = \tfrac{1}{4}(\phi_{i+1,j} + \phi_{i-1,j} + \phi_{i,j+1} + \phi_{i,j-1}) \tag{5-100a}$$

Thus the value of the function at any grid point is the average of its values at the immediately neighboring points.

The equation

$$\frac{\partial^2\phi}{\partial x^2} + \frac{\partial^2\phi}{\partial y^2} = \frac{S}{T}\frac{\partial\phi}{\partial t}$$

with S and T constant over the entire domain, describes unsteady flow in a homogeneous isotropic confined aquifer. For its numerical simulation, we use the grid of Fig. 5-18, with $\Delta x_i = \Delta x$ and $\Delta y_j = \Delta y$. For the left-hand side, we make use of (5-99) written at time t. For the time derivative on the right-hand side, two possibilities exist

$$\text{Forward difference:} \qquad \frac{\partial\phi}{\partial t}\bigg|_{t=k\Delta t} = \frac{\phi^{k+1} - \phi^k}{\Delta t}$$

$$\text{Backward difference:} \qquad \frac{\partial\phi}{\partial t}\bigg|_{t=k\Delta t} = \frac{\phi^k - \phi^{k-1}}{\Delta t}$$

where $t = \sum_{(k)}(\Delta t)_k$, and we have used a superscript, k, to denote time. For example, ϕ at time t, in the case of $(\Delta t)_k = \Delta t =$ constant for all k, is written in the form ϕ^k ($\equiv \phi|_{t=k\Delta t}$). The two possibilities are shown in Fig. 5-20.

Accordingly, the partial differential equation can be simulated by two finite difference schemes:

(i) Using the forward difference simulation for the time derivative, we obtain

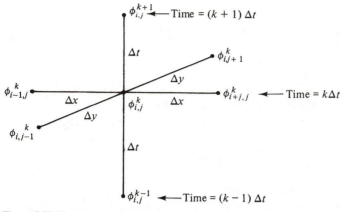

Figure 5-20 Notation for forward and backward simulation of the time derivative.

for constant Δx and Δy

$$\frac{\phi_{i-1,j}^{k} - 2\phi_{i,j}^{k} + \phi_{i+1,j}^{k}}{(\Delta x)^2} + \frac{\phi_{i,j-1}^{k} - 2\phi_{i,j}^{k} + \phi_{i,j+1}^{k}}{(\Delta y)^2} = \frac{S}{T}\frac{\phi_{i,j}^{k+1} - \phi_{i,j}^{k}}{\Delta t} \qquad (5\text{-}101)$$

We note that when all values of $\phi_{i,j}^{k}$ are known at time $k\,\Delta t$, we can obtain $\phi_{i,j}^{k+1}$ for all i, j and for time $(k+1)\,\Delta t$, by solving the *single equation* (5-101) for each node, i, j in the grid. In (5-101), we know the *five values* $\phi_{i-1,j}^{k}$, $\phi_{i,j}^{k}$, $\phi_{i+1,j}^{k}$, $\phi_{i,j-1}^{k}$, $\phi_{i,j+1}^{k}$, and seek the *single value* $\phi_{i,j}^{k+1}$. The head $\phi_{i,j}^{k+1}$ at each node is computed explicitly. The scheme represented by (5-101) is therefore called an *explicit scheme*. It is shown schematically in Fig. 5-21a (for one dimension, x, only).

(ii) Using the backward difference simulation for the time derivative, and writing (5-99) at $t = (k+1)\,\Delta t$, we obtain for constant Δx and Δy

$$\frac{\phi_{i-1,j}^{k+1} - 2\phi_{i,j}^{k+1} + \phi_{i+1,j}^{k+1}}{(\Delta x)^2} + \frac{\phi_{i,j-1}^{k+1} - 2\phi_{i,j}^{k+1} + \phi_{i,j+1}^{k+1}}{(\Delta y)^2} = \frac{S}{T}\frac{\phi_{i,j}^{k+1} - \phi_{i,j}^{k}}{\Delta t}$$

$$(5\text{-}102)$$

Here, assuming that the values $\phi_{i,j}^{k}$ are known at all nodes at time $k\,\Delta t$, (5-102) is a *single* equation containing *five unknowns*

$$\phi_{i-1,j}^{k+1}, \qquad \phi_{i,j}^{k+1}, \qquad \phi_{i+1,j}^{k+1}, \qquad \phi_{i,j-1}^{k+1}, \qquad \phi_{i,j+1}^{k+1}$$

We can, however, write an equation like (5-102) for each node in the flow domain. Then, since there is one unknown value of head (for time $(k+1)\,\Delta t$) at each node, we shall have a system of equations in which the total number is equal to the number of unknowns. We should, therefore, be able to solve the entire set of simultaneous equations, obtaining the new value $\phi_{i,j}^{k+1}$ at each node i, j. This scheme is, therefore, called the *implicit scheme*.

Comparing the two approaches, we note that the implicit scheme involves much more work in solving the set of simultaneous equations. However, as we shall see below, this drawback is offset by the advantage that the implicit scheme is unconditionally stable, regardless of the size of the time step Δt.

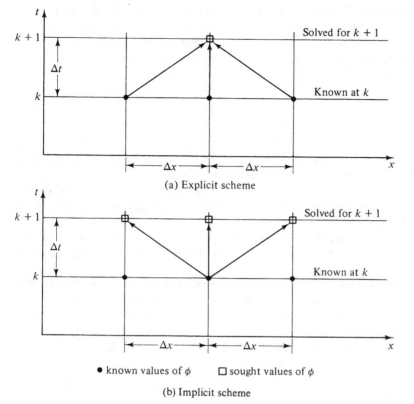

(a) Explicit scheme

● known values of ϕ □ sought values of ϕ

(b) Implicit scheme

Figure 5-21 Explicit and implicit schemes for the heat conduction equation.

Crank–Nicolson Scheme

Another often used numerical approximation is the *Crank–Nicolson scheme*, which is also unconditionally stable. The basic idea here is to use a central difference analog for the time derivative, rather than a forward or a backward one. While in the latter the truncation error is $0(\Delta t)$, in the former it is reduced to $0\{(\Delta t)^2\}$. We obtain the Crank–Nicolson scheme of the aquifer equation by averaging the approximations of $\partial^2\phi/\partial x^2 + \partial^2\phi/\partial y^2$ at k and $k + 1$ time levels. Sometimes a weighted average is used, with a weight λ, $0 < \lambda < 1$. We then obtain for $\partial\phi/\partial t$ evaluated at $k + 1/2$

$$\frac{\phi_{i,j}^{k+1} - \phi_{i,j}^{k}}{\Delta t} = \frac{T}{S}\left\{ \frac{\lambda(\phi_{i-1,j}^{k+1} - 2\phi_{i,j}^{k+1} + \phi_{i+1,j}^{k+1}) + (1 - \lambda)(\phi_{i-1,j}^{k} - 2\phi_{i,j}^{k} + \phi_{i+1,j}^{k})}{(\Delta x)^2} \right.$$

$$\left. + \frac{\lambda(\phi_{i,j-1}^{k+1} - 2\phi_{i,j}^{k+1} + \phi_{i,j+1}^{k+1}) + (1 - \lambda)(\phi_{i,j-1}^{k} - 2\phi_{i,j}^{k} + \phi_{i,j+1}^{k})}{(\Delta y)^2} \right\} \quad (5\text{-}103)$$

Stability

The explicit scheme is very easy to formulate and solve. However, for a numerical solution to be of any value, its solution must converge to that of the corresponding partial differential equation when the finite increments Δx, Δy, and Δt are decreased in size. Analysis has shown (see any text on numerical solutions of partial differential equations, e.g., Rosenberg, 1969) that a very restrictive relationship between the sizes of Δx, Δy and Δt must be satisfied in order to obtain a converging and stable solution. This stability criterion is

$$\frac{T}{S}\left[\frac{\Delta t}{(\Delta x)^2} + \frac{\Delta t}{(\Delta y)^2}\right] \leq \frac{1}{2} \tag{5-104}$$

A simplistic physical interpretation of (5-104) is given after (10-7). It is a serious restriction which often renders the rather simple explicit scheme impractical. For, in order to minimize the truncation error, we have to make Δx and Δy sufficiently small. It then follows that a very small Δt must be used for stability reasons, even when a much larger value could be used. Altogether this means much more computer time.

The implicit backward difference and Crank–Nicolson schemes are unconditionally stable. The latter scheme is usually preferred because of its increased accuracy.

Let us now exemplify the second, physical, approach of deriving a finite difference scheme by considering flow in an inhomogeneous anisotropic aquifer, governed by (5-58). The block-centered finite difference grid is shown in Fig. 5-22 (after Trescott et al., 1976).

By considering the water balance per unit area of the i, j element, and discharge rates at $k\Delta t$ we obtain the implicit equation

$$\frac{1}{\Delta x_i}\left[(T_x)_{i-1/2,j}\frac{\phi^k_{i-1,j} - \phi^k_{i,j}}{\Delta x_{i-1/2}} - (T_x)_{i+1/2,j}\frac{\phi^k_{i,j} - \phi^k_{i+1,j}}{\Delta x_{i+1/2}}\right]$$

$$+ \frac{1}{\Delta y_j}\left[(T_y)_{i,j-1/2}\frac{\phi^k_{i,j-1} - \phi^k_{i,j}}{\Delta y_{j-1/2}} - (T_y)_{i,j+1/2}\frac{\phi^k_{i,j} - \phi^k_{i,j+1}}{\Delta y_{j+1/2}}\right]$$

$$= \frac{S_{i,j}}{\Delta t}(\phi^k_{i,j} - \phi^{k-1}_{i,j}) - N^{k-1/2}_{i,j} \tag{5-105}$$

where: $N^{k-1/2} = (N^k + N^{k-1})/2$,

$$\Delta x_{i\mp1/2} = (\Delta x_{i\mp1} + \Delta x_i)/2, \qquad \Delta y_{j\mp1/2} = (\Delta y_{j\mp1} + \Delta y_j)/2$$

Note that (5-55), when divided by $\delta x\,\delta y\,\delta t$ is actually identical to (5-105), except for $N^{k-1/2}$. However, here we do not continue to perform the limit operation. Also

$$(T_x)_{i\mp1/2,j} = \frac{\Delta x_i + \Delta x_{i\mp1}}{\Delta x_i/T_{i,j} + \Delta x_{i\mp1}/T_{i\mp1,j}}$$

is the value of the transmissivity T_x between nodes i, j and $i \mp 1, j$

$$(T_y)_{i,j\mp1} = \frac{\Delta y_j + \Delta y_{j\mp1}}{\Delta y_j/T_{i,j} + \Delta y_{j\mp1}/T^*_{i,j\mp1}}$$

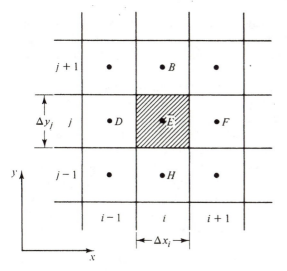

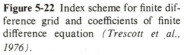

Figure 5-22 Index scheme for finite difference grid and coefficients of finite difference equation *(Trescott et al., 1976)*.

is the value of T_y between nodes i, j and $i, j \mp 1$. In both cases we have taken the harmonic average to ensure continuity across cell boundaries at steady state, and to make the appropriate coefficients zero at no flow boundaries. Equation (5-105) is similar to (10-7), except that here the aquifer is also anisotropic, and the term $N_{i,j}^{k-1/2}$ represents net input (volume per unit area per unit time) into the aquifer (e.g., artificial and natural replenishment minus pumpage) during Δt.

We may rewrite (5-105) in the form (Trescott *et al.*, 1976)

$$F_{i,j}(\phi_{i+1,j}^k - \phi_{i,j}^k) + D_{i,j}(\phi_{i-1,j}^k - \phi_{i,j}^k) + H_{i,j}(\phi_{i,j+1}^k - \phi_{i,j}^k) + B_{i,j}(\phi_{i,j-1}^k - \phi_{i,j}^k)$$

$$= \frac{S_{i,j}}{\Delta t}(\phi_{i,j}^k - \phi_{i,j}^{k-1}) - N_{i,j}^{k-1/2} \quad (5\text{-}106)$$

where the various coefficients can easily be written by comparing (5-105) with (5-106). Explicit forms of (5-105) and (5-106) can be obtained by considering discharge rates at $(k-1)\Delta t$.

For flow in a phreatic aquifer, the governing equation is (5-75), or (5-90). Referring to Kh or to $K(h - \eta)$ as the time-dependent aquifer transmissivity, $T = KB$, we may use (5-105) and (5-106) also for a phreatic aquifer by redefining the transmissivity in them. For example: $(T_x)_{i,j}^k = (K_x)_{i,j} B_{i,j}^k$. In the actual computation, in order to circumvent the non-linearity resulting from terms such as: $(K_x)_{i,j} B_{i,j}^k \phi_{i,j}^k \equiv (K_x)_{i,j}(\phi_{i,j}^k - \eta_{i,j}) \phi_{i,j}^k$, we calculate $(T_x)_{i,j}^k$ in each step, using the value of head $\phi_{i,j}^{k-1}$ known from the previous one.

Equations (5-105) and (5-106) can further be modified to take care of additional components of the water balance. The following modifications may be introduced by replacing the source term $N(x, y, t)$ by a sum of terms which includes: pumpage, $P_{i,j}^k$ (with artificial recharge considered as negative pumpage), natural replenishment, $R_{i,j}^k$, leakage through and release from storage in an upper semipermeable layer, $(q_v)_{i,j}^k$, and evapotranspiration, where present, $E_{i,j}^k$

$$N_{i,j}^k = -P_{i,j}^k + R_{i,j}^k + (q_v)_{i,j}^k - E_{i,j}^k \quad (5\text{-}107)$$

Prickett and Lonnquist (1971) and Trescott *et al.* (1976) discuss these components and show how they are introduced into the finite difference equation (5-105). They also discuss the case of an aquifer which is partly phreatic and partly confined and leakage from streams.

All wells within an element of area $\Delta x_i \, \Delta y_j$ are assumed to operate *at* the node i, j representing it, as if the total pumpage is uniformly spread out over the area of the element. The computed head at a node is also not *at* wells, but an *average* head computed for the element represented by the node. Because of the different kind of head loss which occurs in the case of radially converging flow to a point, a certain modification is often introduced in order to obtain more accurate heads at the nodes (Trescott *et al.*, 1976; see also Bear, 1972, p. 717). This option is of special interest in a phreatic aquifer where, as a result of drawdown, wells dry up when the water table drops below their pumps. A good computer program will include the possibility of indicating when this happens and then eliminate such wells until water levels rise again.

Sometimes, the averaged picture based on averaged regional water levels (or piezometric heads) is insufficient, and we are interested in the accurate distribution of levels in the vicinity of a certain well, or a certain number of wells. We then use the principle of superposition (Sec. 5-7) and a well drawdown equation corresponding to the type of the considered aquifer (Chap. 8). We first solve the regional problem numerically, with all wells assigned to their respective nodes. Then we repeat the procedure in order to subtract the effects of the wells around which we seek a more accurate solution. This is done by solving the same regional problem, but with only the considered wells being assigned to their respective nodes. Finally, we use the drawdown at and around the wells as calculated (analytically) from a well drawdown formula and subtract it from the regional water table obtained after the second step, with each well at its true location (and not at the nearby node). This is a mixed numerical-analytical solution.

Boundary Conditions

Figure 10-4 shows how a real, usually curved, aquifer boundary is approximated by a model boundary that is used in the numerical simulation.

Boundary conditions of the types considered in Sec. 5-5 are reflected in the finite difference equations corresponding to the nodes which are adjacent to the boundaries.

(a) Boundary of specified head The simplest approach is to assign the specified heads (which may be time dependent) to the nodes adjacent to the boundaries (see Fig. 10-4). In this case, however, we reduce the investigated area by half the length of the cell. If we assign the values to a (fictitious) external node (Fig. 10-4), we enlarge the area by half a cell. If the investigated area is very large the error introduced in either case may be minor. One can, of course, attempt to reduce the size of boundary cells in order to reduce the error.

For time-independent head on the boundary, we may assign the specified value for $t = 0$ only, but assign a very large storativity, say $S_{i,j} = 10^{40}$ (to represent $S_{i,j} \to \infty$) to the boundary cell. This will cause the head in this cell to remain unchanged. Trescott *et al.* (1976) assign a negative value of S to constant head nodes that define a constant head boundary. This indicates to the program that these nodes are to be skipped in the computations.

(b) Boundary of specified flux We express the flux by its finite difference approximation, using external fictitious cells. For example, for a boundary node denoted by i, j, we obtain the condition

$$Q^k_{i-1/2,j} = T_{i-1/2,j} \frac{\phi^k_{i-1,j} - \phi^k_{i,j}}{(\Delta x_i + \Delta x_{i-1})} \Delta y_j \tag{5-108}$$

where $Q^k_{i-1/2,j}$ is known.

Other formulations are also possible, depending on the scheme we use. We usually use $T_{i,j} = T_{i-1,j} = T_{i-1/2,j}$.

For an impermeable boundary, $Q^k_{i-1/2,j} = 0$, or

$$\phi^k_{i,j} = \phi^k_{i-1,j}, \qquad \text{for all } k\text{'s} \tag{5-109}$$

As indicated above, these conditions result in a modified equation for a boundary node. It is also possible to write an expression for a boundary node, taking into account the actual irregular shape of the curved boundary in its vicinity (see, for example, Bear, 1972, p. 345).

Alternating Direction Implicit Methods

Another, more efficient unconditionally stable method (actually a family of methods) is the alternating direction implicit (ADI) method, originally proposed by Peaceman and Rachford (1955; see example of application by Pinder and Bredehoeft, 1968). There exists also an iterative alternating direction implicit (IADI) method which allows for large time steps.

The main advantage introduced by the ADI method is that for each time step, Δt, it reduces the large set of simultaneous equations that has to be solved in an ordinary implicit method, down to a number of small sets. This is done by formulating the node equations in two substeps. First, from time level k to time level $k + 1$, with unknowns appearing only in the x direction, using known values for the y direction (in a problem in the xy plane). Then, from time level $k + 1$ to $k + 2$, we write the equations with unknowns appearing only in the y direction. For example, for solving $\nabla^2 \phi = (S/T) \partial\phi/\partial t$, we first write (5-102), replacing $(k + 1)$ by k in the expression for $\partial^2\phi/\partial x^2$, i.e., $\partial^2\phi/\partial x^2$ is expressed explicitly in terms of known values at time level k, while $\partial^2\phi/\partial y^2$ is expressed implicitly in terms of unknown values at $k + 1$. This set of equations is solved for $\phi^{k+1}_{i,j}$ for all values of y. Then, in stepping from the $k + 1$ to the $k + 2$ time level, we rewrite (5-102), with $\partial^2\phi/\partial y^2$ expressed explicitly in terms of the known values at $k + 1$, while $\partial^2\phi/\partial x^2$ is expressed implicitly in terms of the unknown values of $k + 2$ (i.e., with $\partial\phi/\partial t = (\phi^{k+2}_{i,j} - \phi^{k+1}_{i,j})/\Delta t$). This time we solve for $\phi^{k+2}_{i,j}$ for all values of x. In each of the two steps we obtain a tri-diagonal system of equations, the solution of which is very efficient.

In the IADI, the process is repeated for each Δt, using in each iteration updated values for expressing the derivatives. Thus, for example, the single finite difference equation (5-105) is replaced by two equations

$$\frac{1}{\Delta x_i}\left[(T_x)_{i-1/2,j}\frac{\phi_{i-1,j}^{k\,m+1} - \phi_{i,j}^{k,m+1}}{\Delta x_{i-1/2}} - (T_x)_{i+1/2,j}\frac{\phi_{i,j}^{k,m+1} - \phi_{i+1,j}^{k,m+1}}{\Delta x_{i+1/2}}\right]$$

$$+ \frac{1}{\Delta y_j}\left[(T_y)_{i,j-1/2}\frac{\phi_{i,j-1}^{k,m} - \phi_{i,j}^{k,m}}{\Delta y_{j-1/2}} - (T_y)_{i,j+1/2}\frac{\phi_{i,j}^{k,m} - \phi_{i,j+1}^{k,m}}{\Delta y_{j+1/2}}\right]$$

$$= S_{i,j}\frac{\phi_{i,j}^{k,m+1} - \phi_{i,j}^{k-1}}{\Delta t} - N_{i,j}^{k-1/2} \tag{5-110}$$

and

$$\frac{1}{\Delta x_i}\left[(T_x)_{i-1/2,j}\frac{\phi_{i-1,j}^{k,m+1} - \phi_{i,j}^{k,m+1}}{\Delta x_{i-1/2}} - (T_x)_{i+1/2,j}\frac{\phi_{i,j}^{k,m+1} - \phi_{i+1,j}^{k,m+1}}{\Delta x_{i+1/2}}\right]$$

$$+ \frac{1}{\Delta y_j}\left[(T_y)_{i,j-1/2}\frac{\phi_{i,j-1}^{k,m+2} - \phi_{i,j}^{k,m+2}}{\Delta y_{j-1/2}} - (T_y)_{i,j+1/2}\frac{\phi_{i,j}^{k,m+2} - \phi_{i,j+1}^{k,m+2}}{\Delta y_{j+1/2}}\right]$$

$$= S_{i,j}\frac{\phi_{i,j}^{k,m+2} - \phi_{i,j}^{k-1}}{\Delta t} - N_{i,j}^{k-1/2} \tag{5-111}$$

where m denotes the number of iterations.

We note that when we solve (5-110), i.e. for a row, we have only *three* unknowns in each row equation: $\phi_{i-1,j}^{k,m}$, $\phi_{i,j}^{k,m+1}$, and $\phi_{i+1,j}^{k,m+1}$. Similarly, in solving (5-111), for a column, we also have only three unknowns in each column equation: $\phi_{i,j-1}^{k,m+2}$, $\phi_{i,j}^{k,m+2}$, and $\phi_{i,j+1}^{k,m+2}$. In standard matrix notation, a set of equations defined by (5-110), or (5-111), is called a *tri-diagonal matrix*.

In the iterative ADI method, the solution is obtained in the following steps:

(i) Assume $\phi_{i,j}^{k,m}$ for the end of the time interval.

(ii) Solve row equations for $\phi^{k,m+1}$ at the end of the time period.

(iii) With the results of (ii), solve column equations to obtain $\phi_{i,j}^{k,m+2}$.

(iv) Repeat until $\phi_{i,j}^{k,m+2}$ is sufficiently close to $\phi_{i,j}^{k,m+1}$. We then assume that $\phi_{i,j}^{k,m+2} \cong \phi_{i,j}^{k,m+1} \cong \phi_{i,j}^{k}$.

As we have emphasized above, each time we solve for a total number of unknowns equal to the number of cells in a row or in a column, which is much smaller than the total number of nodes. The solution itself is derived by using the Thomas algorithm.

Prickett and Lonnquist (1971), following Peaceman and Rachford (1955), show in detail how the Gauss elimination algorithm is used for solving these equations. They present the complete programs necessary for solving forecasting problems in different types of aquifers.

Trescott *et al.* (1976) discuss, in addition to the iterative ADI, two other solution techniques: the strongly implicit procedure (SIP) and the line successive over-relaxation (LSOR). They also outline the computational algorithms for these three methods, compare results, and provide computer programs. They conclude that in general, the strongly implicit procedure requires less computer time and has fewer numerical difficulties than do the two other ones, especially when solving large complex problems (see also Trescott and Larson, 1977).

Numerical Schemes for a Phreatic Aquifer

The various finite difference approaches presented above are also applicable to flow in phreatic aquifers, as governed by the nonlinear equation (5-75), or the linearized equation (5-82). For example, (5-75) for a homogeneous aquifer, S_y, $K =$ const., may be written in the following (implicit) finite difference form

$$\frac{(h^2)_{i-1,j}^{k+1} - 2(h^2)_{i,j}^{k+1} + (h^2)_{i+1,j}^{k+1}}{2(\Delta x)^2} + \frac{(h^2)_{i,j-1}^{k+1} - 2(h^2)_{i,j}^{k+1} + (h^2)_{i,j+1}^{k+1}}{2(\Delta y)^2}$$

$$= \frac{S_y}{K} \frac{h_{i,j}^{k+1} - h_{i,j}^k}{\Delta t} - \frac{N_{i,j}^{k+1/2}}{K \Delta x \Delta y} \quad (5\text{-}112)$$

where Δx and Δy are constant.

It is possible to linearize the numerical scheme by replacing $h \partial h / \partial x$ and $h \partial h / \partial y$ by $h^k \partial h^{k+1} / \partial x$ and $h^k \partial h^{k+1} / \partial y$, respectively. For an inhomogeneous aquifer we then obtain the general scheme

$$\varepsilon_y \{ \Delta_{x+1/2} [K h^k h^{k+1}] + \Delta_{x-1/2} [K h^k h^{k+1}] \}$$

$$+ (1 - \varepsilon_x) \{ \Delta_{x+1/2} [K h^k h^k] + \Delta_{x-1/2} [K h^k h^k] \}$$

$$+ \varepsilon_y \{ \Delta_{y+1/2} [K h^k h^{k+1}] + \Delta_{y-1/2} [K h^k h^{k+1}] \}$$

$$+ (1 - \varepsilon_y) \{ \Delta_{y+1/2} [K h^k h^k] + \Delta_{y-1/2} [K h^k h^k] \}$$

$$= (S_y)_{i,j} \frac{h_{i,j}^{k+1} - h_{i,j}^k}{\Delta t} + \frac{N_{i,j}^{k+1/2}}{\Delta x \Delta y} \quad (5\text{-}113)$$

where

$$\Delta x_{i \pm 1/2} [K h^k h^{k+1}] = K_{i \pm 1/2, j} h_{i \pm 1/2, j}^k (h_{i \pm 1, j}^{k+1} - h_{i,j}^k)/(\Delta x)^2,$$

$$\Delta x_{i \pm 1/2} [K h^k h^k] = K_{i \pm 1/2, j} h_{i \pm 1/2, j}^k (h_{i \pm 1, j}^k - h_{i,j}^k)/\Delta x^2,$$

$h_{i \pm 1/2, j}^k = (h_{i \pm 1, j}^k + h_{i,j}^k)/2$, and for $K_{i \pm 1/2, j}$ we may take either a harmonic (preferable) or an arithmetic average between the nodes.

Equation (5-113) is explicit when $\varepsilon_x = \varepsilon_y = 0$ and fully implicit when $\varepsilon_x = \varepsilon_y = 1.0$; it is a centered (in time) equation for $\varepsilon_x = \varepsilon_y = \frac{1}{2}$. An ADI scheme is obtained from (5-113) by setting $\varepsilon_x = 1.0$, $\varepsilon_y = 0$ and then $\varepsilon_x = 0$, $\varepsilon_y = 1.0$ at alternating steps.

Method of Finite Elements

Starting about the mid 1960's, another, very powerful numerical technique — generally known as the *finite element method* — has been applied to numerous problems of flow through porous media, groundwater flow, multiphase flow, flow with a phreatic surface, hydrodynamic dispersion, consolidation, and heat and mass flow through porous media. It is a very powerful and extremely flexible method. It can handle any shape of boundary and any combination of boundary conditions, inhomogeneous and anisotropic media, moving boundaries (by continuously changing the grid), free surfaces and interfaces, deformable media, multiphase flows, etc.

In the finite element method, the objective is to transform the partial differential equation into an integral equation which includes derivatives of the first order only. Then the integration is performed numerically over elements into which the considered domain is divided.

Major contributions to the development and application of the method have been made by researchers at the Water Resources Program, Princeton University and at the Energy and Environment Division, Lawrence Berkeley Laboratory, University of California at Berkeley. Pinder and Gray (1977) summarize many of these developments (see also Zienkiewicz, 1971; Remson, 1971; Desai, 1974; Pinder, 1974), Gupta *et al.* (1975) and Gupta and Tanji (1976) present a three-dimensional finite element groundwater model. International Conferences on Finite Elements in Water Resources were held at Princeton (1976) and in London (1978). Their proceedings include many articles on the problems listed above. Numerous publications are also available in the literature.

Calculus of Variations Approach

The method is often presented as an application of the calculus of variations. The starting point is an integral (= a *functional*)

$$I = \iint_{(R)} F(x, y, U_1, U_2, \partial U_1/\partial x, \partial U_1/\partial y, \partial U_2/\partial x, \partial U_2/\partial y)\, dx\, dy \quad (5\text{-}114)$$

where R denotes the considered domain, x, y are two independent variables and $U_1 = U_1(x, y)$, $U_2 = U_2(x, y)$ are two dependent ones. We seek to make I stationary, i.e., to determine U_1 and U_2 which will make I an extremum. This is done by requiring that the *variation* (or differential) of I vanishes, i.e., $\delta I = 0$.

It can be shown (e.g., Hildebrand, 1962; Gelfand and Fomin, 1963) that this requirement holds *only if* the following partial differential equations are satisfied

$$\frac{\partial}{\partial x}\left(\frac{\partial F}{\partial U_{1x}}\right) + \frac{\partial}{\partial y}\left(\frac{\partial F}{\partial U_{1y}}\right) - \frac{\partial F}{\partial U_1} = 0;$$

$$\frac{\partial}{\partial x}\left(\frac{\partial F}{\partial U_{2x}}\right) + \frac{\partial}{\partial y}\left(\frac{\partial F}{\partial U_{2y}}\right) - \frac{\partial F}{\partial U_2} = 0 \quad (5\text{-}115)$$

where (only here!) subscripts x and y denote differentiation with respect to x and y, respectively (that is, $U_{1x} \equiv \partial U_1/\partial x$, etc.)

Equations (5-115) are called the *Euler equations* associated with (5-114). In these equations, U_1, U_{1x}, U_2, U_{2x}, x and y are treated as independent variables. These equations are the necessary conditions for I to be stationary. Sufficiency conditions are often very difficult to establish analytically.

The finite element technique is based on the solution of the variational problem in its original form, (5-114). Once the differential equations describing the problem have been formulated, we seek the functional for which they are the Euler equations. Then, instead of solving the differential equation, we can solve the minimization problem directly. The function to be minimized is an integral over the region of interest (an area integral for the two-dimensional case).

For example, satisfying the partial differential equation

$$\frac{\partial}{\partial x}\left(K_x \frac{\partial \phi}{\partial x} \right) + \frac{\partial}{\partial y}\left(K_y \frac{\partial \phi}{\partial y} \right) = 0 \quad \text{in domain } R \quad (5\text{-}116)$$

that describes steady two-dimensional flow in a nonhomogeneous anisotropic porous medium (x, y principal directions), can be shown to be equivalent to minimizing the functional

$$I = \frac{1}{2} \iint_{(R)} \left[K_x \left(\frac{\partial \phi}{\partial x} \right)^2 + K_y \left(\frac{\partial \phi}{\partial y} \right)^2 \right] dx\, dy \quad (5\text{-}117)$$

Next, the solution domain, R, is divided into *elements*. The shape, size, and distribution of the elements is arbitrary. It is assumed that the value of the dependent variable varies in some manner, say linearly, over each element. This means that the value of the dependent variable at any point within the element is uniquely determined by the values of the variable at the nodes of the element and the position of the point under consideration inside the element.

The contribution of each element to the integral given by (5-114) can be expressed in terms of the values of the dependent variables at the nodes of the element and its geometry. By differentiating this expression with respect to the dependent variable at each node, and adding up the resulting equations for all the elements in the field, a set of simultaneous equations is obtained in which the unknowns are the values of the dependent variables at the nodes, and the coefficients are functions of the coordinates of the nodes. The right-hand side is zero, as we are seeking a stationary point.

Boundary conditions are transposed from conditions along *sides* of an element to conditions at its *nodes*. For example, a given flux along a side is divided (in some proportional way) into two discrete fluxes at the ends of this side. A given value of the dependent variable is represented by its values at the nodes.

In summary, the finite element technique uses the following procedure:

(a) For the partial differential equation which governs the considered flow, derive the associated variational problem.
(b) Divide the field into elements.
(c) Formulate the variational functional within an element.
(d) Take derivatives with respect to the dependent variable at all nodes of the element.
(e) "Assemble" the equations for all elements.
(f) Express the boundary condition in terms of nodal values.
(g) Incorporate the boundary condition into the equations and solve.
(h) The shape and size of the elements is arbitrary. Different shapes (triangles, rectangles, etc.) can be used simultaneously. Smaller elements can be chosen

in regions where there are rapid variations in the properties of the material, or in the values of the dependent variables.

Let us exemplify the above steps by considering the flow described by (5-116) and (5-117). The flow domain is divided into elements, for example, triangular elements as shown in Fig. 5-23. If it is assumed that ϕ varies linearly over the element, then the value of ϕ at any internal point (x, y) is given in matrix notation by

$$\phi(x, y) = [N_i, N_j, N_m] \begin{Bmatrix} \phi_i \\ \phi_j \\ \phi_m \end{Bmatrix} \equiv [N]\{\phi^e\} \tag{5-118}$$

where $N_i = a_i + b_i x + c_i y$

$$a_i = \frac{x_j y_m - x_m y_j}{2\Delta}, \qquad b_i = \frac{y_j - y_m}{2\Delta}, \qquad c_i = \frac{x_m - x_j}{2\Delta}$$

and the other coefficients are obtained by the cyclic permutation $i \to j \to m$. The contribution, E^e, of an element is given by

$$E^e = \frac{1}{2} \iint_{(\Delta)} \left[K_x \left(\frac{\partial \phi}{\partial x} \right)^2 + K_y \left(\frac{\partial \phi}{\partial y} \right)^2 \right] dx\, dy \tag{5-119}$$

Next we take the derivatives of E^e with respect to ϕ_i, ϕ_j, and ϕ_m. For example

$$\frac{\partial E^e}{\partial \phi_i} = \iint_{(\Delta)} \left[K_x \frac{\partial \phi}{\partial x} \frac{\partial}{\partial \phi_i} \left(\frac{\partial \phi}{\partial x} \right) + K_y \frac{\partial \phi}{\partial y} \frac{\partial}{\partial \phi_i} \left(\frac{\partial \phi}{\partial y} \right) \right] dx\, dy \tag{5-120}$$

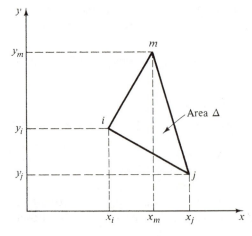

Figure 5-23 A triangular element.

But

$$
\frac{\partial \phi}{\partial x} = \left[\frac{\partial N_i}{\partial x}, \frac{\partial N_j}{\partial x}, \frac{\partial N_m}{\partial x} \right] \{\phi^e\} = [b_i, b_j, b_m] \{\phi^e\}
$$

$$
\frac{\partial \phi}{\partial y} = \left[\frac{\partial N_i}{\partial y}, \frac{\partial N_j}{\partial y}, \frac{\partial N_m}{\partial y} \right] \{\phi^e\} = [c_i, c_j, c_m] \{\phi^e\} \quad \Bigg\} \quad (5\text{-}121)
$$

$$
\frac{\partial}{\partial \phi_i} \left(\frac{\partial \phi}{\partial x} \right) = \frac{\partial N_i}{\partial x} = b_i; \qquad \frac{\partial}{\partial \phi_i} \left(\frac{\partial \phi}{\partial y} \right) = \frac{\partial N_i}{\partial y} = c_i
$$

Hence, (5-120) may be written in the form

$$
\frac{\partial E^e}{\partial \phi_i} = K_x [b_i^2, b_i b_j, b_i b_m] \{\phi^e\} \iint_{(\Delta)} dx\, dy + K_y [c_i^2, c_i c_j, c_i c_m] \{\phi^e\} \iint_{(\Delta)} dx\, dy
$$

$$
= \Delta (K_x [b_i^2, b_i b_j, b_i b_m] + K_y [c_i^2, c_i c_j, c_i c_m]) \{\phi^e\} \qquad (5\text{-}122)
$$

Repeating this procedure for ϕ_j and ϕ_m, we obtain a set of three equations, which can be expressed in matrix form as

$$
\left\{ \frac{\partial E^e}{\partial \phi^e} \right\} = [S^e_{ijm}] \{\phi^e\} \qquad (5\text{-}123)
$$

where $[S^e_{ijm}]$ is the (local) matrix of coefficients. The same procedure is repeated for all elements. Then, all the equations are combined into a set of simultaneous equations, each equal to zero, to make (5-119) stationary. In matrix notation we write

$$
[S] \{\phi\} = 0 \qquad (5\text{-}124)
$$

where $[S]$ is the (global) matrix of coefficients which incorporate the properties of the porous medium and the geometry; $\{\phi\}$ is the vector of the unknown ϕ's at the nodes.

The same approach is also extended to flow in a three-dimensional space and to unsteady flow problems (e.g., Javandel and Witherspoon, 1968).

A boundary condition of prescribed head over a surface area A is easily implemented by assigning the fixed prescribed heads to nodal points. The general boundary condition, e.g., $\mathbf{q} \cdot \mathbf{1n} + \alpha\phi + \beta = 0$, where $\mathbf{1n}$ is an outward normal unit vector to the boundary surface, is handled by adding to the functional (5-110) a second integral, $\iint_{(S)} (\beta\phi + \frac{1}{2}\alpha\phi^2)\, dS$, which on minimization automatically yields the boundary condition; S is area of boundary.

Rather than continue to discuss the finite element method on the basis of minimization of a functional, let us consider another approach in which the finite element approximation is derived directly from the partial differential equations governing the problem. In this approach, the need for finding the functional which is equivalent to the partial differential equation is eliminated.

Weighted Residual Methods

Following Zienkiewicz (1971, p. 39), let us consider a problem of approximately solving a set of partial differential equations in a domain D bounded by a surface S with the unknown variables $\{\phi\}$. We shall write the governing equation and the boundary condition symbolically, using the operators L and C in the form

$$L(\{\phi\}) = 0 \ \text{in} \ D; \qquad C(\{\phi\}) = 0 \ \text{on} \ S \tag{5-125}$$

Let a trial solution which satisfies the boundary conditions be written in the general form

$$\{\phi\}_a = [N] \{\Phi\}, \tag{5-126}$$

in which, as in (5-118), $[N]$ are prescribed functions of coordinates and $\{\Phi\}$ is a set of n parameters. Then, in general

$$L(\{\phi\}_a) = R \neq 0 \tag{5-127}$$

where R denotes a *residual*. The best solution will be one which reduces R to a least value at all points of D.

One way to achieve this is to make use of the fact that if R is made identically zero everywhere, then

$$\iiint_{(D)} WR \, dU = 0 \tag{5-128}$$

where W is any function of the coordinates. If the number of unknown parameters $\{\Phi\}$ is n, then if, n, linearly independent, functions W_i are chosen, we can write a suitable number of simultaneous equations as

$$\iiint_{(D)} W_i R \, dU = \iiint_{(D)} W_i L([N] \{\Phi\}) \, dU = 0 \tag{5-129}$$

from which $\{\Phi\}$ can be found. These processes are known as *weighted residual methods* and W_i is the *weighting function*. Different techniques can be implemented, depending on the choice of the weighting function. Here we shall focus our attention on the *Galerkin method* which is extensively used in recent years in single and multiphase flow through porous media, in reservoir engineering and in groundwater hydrology.

Galerkin Method

In this method $W_i = N_i$, i.e., the weighting function is made equal to the shape function defining the approximation. Zienkiewicz (1971) adds the following comments.

(i) The integration as described above is performed over elements independently, and then summed to give the total contribution.

(ii) The differential operator, L, appears directly in the integral of the weighted residual process, and in this, higher order differentials generally exist than in the variational functional I. This requires a higher order of continuity in the shape function definition. To overcome this difficulty it is possible to transform the integrals in (5-129), using integration by parts or its more general forms of Green's or Stokes' transformations.

Let us exemplify the Galerkin method by applying it to the basic equation of a leaky confined aquifer

$$L(\phi) = \frac{\partial}{\partial x}\left(T_{xx}\frac{\partial \phi}{\partial x} \right) + \frac{\partial}{\partial y}\left(T_{yy}\frac{\partial \phi}{\partial y} \right) - S\frac{\partial \phi}{\partial t} - Q - \frac{\phi - \phi_w}{\sigma'} = 0 \qquad (5\text{-}130)$$

where T_{xx} and T_{yy} are principal values of T, $Q = \sum_{k=1}^{m} Q_w(x_k, y_k)\,\delta(x - x_k, y - y_k)$ represents the strength of sinks at m points (x_k, y_k) and Q_w is the volumetric discharge from the aquifer; σ' is the resistance of the semipervious layer, and ϕ_w is the head in the adjacent aquifer (following Pinder and Frind, 1972).

To solve $L(\phi) = 0$ by the Galerkin method, we assume a trial solution of the form

$$\hat\phi(x, y, t) = \sum_{i=1}^{n} C_i(t)\, v_i(x, y) \qquad (5\text{-}131)$$

where $v_i(x, y)$, $i = 1, 2, \ldots, n$, is a system of functions (called *basis functions*) chosen beforehand and satisfying the boundary conditions imposed on (5-130). The functions $v_i(x, y)$, $i = 1, 2, \ldots, n, \ldots$, are assumed to be linearly independent and to represent the first n functions of some system of functions $v_i(x, y)$, $i = 1, 2, \ldots, n$, which is *complete* in the given region. In the subdivided domain (Fig. 5-24), n nodes are chosen, usually at element vertices and at points of interest. The trial solution is represented in a piecewise fashion across the domain, element by element, in terms of the space variables and the nodal values of the solution and its derivatives. The functions $C_i(t)$ in (5-131) are undetermined coefficients that will be shown below to be the solution of (5-130) at specified points (nodes) in the domain D (Fig. 5-24).

The approximation function $\hat\phi$ will be an exact solution of (5-130) only if $L(\hat\phi) \equiv 0$. Recalling that the condition for two functions $g_m(x)$ and $g_n(x)$ to be orthogonal in the region $a \le x \le b$ is $\int_a^b g_m(x)\,g_n(x)\,dx = 0$, one sees that this condition is equivalent to the requirement of the orthogonality of the expression

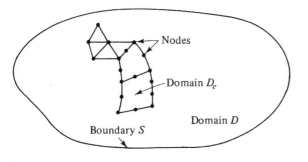

Figure **5-24** Domain divided into finite elements.

$L(\hat{\phi})$ to all the basis functions $v_i(x, y)$, $i = 1, 2, \ldots, n, \ldots$, or

$$\iint_{(D)} L(\hat{\phi}[x, y, t]) v_i(x, y) \, dx \, dy = 0; \qquad i = 1, 2, \ldots, n, \ldots \qquad (5\text{-}132)$$

Because we have selected only n basis functions, there are n undetermined coefficients $C_i(t)$, $i = 1, 2, \ldots, n$, and we can satisfy only n conditions of orthogonality. These conditions are

$$\iint_{(D)} L(\hat{\phi}[x, y, t]) v_i(x, y) \, dx \, dy$$

$$= \iint_{(D)} L\left(\sum_{j=1}^{n} C_j[t] v_j[x, y] \right) v_i(x, y) \, dx \, dy = 0; \qquad i = 1, 2, \ldots, n \qquad (5\text{-}133)$$

Assuming the appropriate integrations were performed, the coefficients $C_i(t)$ could be determined. Then, substituting these values into (5-131) will lead to the desired solution of (5-130).

We still have to select the basis functions v_i. Pinder and Frind (1972) remark that the suitability of the Galerkin approximation for computer application is largely a result of the choice of the basis functions. Efficient numerical schemes can be developed when continuous polynomial functions are used. In selecting these functions, a series of nodes (Figs. 5-24 and 5-25) are chosen in the domain D, and the basis functions are defined such that $v_i(x, y)$ is unity at node i and zero at all other nodes. Figure 5-26 gives an example of a piecewise linear basis function. When functions are chosen in this way, the undetermined coefficients $C_i(t)$, $i = 1, 2, \ldots n$, are equal to the required function $\hat{\phi}(x, y)$ at the n node points. The basis functions themselves may be piecewise linear, over triangular elements or higher order functions (over square or rectangular elements). *Isoparametric quadrilateral* elements have been introduced by Ergatoudis *et al.* (1968) with sides which may be distorted in any way (Fig. 5-25). Linear, quadratic, or cubic polynomials may be used, depending on the shape of the boundaries or the ex-

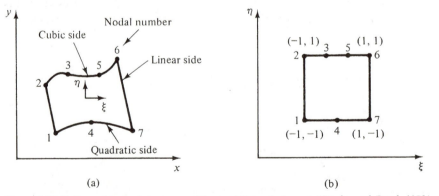

Figure 5-25 Deformed, mixed, isoparametric quadrilateral element *(Pinder and Frind, 1972)*. (a) In global coordinates xy. (b) In local coordinates $\xi\eta$.

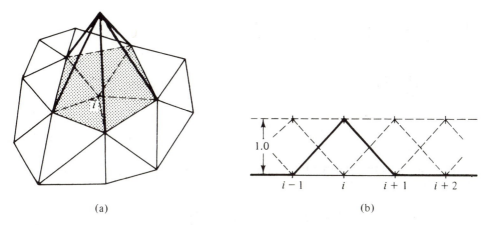

(a)   (b)

Figure 5-26 Piecewise linear basis function. (a) In the xy plane. (b) In one dimension.

pected shape of the unknown function. To facilitate integration. they introduce a dimensionless and curvilinear local $\xi\eta$ coordinate system (Fig. 5-25b), in which the element appears as a square with the side nodes located at midpoints (quadratic side) or one-third points (cubic side). The functions $v_i(x, y)$ are written in terms of ξ and η and are selected such that in addition to fulfilling the basic requirements of a basis function, they also relate the global and the local coordinate systems. Pinder and Frind (1972) give details of $v_i(\xi, \eta)$ at the nodes.

After generating the appropriate basis function for each of the n nodes in D, it is necessary to solve (5-133) for the undetermined coefficients $C_i(t)$. By substituting (5-130) and (5-131) into (5-133), we obtain

$$\iint_{(D)} \left\{ \left[\frac{\partial}{\partial x}\left(T_{xx}\frac{\partial}{\partial x} \right) + \frac{\partial}{\partial y}\left(T_{yy}\frac{\partial}{\partial y} \right) \right] \cdot \sum_{j=1}^{n} C_j v_j - S\frac{\partial}{\partial t}\sum_{j=1}^{n} C_j v_j - Q \right.$$

$$\left. - \frac{1}{\sigma'}\left(\sum_{j=1}^{n} C_j v_j - \phi_w \right) \right\} v_i \, dx \, dy = 0; \qquad i = 1, 2, \ldots, n \quad (5\text{-}134)$$

where D is composed of all the elements over which the ith basis is defined. To eliminate the second derivative, Pinder and Frind (1972) apply Green's theorem in the form

$$\iint_{(D)} \psi \, \nabla'^2 \phi \, dx dy = - \iint_{(D)} \nabla'\psi \cdot \nabla'\phi \, dx \, dy + \int_{(S)} \psi \, \nabla\phi \cdot d\mathbf{S} \quad (5\text{-}135)$$

and $\partial\phi/\partial n = \nabla'\phi \cdot \mathbf{1n}$, where $\mathbf{1n}$ is the outward normal to the boundary S.

By assuming transmissivity to be constant over each area of integration, recalling that C_i is a function of time only, and introducing (5-135) into (5-134), we obtain

$$\iint_{(D)} \sum_{j=1}^{n} \left(T_{xx} \frac{\partial v_i}{\partial x} \frac{\partial v_j}{\partial x} + T_{yy} \frac{\partial v_i}{\partial y} \frac{\partial v_j}{\partial y} + \frac{1}{\sigma'} v_i v_j \right) C_j \, dx \, dy$$

$$+ \iint_{(D)} S v_i \sum_{j=1}^{n} v_j \frac{dC_j}{dt} \, dx \, dy - \iint_{(D)} v_i \left(\frac{\phi_w}{\sigma'} - Q \right) dx \, dy$$

$$- \int_{(S)} v_i \sum_{j=1}^{n} \left(T_{xx} \frac{\partial v_j}{\partial x} l_x + T_{yy} \frac{\partial v_j}{\partial y} l_y \right) C_j \, dS = 0; \qquad i = 1, 2, \ldots, n \quad (5\text{-}136)$$

where l_x and l_y are the direction cosines of $\mathbf{1n}$. The n equations of (5-136) can be written in matrix form as

$$[H]\{C\} + [P]\left\{ \frac{dC}{dt} \right\} + \{F\} = 0 \qquad (5\text{-}137)$$

where $[H]$ and $[P]$ are $n \times n$ matrices, with

$$H_{ij} = \iint_{(D)} \left(T_{xx} \frac{\partial v_i}{\partial x} \frac{\partial v_j}{\partial x} + T_{yy} \frac{\partial v_i}{\partial y} \frac{\partial v_j}{\partial y} + \frac{v_i v_j}{\sigma'} \right) dx \, dy$$

$$P_{ij} = \iint_{(D)} S v_i v_j \, dx \, dy$$

$\{F\}$ is a vector in which

$$F_i = - \iint_{(D)} v_i \left(\frac{\phi_w}{\sigma'} - Q \right) dx \, dy - \int_{(S)} v_i \sum_{j=1}^{n} \left(T_{xx} \frac{\partial v_j}{\partial x} l_x + T_{yy} \frac{\partial v_j}{\partial y} l_y \right) C_j \, dS$$

$$(5\text{-}138)$$

The last term in (5-138) incorporates the prescribed flux boundary condition (which in the calculus of variation is called *natural boundary condition*) $\mathbf{T} \cdot \nabla' \phi \cdot \mathbf{1n} = q_n$ which is the flux of water into the element per unit length of boundary S. When $q_n \neq 0$, this last term therefore reduces to $\int_{(S)} v_i q_n \, dS$.

We do not generate (5-136) at nodes where the boundary condition is one of prescribed head.

Because of the way of selecting the v_i's so that they are nonzero only over elements on which they are located, the domain (D) in each case consists of from one to four elements, depending on the location of the ith node in the element array. As a result, the matrices $[H]$ and $[P]$ are sparse and banded, and the computation time and computer storage requirements are greatly reduced. Pinder and Frind (1972) also give the details of the integration for equilateral isoparametric elements in terms of the local coordinates ξ and η.

Once the matrices of (5-137) have been determined, we still have to solve for the n undetermined coefficients $C_i(t)$. Pinder and Frind (1972) use for this purpose a recurrence formula suggested by Zienkiewicz and Parekh (1970). Given the initial values $C_i(t)$, a finite difference approximation for the time derivative is introduced into (5-137). For the mid-interval in time, $t + \Delta t/2$, we obtain

$$[H]\{C\}\bigg|_{t+\Delta t/2} + [P]\left(\{C\}\bigg|_{t+\Delta t} - \{C\}\bigg|_{t}\right)\bigg/\Delta t + \{F\} = 0 \qquad (5\text{-}139a)$$

Introducing

$$\{C\}\bigg|_{t+\Delta t/2} \approx \left(\{C\}\bigg|_{t} + \{C\}\bigg|_{t+\Delta t}\right)\bigg/2$$

in (5-139a) gives

$$([H] + 2[P]/\Delta t)\{C\}\bigg|_{t+\Delta t/2} = 2[P]\{C\}\bigg|_{t}\bigg/\Delta t - \{F\} \qquad (5\text{-}139b)$$

from which we can find $\{C\}|_{t+\Delta t/2}$ from the known values of $\{C\}|_{t}$. The solution of (5-137) at time $t + \Delta t$ is obtained from

$$\{C\}\bigg|_{t+\Delta t} \approx 2\{C\}\bigg|_{t+\Delta t/2} - \{C\}\bigg|_{t} \qquad (5\text{-}140)$$

Other schemes for expressing $\{C\}|_{t+\Delta t/2}$ in terms of $\{C\}|_{t}$ and $\{C\}|_{t+\Delta t}$ are also possible (Pinder and Gray, 1977).

A rather large number of articles on the application of the finite element method, and especially of the Galerkin method, to the solution of two- and three-dimensional flow problems has been published in the literature of water resources and presented in special symposia, especially since the late 1960's. They are too many to be listed here.

Among the advantages given in the literature for preferring the finite element technique to the finite difference one, although both cases eventually lead to similar (and even identical in certain cases) sets of difference equations, are as follows.

(a) Boundary conditions, especially the flux boundary conditions are handled naturally. There is no need for special formulae, as in the method of finite differences.
(b) The size of the elements can be varied readily, with small elements used in areas of rapid changes, or where more detailed results are required. At the same time, very large elements may be used in other parts of the flow domain.
(c) It can easily handle flow in inhomogeneous and/or anisotropic domains. A discontinuity in hydraulic conductivity or in transmissivity is handled in a routine manner as nodes are located on the surface of discontinuity and the properties are those of the elements and not at the nodes. In the case of an-isotropy, no particular problem arises when the principal axes vary from one element to the next. All we have to know is the angles that these axes make with the local coordinates in each element.

Pinder and Gray (1976) compare the two methods and conclude that the finite element method may be considered as a numerical scheme wherein the algebraic equations represent spatially averaged derivative approximations. They emphasize that while finite difference formulae are usually written for a *point* in the flow domain, finite element formulae are approximate spatial integrations of point formulae.

Our intention in the present section was not to give details of solving the forecasting problem, but mainly to indicate (a) that numerical techniques are nowadays the main tool for solving this problem, and (b) that these techniques have been brought to the point where they are accessible to those who have to solve such problems as part of a management procedure.

5-7 SUPERPOSITION

The fact that the continuity equations for flow in aquifers developed in Secs 5-2 and 5-4 (except for (5-57)) and the boundary conditions discussed in Sec. 5-5 are *linear*, enables us to employ a powerful tool—the *principle of superposition*—in order to facilitate their solution. This is of special interest where mixed or inhomogeneous boundary conditions occur.

By applying this principle, the solution of a single complicated problem is obtained through the solution of a number of simpler ones. In Sec. 8-9 we shall use this principle in order to determine the drawdown in a multiple well system. In Chap. 12 we shall show how this principle enables us to make use of the linear programming technique in order to solve groundwater management problems.

Briefly, the principle of superposition states that if $\phi_1 = \phi_1(x, y, z, t)$ and $\phi_2 = \phi_2(x, y, z, t)$ are two general solutions of a homogeneous linear partial differential equation $L(\phi) = 0$, where L represents a linear operator (e.g., in (5-56) $L \equiv T\{\partial^2(\ )/\partial x^2 + \partial^2(\ )/\partial y^2\} + (\partial T/\partial x)\partial(\ )/\partial x + (\partial T/\partial y)\partial(\ )/\partial y - S\partial(\ )/\partial t$), then their sum $\phi_1 + \phi_2$, or, in general, any linear combination of ϕ_1 and ϕ_2

$$\phi = C_1\phi_1 + C_2\phi_2 \tag{5-141}$$

where C_1 and C_2 are constants, is also a solution of $L(\phi) = 0$. Or, in general, if $\phi_i = \phi_i(x, y, z, t)$, $i = 1, 2, \ldots, n$, are particular solutions of $L(\phi) = 0$, then

$$\phi = \sum_{i=1}^{n} C_i\phi_i \tag{5-142}$$

where C_i's are constants, is also a solution of this equation.

In each case the constants are determined by requiring that ϕ should also satisfy the prescribed boundary conditions. For example, let the boundary condition on a surface S be $a_0\phi + a_1\partial\phi/\partial x + a_2\partial\phi/\partial y + a_3\partial\phi/\partial z = f(x, y, z, t)$. By requiring that both ϕ_1 and ϕ_2 satisfy this condition, but with f_1 for ϕ_1 and f_2 for ϕ_2, we obtain

$$a_0\phi_1 + a_1\partial\phi_1/\partial x + a_2\partial\phi_1/\partial y + a_3\partial\phi_1/\partial z = f_1(x, y, z, t)$$

$$a_0\phi_2 + a_1\partial\phi_2/\partial x + a_2\partial\phi_2/\partial y + a_3\partial\phi_2/\partial z = f_2(x, y, z, t)$$

$$C_1 f_1(x, y, z, t) + C_2 f_2(x, y, z, t) = f(x, y, z, t) \qquad (5\text{-}143)$$

We then determine C_1 and C_2 from the last equation (if an explicit solution is possible). Sometimes the sum in (5-142) must be extended to infinity in order to satisfy the boundary conditions.

The solution ϕ in (5-141) and (5-142), with coefficients determined so that boundary conditions are satisfied, is called the *complete solution of the homogeneous equation*.

Sometimes the equation we wish to solve is nonhomogeneous (i.e., it includes terms, such as $N(x, y, t)$ and $P(x, y, t)$, e.g., in (5-81), that do not involve ϕ). If $\phi = \phi_0$ is a solution (no matter how special) of the nonhomogeneous equation, then any solution of the nonhomogeneous equation can be written in the form

$$\phi = \phi_0 + C_1\phi_1 + C_2\phi_2 \qquad (5\text{-}144)$$

ϕ_0 is called the particular integral of the nonhomogeneous equation, while $C_1\phi_1 + C_2\phi_2$, which is the complete solution of the associated homogeneous equation, obtained from the given nonhomogeneous one by deleting the terms that do not contain ϕ, is called the complementary solution of the nonhomogeneous equation.

Several examples of the use of the principle of superposition are considered below. In all cases one should make sure that in addition to the equation, both initial and boundary conditions are also satisfied.

Example 1 *Steady flow.* Let the boundary C of a flow domain D be composed of n segments $C^{(1)}, C^{(2)}, \ldots, C^{(n)}$, on each of which we have the boundary condition of a constant potential, that is, $\phi = \phi^{(i)} = \text{const.}$ on $C^{(i)}$, for $i = 1, \ldots n$ (Fig. 5-27). The flow in D is governed by the Laplace equation

$$\nabla^2\phi \equiv \partial^2\phi/\partial x^2 + \partial^2\phi/\partial y^2 = 0 \qquad (5\text{-}145)$$

Let $\phi_1 = \phi_1(x, y)$ be a solution of (5-145) satisfying the boundary conditions $\phi_1 = \phi^{(1)} \neq 0$ on $C^{(1)}$, $\phi_1 = 0$ elsewhere, that is, on $C - C^{(1)}$. Similarly, let $\phi_2 = \phi_2(x, y)$ satisfy (5-145), with $\phi_2 = \phi^{(2)}$ on $C^{(2)}$ and $\phi_2 = 0$ on $C - C^{(2)}$, etc. It can be easily be verified that $\phi = \phi_1 + \phi_2 + \cdots + \phi_n$ is a solution of (5-145) that satisfies the given boundary conditions of C.

There is no difficulty in extending this example to the case where the boundary condition to be satisfied on certain portions of the boundary is of the second kind. This case is demonstrated in Fig. 5-28. It can easily be verified that $\phi = \phi_1 + \phi_2 + \phi_3 + \phi_4$ is indeed a solution of the given problem.

The principle of superposition means that (1) the presence of one boundary condition does not affect the response produced by the presence of other boundary conditions (and, as we shall see below, of the initial conditions as

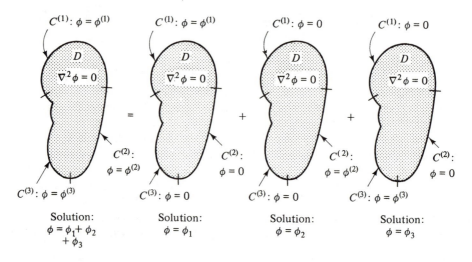

Figure 5-27 Decomposition of a regional flow problem into subproblems with simpler boundary conditions.

well), and (2) there are no interactions among the responses produced by the different boundary conditions. Therefore, to determine the combined effect of a number of boundary conditions, we may first solve for the effect of each individual boundary and then combine the results.

The advantage of the decomposition into subproblems is that, in general, the solution of each subproblem (by any of the methods described in Sec. 5-6) is simpler.

The principle of superposition is most useful in the case of multiple well systems (considered as point sources or sinks). This subject is discussed in detail in Secs. 8-9 through 8-11. The following example demonstrates this application.

Example 2 *Steady flow with wells.* Let two wells pumping at rates $Q = A$ and $Q = B$ be located at points P_1 and P_2, respectively, inside a flow domain D (Fig. 5-29). The flow is again described by (5-145). Using the principle of superposition, the problem may be decomposed into three subproblems. We first solve a case without wells, obtaining the solution $\phi = \phi_1(x, y)$. This means that $\nabla^2 \phi_1 = 0$ in D, $\phi_1 = \phi^{(1)}$ on $C^{(1)}$, and $\phi_1 = \phi^{(2)}$ on $C^{(2)}$. We then solve two more cases: (1) $Q = 1$ at P_1, $Q = 0$ at P_2, $\phi = 0$ on $C^{(1)} + C^{(2)}$, leading to a solution ϕ_2, and (2) $Q = 0$ at P_1, $Q = 1$ at P_2, $\phi = 0$ on $C^{(1)} + C^{(2)}$, leading to a solution ϕ_3. Then, the solution of the given problem is $\phi = \phi_1 + A\phi_2 + B\phi_3$. Since $\nabla^2 \phi_2 = 0$ and $\nabla^2 \phi_3 = 0$, we also have $\nabla^2 \phi = 0$. Similarly it can be checked that the boundary conditions on $C^{(1)}$ and $C^{(2)}$ are satisfied and that $Q = A$ and $Q = B$ at P_1 and P_2, respectively. Actually we could have solved the two last subproblems with discharge rates A and B rather than unity, leading to solutions $A\phi_2$ and $B\phi_3$, respectively (Fig. 5-29). However, by deriving solutions for discharge rates of one

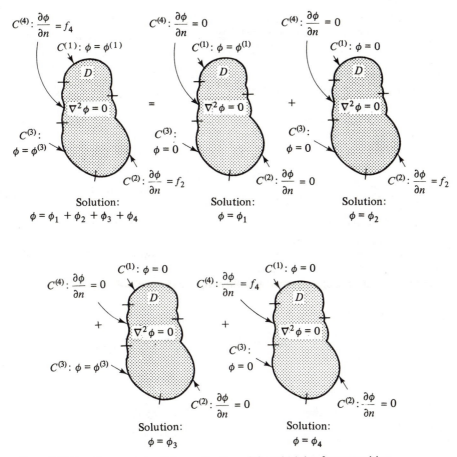

$C^{(4)}: \dfrac{\partial \phi}{\partial n} = f_4$

$C^{(1)}: \phi = \phi^{(1)}$

D

$\nabla^2 \phi = 0$

$=$

$C^{(3)}: \phi = \phi^{(3)}$

$C^{(2)}: \dfrac{\partial \phi}{\partial n} = f_2$

Solution:
$\phi = \phi_1 + \phi_2 + \phi_3 + \phi_4$

$C^{(4)}: \dfrac{\partial \phi}{\partial n} = 0$

$C^{(1)}: \phi = \phi^{(1)}$

D

$\nabla^2 \phi = 0$

$C^{(3)}: \phi = 0$

$C^{(2)}: \dfrac{\partial \phi}{\partial n} = 0$

Solution:
$\phi = \phi_1$

$+$

$C^{(4)}: \dfrac{\partial \phi}{\partial n} = 0$

$C^{(1)}: \phi = 0$

D

$\nabla^2 \phi = 0$

$C^{(3)}: \phi = 0$

$C^{(2)}: \dfrac{\partial \phi}{\partial n} = f_2$

Solution:
$\phi = \phi_2$

$+$

$C^{(4)}: \dfrac{\partial \phi}{\partial n} = 0$

$C^{(1)}: \phi = 0$

D

$\nabla^2 \phi = 0$

$C^{(3)}: \phi = \phi^{(3)}$

$C^{(2)}: \dfrac{\partial \phi}{\partial n} = 0$

Solution:
$\phi = \phi_3$

$+$

$C^{(4)}: \dfrac{\partial \phi}{\partial n} = f_4$

$C^{(1)}: \phi = 0$

D

$\nabla^2 \phi = 0$

$C^{(3)}: \phi = 0$

$C^{(2)}: \dfrac{\partial \phi}{\partial n} = 0$

Solution:
$\phi = \phi_4$

Figure 5-28 Another example of the application of the principle of superposition.

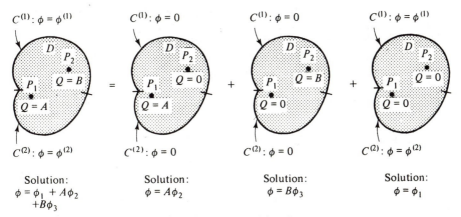

$C^{(1)}: \phi = \phi^{(1)}$

D P_2

P_1 $Q = B$

$Q = A$

$C^{(2)}: \phi = \phi^{(2)}$

Solution:
$\phi = \phi_1 + A\phi_2$
$+ B\phi_3$

$=$

$C^{(1)}: \phi = 0$

D P_2

P_1 $Q = 0$

$Q = A$

$C^{(2)}: \phi = 0$

Solution:
$\phi = A\phi_2$

$+$

$C^{(1)}: \phi = 0$

D P_2

P_1 $Q = B$

$Q = 0$

$C^{(2)}: \phi = 0$

Solution:
$\phi = B\phi_3$

$+$

$C^{(1)}: \phi = \phi^{(1)}$

D P_2

P_1 $Q = 0$

$Q = 0$

$C^{(2)}: \phi = \phi^{(2)}$

Solution:
$\phi = \phi_1$

Figure 5-29 Use of superposition for a flow regime with wells.

unit at each well, we can now use the results ϕ_2 and ϕ_3 to obtain solutions for any other rates simply by multiplying the values of ϕ_2 and ϕ_3 by the actual discharge rates. This stems directly from the principle of superposition.

The same procedure can easily be extended to a larger number of wells and to recharging wells which may be considered as negative pumping wells.

Example 3 *Unsteady flow with boundary conditions independent of time.* Consider the case of unsteady flow described by the continuity equation (5-60) with $N = 0$

$$\partial^2\phi/\partial x^2 + \partial^2\phi/\partial y^2 = (S/T)\,\partial\phi/\partial t \quad \text{in } D$$

with initial conditions $\phi = f(x, y)$ at $t = 0$ in D (5-146)

and boundary conditions $L(\phi) = g(x, y)$ at $t \geq 0$ on C

where $L(\phi) = a_0\phi + a_1\partial\phi/\partial x + a_2\partial\phi/\partial y$; a_0, a_1 and a_2 are constants. This means that the boundary conditions here are independent of time. We may decompose this problem into the following two subproblems.

(a) $\partial^2\phi/\partial x^2 + \partial^2\phi/\partial y^2 = 0$ at $t \geq 0$ in D

         $L(\phi) = g(x, y)$ at $t \geq 0$ on C

     Solution: $\phi = \phi^{(1)}(x, y)$

(b) $\partial^2\phi/\partial x^2 + \partial^2\phi/\partial y^2 = (S/T)\,\partial\phi/\partial t$ at $t \geq 0$ in D

          $\phi = f(x, y) - g(x, y)$ at $t = 0$ in D

        $L(\phi) = 0$ at $t \geq 0$ on C

     Solution: $\phi = \phi^{(2)}(x, y, t)$

where subproblem (a) describes steady flow. It can easily be verified that $\phi = \phi^{(1)} + \phi^{(2)}$ is the sought solution as it satisfies both the equation and the initial and boundary conditions of the original problem. The latter has been reduced and two subproblems (a) steady flow with prescribed boundary conditions and (b) unsteady flow with prescribed initial conditions and with *homogeneous boundary conditions* $(L(\phi) = 0)$.

Example 4 *Unsteady flow with accretion.* Consider the problem described by (5-60) with $N \neq 0$.

$$T\nabla^2\phi + N(x, y, t) = S\,\partial\phi/\partial t \quad \text{at } t \geq 0 \text{ in } D$$

$$\phi = f(x, y) \quad \text{at } t = 0 \text{ in } D \qquad (5\text{-}147)$$

$$L(\phi) = g(x, y) \quad \text{at } t \geq 0 \text{ on } C$$

The problem may be decomposed into the following two subproblems.

(a) $T\nabla^2\phi = S\,\partial\phi/\partial t$ (b) $T\nabla^2\phi + N(x, y, t) = S\,\partial\phi/\partial t$ at $t \geq 0$ in D

        $\phi = f(x, y)$ $\phi = 0$ at $t = 0$ in D

        $L(\phi) = g(x, y)$ $L(\phi) = 0$ at $t = 0$ on C

Solution: $\phi = \phi_1(x, y, t)$ Solution: $\phi = \phi_2(x, y, t)$

The solution of the original problem is $\phi = \phi_1 + \phi_2$. Problem (a) may be further decomposed into two subproblems as in Example 3.

We have thus separated the effect of accretion, N (negative or positive), from that of the boundary and the initial conditions. If now we have another problem with $N' = \alpha N$, the corresponding solution ϕ_2' to subproblem (b) would be $\phi_2' = \alpha\phi_2$.

Let the flow domain D be subdivided into n subdomains $D^{(1)}, D^{(2)}, \ldots$, $D^{(n)}$, in which the accretion is $N^{(1)}, N^{(2)}, \ldots, N^{(n)}$, respectively. We may then decompose problem (b) into n subproblems such that $N^{(i)} \neq 0$ in $D^{(i)}$ and $N^{(i)} = 0$ in $D - D^{(i)}$, $i = 1, 2, \ldots, n$. If the respective solutions are $\phi_2^{(1)}, \phi_2^{(2)}, \ldots$, $\phi_2^{(n)}$, then the combined effect is

$$\phi_2 = \phi_2^{(1)} + \phi_2^{(2)} + \cdots + \phi_2^{(n)}$$

Of special interest is the case where all $N^{(i)}$'s are constants (i.e., independent of x, y, t). Then we may again solve for unit values of accretion in each subdomain, and obtain the total effect by superposition.

The term $N(x, y, t)$ in the continuity equation (5-60) may also represent pumping in wells; it will then have a negative sign. The solution of subproblem (b) then represents the effect of the wells on the piezometric head. Accordingly, whenever we have to determine the flow regime in the presence of pumping wells, we can first solve for the flow regime without the wells (subproblem (a)) and then obtain the effect of the wells under homogeneous initial and boundary conditions. This effect takes the form of drawdown, or lowering of the piezometric surface with respect to that obtained in the absence of the wells. An example is presented in Fig. 5-30 which shows a vertical cross section of an aquifer. We obtain the piezometric head in the presence of pumping wells by subtracting the sum of drawdowns caused by the individual wells from the piezometric head in the absence of wells. The same procedure is applicable to build-up produced by artificial recharge. Obviously when both recharge $N(x, y, t)$ and pumping $-P(x, y, t)$ are present, we may decompose subproblem (b), study the effect of each parameter separately, and then combine the results.

Example 5 *Unsteady flow with time-dependent boundary conditions.* Let $\phi = F(x, y, t)$ be the solution of

$$T\nabla^2\phi = S\,\partial\phi/\partial t \quad \text{at } t \geq 0 \text{ in } D$$

with initial conditions: $\phi(x, y, 0) = 0$ at $t = 0$ in D (5-148)

and boundary conditions: $\phi = 1$ at $t \geq 0$ on C

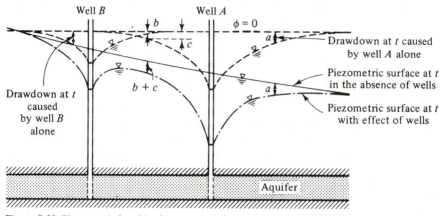

Figure 5-30 Piezometric head in the presence of pumping wells, obtained by superposition.

Then the solution of

$$T \nabla^2 \phi = S \, \partial \phi / \partial t \quad \text{at } t \geq 0 \text{ in } D$$

and initial conditions: $\quad \phi(x, y, 0) = 0 \qquad \text{at } t = 0 \text{ in } D$ $\qquad$ (5-149)

and boundary conditions: $\qquad \phi = g(x, y, t) \text{ at } t \geq 0 \text{ on } C$

is given by

$$\phi(x, y, t) = \int_0^t g(x, y, \tau) \frac{\partial}{\partial t} F(x, y, t - \tau) \, d\tau \qquad (5\text{-}150)$$

This integral is known as *Duhamel's* (or *superposition*) *integral*. Here F has the meaning of a step response.

If now we have a problem where the initial conditions are $\phi(x, y, 0) = f(x, y)$ at $t = 0$ in D, and the boundary conditions are $\phi = g(x, y, t)$ on C, we first decompose the problem into two subproblems with a solution ϕ_1 which satisfies the continuity equation, is equal to $f(x, y)$ at $t = 0$, and vanishes everywhere on C, and ϕ_2 which satisfies the same equation, vanishes everywhere at $t = 0$ and satisfies $\phi = g(x, y, t)$ on C. The sought solution is the sum $\phi = \phi_1 + \phi_2$. An example of the application of superposition in groundwater management is given in Chap. 12.

The decomposition here expresses the fact that responses produced by several separate excitations operating simultaneously are independent. Hence, the resulting response may be obtained as an algebraic sum of the individual responses resulting from the various individual excitations operating one at a time.

Equation (5-75), which is used to describe groundwater flow in a phreatic aquifer, is nonlinear with respect to h. Therefore we cannot apply the principle of superposition to it. This means that *superposition is not applicable to phreatic aquifers*. However, as already mentioned in Sec. 5-4, when variations in h are

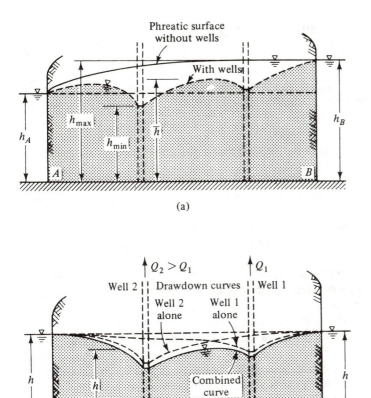

(a)

(b)

Figure 5-31 Superposition in a phreatic aquifer with wells.

small with respect to h (or some average $\bar{h}$) we employ the linearized equation (5-81) to describe the flow. This is a linear equation which enables us to use superposition with respect to h. An example is shown in Fig. 5-31. Assuming that $(h_{max} - h_{min}) \ll \bar{h}$, we may use the linearized equation (5-81) to determine the initial water table without the wells, then the drawdown caused by the wells alone (using superposition to combine the effect of the individual wells). The water table in the presence of wells is obtained by subtracting the drawdown caused by the wells from the initial water table (or adding build-up produced by recharging wells).

In steady flow in a phreatic aquifer, the right-hand side of (5-82) vanishes, and we have a linear equation, but in h^2. If we then have also boundary conditions which are linear in h^2, we may employ the principle of superposition with respect to h^2 (see Sec. 8-9).

Superposition can be applied (when necessary) to most groundwater flow problems by following the procedures outlined in the examples presented above.

5-8 HYDROLOGIC MAPS AND FLOW NETS

Hydrologic Maps

Groundwater mapping is a compact, visual form of recording the results of both field and desk investigations of aquifer regimes, as expressed by water table elevations, $h = h(x, y \cdot t)$. in a phreatic aquifer or by piezometric surface elevations, $\phi = \phi(x. y \cdot t)$, in a confined or a leaky one.

Underlying the drawing of hydrologic maps and their utilization in practice is the assumption (Sec. 2-4) that flow in an aquifer is essentially horizontal and, therefore, equipotential surfaces are vertical (see introductory comments to Chap. 8). The hydrologic map is drawn for a specific aquifer and for a specified instant of time.

UNESCO (1977), jointly with WMO, has published a summary on hydrologic maps. A detailed discussion of contour maps and their interpretation is also given by Castany (1966).

Using water levels measured simultaneously at a certain time in all observation wells of a given aquifer (and one should be careful not to mix observations in wells tapping different aquifers), we use interpolation of one kind or another (e.g., linear), to draw contours of the water table or of the piezometric surface at desirable intervals. In the latter case, the term equipotentials is often used. Figure 5-32 shows an example of a contour map (= hydrologic map) obtained in this way. When the objective of a map is to give a regional picture of the flow regime in an aquifer, it should exclude local cones of depression occurring around pumping wells operating in the area. To eliminate this local effect, pumping wells (actually also recharging ones) are shut off a sufficient period of time (often 24–48 h) prior to measuring the water levels in them (if they are used as observation wells) and/or in observation wells in their vicinity. The period of time should be sufficient for complete recovery of water levels, so that the measured ones correspond to the regional water table (or piezometric surface).

Nowadays, programs are available for for drawing contour maps by computers, using library or specially developed programs.

The forecasts of water levels by solving aquifer continuity equations (Sec. 5-4) analytically or numerically, may also be represented in the form of contour maps.

The flow regime in a considered aquifer can be investigated, using the contour maps, as following.

(a) Assuming that the aquifer is isotropic, the direction of flow (of water and pollutants) at every point is perpendicular to the equipotentials.

An example is shown in Fig. 5-33. We shall explain below why, in general, we draw only short arrows on a contour map to indicate flow direction and not complete streamlines.

(b) For an isotropic aquifer, the rate of flow $\mathbf{Q}'$ is given by $\mathbf{Q}' = -T\nabla\phi = T\mathbf{J}$. The gradient itself can be determined from $J = \Delta\phi/\Delta n$ by reading $\Delta\phi$ and Δn on the map (Fig. 5-33, where $\Delta\phi = 123 - 122 = 1.0$ m and $\Delta n = 350$ m at P, according to the scale of the map).

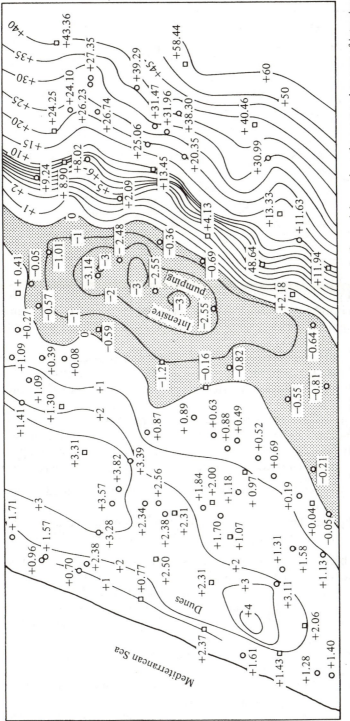

Figure 5-32 A groundwater contour map of a coastal aquifer (Israel) indicating increased recharge in the sand dunes along the coast and an area of intensive pumping. (*Courtesy of Hydrological Service, Israel.*)

161

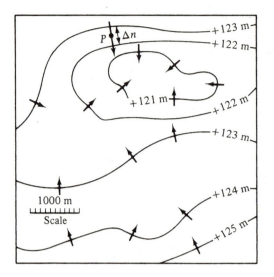

Figure 5-33 Contour map with arrows indicating directions of flow.

In a phreatic aquifer, we may wish to use the Dupuit assumption and the Dupuit–Forcheimer discharge formula (4-64) for determining $\mathbf{Q}'$. In this case, information on the elevations of the impervious bottom of the aquifer is also needed.

(c) In an anisotropic aquifer ($T_x \neq T_y$; x, y principal directions), the flow is no more perpendicular to the equipotentials. We have to calculate $\mathbf{Q}'$ from $Q'_x = T_x J_x$, $Q'_y = T_y J_y$, where J_x ($=\Delta\phi/\Delta x$) and J_y ($=\Delta\phi/\Delta y$) can be read from the map.

(d) We can use two superimposed contour maps at two different times to draw a *map of water table changes*. Given aquifer storativity, this information is used to determine changes in the volume of water stored in an aquifer during the time interval between dates for which the two maps have been drawn.

In Sec. 3-10 we have shown how the above results are used in setting up a regional water balance.

By analyzing the pattern of equipotentials, one learns about the general features of the flow pattern in a given aquifer, as shown by the following examples.

(a) In the absence of sources and sinks, a contour map cannot show a minimum and/or a maximum anywhere *inside* a flow domain. These may occur only along the boundaries of the domain. If a minimum in water levels, indicated by a closed contour curve, does occur in a flow domain, this means that a sink (e.g., due to intensive pumping, a spring, or marshland losing water by evapotranspiration) occurs in that area (Fig. 5-34a).

Similarly a maximum (also indicated by a closed contour) indicates a zone of recharge (Fig. 5-34b), both natural and artificial.

One should note that sources and sinks are not only those visible ones at the ground surface (e.g., pumping, natural and artificial replenishment, springs), but may include hidden underground ones, such as a zone of leakage into or

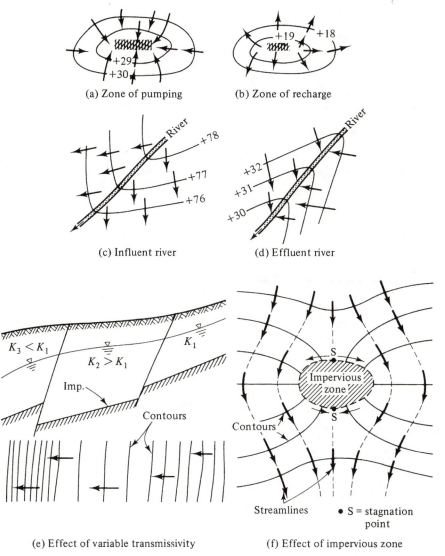

(a) Zone of pumping

(b) Zone of recharge

(c) Influent river

(d) Effluent river

(e) Effect of variable transmissivity

(f) Effect of impervious zone

Figure 5-34 Typical features of contour maps.

out of an aquifer from underlying or overlying aquifers through semipervious strata or fissures, or local direct contact with an adjacent aquifer.

In unsteady flow, maxima and minima may exist *inside* a flow domain. In this case the interpretation may be somewhat different, namely, that water is taken out of storage in an aquifer (say in the case of a decaying mound, following a period of artificial recharge), or added to it (as where an area of depression in the water table is being filled up once pumping has stopped).

(*b*) Contours can indicate river–aquifer relationships (Figs 5-34c and d). No

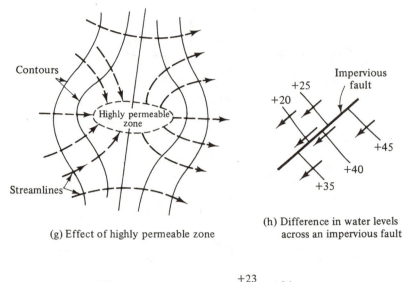

(g) Effect of highly permeable zone

(h) Difference in water levels
across an impervious fault

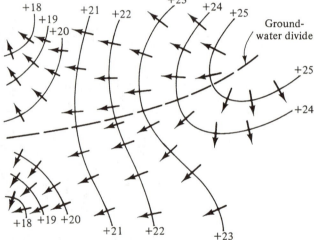

(i) A groundwater divide

Figure 5-34 (cont.)

further explanation seems necessary. Note, however, that the streams in Figs
5-34c and d are not equipotentials.

(c) Gradients may increase (Fig. 5-34e) or decrease in the direction of the flow.
In the absence of sources and/or sinks, this is caused by either a reduction
in aquifer thickness and/or its hydraulic conductivity. (See discussion below
on flow nets.)

(d) Impervious zones or lines (e.g. faults) force the flow to change direction.
These obstacles may not be visible at ground surface, but they are indicated
by studying the contour map (Fig. 5-34f).

(*e*) Zones of very high transmissivity may behave practically as equipotential (or nearly so) open water bodies. Old buried rivers may produce in their vicinity flow patterns similar to those shown in 5-34c and d, although no stream is visible at the ground surface.

(*f*) Different water levels may occur across an impervious fault (Fig. 5-34h). One should remember that an impervious boundary is also a streamline. If a fault zone is semipervious, water may flow through, indicating a local loss of head across the fault.

(*g*) A *groundwater divide* is a curve in the horizontal plane which separates the flow domain into subdomains, such that all groundwater from a subdomain will eventually drain out through a separate outlet (a lake, spring, river, etc.). Figure 5-34i shows a segment of a groundwater divide. This is an important feature in pollution studies as a spring, a lake, or a river can only be polluted by polluted groundwater originating in the subdomain drained into it. It is of interest to note that unlike a topographic water divide, the groundwater one may shift with time. It can be made to shift by manipulating groundwater levels.

An important conclusion from the above examples (and this is by no means a complete list of the possible special features observed on contour maps) is that serious mistakes in a hydrologic map may be made if contours are constructed by mechanical interpolation of observed water levels only, without taking into account all available geological information on faults, variations in aquifer thickness and hydraulic conductivity, depth of water table below ground surface (which may indicate area of groundwater loss in the form of a spring, or evaporVtranspiration from marshes, or from a water table which is very close to the ground surface), etc.

We have already mentioned above that another source of errors is the mixing of data on water levels observed at wells tapping different aquifers.

Flow Nets

A *streamline* is a curve that is everywhere tangent to the specific discharge vector, **q**. Thus, streamlines indicate the direction of flow at every point in a flow domain. A stream tube is formed by adjacent streamlines. Since, by definition, no flow can cross a streamline, the flow rate along a stream tube is constant. This statement is true provided that the flow is steady (i.e., no water is taken out of, or added to storage) and no distributed sources and sinks exist in the flow domain. Point sources and sinks (e.g., wells) may exist, but they can be excluded from the flow domain, leaving one with no sources and sinks.

By the above definition, the equation of a streamline is

$$\mathbf{q} \times \mathbf{ds} = 0 \tag{5-151}$$

where $\times$ denotes a cross product and **ds** is an element of length along the streamline (Fig. 5-35a). In cartesian coordinates, (5-151) is equivalent to

$$q_y dx - q_x dy = 0 \tag{5-152}$$

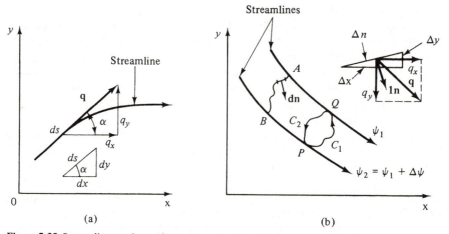

Figure 5-35 Streamlines and specific discharge in plane flow.

We can now define a function $\Psi = \Psi(x, y)$ which is a constant along a streamline (see below). Hence, along a streamline

$$d\Psi = 0 = \frac{\partial \Psi}{\partial x} dx + \frac{\partial \Psi}{\partial y} dy \qquad (5\text{-}153)$$

By comparing (5-152) and (5-153), we find that

$$q_x = -\frac{\partial \Psi}{\partial y}, \qquad q_y = \frac{\partial \Psi}{\partial x} \qquad (5\text{-}154)$$

The function Ψ is called a *stream function* (dims. L^2/T), or *Lagrange stream function*. The physical interpretation of the *generating function* Ψ (as according to (5-154) it can be used to generate the specific discharge) may be obtained by considering the integral of Ψ between two points, say A and B in Fig. 5-35b. Before determining the value of this integral, let us consider the two integrals

$$\int_P^Q \rho\mathbf{q} \cdot \mathbf{dn} = \text{mass flow rate across } C_1$$

$$\int_Q^P \rho\mathbf{q} \cdot \mathbf{dn} = \text{mass flow rate across } C_2$$

where $\mathbf{dn} = \Delta n \, \mathbf{1n}$ is shown in Fig. 5-35b. If the fluid and medium are incompressible ($=$ steady flow) and no sources and sinks are present within the area bounded by C_1 and C_2, the mass of fluid in the area bounded by C_1 and C_2 remains constant and therefore

$$\int_P^Q \rho\mathbf{q} \cdot \mathbf{dn} \bigg|_{\text{along } C_1} + \int_Q^P \rho\mathbf{q} \cdot \mathbf{dn} \bigg|_{\text{along } C_2} = 0 \qquad (5\text{-}155)$$

From (5-155) it follows that

$$\int_P^Q \rho\mathbf{q} \cdot \mathbf{dn}\bigg|_{\text{along } C_1} = \int_P^Q \rho\mathbf{q} \cdot \mathbf{dn}\bigg|_{\text{along } C_2} \qquad (5\text{-}156)$$

i.e. we may choose any arbitrary curve between points P and Q on the two streamlines.

Accordingly we may choose any path of integration between points A and B. Since the differential of $\Psi_B - \Psi_A \ (\equiv \Psi_2 - \Psi_1)$ depends only on the endpoints of the integration, we obtain

$$\int_{\Psi_A}^{\Psi_B} d\Psi = \Psi_B - \Psi_A = \Psi_2 - \Psi_1 \qquad (5\text{-}157)$$

On the other hand, the total flow through the streamtube is given by

$$Q_{AB} = \int_A^B \mathbf{q} \cdot \mathbf{dn} = \int_A^B (q_x\,dy - q_y\,dx) = \int_A^B \left(-\frac{\partial \Psi}{\partial y}\,dy - \frac{\partial \Psi}{\partial x}\,dx \right)$$

$$= -\int_A^B d\Psi = \Psi_A - \Psi_B \qquad (5\text{-}158)$$

Hence the total discharge (in terms of volume per unit width normal to the xy plane per unit time) between two streamlines is equal to the difference in the stream functions corresponding to these lines. Note that according to our sign convention here (where $\nabla\Psi$ is obtained from $\nabla\phi$ by a counterclockwise rotation), $\Psi_B > \Psi_A$. If points A and B are on the same streamline, $Q_{AB} = 0$, $d\Psi = 0$ and $\Psi = \text{constant}$. We must emphasize again that the discussion above is valid for incompressible flow and in the absence of sources and sinks.

For an isotropic domain, we have *Cauchy-Riemann's conditions*

$$q_x = -K\frac{\partial \phi}{\partial x} = -\frac{\partial \Psi}{\partial y}; \qquad q_y = -K\frac{\partial \phi}{\partial y} = \frac{\partial \Psi}{\partial x} \qquad (5\text{-}159)$$

Hence, by multiplying the two equations, we obtain

$$\frac{\partial \phi}{\partial x}\frac{\partial \Psi}{\partial x} + \frac{\partial \phi}{\partial y}\frac{\partial \Psi}{\partial y} \equiv \nabla'\phi \cdot \nabla'\Psi = 0 \qquad (5\text{-}160)$$

that is, the family of equipotentials is perpendicular everywhere to the family of streamlines. The two families of curves are not orthogonal when the medium is anisotropic.

Finally, it can be shown (Bear, 1972, p. 230), that Ψ satisfies the following partial differential equation.

For a homogeneous isotropic medium

$$\nabla^2\Psi = 0 \qquad (5\text{-}161)$$

For a nonhomogeneous, isotropic medium

$$K\nabla^2\Psi - \nabla'K \cdot \nabla'\Psi = 0 \qquad (5\text{-}162)$$

For a nonhomogeneous nonisotropic medium (x, y principal directions)

$$K_x \frac{\partial^2 \Psi}{\partial x^2} + K_y \frac{\partial^2 \Psi}{\partial y^2} - \frac{K_x}{K_y} \frac{\partial K_y}{\partial x} \frac{\partial \Psi}{\partial x} - \frac{K_y}{K_x} \frac{\partial K_x}{\partial y} \frac{\partial \Psi}{\partial y} = 0 \qquad (5\text{-}163)$$

For a homogeneous nonisotropic medium

$$K_x \frac{\partial^2 \Psi}{\partial x^2} + K_y \frac{\partial^2 \Psi}{\partial y^2} = 0 \qquad (5\text{-}164)$$

These equations (with appropriate boundary and initial conditions on Ψ) can be used to determine $\Psi = \Psi(x, y)$ in the given domain. Of special interest is (5-161) which is the same Laplace equation as that satisfied by ϕ in steady flow in an isotropic homogeneous domain in the absence of distributed sources and sinks.

In two-dimensional plane flow, a plot of the two families of curves: equipotentials $\phi = \phi(x, y) = $ const. and streamlines $\Psi = \Psi(x, y) = $ const., is called a flow net. In an isotropic domain, the two families are orthogonal to each other.

The use of Ψ is meaningful only in steady flow and in the absence of distributed sources and sinks in the flow domain. This, in fact, is the reason for not plotting streamlines on a hydrologic map. Point sources and sinks introduce no difficulty as we may exclude them from the flow domain; they become part of the domain's boundary.

Figure 5-36 shows a portion of a flow net with three streamlines in a homogeneous isotropic medium. It is customary to draw the flow net such that the difference $\Delta\phi$ between any two adjacent equipotentials is constant. The same is true for $\Delta\Psi$ between adjacent streamlines. For each of the streamtubes, we have

$$\Delta Q = K\Delta n_1 (\Delta\phi/\Delta s_1) = K\Delta n_2 (\Delta\phi/\Delta s_2) \qquad (5\text{-}165)$$

where Q is the discharge per unit thickness. We may replace K by the transmissivity T and then Q is the flow through the entire thickness of an aquifer. From (5-165) we obtain

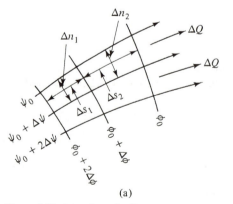

(a)   (b)

Figure 5-36 A portion of a flow net.

$$\Delta n_1/\Delta s_1 = \Delta n_2/\Delta s_2 \tag{5-166}$$

that is, in a homogeneous medium the ratio $\Delta n/\Delta s$ must remain constant throughout the flow net.

For an inhomogeneous medium

$$K_1 \Delta n_1 (\Delta\phi/\Delta s_1) = K_2 \Delta n_2 (\Delta\phi/\Delta s_2); \quad K_1 \frac{\Delta n_1}{\Delta s_1} = K_2 \frac{\Delta n_2}{\Delta s_2} \tag{5-167}$$

that is, the ratio $\Delta n/\Delta s$ varies. When streamlines are approximately parallel (that is, $\Delta n_1 \approx \Delta n_2$), we have $(K_1/K_2) \cong (\Delta\phi/\Delta s_2)/(\Delta\phi/\Delta s_1)$ that is, the hydraulic conductivity is inversely proportional to the hydraulic gradient; contours will be closely spaced in regions of low hydraulic conductivity.

It is convenient to draw the flow net for a homogeneous isotropic medium so that approximate curvilinear squares are formed (Fig. 5-36b). For this case $\Delta s = \Delta n$ and $Q = K\Delta\phi$. In certain cases, however, it is more convenient to draw the flow net for a given domain such that we have m streamtubes, each carrying the same discharge $\Delta Q = Q_{\text{total}}/m$, and n equal drops in piezometric head, $\Delta\phi\,(=(\phi_{\max} - \phi_{\min})/n)$. Then

$$Q_{\text{total}} = m\,\Delta Q = mK\,\Delta n\,\frac{\Delta\phi}{\Delta s} = mK\,\Delta n\,\frac{\phi_{\max} - \phi_{\min}}{n\,\Delta s} = \frac{m\,\Delta n}{n\,\Delta s}\,K(\phi_{\max} - \phi_{\min})$$

$$\tag{5-168}$$

Several methods are described in the literature for drawing a flow net for a given flow domain. In fact, the drawing of a flow net is a graphical way of solving the Laplace equation. Figures 8-20, 8-24, 8-25, and 8-29 through 8-31 show examples of flow nets in the xy plane. Figure 5-37 shows an example of a flow net in the vertical plane (Toth, 1962). All these flow nets are drawn from the distributions of ϕ and Ψ obtained by solving the partial differential equations for ϕ and Ψ, respectively, graphically, numerically, or by using an analog (Sec. 5-6).

5-9 RELATIONSHIPS BETWEEN FLOWS IN ISOTROPIC AND ANISOTROPIC AQUIFERS

In Sec. 5.2, the various continuity equations were presented for both isotropic and anisotropic media. In general, the solution of problems of flow in anisotropic media is somewhat more difficult. In what follows, we shall show (following Bear and Dagan, 1965; see also Bear, 1972, p. 290) that instead of solving a flow problem in a given anisotropic domain, one can use certain relationships by means of which the problem may be transformed into one of flow in an *equivalent isotropic domain*. Once a solution is obtained for this domain (by analytical, numerical, or analog methods), it is transformed back to the original domain by using the same relationships. In each case, both the equations and the geometry of the boundary and the boundary conditions on it must be transformed. The analysis is based on the homogeneity of the equations describing the flow. We shall also show how the same method can be used to derive dimensionless variables and parameters for any given problem.

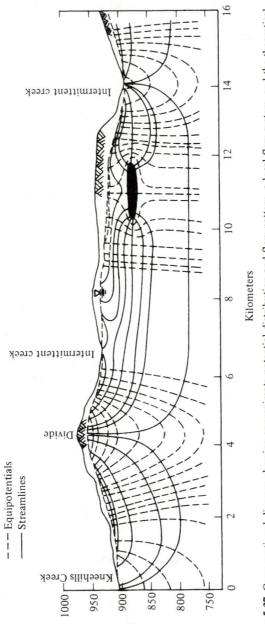

Figure 5-37 Cross-sectional diagram showing approximate potential distribution and flow pattern, a local flow system, and the theoretical effect of a highly permeable body (in black) on the flow system across two adjacent valley sides *(after Toth, 1962)*.

In what follows, we shall assume that the principal directions of anisotropic permeability coincide with the axes of the cartesian (horizontal–vertical) xyz coordinate system. When the principal directions (say, x', y', z') of k do not coincide with the (horizontal–vertical) xyz cartesian system of coordinates, the entire analysis described below is carried out in terms of x', y', z'. However, one should be careful when a phreatic surface is present (see Bear, 1972, p. 269).

The starting point is always the well-posed mathematical statement of the flow problem on hand. Of particular interest for our purpose is the partial differential equation governing the potential, or water level, distribution.

Let subscripts p and e denote the (given) *anisotropic prototype domain* and *equivalent isotropic domain*, respectively. For these two domains, we write the complete mathematical statement of the considered flow problem. For example, using (5-29) with ϕ replacing ϕ^*, we write

$$\frac{\partial}{\partial x_p}\left(K_{xp}\frac{\partial\phi_p}{\partial x_p}\right) + \frac{\partial}{\partial y_p}\left(K_{yp}\frac{\partial\phi_p}{\partial y_p}\right) + \frac{\partial}{\partial z_p}\left(K_{zp}\frac{\partial\phi_p}{\partial z_p}\right) = S_{0p}\frac{\partial\phi_p}{\partial t_p}, \qquad (5\text{-}169)$$

$$\frac{\partial}{\partial x_e}\left(K_e\frac{\partial\phi_e}{\partial x_e}\right) + \frac{\partial}{\partial y_e}\left(K_e\frac{\partial\phi_e}{\partial y_e}\right) + \frac{\partial}{\partial z_e}\left(K_e\frac{\partial\phi_e}{\partial z_e}\right) = S_{0e}\frac{\partial\phi_e}{\partial t_e} \qquad (5\text{-}170)$$

If we are interested also in velocities, then we add

$$V_{xp} = \frac{dx_p}{dt_p} = -\frac{K_{xp}}{n_p}\frac{\partial\phi_p}{\partial x_p}, \quad \text{and similar equations for } V_{yp} \text{ and } V_{zp} \qquad (5\text{-}171)$$

$$V_{xe} = \frac{dx_e}{dt_e} = -\frac{K_e}{n_e}\frac{\partial\phi_e}{\partial x_e}, \quad \text{and similar equations for } V_{ye} \text{ and } V_{ze} \qquad (5\text{-}172)$$

If volume of water is also of interest, we add

$$dU_p = Q_p dt_p = n_p dx_p dy_p dz_p \qquad (5\text{-}173)$$

and

$$dU_e = Q_e dt_e = n_e dx_e dy_e dz_e \qquad (5\text{-}174)$$

Step 1

Writing all the equations related to the problem on hand is always the first step.

Step 2

Use subscript r to denote the ratio between the values of corresponding variables or parameters in the two systems

$$\phi_r = \frac{\delta\phi_e}{\delta\phi_p}; \quad x_r = \frac{\delta x_e}{\delta x_p}; \quad y_r = \frac{\delta y_e}{\delta y_p}; \quad z_r = \frac{\delta z_e}{\delta z_p}; \quad K_{xr} = \frac{K_e}{K_{xp}};$$

$$K_{yr} = \frac{K_e}{K_{yp}}; \quad K_{zr} = \frac{K_e}{K_{zp}}; \quad n_r = \frac{n_e}{n_p}; \quad t_r = \frac{\delta t_e}{\delta t_p}; \quad S_{0r} = \frac{S_{0e}}{S_{0p}};$$

$$V_{xr} = \frac{V_{xe}}{V_{xp}}; \ldots \quad U_r = \frac{U_e}{U_p}; \quad Q_r = \frac{Q_e}{Q_p}; \ldots \qquad (5\text{-}175)$$

This is done for all dependent variables, independent variables, and parameters of the two systems, as included in the mathematical statement of the problem on hand. These ratios may be considered as *scales* between the two systems.

Step 3

Express all values of the *e*-system in terms of the corresponding values of the *p*-system and the scales

$$\delta\phi_e = \phi_r \delta\phi_p; \qquad \delta x_e = x_r \delta x_p; \qquad \delta y_e = y_r \delta y_p; \qquad \delta z_e = z_r \delta z_p$$

$$K_e = K_{xr}K_{xp} = K_{yr}K_{yp} = K_{zr}K_{zp}; \qquad \delta t_e = t_r \delta t_p; \qquad S_{0e} = S_{0r}S_{0p}$$

$$n_e = n_r n_p; \dots \tag{5-176}$$

and insert these values in the equations for the *e*-system. For example, (5-170), (5-172), and (5-174) become

$$\frac{K_{xr}\phi_r}{x_r^2} \frac{\partial}{\partial x_p}\left(K_{xp}\frac{\partial\phi_p}{\partial x_p}\right) + \frac{K_{yr}\phi_r}{y_r^2}\frac{\partial}{\partial y_p}\left(K_{yp}\frac{\partial\phi_p}{\partial y_p}\right) + \frac{K_{zr}\phi_r}{z_r^2}\frac{\partial}{\partial z_r}\left(K_{zp}\frac{\partial\phi_p}{\partial z_p}\right)$$

$$= S_{0r}\frac{\phi_r}{t_r}S_{0p}\frac{\partial\phi_p}{\partial t_p}; \tag{5-177}$$

$$V_{xr}V_{xp} = \frac{x_r}{t_r}\frac{dx_p}{dt_p} = -\frac{K_{xr}\phi_r}{n_r x_r}\frac{K_{xp}}{n_p}\frac{\partial\phi_p}{\partial x_p}, \dots \tag{5-178}$$

$$U_r dU_p = Q_r t_r Q_p dt_p; \qquad U_r dU_p = n_r x_r y_r z_r n_p dx_p dy_p dz_p \tag{5-179}$$

Step 4

We now compare the transformed *e*-system equations—here (5-177) to (5-179) —with those for the (given) *p*-system and observe that the two systems of equations become identical (term for term) if the following relationships between the scales are satisfied

$$\frac{K_{xr}\phi_r}{x_r^2} = \frac{K_{yr}\phi_r}{y_r^2} = \frac{K_{zr}\phi_r}{z_r^2} = S_{0r}\frac{\phi_r}{t_r} \tag{5-180}$$

$$V_{xr} = \frac{x_r}{t_r} = \frac{K_{xr}\phi_r}{n_r x_r}; \qquad V_{yr} = \frac{y_r}{t_r} = \frac{K_{yr}\phi_r}{n_r y_r}; \qquad V_{zr} = \frac{z_r}{t_r} = \frac{K_{zr}\phi_r}{n_r z_r}; \tag{5-181}$$

$$U_r = Q_r t_r = n_r x_r y_r z_r \tag{5-182}$$

Step 5

These, then, are the relationships among the scales which can be used to transform the given anisotropic system into *an equivalent isotropic* one. We count the number of scales we have and the number of *independent* relationships we have established among them (recalling that for the equivalent isotropic system, we have only a single K_e so that of K_{xr}, K_{yr}, and K_{zr}, only one is independent of the others). The difference between the two gives the number of scales which may be chosen arbitrarily, provided we do not violate the relations among scales (e.g., by choosing $Q_r t_r$ and U_r independently).

In the example above, we have 13 scales, ϕ_r, K_{xr}, K_{yr}, K_{zr} (the last three depending on a single parameter K_e), x_r, y_r, z_r, t_r, n_r, S_{0r}, V_{xr}, V_{yr}, V_{zr}, U_r, Q_r. For these we have nine independent relationships included in (5-180) through (5-182), noting that certain relationships are repeated in these equations. This means that four of the scales may be chosen arbitrarily (of course, without violating any of the equations) while the remaining ones will be determined from the above relationships. For example, choosing ϕ_r, Q_r, t_r, and n_r, we obtain

$$x_r = K_{xr}^{1/2}(\phi_r t_r/n_r)^{1/2} = (K_{yp}K_{zp}/K_{xp}^2)^{1/6}(Q_r t_r/n_r)^{1/3}$$

$$y_r = K_{yr}^{1/2}(\phi_r t_r/n_r)^{1/2} = (K_{xp}K_{zp}/K_{yp}^2)^{1/6}(Q_r t_r/n_r)^{1/3}$$

$$z_r = K_{zr}^{1/2}(\phi_r t_r/n_r)^{1/2} = (K_{xp}K_{yp}/K_{zp}^2)^{1/6}(Q_r t_r/n_r)^{1/3}$$

$$K_e = (K_{xp}K_{yp}K_{zp})^{1/3}(Q_r t_r/n_r)^{2/3}(n_r/t_r\phi_r)$$

$$S_{0r} = n_r/\phi_r$$

$$U_r = Q_r t_r$$

$$V_{xr} = (K_{yp}K_{zp}/K_{xp}^2)^{1/6}(Q_r/n_r t_r^2)^{1/3}$$

$$V_{yr} = (K_{xp}K_{zp}/K_{yp}^2)^{1/6}(Q_r/n_r t_r^2)^{1/3}$$

$$V_{zr} = (K_{xr}K_{yr}/K_{zr}^2)^{1/6}(Q_r/n_r t_r^2)^{1/3}$$

If we choose $n_r = \phi_r = t_r = Q_r = 1$ (i.e., the values of these parameters are identical in the two systems), then $S_{0r} = 1$, $U_r = 1$ and

$$K_e = (K_{xp}K_{yp}K_{zp})^{1/3} \tag{5-183}$$

$$x_r = (K_{yp}K_{zp}/K_{xp}^2)^{1/6} \qquad y_r = (K_{xp}K_{zp}/K_{yp}^2)^{1/6} \qquad z_r = (K_{xp}K_{yp}/K_{zp}^2)^{1/6}$$

$$\tag{5-184}$$

Step 6

Using these scales, the original problem is first transformed into an equivalent isotropic system (including the transformation of the domain's boundaries) and then solved (analytically or numerically) in that system. Once the solution is obtained, it is then transformed by the same relationships back to the original, anisotropic system.

As a second example, consider the case of flow in a confined aquifer described by (5-58)

$$\frac{\partial}{\partial x}\left(T_x\frac{\partial\phi}{\partial x}\right) + \frac{\partial}{\partial y}\left(T_y\frac{\partial\phi}{\partial y}\right) + N = S\frac{\partial\phi}{\partial t} \tag{5-185}$$

and

$$Q = -T_x\delta y\frac{\partial\phi}{\partial x} \tag{5-186}$$

Following the procedure outlined above, we shall obtain

$$\frac{T_{xr}\phi_r}{x_r^2} = \frac{T_{yr}\phi_r}{y_r^2} = N_r = S_r\frac{\phi_r}{t_r}; \qquad Q_r = T_{xr}y_r\phi_r/x_r \tag{5-187}$$

as four equations for x_r, y_r, T_e, ϕ_r, N_r, S_r, t_r, Q_r. This means that four of them may be given arbitrary values. Let these be ϕ_r, t_r, N_r, Q_r. Choosing $\phi_r = t_r = N_r = Q_r = 1$, we obtain

$$S_r = 1; \qquad x_r = (T_{yp}/T_{xp})^{1/4}$$

$$T_e = (T_{xp}T_{yp})^{1/2}; \qquad y_r = (T_{xp}/T_{yp})^{1/4} \tag{5-188}$$

Obviously, $Q_{xr} = Q_{yr} = Q_r = U_r/t_r$.

Consider now the case of flow to a single well pumping at a constant rate, Q_w, from this aquifer (Sec. 8-5). The drawdown $s(x, y, t) = \phi(x, y, 0) - \phi(x, y, t)$ in an isotropic aquifer is given by (Theis, 1935)

$$s_e = (Q_w/4\pi T_e)\, W\left[S_e(x_e^2 + y_e^2)/4t_e T_e\right]; \qquad W(u) = \int_u^\infty (e^{-x}/x)\, dx$$

$$r_e^2 = x_e^2 + y_e^2 \tag{5-189}$$

obtained by solving (5-185) with $N = 0$; $W(u)$ is the *well function for a confined aquifer* (Sec. 8-5).

We may use (5-188) in order to obtain the solution for an anisotropic aquifer where the flow is described by (5-185): $(Q_w)_e = (Q_w)_p$; $s_e = s_p$, $S_e = S_p$, $t_e = t_p$, $(\delta x)_e = (\delta x)_p (T_{yp}/T_{xp})^{1/4}$, $(\delta y)_e = (\delta y)_p (T_{xp}/T_{yp})^{1/4}$, $T_e = (T_{xp}T_{yp})^{1/2}$. We obtain

$$s_p = \frac{(Q_w)_p}{4\pi(T_{xp}T_{yp})^{1/2}}\, W\left[\frac{S_p\{x_p^2(T_{yp}/T_{xp})^{1/2} + y_p^2(T_{xp}/T_{yp})^{1/2}\}}{4t_p(T_{xp}T_{yp})^{1/2}}\right]$$

Or, omitting the subscript p

$$s = \left[Q_w/4\pi(T_xT_y)^{1/2}\right] W\left[S(x^2T_y + y^2T_x)/4tT_xT_y\right] \tag{5-190}$$

Equation (5-190) may be rewritten for an anisotropic domain in the form

$$s(r, t) = (Q_w/4\pi T_e)\, W(Sr^2/4tT_q) \tag{5-191}$$

where $T_e = (T_xT_y)^{1/2}$ and $T_q = T_x/[\cos^2\theta + (T_x/T_y)\sin^2\theta]$ is the directional transmissivity in the direction of the flow (radial direction); $\theta = \tan^{-1}(y/x)$.

To summarize, the procedure described above involves the following steps:

Step 1: Write down *all* the equations (sometimes these are approximate equations) describing the flow in the anisotropic system and in the equivalent isotropic one.

Step 2: Introduce scales between *all* the variables and parameters appearing in the two systems of equations.

Step 3: Use the scales to rewrite the equations for the prototype in terms of the equivalent system's parameters.

Step 4: Derive the conditions among scales required to make the two sets of equations identical. Examine, and leave only independent conditions.

Step 5: The difference between the number of parameters (constants and variables) involved in the equivalent system and the number of independent
conditions indicates the number of parameters (or their scales) that may
be chosen arbitrarily. The choice of parameters is immaterial.

Step 6: Use the derived scales to convert the anisotropic system (including the
geometry of its boundaries) into an equivalent isotropic one, and then
again to transform the solution back to the given anisotropic system.

Dimensionless Variables

It is often convenient to solve a problem, and present its results, in terms of
generalized *dimensionless variables and parameters*. This is especially advantageous
when solutions are required for aquifer parameters (like T and S) taking on a
number of possible values. When a problem is transformed and solved in terms
of these dimensionless variables and parameters, the results can be used for a
large number of *similar* cases, by inserting in them the values of the parameters
corresponding to specific cases of interest.

The relationships among scales (Step 4) can be used to derive the dimensionless variables and parameters.

However, in this case, the equivalent system is also anisotropic so that
$K_{xr} = K_{xe}/K_{xp}$, $K_{yr} = K_{ye}/K_{yp}$, $K_{zr} = K_{ze}/K_{zp}$. The methodology is identical
to the one described above. We shall demonstrate this by some examples.

From (5-181) and (5-182), or directly from Darcy's law written in the form
$Q = -K_x(\delta y \, \delta z) \, \delta\phi/\delta x$, we obtain

$$Q_r = K_{xr} y_r z_r \phi_r / x_r \tag{5-192}$$

or

$$\left(\frac{Q \, \delta x}{K_x \delta y \, \delta z \, \delta\phi} \right)_p = \left(\frac{Q \, \delta x}{K_x \delta y \, \delta z \, \delta\phi} \right)_e \tag{5-193}$$

The group $Q \, \delta x/K_x \delta y \, \delta z \, \delta\phi$ is the dimensionless discharge. We may represent
δx and δy by some characteristic length L and width W of the problem, δz by some
characteristic depth, e.g., the thickness B of the aquifer, $\delta\phi$ by some characteristic
$(\Delta\phi)_0$ of the problem, and obtain a dimensionless discharge $QL/K_x(\Delta\phi_0) \, WB$.

In a similar way, it follows from (5-181) that

$$n_r x_r/t_r = K_{xr} \phi_r / x_r \qquad \text{or} \qquad [K_x\delta t \, \delta\phi/n(\delta x)^2]_p = [K_x\delta t \, \delta\phi/n(\delta x)^2]_e \tag{5-194}$$

This leads to the dimensionless time $t K_x(\Delta\phi)_0/nL^2$.

Thus, we obtain the dimensionless variables by requiring that the dimensionless groups be identical in the two (and hence in any) considered systems.

It should be noted that the procedure presented in the present paragraph
can also be used for determining the scales of laboratory models and analogs,
if these are used for solving groundwater flow problems (Sec. 5-6; Bear, 1972,
Chapter 11).

5-10 ANALYTICAL SOLUTIONS OF ONE-DIMENSIONAL STEADY FLOWS IN AQUIFERS

The complete statement of the groundwater flow problem (= forecasting problem) is presented in Sec. 5-5. In general, this is an unsteady one in the xy plane, with $\phi(x, y, t)$ or $h(x, y, t)$ as the unknown dependent variables. Because of the often irregular shape of the boundary and/or inhomogeneity of aquifer parameters, analytical solutions are impractical. Numerical techniques (Sec. 5-6) are usually employed for solving regional problems of practical interest.

On the other hand, one-dimensional steady flow problems, where $\phi = \phi(x)$ or $h = h(x)$ is the sought solution, can often be solved easily by direct integration of the differential equation of continuity. Although one would tend, at first, to regard such cases as of little relevance to the practice, yet many cases of practical interest involve two-dimensional flow (in a three-dimensional space) in the vertical plane which can be reduced to ones of one-dimensional flow when the assumptions of essentially horizontal flow in an aquifer (Sec. 2-4) are justified. Some steady one-dimensional flow cases in which a complete analytical solution is possible (and relatively easy) are presented in this section in order to demonstrate the complete statement and corresponding solution of problems. This is by no means a complete survey of cases where an analytical solution is possible (and relatively easy). Our objective is just to give a selection of examples which demonstrates the analytical approach.

Unsteady flow problems require the solution of partial differential equations of continuity, such as $T \,\partial^2\phi/\partial x^2 = S \,\partial\phi/\partial t$ for a confined aquifer, or a phreatic one assuming $T = K\bar{h} = \text{const.}$, and $T \,\partial^2\phi/\partial x^2 - \phi/\lambda^2 = S \,\partial\phi/\partial t$ for a leaky one. Numerous analytical solutions of these equations for specific cases are available in the literature. Because the first of these equations describes also heat conduction, we can also borrow solutions available in that area (see, for example, Carslaw and Jaeger, 1946).

(a) Flow in a Confined Aquifer

Let Fig. 5-38 represent a cross section through an aquifer. We assume that the geometry (i.e., thickness to length ratio) is such that we have essentially horizontal flow in the $+x$ direction. Both thickness, B, and hydraulic conductivity, K, vary with distance x. At $x = 0$ and $x = L$ the piezometric heads are ϕ_A and ϕ_B, respectively.

The first step is to establish the continuity equation, either by reducing (5-56) to steady one-dimensional flow, or by constructing the continuity equation, using the control box shown in Fig. 5-38

$$Q\bigg|_{x-\Delta x/2} - Q\bigg|_{x+\Delta x/2} = 0; \qquad \frac{dQ}{dx} = 0; \qquad Q = -K(x)\,B(x)\,\frac{d\phi}{dx};$$

$$\frac{d}{dx}\left[T(x)\,\frac{d\phi}{dx}\right] = 0 \tag{5-195}$$

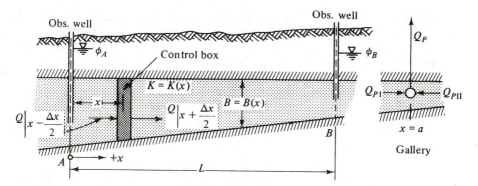

Figure 5-38 Flow in an inhomogeneous confined aquifer.

where Q is discharge per unit width normal to the cross section. By integrating (5-195) twice, we obtain

$$T(x)\frac{d\phi}{dx} = C_1; \qquad \frac{d\phi}{dx} = \frac{C_1}{T(x)}; \qquad d\phi = C_1\frac{dx}{T(x)} \qquad (5\text{-}196)$$

where C_1 is a constant, and

$$\int_{\phi'=\phi_A}^{\phi(x)} d\phi' = \phi(x) - \phi_A = C_1\int_{x'=0}^{x}\frac{dx'}{T(x')}; \qquad \phi_B - \phi_A = C_1\int_{x'=0}^{L}\frac{dx'}{T(x')} \qquad (5\text{-}197)$$

Hence

$$\frac{\phi(x) - \phi_A}{\phi_B - \phi_A} = \frac{\displaystyle\int_{x'=0}^{x}\frac{dx'}{T(x')}}{\displaystyle\int_{x=0}^{L}\frac{dx'}{T(x')}} \qquad (5\text{-}198)$$

from which we can obtain $\phi = \phi(x)$ given $T = T(x)$. For $T(x) = \text{const.}$, we obtain

$$\frac{\phi_A - \phi(x)}{\phi_A - \phi_B} = \frac{x}{L} \qquad (5\text{-}199)$$

that is *independent* of T. It is of interest to note that

$$Q = -T(x)\frac{d\phi}{dx} = -C_1 = \text{const.}$$

For a given $T(x)$ we can find C_1 (and hence Q) from the second equation in (5-197)

$$Q = \frac{\phi_A - \phi_B}{\displaystyle\int_{x'=0}^{L} dx'/T(x')} \qquad (5\text{-}200)$$

When water is extracted at a rate Q_P (per unit length), say, by a horizontal gallery at $x = a$ (or by an array of closely spaced wells, the effect of which at distances $x < a - m$ and $x > a + m$, where m is the distance between the wells in the array (Fig. 8-20), is practically the same as that of continuous gallery), we have to divide the domain $0 \le x \le L$ into two: I: $0 \le x \le a$; II: $a < x \le L$. At $x = a$, $\phi = \phi_a$ when approached from both sides and $Q_P = Q_{PI} + Q_{PII}$. Then

I
$$\phi_A - \phi(x) = C_{11} \int_{x'=0}^{x} \frac{dx'}{T(x')}; \qquad \phi_a - \phi_A = C_{11} \int_{x'=0}^{a} \frac{dx'}{T(x')}$$

$$Q_{PI} = \frac{\phi_A - \phi_a}{\int_{x'=0}^{a} dx'/T(x')} \tag{5-201}$$

II
$$Q_{PII} = \frac{\phi_B - \phi_a}{\int_{x'=a}^{L} dx'/T(x')} \tag{5-202}$$

Hence

$$Q_P = Q_{PI} + Q_{PII} = \frac{\phi_A - \phi_a}{\int_0^a dx'/T(x')} + \frac{\phi_B - \phi_a}{\int_a^L dx'/T(x')} \tag{5-203}$$

For $T(x) = T = $ const., we obtain

$$Q_P = T\left(\frac{\phi_A - \phi_a}{a} + \frac{\phi_B - \phi_a}{L - a} \right) \tag{5-204}$$

which serves also as a relationship between ϕ_a and Q_P at the gallery. Obviously we assume that the assumption of essentially horizontal flow holds in spite of the fact that in the close proximity of the gallery, the flow may deviate appreciably from this assumption.

(b) Flow in a Leaky Confined Aquifer

The cross section is given in Fig. 5-39. It is assumed that the piezometric head in the leaky phreatic aquifer remains $\phi_0 = $ const., independent of variations of $\phi(x)$ in the main aquifer.

The continuity equation is obtained either from (5-70), or from a balance for the control box shown in Fig. 5-39

$$Q\Big|_{x-\Delta x/2} + q_v \Delta x - Q\Big|_{x+\Delta x/2} = 0; \qquad q_v = \frac{\phi_0 - \phi(x)}{B'/K'} = \frac{\phi_0 - \phi(x)}{\sigma'};$$

$$-\frac{dQ}{dx} + q_v = 0; \qquad \frac{d}{dx}\left(T\frac{d\phi}{dx} \right) + \frac{\phi_0 - \phi(x)}{\sigma'} = 0; \qquad \frac{d^2\phi}{dx^2} + \frac{\phi_0 - \phi(x)}{\lambda^2} = 0$$

$$\tag{5-205}$$

where $\lambda^2 = T\sigma'$ and $\sigma' = B'/K'$. The solution of (5-205) is

$$\phi(x) = C_1 \exp(-x/\lambda) + C_2 \exp(x/\lambda) + \phi_0 \tag{5-206}$$

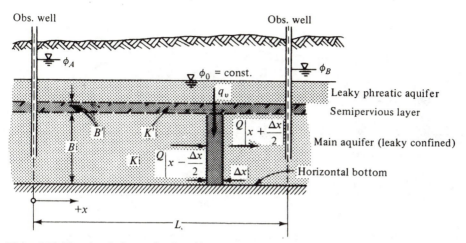

Figure 5-39 Flow in a leaky confined aquifer.

where C_1 and C_2 are constants to be derived from the boundary conditions: $\phi = \phi_A$ at $x = 0$, $\phi = \phi_B$ at $x = L$. We obtain

$$\phi(x) - \phi_0 = \frac{1}{\sinh L/\lambda} \left\{ (\phi_A - \phi_0) \sinh \frac{L-x}{\lambda} + (\phi_B - \phi_0) \sinh \frac{x}{\lambda} \right\} \quad (5\text{-}207)$$

The discharge $Q(x)$ can be found from $Q(x) = -T d\phi/dx$.

When at $x = 0$, $\phi = \phi_A$, while the aquifer extends to infinity, with $x \to \infty$, $\phi \to \phi_0$, we obtain

$$C_2 = 0, \qquad C_1 = \phi_A - \phi_0$$

Hence

$$\phi(x) - \phi_0 = (\phi_A - \phi_0) \exp(-x/\lambda) \quad (5\text{-}208)$$

$$Q(x) = -T\frac{d\phi}{dx} = \frac{T}{\lambda}(\phi_A - \phi_0) \exp\left(-\frac{x}{\lambda}\right) = \frac{T(\phi - \phi_0)}{\lambda} \quad (5\text{-}209)$$

It is of interest to note that from (5-209) it follows that $Q(x)$ is the same as that obtained in a confined aquifer with $\phi = \phi_0$ at $x = 0$ and $\phi = \phi(x)$ at $x = \lambda$.

When a gallery is placed at $x = a \, (<L)$ and water is extracted from it at a constant rate Q_P per unit length, we have, again, to divide the domain $0 \leq x \leq L$ into two subdomains, solve for ϕ_a at $x = a$, separately for the two subdomains and require that ϕ_a be the same when calculated for $x = a$ in both cases.

(c) Flow in a Phreatic Aquifer

Several examples of this case have already been presented in Sec. 4-5 (see Figs 4-7, 4-9, and 4-10). Another example, shown in Fig. 5-40, is a strip of land of width L between two parallel rivers. Although the rivers only partly penetrate the water

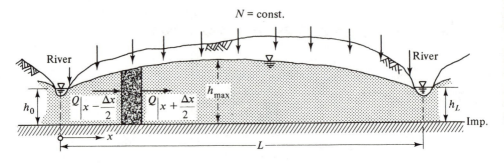

Figure 5-40 Flow in a phreatic aquifer with accretion.

bearing formation we assume that at $x = 0$ and $x = L$ we have vertical equipotentials $\phi = h_0 \, (\ll L)$ and $\phi = h_L \, (\ll L)$, respectively, and that everywhere the flow is essentially horizontal. We know that in the vicinity of the water table peak and under the streams this assumption is incorrect (in fact the flow in these places is along the vertical), yet the regions of error are relatively small and the results based on the assumption of horizontal flow should be considered a good estimate for all practical purposes.

The continuity equation is obtained either from (5-70), or from a water balance written for the control box shown in Fig. 5-40

$$Q\bigg|_{x-\Delta x/2} + N\,\Delta x - Q\bigg|_{x+\Delta x/2} = 0; \qquad -\frac{dQ}{dx} + N = 0;$$

$$K\frac{d}{dx}\left(h\frac{dh}{dx}\right) + N = 0 \qquad (5\text{-}210)$$

By integration, we obtain

$$\frac{Kh^2}{2} + \frac{Nx^2}{2} + C_1 x + C_2 = 0 \qquad (5\text{-}211)$$

Using the boundary conditions $x = 0$, $h = h_0$; $x = L$, $h = h_L$, we obtain

$$C_2 = -\frac{K}{2}h_0^2; \qquad C_1 = -\frac{K}{2L}(h_L^2 - h_0^2) - \frac{NL}{2}$$

Hence

$$K(h^2 - h_0^2) - Nx(L - x) + K\frac{x}{L}(h_0^2 - h_L^2) = 0 \qquad (5\text{-}$$

gives the shape of the water table $h = h(x)$. By differentiating (5-212), we

$$Kh\frac{dh}{dx} \equiv -Q(x) = N\left(\frac{L}{2} - x\right) - \frac{K}{2L}(h_0^2 - h_L^2)$$

Hence

$$Q\bigg|_{x=0} = -\frac{NL}{2} + \frac{K}{2L}(h_0^2 - h_L^2)$$

$$Q\bigg|_{x=L} = +\frac{NL}{2} + \frac{K}{2L}(h_0^2 - h_L^2)$$

(5-213)

where positive Q means flow in the $+x$ direction. The water table peak is obtained from $dh/dx = 0$ (or $Q(x) = 0$); it occurs at

$$x\bigg|_{h=h_{max}} = \frac{L}{2} - \frac{K}{2NL}(h_0^2 - h_L^2)$$

(5-214)

When a gallery is placed at $x = a$, $0 < a < L$, and water is withdrawn from it at a rate Q_P (per unit length), producing there a water table at elevation h_a, we follow the usual procedure of solving once for the region $0 \le x < a$ and then for $a \le x \le L$. We require that at $x = a$, $h = h_a$ when approached from both sides and that the total withdrawal be equal to the sum of the flows to the gallery from both sides.

(d) Flow in a Leaky Phreatic Aquifer

Consider the case shown in Fig. 5-41a. From a water balance written for the control box shown in this figure, we obtain the continuity equation

$$Q\bigg|_{x-\Delta x/2} + N\,\Delta x - Q\bigg|_{x+\Delta x/2} - q_v(x)\,\Delta x = 0; \qquad \frac{\partial Q}{\partial x} - N + q_v(x) = 0,$$

$$Q = -Kh\frac{dh}{dx}; \qquad q_v = \frac{h + B'}{\sigma'}; \qquad \frac{K}{2}\frac{d^2h^2}{dx^2} + N - \frac{h + B'}{\sigma'} = 0$$

(5-215)

We have assumed $p = 0$ along the bottom of the semipervious layer. Equation (5-215) can also be written as

$$\frac{d^2h^2}{dx^2} - 2Ah - 2B = 0; \qquad A = \frac{1}{K\sigma'}; \qquad B = \frac{B' - N\sigma'}{K\sigma'}$$

(5-216)

Although this equation is nonlinear in h, integration in *this special case* is possible as is shown below. We multiply both sides of (5-216) by $h\,dh/dx$ and obtain

$$2\left(h\frac{dh}{dx}\right)\frac{d}{dx}\left(h\frac{dh}{dx}\right) - 2Ah\left(h\frac{dh}{dx}\right) - 2Bh\frac{dh}{dx} = 0$$

$$\frac{d}{dx}\left(h\frac{dh}{dx}\right)^2 - \frac{2}{3}A\frac{d}{dx}(h^3) - B\frac{d}{dx}(h^2) = 0$$

which, when integrated, yields

$$\left(h\frac{dh}{dx}\right)^2 = \tfrac{2}{3}Ah^3 + Bh^2 + C, \qquad h\frac{dh}{dx} = -(\tfrac{2}{3}Ah^3 + Bh^2 + C)^{1/2}$$

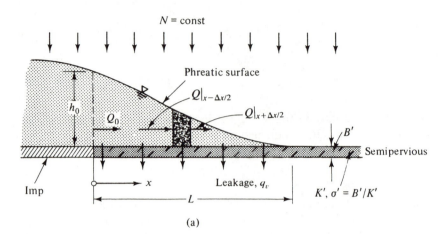

(a)

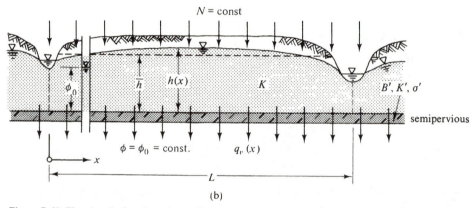

(b)

Figure 5-41 Flow in a leaky phreatic aquifer.

The minus sign of the square root is chosen because dh/dx is negative. We then obtain the constant C from the boundary condition $Q = -Kh(dh/dx) = 0$ at $x = L$, where $h = 0$; $C = 0$. Hence

$$\frac{dh}{dx} = -(\tfrac{2}{3}Ah + B)^{1/2}; \quad dx = -\frac{dh}{(\tfrac{2}{3}Ah + B)^{1/2}},$$

$$x + C_1 = -\frac{3}{A}\left(\frac{2}{3}Ah + B\right)^{1/2}$$

with $x = 0$, $h = h_0$, $C_1 = -(3/A)\{(2/3)Ah_0 + B\}^{1/2}$. We therefore obtain

$$h(x) = h_0 + \frac{Ax^2}{6} - \left(\frac{2}{3}Ah_0 + B\right)^{1/2} x \qquad (5\text{-}217)$$

$$\left. \begin{array}{l} Q(x) = -Kh\dfrac{dh}{dx} = Kh\left(\dfrac{2Ah}{3} + B\right)^{1/2} \\[4mm] Q\bigg|_{x=0} = Kh_0\left(\dfrac{2Ah_0}{3} + B\right)^{1/2} \end{array} \right\} \qquad (5\text{-}218)$$

From $x = L$, $h = 0$, $Q|_L = 0$, we obtain

$$L = \frac{3}{A}\left[\left(\frac{2}{3}Ah_0 + B\right)^{1/2} - B^{1/2}\right] \qquad (5\text{-}219)$$

Phreatic flow conditions with leakage lead always to a nonlinear continuity equation in h. As has already been mentioned in Sec. 5-4(c), the equation can be linearized by assuming some constant average thickness of the saturated flow domain. In the one-dimensional case, the linearized equation can then be integrated rather easily. The error introduced by the linearization is smaller as the deviation of h from $\bar{h}$ becomes smaller.

As an example, consider the case shown in Fig. 5-41b. One should note that although we have indicated a downward direction for the leakage q_v, the actual direction is determined by the relationship between h above the semipervious layer and ϕ_0 (= const.) below it.

The continuity equation is the same as (5-215), except that there $\phi_0 = 0$

$$\frac{K}{2}\frac{d^2(h^2)}{dx^2} + N - \frac{h - \phi_0}{\sigma'} = 0 \qquad (5\text{-}220)$$

Emich (1962) presents an exact analytical solution of this equation, using elliptic integrals. For practical engineering problems, let us linearize this equation by assuming that we have a constant transmissivity $T = K\bar{h}$, where $\bar{h}$ is some average thickness of the flow domain. Then, (5-220) reduces to

$$T\frac{d^2h}{dx^2} - \frac{h - \phi_0}{\sigma'} + N = 0, \qquad \text{or:} \qquad \frac{d^2h}{dx^2} - \frac{h - \phi_0}{\lambda^2} + \frac{N}{T} = 0 \qquad (5\text{-}221)$$

The solution is

$$h = \phi_0 + C_1 \exp(-x/\lambda) + C_2 \exp(x/\lambda) + N\lambda^2/T \qquad (5\text{-}222)$$

The constants C_1 and C_2 are derived from the boundary conditions $x = 0$, $h = h_0$, $x = L$, $h = h_L$.

When $h(x)$ varies appreciably with respect to the average value $\bar{h}$, it is possible to divide the flow domain $0 < x < L$ into several subdomains with $\bar{h}_1, \bar{h}_2$, etc., and corresponding transmissivities T_1, T_2, etc., and solve the problem as one in an inhomogeneous domain.

(e) Radial Flow to a Well in a Phreatic Aquifer

The whole of Chap. 8, on well hydraulics, is devoted to analytical solutions of problems of radially converging or diverging flows in different types of aquifers.

5-11 LAND SUBSIDENCE

We have mentioned several times before that *land subsidence* (or *consolidation*), like water levels and solute concentrations, is part of the aquifer's response to planned activities. Maximum permissible consolidation may also serve as a constraint in a groundwater management problem. It is, therefore, necessary to predict the subsidence of the ground surface overlying an exploited aquifer as a result of pumping, whenever appreciable subsidence is suspected.

The basic cause for subsidence has been presented already in Sec. 5-1. We have seen there that pumping from an aquifer produces a reduction in water pressure and therefore, at the same time, an increase in the effective stress in the solid skeleton of the porous medium. The latter produces consolidation. In general, an increase in effective stress may also be produced by increasing the load on the ground surface; however, this is not the case here. The degree and rate of consolidation depend on the stress–strain relationship of the porous material comprising the aquifer.

There exist many examples of the occurrence of subsidence of large areal extent as a result of pumping groundwater. The most spectacular one in an urban area is in Mexico City, where almost the entire metropolitan area has subsided more than 3 m (with up to 8 m at some locations). The San Joaquin Valley in California has experienced subsidence at a rapid rate of 30–40 cm per year, reaching a total subsidence of 9 m at some locations. Other places in which significant subsidence has occurred are the Taipei Basin, Tokyo, the Texas Gulf Coast, London, and Bangkok. In almost all these places, the aquifers contain (or a sequence of confined aquifers are separated from each other by) compressible clay strata.

Basically, consolidation is a three-dimensional phenomenon. De Josselin de Jong (1963), Verruijt (1965, 1969), among many others, present theories and examples of three-dimensional consolidation (see also Bear (1972, p. 208)). In general, two main approaches exist to the analysis of consolidation.

(a) The approach originated by Biot (1941) which regards consolidation as an elasticity problem with variable body forces. Strain and flow in a three-dimensional space are considered simultaneously and the redistribution of total stress is accounted for. A simultaneous solution is sought for two dependent variables: pressure in the water and strain in the solid. By integrating the vertical strain over the thickness of the aquifer, we obtain the vertical displacement, or subsidence (De Josselin de Jong, 1963; Verruijt, 1965, 1969).

(b) A generalization of Terzaghi's (1925) theory to include three-dimensional dissipation of water pressure, without considering the state of strain in the solid skeleton at the same time. The redistribution of total stress and its effect on water pressure is ignored when the latter is solved for by solving a single partial differential equation in terms of pressure only. Once the pressure distribution is known, effective stress, accompanying strain and vertical settlement, can be found.

In the practice of soil mechanics, the second approach is more commonly employed; it is considered a rather fair approximation to the corresponding Biot solution. Actually in most consolidation analyses carried out in soil mechanics, it is assumed that both flow and soil deformation occur mainly in the vertical direction, thus ignoring any lateral deformation. The problem then reduces to one-dimensional consolidation in the vertical direction. In the latter case, the two approaches coincide. The reader is referred to standard texts on soil mechanics (e.g., Scott, 1963; Harr, 1966) for more details on consolidation theory and practice. Of special interest should be the Proceedings of the IASH/UNESCO Symposium on Land Subsidence held in Tokyo, Japan, in 1969.

In the following, we shall briefly review the relatively simple second approach, which may be useful for predicting land subsidence due to pumping.

We may start from the three-dimensional mass conservation equation (5-13)

$$\nabla \cdot \rho \mathbf{q} + \partial(n\rho/\partial t) = 0 \qquad (5\text{-}223)$$

and develop it to a continuity equation in terms of the dependent variable pressure, $p = p(x, y, z, t)$. The resulting equation depends on the assumptions we make. For example, using certain simplifying assumptions we obtain (5-26). Assuming that the water is homogeneous, and practically incompressible, leads to

$$\nabla \cdot \mathbf{q} + \partial n/\partial t = 0 \qquad (5\text{-}224)$$

Overlooking the fact that Darcy's law describes flow with respect to the solids, we obtain from (5-7) and (5-224)

$$\nabla \cdot (\mathbf{K} \cdot \nabla \phi^*) = - \partial n/\partial t = \alpha(1 - n)\partial p/\partial t \qquad (5\text{-}225)$$

If we do take into account the solid's velocity $\mathbf{V}_s$, with $\mathbf{q} - n\mathbf{V}_s = -\mathbf{K} \cdot \nabla \phi^*$, we obtain from (5-224)

$$-\nabla \cdot (\mathbf{K} \cdot \nabla \phi^*) - \partial n/\partial t - \nabla \cdot n\mathbf{V}_s = -(\partial n/\partial t + \mathbf{V}_s \cdot \nabla n)$$

$$-n\nabla \cdot \mathbf{V}_s = -d_s n/dt - n\nabla \cdot \mathbf{V}_s \qquad (5\text{-}226)$$

where $d_s n/dt$ is the *material derivative* of n with respect to $\mathbf{V}_s$. Volume conservation of the solid is expressed by

$$\partial(1 - n)/\partial t + \nabla \cdot [(1 - n)\mathbf{V}_s] = 0$$

from which we obtain

$$\nabla \cdot \mathbf{V}_s = \frac{1}{1 - n}\frac{d_s n}{dt} = -\frac{1}{1 - n}\frac{d_s(1 - n)}{dt} = -\alpha'\frac{d_s \sigma'}{dt} = \alpha'\frac{d_s p}{dt} \quad (5\text{-}227)$$

where α' is the coefficient of matrix compressibility of a moving solid. Inserting (5-227) in (5-226) gives

$$\nabla \cdot (\mathbf{K} \cdot \nabla \phi^*) = \frac{1}{1 - n}\frac{d_s n}{dt} \approx \frac{1}{1 - n}\frac{\partial n}{\partial t} = \frac{1}{1 + e}\frac{\partial e}{\partial t} \qquad (5\text{-}228)$$

$$\nabla \cdot (\mathbf{K} \cdot \nabla \phi^*) = \alpha' d_s p/dt \approx \alpha' \partial p/\partial t \qquad (5\text{-}229)$$

where the approximations result from the assumptions $\partial n/\partial t \gg \mathbf{V}_s \cdot \nabla n$ and $\partial p/\partial t \gg \mathbf{V}_s \cdot \nabla p$; $e = n/(1 - n)$ is the *void ratio* (= volume of voids to volume of solids). Note that $\alpha' = [1/(1 - n)] \, d_s n/dp \approx [1/(1 - n)] \, \partial n/\partial p = \alpha$ as defined by (5-7).

By removing the assumption of water incompressibility, but with $\mathbf{q} \cdot \nabla \rho \ll n \, \partial \rho/\partial t$, we obtain

$$\nabla \cdot (K \cdot \nabla \phi^*) \cong (n\beta + \alpha') \, \partial p/\partial t = \rho g(n\beta + \alpha') \, \partial \phi^*/\partial t \qquad (5\text{-}230)$$

to be compared with (5-26) which neglects $\mathbf{V}_s$.

At this point it is usually assumed that K is independent of variations in ρ (actually also in μ and n) due to pressure changes. For a homogeneous non-isotropic medium, with x, y, z principal directions, we obtain from (5-230)

$$K_x \frac{\partial^2 p}{\partial x^2} + K_y \frac{\partial^2 p}{\partial y^2} + K_z \frac{\partial^2 p}{\partial z^2} = \rho g(n\beta + \alpha') \frac{\partial p}{\partial t} \qquad (5\text{-}231)$$

In soil mechanics, $(n\beta + \alpha')$ is often replaced by $(\beta e + \partial e/\partial p)/(1 + e) \approx (\beta e + \partial e/\partial p)/(1 + e_0)$, where e_0 is the initial value of e.

Figure 5-42 shows typical experimental relationships between void ratio and effective stress as defined in Sec. 5-1. Texts on soil mechanics (e.g., Scott, 1963) discuss these relationships in detail. For granular soil, the approximation usually employed is (Fig. 5-42a)

$$e = e_0 - a_v \sigma' \quad \text{or} \quad e = e_0 - a_v(\sigma' - \sigma'_0) \qquad (5\text{-}232)$$

where a_v (>0) is a *coefficient of compressibility*; e_0 is the void ratio at the effective stress σ'_0.

For cohesive soils, experimental results usually lead to the relationship

$$e = e_0 - C_c \log(\sigma'/\sigma'_0) \qquad (5\text{-}233)$$

where C_c is called the *compression index*. However, for small changes in e, (5-231) may also be used for cohesive soils like clay.

From (5-232), we obtain

$$\partial e/\partial \sigma' = -a_v \qquad (5\text{-}234)$$

It is usually assumed that for small changes in e, a_v may be considered a constant (thus giving the slope of the straight line approximation in Fig. 5-42a).

As in Secs. 5-1 through 5-3, we assume that the stress, σ, due to an external load remains unchanged. Then

$$\partial e/\partial p = -\partial e/\partial \sigma' = a_v = \alpha/(1 - n) \cong \alpha'(1 + e) \qquad (5\text{-}235)$$

where α is defined by (5-7). Then (5-231) for $\beta = 0$ becomes

$$K_x \frac{\partial^2 p}{\partial x^2} + K_y \frac{\partial^2 p}{\partial y^2} + K_z \frac{\partial^2 p}{\partial z^2} = \rho g \alpha' \frac{\partial p}{\partial t} \cong \frac{\rho g a_v}{1 + e_0} \frac{\partial p}{\partial t} \qquad (5\text{-}236)$$

where $\rho g a_v/(1 + e_0)$ is a constant coefficient. A similar equation may be obtained

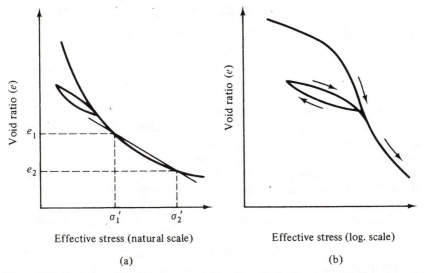

Figure 5-42 Typical relationships between void ratio and effective stress. (a) Granular soil. (b) Clay.

from (5-231). In a homogeneous isotropic medium, the group $K(1 + e_0)/\rho g a_v$, denoted by C_v, is often called the *coefficient of consolidation* $(=k(1 + e_0)/\mu a_v)$.

In consolidation investigations in soil mechanics, the water pressure, $p(x, y, z, t)$, is usually divided into two parts: a steady state pressure $p_0(x, y, z)$ and a transient (pore) water pressure excess $u = u(x, y, z, t) = p - p_0$ over the steady state pressure. Hence, the partial differential equations are usually written in terms of u, assuming that p_0, which produces no consolidation, is known.

Once we have a partial differential equation, we solve it, for a given aquifer domain, subject to specified initial and boundary conditions to obtain $p(x, y, z, t)$, or $u(x, y, z, t)$.

It should be obvious by now that what we have actually been doing above is rewriting (5-26) in terms of p, subject to additional assumptions, e.g., $\mathbf{V}_s \neq 0$, and giving a different interpretation to S_0 in it, again for different sets of assumptions. For example in (5-236), $a_v/(1 + e_0)$ gives the volume of water drawn out of storage per unit volume of porous medium, per unit decline of pressure. In fact, this is the drainage which accompanies consolidation, for in the course of consolidation, water drains out of the soil until the excess pore pressure becomes zero. Then the entire stress is transferred to effective stress.

Assuming vertical displacement only, the total settlement $\delta(x, y, t)$ of an aquifer of thickness B is equal to the volume of water drained out of a column of unit horizontal area

$$\delta(x, y, t) = \frac{a_v}{1 + e_0} \int_0^B \left[p(x, y, z, t) - p_0(x, y, z) \right] dz \qquad (5\text{-}237)$$

Equation (5-237) may also be obtained by noting that $e = e_0 + \Delta e$ and

$$\frac{\Delta e}{1 + e_0} = \left\{ \begin{array}{l} \text{Volume of water drained per} \\ \text{unit volume of porous medium} \end{array} \right\} = \frac{d\delta \, dx \, dy}{dx \, dy \, dz} = \frac{d\delta}{dz}$$

$$\delta(x, y, t) = \int_0^B \frac{\Delta e}{1 + e_0} \, dz = \int_0^B \frac{-a_v \Delta \sigma'}{1 + e_0} \, dz = \int_0^B \frac{a_v \Delta p}{1 + e_0} \, dz \qquad (5\text{-}238)$$

Obviously, in an inhomogeneous domain, we may have $a_v = a_v(z)$ and $e_0 = e_0(z)$.

In the way presented above, we have reduced the problem of forecasting consolidation to two steps. First, we determine the pressure (or excess pressure) distribution, and then use (5-238) to determine $\delta(x, y, t)$.

Before leaving this subject, we would like to make two more comments.

First, we have stated at the beginning of this section that our objective is to forecast consolidation and resulting land subsidence due to pumpage. Hence, we have to add sink terms representing pumpage to the various partial differential equations of continuity. For a distributed sink, we add $-P(x, y, z, t)$, where P (dims T^{-1}) is the rate at which water is withdrawn per unit volume of medium per unit time. For point sinks, we may write $-\sum_{(i)} P_w(x_i, y_i, z_i, t) \times \delta(x - x_i, y - y_i, z - z_i)$, where $P_w(x_i, y_i, z_i, t)$ is the discharge of a sink (dims $L^3 T^{-1}$) located at point (x_i, y_i, z_i). For example, the basic equation (5-18) then becomes

$$-\operatorname{div}(\rho \mathbf{q}) - \sum_{(i)} \rho P_w(x_i, y_i, z_i, t) \, \delta(x - x_i, y - y_i, z - z_i) = \partial \rho n / \partial t \qquad (5\text{-}239)$$

A second comment concerns the fact that the entire discussion so far has been related to three-dimensional flow; we have sought a solution for $p(x, y, z, t)$. However, when we consider land subsidence over an extended area due to pumpage from an underlying aquifer, the flow in the aquifer is essentially two dimensional in the horizontal plane, with practically vertical equipotentials. Following the routine presented in Sec. 5-4, this means that we have to integrate, or average, any of the continuity equations in terms of $p(x, y, z, t)$ in order to obtain an average equation in terms of an average pressure, $\tilde{p}(x, y, t)$ over the thickness, B.

Actually, we do not have to repeat the developments, as we may use any of the averaged equations already developed in Sec. 5-4 (with an appropriate interpretation of S_0 and neglecting the effect of consolidation on transmissivity). Once we have $\tilde{\phi}$ we may obtain $\tilde{p}$ from $\tilde{p} = \rho g(\tilde{\phi} - \tilde{z})$, where $\tilde{z}$ is the average elevation of the aquifer (and we assume $\rho g = \text{const.}$). We then note that (5-237) actually states that

$$\delta(x, y, t) = \frac{a_v B}{1 + e_0} \left[\tilde{p}(x, y, t) - \tilde{p}_0(x, y) \right] \qquad (5\text{-}240)$$

In the procedure, described above, based on Terzaghi's (1925) theory, only the linearly elastic part of the consolidation, often called *primary consolidation*, is taken into account. The departure of observed consolidation from that predicted by Terzaghi's theory is called *secondary consolidation*. This part of the consolidation, which is significant when silt and clay lenses are present within any aquifer, is due to the time lag between measured changes in the piezometric head in the coarse-grained portions of an aquifer and the resulting observed compaction. Gambolati and Freeze (1973), Gambolati et al. (1974), Helm (1975), and Brutsaert and Corapcioglu (1976), have presented models that account for secondary consolidation. In a later paper, Corapcioglu and Brutsaert (1977) make use of a viscoelastic model. In this model, the time-dependent strain, $\varepsilon(t)$, is given by

$$\varepsilon(t) = x_1\sigma'(t) + \frac{1}{q_1''} \int_0^t \sigma'(\tau)\exp[-(t-\tau)/\alpha_2 q_1'']\,d\tau \qquad (5\text{-}241)$$

where α_1 and α_2 are primary and secondary compressibilities, respectively, q_1'' is the viscosity of the model and $\sigma'(t)$ is the stress.

The corresponding governing equation for saturated horizontal groundwater flow in a viscoelastic homogeneous confined isotropic aquifer, assuming that horizontal displacements of the aquifer skeleton are negligible. is (Corapcioglu and Brutsaert, 1977)

$$\frac{K}{\rho g}\nabla'^2 s_p = (\alpha_1 + n\beta)\frac{\partial s_p}{\partial t} + \frac{1}{q_1''}\frac{\partial}{\partial t}\int_0^t s_p(\tau)\exp[-(t-\tau)/\alpha_2 q_1'']\,d\tau \qquad (5\text{-}242)$$

where K is hydraulic conductivity, β is water compressibility, and s_p is the drawdown in water pressure below its initial value. The second term on the right-hand side of (5-242) expresses the nonelastic part of the consolidation. Note that in (5-242), the aquifer's elastic storativity is expressed by $\alpha_1 + n\beta$, that is, taking into account the movement of the upper bound of the aquifer (see comment following (5-67)), or $V_s = 0$.

SIX

FLOW IN THE UNSATURATED ZONE

The previous chapters dealt with the hydraulics of groundwater flow in aquifers, i.e., in the saturated zone, and did so mainly from the regional point of view. However, certain flow processes which take place in the unsaturated zone are highly important also from the regional point of view and should be incorporated in our considerations.

The first example is the infiltration process. As discussed in Chap. 3, a phreatic aquifer is replenished from above by water from various sources: precipitation, irrigation, artificial recharge by surface spreading techniques, etc. In all these cases, water moves downward, from the ground surface to the phreatic surface, through the unsaturated zone (or zone of aeration; Fig. 2-3). The understanding of, and consequently the ability to calculate and predict, the movement of water in the unsaturated zone is, therefore, essential when we wish to determine the (total) replenishment of a phreatic aquifer (as part of our groundwater flow model).

The second example is related to water quality (Chap. 7). Pollutants applied in various forms at the ground surface, e.g., solid waste in landfills, septic tanks, fertilizers, and pesticides, over extended areas, are dissolved in water applied to the land surface. The infiltrating water will then carry pollutants as it moves downward towards the water table. As pollutants travel downward with the infiltrating water, various phenomena, e.g., dispersion and adsorption, take place. These affect the concentration of pollutants in the water which eventually reaches the phreatic surface.

As in the case of aquifer replenishment, the understanding of, and ability to calculate and forecast, the movement and accumulation of pollutants in the unsaturated zone is needed if we wish to determine the rate at which a pollutant reaches the water table. However, one cannot study the movement of a pollutant carried by the water without information on the movement of the water itself.

In the following paragraphs, a brief review is presented on the motion and continuity equations of unsaturated flow. Only those concepts which are directly related to the hydrology of groundwater are reviewed. For additional details on unsaturated flow, the reader is referred to texts and reviews on unsaturated flow (e.g., Wesseling, 1961; Stallman, 1967; Childs, 1967, 1969; Swartzendruber, 1969; Philip, 1970; Bear, 1972, Chap. 9; Kirkham and Powers, 1972; Morel-Seytoux, 1973).

6-1 CAPILLARY PRESSURE AND RETENTION CURVES

Moisture Content

In unsaturated flow, the void space is partly filled by air and partly by water. Two variables may be used to define the relative quantity of water at a certain time in the vicinity of a point in a porous medium domain (i.e., in an REV for which this point is a centroid)

$$\theta_w = \frac{\text{Volume of water in REV}}{\text{Bulk volume of REV}}; \qquad 0 \le \theta_w \le n \tag{6-1}$$

$$S_w = \frac{\text{Volume of water in REV}}{\text{Volume of voids in REV}}; \qquad 0 \le S_w \le 1 \tag{6-2}$$

θ_w is called the *water (or moisture) content*; S_w is called the *water saturation*. Obviously, the two definitions are related to each other by

$$\theta_w = nS_w \tag{6-3}$$

where n is the porosity at the considered point.

Contact Angle and Wettability

When a liquid is in contact with another substance (another liquid immiscible with the first, a gas, or solid), there is *free interfacial energy* present between them. The interfacial energy arises from the difference between the inward attraction of the molecules in the interior of each phase and those at the surface of contact. Since a surface possessing free energy contracts, if it can do so, the free interfacial energy manifests itself as an *interfacial tension*. Thus, the interfacial tension σ_{ik} for a pair of substances i and k is defined as the amount of work that must be performed in order to separate a unit area of substance i from substance k. For air and water, $\sigma_{ik} = 72.5 \, \text{erg/cm}^2$ at $20°\text{C}$. The interfacial tension σ_i between a substance and its own vapor is sometimes called *surface tension*; however, the term surface tension is often also used to indicate the interfacial tension between two phases.

Figure 6-1 shows two immiscible fluids in contact with a solid plane surface. The angle θ, called *contact angle*, denotes the angle between the solid surface and the liquid–gas, or liquid–liquid interface, measured through the denser fluid. Equilibrium requires that

$$\sigma_{LG} \cos \theta + \sigma_{SL} = \sigma_{GS} \qquad \text{or} \qquad \cos \theta = (\sigma_{GS} - \sigma_{SL})/\sigma_{LG} \tag{6-4}$$

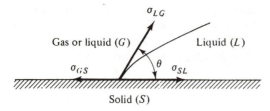

Figure 6-1 Interfacial tension.

Equation (6-4), called *Young's equation*, states that $\cos \theta$ is defined as the ratio of the energy released in forming a unit area of interface between a solid S and a liquid L, instead of between a solid S and a fluid G, to the energy required to form a unit area between the liquid L and the fluid G.

Sometimes a factor is introduced in (6-4) to account for the roughness of the solid. From (6-4) it follows that no equilibrium is possible if $(\sigma_{GS} - \sigma_{SL})/\sigma_{LG} > 1$. In this case, the liquid L will spread indefinitely over the surface. This leads to the *concept of wettability* of a solid by a liquid.

The product $\sigma_{LG} \cos \theta$, called adhesion tension, determines which of the two fluids (L or G) will preferentially wet the solid, i.e., adhere to it and spread over it.

When $\theta < 90°$, the fluid (e.g., L in Fig. 6-1) is said to wet the solid and is called a *wetting fluid*. When $\theta > 90°$, the fluid (G in Fig. 6-1) is called *nonwetting fluid*. In any system similar to that shown in Fig. 6-1, it is possible to have either a fluid L-wet or a fluid G-wet solid surface, depending on the chemical composition of the two fluids and the solid. In the unsaturated (air–water) zone in soil, water is the wetting phase, while air is the nonwetting one.

Interfacial tension and wettability may be different when a fluid–fluid interface (e.g., an air–water interface) is advancing or receding on a solid surface. This phenomena is called *hysteresis* (see Fig. 6-10).

With the concept of wettability as defined above, and for the air–water system considered here, we may distinguish three ranges of water saturation between the limits of 0% and 100%. Figure 6-2 shows water in a water wet granular soil (e.g., sand). At a very low saturation (Fig. 6-2a), water forms rings, called *pendular rings*, around the grain contact points. The air–water interface has the shape of a "saddle". Sometimes a number of adjacent pendular rings coalesce.

At this low saturation, the rings are isolated and do not form a continuous water phase, except for a very thin film of water of nearly molecular thickness (and which does not behave as ordinary liquid water) on the solid surfaces. Practically no pressure can be transmitted from one ring to the next through the water phase. Figure 6-2b shows a pendular ring between two spheres. For this idealized case, it is possible to relate the volume of the ring to the radius of curvature of the air–water interface; the latter, in turn is related, as we shall see below, to the capillary pressure.

As water saturation increases, the pendular rings expand until a continuous

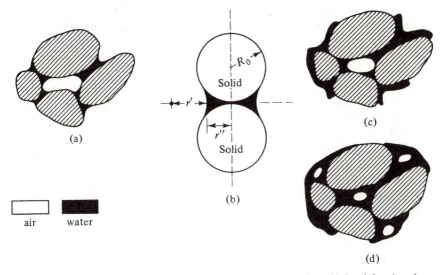

Figure 6-2 Possible water saturation states. (a) Pendular saturation. (b) Pendular rings between two spheres. (c) Funicular saturation. (d) Insular air saturation.

water phase is formed. The saturation at which this occurs is called *equilibrium water saturation*. Above this critical saturation, the saturation is called *funicular* and flow of water is possible (Fig. 6-2c). Both water and air phases are continuous. As the water saturation increases, a situation develops in which the air (= nonwetting phase) is no longer a continuous phase; it breaks into individual droplets lodged in the larger pores (Fig. 6-2d). The air is then said to be in a *state of insular saturation*. A globule of air can move only if a pressure difference sufficient to squeeze it through a capillary size restriction is applied across it in the water. Obviously, if air is not present at all in the void space, we have complete saturation.

Sometimes the term *adsorbed stage* is used for water present in the pore space at a very low saturation, such that it forms continuous, or discontinuous, films of one or more molecular layers (Stallman, 1964) on adsorption sites on the solid.

Capillary Pressure

When two immiscible fluids (here air and water) are in contact, a discontinuity in pressure exists across the interface separating them. This is a consequence of the interfacial tension which exists between the two phases in contact. The magnitude of the pressure difference depends on the interface curvature at that point (which, in turn, depends on the saturation). Here "point" is a *microscopic* point on the air–water interface inside the void space. The difference in pressure is called *capillary pressure* (p_c)

$$p_c = p_{air} - p_{water} \qquad (6\text{-}5)$$

where the pressures are taken in the two phases as the interface is approached from their respective sides. In a general two-phase system, p_c is the difference between the pressure on the nonwetting side of the interface and that on the wetting one.

There are several ways of determining the relationship between the curvature of the air–water interface (or any two phases), the interfacial tension and the pressures, or more generally the stresses, in the two fluids separated by the interface. In all of them, the interface is considered as a (two-dimensional) material body (actually, surface) which has rheological properties of its own. Its behavior is similar to a stretched membrane under tension in contact with the adjacent two fluids. In fact, with this assumption, the consideration of equilibrium surface (= interfacial) tension leads to the conclusion that the normal component of fluid stress, or pressure, is discontinuous at a curved interface. Scriven (1960) and Slattery (1967) present detailed analyses of interface behavior. However, for our purpose here, a much simpler approach will suffice.

Figure 6-3 shows an infinitesimal element of a curved air–water interface. Assuming the interfacial tension, σ_{12}, to be constant, one can show that equilibrium of forces on the element requires that

$$p_c = p_2 - p_1 = \sigma_{12}(1/r' + 1/r'') = 2\sigma_{12}/r^* \tag{6-6}$$

where r^* is the mean radius of curvature defined by $2/r^* = 1/r' + 1/r''$. Equation (6-6) is known as the Laplace formula for capillary pressure. The capillary pressure is thus a measure of the tendency of the partially saturated porous medium to suck in water or to repel air. In soil science, the negative value of the capillary pressure head is often called *suction* or *tension* (denoted by ψ).

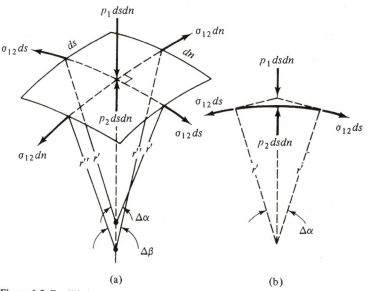

(a)   (b)

Figure 6-3 Equilibrium at a curved air–water interface.

Equation (6-6) is developed as a balance of forces at a point on the curved interface. The interface itself acts as a common boundary to the two subdomains of the void space, one occupied by air and the other by water. According to our continuum approach, we are interested in the average pressure in each of the two phases. If we assume the air in the void space to be at atmospheric pressure, then the water in the void space is at a pressure, p_w, less than atmospheric. In the more general case, $p_{air} \neq 0$. A simple model explaining what happens in the void space is the water in a capillary tube (simulating the narrow opening between grains) shown in Fig. 6-4. In the water just below the meniscus, the pressure p_w is given by

$$p_c = p_{air} - p_w; \qquad p_w = -p_c; \qquad p_{air} = 0 \qquad (6\text{-}7)$$

Pressure is also negative along the tube down to the surface $p_w = 0$.

Figure 6-5 shows what happens at the contact of the air–water interface and the soil (compare with the air–water–glass contact in Fig. 6-4). Figures 6-6 and 6-7 show how the negative (i.e., less than atmospheric) pressure in the water can be determined. In Fig. 6-6a, the unsaturated soil sample is placed on a porous membrane (or porous plate) which has very small openings such that air cannot be sucked through them into the manometer, even through the largest openings (remember that $p_c = 2\sigma/r^*$). After some time, equilibrium is reached between the water in the soil and that in the manometer. The manometer reads an *average*

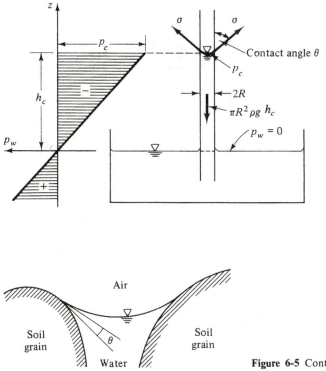

Figure 6-4 A capillary tube.

Figure 6-5 Contact angle in a pendular ring.

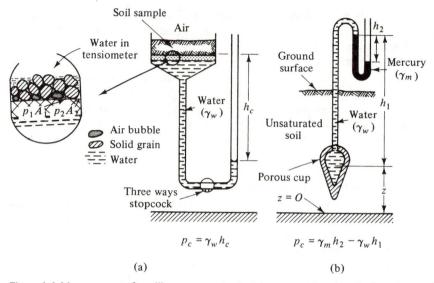

$$p_c = \gamma_w h_c$$

$$p_c = \gamma_m h_2 - \gamma_w h_1$$

(a)   (b)

Figure 6-6 Measurement of capillary pressure in the laboratory (a) and in the field (b).

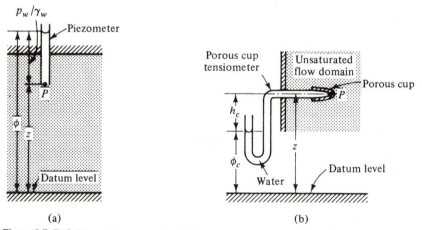

(a)   (b)

Figure 6-7 Definition diagram for ϕ and ϕ_c.

pressure over the area of contact between the water in the soil and that in the manometer (see Bear, Zaslavsky, and Irmay, 1968, p. 46). For higher suctions, porous plates with smaller openings are used (e.g., unglazed earthenware or porcelain). The porous plate, which is thus permeable to water but impermeable to air, is necessary to establish hydraulic contact between the water in the soil and that in the manometer, without air being sucked into the latter.

The instrument used for measuring the capillary pressure in an unsaturated zone is called *tensiometer* (a name introduced by Richards and Gardner, 1936).

The contact between the water in the tensiometer and that in the soil is established through a porous cap (sometimes a porous plate is used).

In using a tensiometer, one has to make sure that equilibrium has been reached, as sometimes this may take a very long time. Also, that suction is such that air is not drawn through the porous plate at any point. The pressure at which air will enter is called *bubbling pressure* (or *air entry* value or *threshold pressure*). Membrane materials are available with bubbling pressures of 20–50 atmospheres.

By analyzing the forces acting on the water column in the capillary tube of Fig. 6-4, we find

$$h_c \pi R^2 \rho g = 2 \pi R \sigma \cos \theta; \qquad h_c = 2\sigma \cos \theta / T \rho g \qquad (6\text{-}8)$$

where σ denotes surface tension and ρ is the water density ($\equiv \rho_w$). With the average radius of curvature, r^*, of the meniscus equal to $R/\cos \theta$, we obtain from (6-6)

$$h_c = p_c / \rho g \qquad (6\text{-}9)$$

called capillary pressure head. For $p_{air} = 0$, we have $h_c = -p_w / \gamma_w$.

Another term often used by soil scientists is pF ($= \log_{10} h_c$), where h_c is in centimeters. This logarithmic scale facilitates the presentation of a wide range of h_c in a single diagram.

As in saturated flow of water, we may define also in unsaturated flow a piezometric head $\phi = z + p_w / \gamma_w$ at every point of the flow domain. Often the term *capillary head* (symbol ϕ_c) is used to denote the piezometric head in unsaturated flow (Fig. 6-7). For $p_{air} = 0$

$$\phi_c = z + p_w / \gamma_w = z - p_c / \gamma_w = z - h_c: \qquad p_c > 0 \qquad (6\text{-}10)$$

Many authors use the symbol ψ for the capillary pressure head so that for $p_{air} = 0$, $\psi \equiv -p_w / \gamma_w > 0$. As in saturated flow, the concepts of ϕ_c, ψ, h_c should be used only when $\rho_w = $ const.

Once we understand the phenomenon of capillary pressure at the microscopic level, we have to describe it at the macroscopic, averaged one. For example, this means that the averaged capillary pressure in the soil (as measured by a tensiometer, except for some inaccuracy introduced by the fact that the latter gives an areal average rather than a volumetric one) is equal to the difference between the average pressure in the water and that in the air. To save on symbols, we shall use (6-5), (6-9), and similar equations as macroscopic relationships, with p_w, p_a, etc., denoting intrinsic phase averages as defined by Eq. (A-2). In each case it will be obvious whether p_w and p_a denote microscopic or macroscopic values.

Moisture Retention

Consider a saturated sample placed on the porous plate of the tensiometer shown in Fig. 6-6a. Then water is drained by lowering the manometer limb, or through the stopcock. Figure 6-8 (after Childs, 1969) shows several successive stages of drainage. The initial stage is denoted by 1. As water is drained, inter-

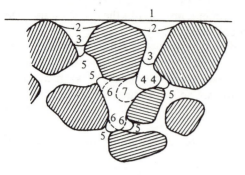

Figure 6-8 Stages of water drainage (1-5) and re-entry (6-7) *(after Childs, 1969).*

faces (meniscii) are formed, 2. The radius of curvature of the interfaces depends on the suction. As water is drained and the interfaces are drawn further down, the curvature becomes sharper and the suction increases. At every stage, the greatest suction that can be maintained by the interface corresponds to the sharpest curvature that can be accommodated in the channel through which the interface is being withdrawn, and the sharpest curvature occurs at the narrowest part (e.g., interface 3 in Fig. 6-8). As drainage progresses, the interface retreats into channels which support a curvature of greater radius (e.g., interface 4). However, since this means reduced suction, this is a nonequilibrium stage and the water will continue to retreat until the interfaces have taken up a position of equilibrium in channels which are sufficiently narrow to support interfaces of sharper curvature. Obviously, if all channels are equal and large, at a certain suction no equilibrium can be maintained any more and a sudden, almost complete, withdrawal of the water from the soil will be observed. We say "almost" because some water will remain as isolated pendular rings. Within such rings which are completely isolated, the pressure is independent of the pressure in the remaining, continuous water body in the void space (except for dependence through the water vapor phase).

In general, the pores have different dimensions and, therefore, will not empty at the same suction. The large pores (or those with larger channels of entry) will empty at low suctions, while those with narrow channels of entry, supporting interfaces of sharper curvature, will empty at higher suctions.

Let us now reverse the process, and rather than increase the suction in order to empty more pores, reduce it in an attempt to refill the pore space. The transition now is from stage 5 to stages 6 and 7 in Fig. 6-8. The interface curvature becomes progressively less sharp.

Figure 6-9 shows typical examples of $p_c = p_c(S_w)$, or $h_c = h_c(S_w)$, curves during drainage. Here p_c has its macroscopic interpretation. Note the different abscissas and ordinates used in the various curves. In soil science, these curves are called *retention curves*, as they show how water is retained in the soil by capillary forces against gravity. Some authors refer to the drainage retention curve as a *desorption curve* and to the imbibition curve as a *sorption curve*. Point A in Fig. 6-9a is the *critical capillary head*, h_{cc}. If we start from a saturated sample, say, in the apparatus shown in Fig. 6-6a, and produce a small capillary head h_c, almost no water will leave the sample (i.e., no air will penetrate the sample) until the

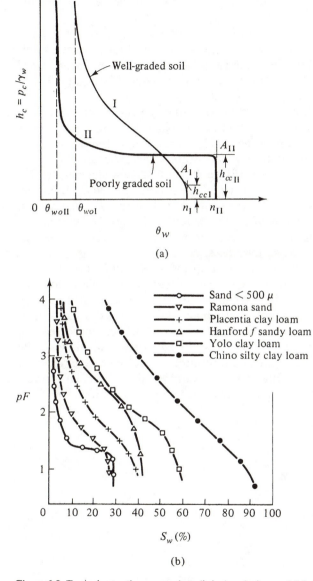

(a)

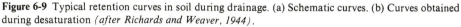

(b)

Figure 6-9 Typical retention curves in soil during drainage. (a) Schematic curves. (b) Curves obtained during desaturation *(after Richards and Weaver, 1944)*.

critical capillary head is reached. When expressed in terms of pressure, the critical value (point A of Fig. 6-9a) is called the bubbling pressure. As the value of h_c is increased, an initial small reduction in θ_w, associated with the retreat of the air–water meniscii into the pores at the external surface of the sample, is observed. Then, at the critical value h_{cc}, the larger pores begin to drain.

The shape of the retention curve, and hence also the threshold pressure, depends on the pore-size distribution and pore shapes of the porous medium (Fig. 6-9).

As drainage progresses, we observe that a certain quantity of water remains (in the form of isolated pendular rings and immobile thin films) in the sample even at very high capillary pressures. This value of θ_w, denoted by θ_{w0}, is called irreducible water content. In terms of saturation, it is denoted by S_{w0} ($= \theta_{w0}/n$) and called irreducible water saturation (Fig. 6-9a).

Upon rewetting (or imbibition), we observe that the retention curve $h_c = h_c(\theta_w)$ differs from that obtained during drainage. We have here the phenomenon of hysteresis resulting from two phenomena. The first, called *the ink-bottle effect* (Fig. 6-10a), results from the fact that as water reenters narrow channels, a local increase of suction is required. In the soil (Fig. 6-8) at this stage we have instability and the interface cannot advance until a neighboring pore is filled. Equilibrium at a given suction may be obtained with somewhat different θ_w. The second effect, sometimes called *the raindrop effect* (Fig. 6-10b), is due to the fact that the contact angle at an advancing interface differs from that at a receding one. En-trapped air is another factor causing hysteresis.

Figure 6-11 shows hysteresis in the relationship $h_c = h_c(\theta_w)$. The drainage and imbibition curves form a closed loop. In fine-grained soil, the effect indicated in Fig. 6-11, as caused by entrapped air, is caused also by subsidence or shrinkage. It is possible to start the imbibition process from any point on the drainage curve, or to start the drainage process from any point on the imbibition curve (leading to the dashed lines, called *drying* and *wetting scanning curves*). In this way, the relationship between capillary pressure and saturation (expressed by the retention curve) depends also on the wetting–drying history of the particular sample under consideration. For a given capillary pressure, a higher saturation is obtained

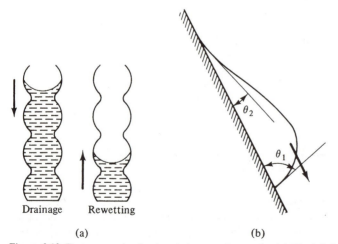

Drainage   Rewetting

(a)   (b)

Figure 6-10 Factors causing hysteresis in retention curve. (a) The ink-bottle effect. (b) The raindrop effect.

when a sample is being drained than during imbibition. As long as the soil remains stable (i.e., no consolidation), the hysteresis loop can be repeatedly traced.

For a given soil, the retention curve $p_c = p_c(S_w)$, or $h_c = h_c(\theta_w)$, can be obtained in the laboratory using an apparatus similar to that described in Fig. 6-6a. Obviously, the capillary pressure is an averaged one in the sense of the continuum approach.

At equilibrium, with no flow taking place, the piezometric head, ϕ, is the same for all points of an unsaturated zone. Consider points 1 and 2 with p_{c1}, ϕ_1, and p_{c2}, ϕ_2, respectively. We have

$$\phi_1 = z_1 + \frac{p_{w1}}{\gamma_w}; \qquad \phi_2 = z_2 + \frac{p_{w2}}{\gamma_w}; \qquad \gamma_w = \text{const.} \qquad (6\text{-}11)$$

From $\phi_1 = \phi_2$, and since the air phase is taken at atmospheric pressure, it follows that

$$z_1 - z_2 = (p_{w2} - p_{w1})/\gamma_w = (p_{c1} - p_{c2})/\gamma_w \qquad (6\text{-}12)$$

If $z_2 = 0$ is chosen as a point on the phreatic surface, $p_{c2} = 0$, and we denote $z_1 \equiv z, p_{c1} \equiv p_c$, we obtain

$$z = p_c(S_w)/\gamma_w \qquad \text{or} \qquad z = z(S_w) = h_c \qquad (6\text{-}13)$$

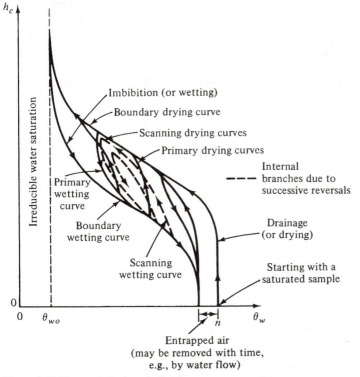

Figure 6-11 Hysteresis in $h_c = h_c(\theta_w)$ for a coarse material.

We have used the subscript w above to denote the water phase, as sometimes the simultaneous flow of the air and the water is being considered.

Equation (6-13) gives the moisture distribution above the phreatic surface. As z increases upward, so does p_c. This means that S_w decreases with height above the phreatic surface. Figure 4-5b shows two curves $S_w = S_w(z)$. Comparing these figures with the retention curves of Fig. 6-9a, it becomes obvious that these are actually the same curves. This supplements the discussion on the capillary fringe presented in Sec. 4-5.

From the discussion above, it follows that immediately above a water table ($p = 0$) we have a zone that is saturated with water, or nearly so, because a certain suction must be reached before any substantial reduction in water content can be produced. Then, above this zone, there is a marked drop in the water content with a relatively small rise in the capillary pressure. This zone contains most of the water present in the zone of aeration. From Figs 6-9a and b it is clear that this statement better describes the situation for poorly graded or coarse-textured soil (sand, gravel, etc.), but is also valid for fine-textured or well-graded soils when the water table is sufficiently deep below the ground surface. As this phenomenon is analogous to the rise in a capillary tube where the water rises to a certain height above the free water surface, with a fully saturated tube below the meniscus and zero saturation above it, the nearly saturated zone above the phreatic surface, when it occurs, is called the *capillary fringe*, or *capillary rise* (see Fig. 2-3 and Sec. 4-5). Thus, in Fig. 6-9a, h_{cc} is the capillary rise for the poorly graded soil.

The capillary fringe is thus an approximate practical concept that is very useful and greatly simplifies the treatment of phreatic flows when we wish to take into account the fact that a certain saturated (or nearly so) zone is present above a phreatic surface.

Further comments, with some estimates of the thickness of the capillary fringe, are given at the beginning of Sec. 4-5.

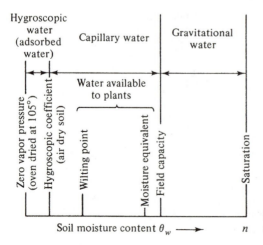

Figure 6-12 Soil moisture classes and equilibrium points.

Figure 6-12 gives some of the terms used for levels of water (or moisture) content in unsaturated flow studies. Obviously, the word "point," used to define such moisture levels as "wilting point," actually indicates a range rather than a well-defined point.

Field Capacity

Field capacity is usually defined as that value of water content remaining in a unit volume of soil after downward gravity drainage has ceased, or materially done so, say, after a period of rain, or excess irrigation. A difficulty inherent in this definition is that no quantitative specification of what is meant by "materially ceased" is given. Although, according to this definition, field capacity is a property of a unit volume of soil (depending on the soil structure, grain-size distribution, etc.), it is obvious from any of the curves describing the moisture distribution above the water table (e.g., Fig. 4-5) that the amount of water retained in a unit volume of soil at equilibrium under field conditions depends on the elevation of this unit volume above the water table. In addition, in the soil-water zone adjacent to the ground surface (Fig. 2-3), equilibrium is seldom reached as water in this zone constantly moves up or down and the water content is also being reduced by plant uptake. From these observations, it follows that the above definition of field capacity should be supplemented by the constraint that the soil sample should be at a point sufficiently high above the water table. Returning to the relationship $p_c = p_c(S_w)$, the notion of field capacity of unsaturated flow is identical to the notion of irreducible wetting fluid saturation in Figs 6-9a and 6-11. The field capacity θ_{w0} is shown in Fig. 6-13. The complement of the field capacity,

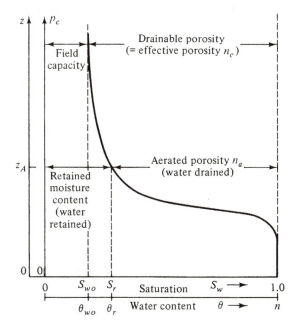

Figure 6-13 Field capacity and effective porosity.

i.e., volume of water drained by gravity from a unit volume of saturated soil, is called *effective porosity* and is denoted by $n_e (= n - \theta_{w0})$.

At any point, once gravity drainage has materially stopped, a certain amount of moisture is retained in the soil (per unit volume of soil). This amount can be used to define the *retained moisture content* $\theta_r(z, t)$ and the *aerated porosity* $n_a(z, t)$ (Fig. 6-13). Both θ_r and n_a vary with elevation above the water table, and when the water table is displaced, also with time. (For further details, see discussion by Zaslavsky in Bear, Zaslavsky, and Irmay, 1968.)

Specific Yield

Specific yield is another unsaturated flow concept employed in investigations of drainage of agricultural lands and in groundwater hydrology. It is defined as the volume of water per unit area of soil, drained from a soil column extending from the water table to the ground surface, per unit lowering of the water table. The corresponding amount of water retained in the soil against gravity when the water table is lowered is called specific retention. When expressed in terms of moisture content, we obtain for every instant

$$\theta_{wy} + \theta_{wr} = n \tag{6-14}$$

where θ_{wy} is specific yield and θ_{wr} is specific retention. By dividing (6-14) by porosity, n, we obtain the same relationship in terms of saturation

$$S_{wy} + S_{wr} = 1 \tag{6-15}$$

Note that in Fig. 5-4, in (5-12), and, in general, in the definition of specific yield as equivalent to the storativity of a phreatic aquifer, the specific yield (and also the specific retention) are in terms of moisture content. Thus (6-14) is actually identical to (5-12), with $S_y = \theta_{wy} (= nS_{wy})$.

Thus, specific retention is a field concept, obtained by averaging what actually happens in a natural soil in the zone of aeration when the water table is lowered. In Sec. 5-1, S_y is defined as the storativity, or specific yield, of a phreatic aquifer. Figure 6-14 shows the effect of depth and time on the specific yield. With the nomenclature of this figure, we have per unit area

$$S_y(d', d'') = \cfrac{\cfrac{\text{Volume of water drained}}{(d'' - d')}}{} =$$

$$= \frac{1}{(d'' - d')} \left[n(d'' - d') + \int_{z'=0}^{z'=d'} \theta_w'(z', t) \, dz' - \int_{z''=0}^{z''=d''} \theta_w''(z'', t) \, dz'' \right] \tag{6-16}$$

where $S_y \equiv \theta_{wy} = nS_{wy}$. The volume of water drained is indicated by the shaded area in Fig. 6-14. For a homogeneous isotropic soil, the two curves $\theta_w' = \theta_w'(z', t')$ and $\theta_w'' = \theta_w''(z'', t'')$ are identical in shape. If, in addition, both water table positions are sufficiently deep below the ground surface, the two curves will merge at $\theta = \theta_{w0}$. Hence, we have for very large d' and d''

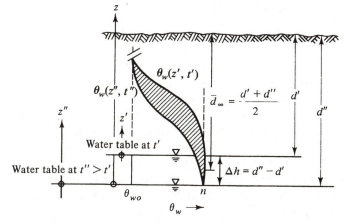

(a) Deep water table

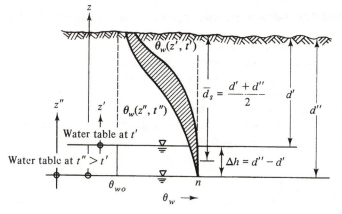

(b) Shallow water table

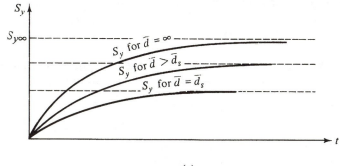

(c)

Figure 6-14 Effect of depth and time on specific yield.

$\theta = \theta_{w0}$. Hence, we have for very large d' and d''

$$S_{y\infty} \equiv S_y\Big|_{\bar{d} \to \,\prime} = n - \theta_{w0} = n(1 - S_{w0}) \qquad (6\text{-}17)$$

It is thus apparent that for a homogeneous isotropic soil and very deep water table, the specific retention is identical to the field capacity. For such conditions, Fig. 5-4 shows the relationship between specific yield and specific retention for various soils. Note again that the phreatic aquifer storativity $S \equiv S_y$ (Sec. 5-1) is equal to $S_{y\infty} (= nS_{wy\infty})$. However, when the soil is inhomogeneous (e.g., composed of layers), or when the water table is at a shallow depth, the moisture distribution curves, corresponding to the two water table positions, are no longer parallel, and the identities presented above are no longer valid; we must distinguish between field capacity and specific retention.

When the time lag is also taken into account, as it takes time for drainage to be completed, we obtain a specific yield that is time dependent and that approaches asymptotically the values corresponding to the depths considered (Figs 5-5 and 6-14c). The term *drainable porosity* is sometimes used to denote the instantaneous specific yield.

When the water table is lowered instantaneously (or relatively fast), say by a vertical distance Δh, as a result of drainage, the corresponding changes in the moisture distribution lag behind and reach a new equilibrium (or practically so) only after a certain time interval that depends on the type of soil. A time lag will also take place when infiltration causes the water table to rise. When the water table is rising or falling slowly, the changes in moisture distribution have sufficient time to adjust continuously and the time lag practically vanishes.

Conditions Between Two Media

Finally, at an interface between two media (say, a coarse sand and a fine one), we require that the pressure, p_w, in the water (actually also that in the air, p_a) be the same as the surface of separation is approached from both sides. Denoting the two media by subscripts 1 and 2, this means that since we assume $p_{a1} = p_{a2} = 0$

$$p_{w1} = p_{w2}, \qquad \text{or } p_{c1} = p_{c2} \qquad (6\text{-}18)$$

Figure 6-15 shows, schematically, the two retention curves. It is obvious that we have a jump in the water saturation across the surface separating the two media; $S_{w1} < S_{w2}$.

6-2 THE MOTION EQUATION

In principle, both the water and the air (= two immiscible fluids) move simultaneously in the void space. Movement of water vapor may also take place. Variations in temperature and in the concentration of dissolved solids will

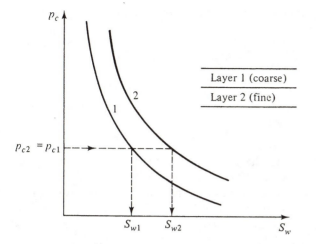

Figure 6-15 Water saturation discontinuity between layers of different retention curves.

affect the vapor pressure of the water, and phase changes and transport of water vapor by diffusion will take place.

In what follows, we shall ignore the movement of water in the vapor phase. We shall also ignore movement due to pressure differences resulting from variations in salt concentration (*osmotic effect*) and movement due to temperature variations (*thermo-osmotic effect*). We shall assume that the solid matrix is rigid and stable (i.e., no consolidation or subsidence). Some of the above phenomena, however, may be important in certain situations and should not be ignored then.

Thus, we shall consider flow resulting only from variations in piezometric head or in pressure in the water; the density, ρ_w, may vary.

Many investigators conclude from experiments that when two immiscible fluids flow simultaneously through a porous medium, each fluid establishes its own tortuous paths. They assume that a unique (or nearly so) set of channels corresponds to every degree of saturation. From the discussion in Sec. 6-1, it follows that as the degree of saturation of a nonwetting fluid is reduced, the channels of that fluid tend to break down until only isolated regions of it remain at residual nonwetting fluid saturation. Similarly, when the saturation of the wetting fluid is reduced, it becomes discontinuous at the irreducible wetting fluid saturation. When any of these fluids becomes discontinuous throughout the flow domain, no flow of that fluid can take place. The same ideas can be extended also to three phases or more (e.g., oil, water, and gas).

With these ideas in mind, it seems natural to apply the concept of permeability, established for the saturated flow of a single fluid, modifying its value owing to the presence of the second phase which occupies part of the void space. Accordingly, we may now apply (4-9), or (4-14), for an isotropic medium, or (4-37) for an anisotropic one, to describe, separately, the flow of each of the two phases—the wetting one (here water) and the nonwetting one (here air). The difference, however, is that in this case, the permeability for each of the phases is a function of the degree of saturation.

Accordingly, the motion equations for the air (nonwetting phase) and the water (wetting phase) in unsaturated flow in an anisotropic porous medium are

$$q_{iw} = -\frac{k_{ijw}(S_w)}{\mu_w}\left[\frac{\partial p_w}{\partial x_j} + \rho_w g \frac{\partial z}{\partial x_j}\right] = -k_{ij}\frac{k_{rw}(S_w)}{\mu_w}\left[\frac{\partial p_w}{\partial x_j} + \rho_w g \frac{\partial z}{\partial x_j}\right],$$

$$q_{inw} = -\frac{k_{ijnw}(S_{nw})}{\mu_{nw}}\left[\frac{\partial p_{nw}}{\partial x_j} + \rho_{nw} g \frac{\partial z}{\partial x_j}\right] = -k_{ij}\frac{k_{rnw}(S_{nw})}{\mu_{nw}}\left[\frac{\partial p_{nw}}{\partial x_j} + \rho_{nw} g \frac{\partial z}{\partial x_j}\right]$$

$$i, j = 1, 2, 3 \qquad (6\text{-}19)$$

where subscripts *nw* and *w* denote the nonwetting and the wetting fluids, respectively, and the summation convention is employed. In (6-19), $k_{ijw}(S_w)$ and $k_{ijnw}(S_{nw})$ are the unsaturated permeabilities of the two fluids (= effective permeabilities), respectively, while k_{ij} is the permeability at saturation; $q_{iw}(=\theta_w V_{iw})$ and $q_{inw}(=\theta_{nw}V_{inw})$ are the *i*th components of the specific discharge vectors of the two fluids, respectively, and $k_{rw}(=k_{ijw}/k_{ij})$ and $k_{rnw}(=k_{ijnw}/k_{ij})$ are the relative permeabilities to the two fluids, respectively. The densities ρ_w and ρ_{nw} are not necessarily constant. Although this is questionable, it is assumed that the relative permeability is the same function of saturation for all permeability components of an anisotropic porous medium. The effective permeability and the relative permeability are discussed in detail in the following section. Finally, p_w and p_{nw} are the pressures in the two fluids, respectively, each being an average value over the volume occupied by that fluid within the REV around the considered point in the flow domain.

For constant densities of the two fluids, ρ_w and ρ_{nw}, we may rewrite (6-19) in terms of the gradient of the piezometric head

$$\left.\begin{aligned}\mathbf{q}_w &= -\frac{k_w(S_w)}{\mu_w}\cdot\nabla(p_w + \gamma_w z) = -\frac{k_w(S_w)\gamma_w}{\mu_w}\cdot\nabla\phi_w; \qquad \phi_w = z + \frac{p_w}{\gamma_w}, \\[2ex] \mathbf{q}_{nw} &= -\frac{k_{nw}(S_{nw})}{\mu_{nw}}\cdot\nabla(p_{nw}+\gamma_{nw}z) = -\frac{k_{nw}(S_{nw})\gamma_{nw}}{\mu_{nw}}\cdot\nabla\phi_{nw}; \quad \phi_{nw} = z + \frac{p_{nw}}{\gamma_{nw}}\end{aligned}\right\} \quad (6\text{-}20)$$

In (6-20) we have used the more compact vector notation instead of the indicial notation employed in (6-19).

If we assume that the air is stationary, or we ignore its movement, that is, $\phi_{nw} \equiv \phi_{air} = \text{const.}$, we have for the water flow

$$\mathbf{q}_w = -\frac{k_w(S_w)}{\mu_w}\cdot\nabla(p_w + \gamma_w z) = -\frac{k_w(S_w)}{\mu_w}\cdot\nabla(-p_c + \gamma_w z) \qquad (6\text{-}21)$$

where $p_c = -p_w$ and $k_w(S_w)$ is the effective permeability to the water. We may also use the concept of effective hydraulic conductivity $K_w(S_w) = k_w(S_w)\gamma_w/\mu_w$. Then, for $\rho_w = \gamma_w/g = \text{const.}$

$$\mathbf{q}_w = -K_w(S_w)\cdot\nabla\phi_w = -\frac{K_w(S_w)}{\gamma_w}\cdot\Delta p_w - K(S_w)\cdot\mathbf{1z} \qquad (6\text{-}22)$$

Another form of the motion equation for water of constant density and $p_{air} = 0$ is

$$\mathbf{q}_w = K_w(S_w)\cdot\nabla\psi - K_w(S_w)\cdot\mathbf{1z}; \qquad \gamma_w = \text{const.} \qquad (6\text{-}23)$$

where $\psi = -p_w/\gamma_w > 0$ is the capillary pressure head. Obviously, we could replace S_w by θ_w in all the above equations. Also $K_w(S_w)$ can be replaced by $K_w(\psi)$.

For horizontal flows, the term $K_w(S_w) \cdot \mathbf{1z}$ vanishes in (6-22) and (6-23).

It is important to recall that the relationship $p_c = p_c(\theta_w)$ is not a unique one because of hysteresis; hence, the history of wetting and drying may play an important role in the analysis of flow problems.

Nevertheless, assuming that the relationship $p_c = p_c(\theta_w)$ is a unique one (as when dealing with a problem of only drainage), (6-23) written for an isotropic medium and $\gamma_w = $ const., becomes

$$\mathbf{q}_w = K_w(\theta_w)\frac{d\psi}{d\theta_w}\nabla\theta_w - K_w(\theta_w)\,\mathbf{1z} = \left[\frac{K_w(\theta_w)}{d\theta_w/d\psi}\right]\nabla\theta_w - K_w(\theta_w)\,\mathbf{1z} \quad (6\text{-}24)$$

where we have replaced S_w by θ_w and $\psi = p_c/\gamma_w = \psi(\theta_w)$ is also assumed unique. Klute (1952) calls the group $D(\theta_w) = -K_w(\theta_w)\,d\psi/d\theta_w = -K_w(\theta_w)/[d\theta_w/d\psi] = (K_w(\theta_w)/\gamma_w)[d\theta_w/dp_w]$ *coefficient of diffusivity* or *capillary diffusivity* (dims L^2T^{-1}). Then, (6-24) becomes

$$\mathbf{q}_w = -D(\theta_w)\nabla\theta_w - K(\theta_w)\,\mathbf{1z} \quad (6\text{-}25)$$

For horizontal, two-dimensional flow in the xy plane

$$\mathbf{q}'_w = -D(\theta_w)\nabla'\theta_w; \qquad \nabla'(\ ) = [\partial(\ )/\partial x]\,\mathbf{1x} + [\partial(\ )/\partial y]\,\mathbf{1y} \quad (6\text{-}26)$$

For vertical flow

$$q_{wz} = -D(\theta_w)\,\partial\theta_w/\partial z - K(\theta_w) \quad (6\text{-}27)$$

Sometimes the definitions $C = d\theta_w/dp_w$ ($= water\ capacity$) and $D = -K_w/(d\theta_w/dp_w)$ are introduced. The similarity between (6-26) and Fick's law of diffusion explains why the term diffusivity is used here.

The second term on the right-hand side of (6-25) gives the effect of gravity. Sometimes this term is neglected, reducing (6-21) to $\mathbf{q}_w = -D(\theta_w)\nabla\theta_w$.

The dependence of D on θ, or of K on ψ, introduces a *nonlinearity* into the equations of motion presented above, and hence also into the continuity equation (Sec. 6-4). The gravity term in the motion equation also makes the continuity equation a rather difficult one for exact solution by analytical methods. Without the assumptions of uniqueness, $K(\theta), p_c(\theta)$ and $D(\theta)$ are subject to hysteresis.

One should note that although the relationship $p_c = p_c(\theta_w)$ is usually obtained from a static test (i.e., in the absence of flow), we use it in the motion equation. We assume that the relationship remains unchanged also under dynamic conditions. Another interesting observation is that in the motion equations presented above, say (6-19), we have assumed no interaction (e.g., momentum transfer) between the two phases across their common interfaces inside the pore space.

6-3 RELATIVE PERMEABILITY

In this section, we shall assume that the soil is isotropic. The discussion can easily be extended to anisotropic soils.

In (6-19) through (6-21), $k_w(\theta_w)$ is the *effective permeability* of the soil in unsaturated flow of water; $k_{rw}(\theta_w)$ in (6-23) is the *relative permeability*, where "relative" is to that at full saturation. Obviously, we have also $k_a(\theta_a)$ and $k_{ra}(\theta_a)$ for the air phase, to be used when the air motion is taken into account.

Figure 6-16 shows the variations of relative permeability of the water, k_{rw}, with saturation, S_w, according to experiments by Wyckoff and Botset (1936). As the saturation decreases, the large pores drain first so that the flow takes place through the smaller ones. This causes both a reduction in the cross-sectional area available for the flow, and an increase in tortuosity of the flow paths. The combined effect causes a rather rapid reduction in the permeability as the moisture content decreases. As the water films become thinner, certain phenomena at the solid–water interface come into play, causing a further reduction in permeability. Among such factors, we may mention an increase of viscosity of the water in close proximity to the solid surfaces. Point A in Fig. 6-16 indicates the irreducible water (= wetting fluid) saturation, S_{w0}. The corresponding irreducible water content will be denoted by θ_{w0} ($= nS_{w0}$). At point A, the water phase becomes discontinuous, existing only in the very small pores, as very thin films on the solid, or as isolated rings.

Several authors suggest relationships between the permeability, k_w, conductivity, K_w, and saturation, S_w (or water content θ_w). Childs and Collis-George (1950) assume

$$K_w(\theta_w) = B\theta_w^3/M^2 \qquad (6\text{-}28)$$

where M is the specific surface area of the soil phase and B is a constant.

Irmay (1954) derives a similar relationship assuming that the resistance to flow offered by the solid matrix is proportional to the solid–liquid interfacial area. The hydraulic conductivity K_w then becomes proportional to the hydraulic radius R (= volume of voids, U_v, divided by wetted area of solid, A_s)

$$R = U_v/A_s = nU_b/M(1-n)\,U_b = n/M(1-n), \qquad A_s = MU_s$$

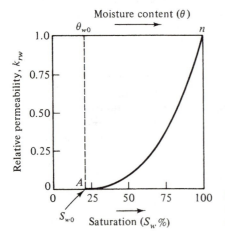

Moisture content (θ)

Figure 6-16 Relative permeability of unsaturated sand according to experiments by Wyckoff and Botset (1936) and theoretical analysis by Irmay (1954).

where subscripts v, b, and s denote voids, bulk, and solids, respectively. For a cubic arrangement of spheres of diameter d in saturated flow, the specific surface is $M = 6/d$. This model leads to the cubic parabola

$$K_w(\theta_w) = K_0 \left(\frac{\theta_w - \theta_{w0}}{n - \theta_{w0}} \right)^3 = K_0 \left(\frac{S_w - S_{w0}}{1 - S_{w0}} \right)^3 \tag{6-29}$$

where K_0 is the hydraulic conductivity at saturation. The experimental curve of Fig. 6-16 fits such a cubic parabola. Experiments by several authors with soils of uniform grain size seem to agree with the relationship in (6-29).

Corey (1957) finds that for many consolidated rocks, a relative permeability proportional to S_w^4 describes unsaturated flow. Irmay's model, like most other models, fails to consider an additional resistance that might be introduced at the (microscopic) air–water interfaces present in the pores. Bear (1972, p. 463) reviews several models which lead to expressions for $k_{rw}(S_w)$ and $k_{rnw}(S_{nw})$.

In general, we have to consider also the permeability to the nonwetting phase. Figure 6-17a shows typical $k_{rw}(S_w)$ and $k_{rnw}(S_{nw})$ curves; the dashed portions of the curves correspond to the case when we start from complete saturation of the considered phase. Otherwise, we cannot obtain saturations $S_w > 100 - S_{nw0}$ and $S_{nw0} > 100 - S_{w0}$. Thus, k_{rw} under field conditions is at most k_{rw}/S_{nw0}.

We have seen above that, due to hysteresis, we may have different flow channels at the same saturation during wetting and drying. This leads to a certain amount of hysteresis also in the relationship $k_{rw}(S_w)$ and $k_{rnw}(S_{nw})$. Figure 6-17b shows a typical example of this phenomenon.

Hydraulic conductivity may also be presented as a function of p_c or of the capillary pressure head ψ. However, the relationship $K_w(\psi)$ shows much hysteresis, probably because of the large hysteresis in the function $\psi(\theta_w)$. The effect of hysteresis is also shown in Fig. 6-18. The figure also shows how, upon rewetting to

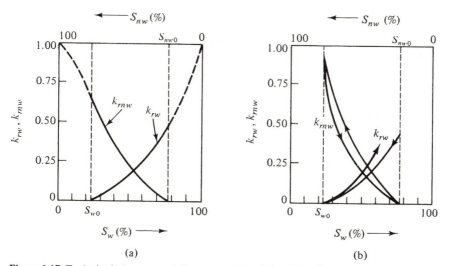

Figure 6-17 Typical relative permeability curves (a) and the effect of hysteresis (b).

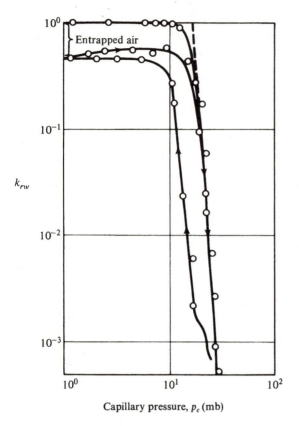

Figure 6-18 Relative permeability as a function of capillary pressure for Loveland fine sand *(Klute, 1967 as adapted from Brooks and Corey, 1964)*.

zero pressure head, the relative hydraulic conductivity is less than that obtained at full saturation owing to the air entrapped in the pores. Childs (1967, p. 206) discusses Darcy's law with hysteresis.

Among other expressions for permeability, or hydraulic conductivity, of unsaturated flow suggested by various authors, we may mention

Gardner (1958): $K = a/(b + \psi^m)$, $\qquad \psi = p_c/\gamma_w = -p_w/\gamma_w$ (6-30)

where a, b, and m are constants, with $m \approx 2$ for heavy clay soil and $m \approx 4$ for sand; K_0 is the hydraulic conductivity at saturation

Gardner (1958): $\qquad\qquad K = K_0 \exp(-a\psi)$ (6-31)

which does not fit experimental data so well, but is sometimes more convenient for analytical purposes

Brooks and Corey (1966): $\quad k = k_0 \qquad\qquad\quad \text{for } p_c \leq p_b$ ⎫

$\qquad\qquad\qquad\qquad\quad k = k_0(p_b/p_c)^n \quad \text{for } p_c \geq p_b$ ⎬ (6-32)

where k is the permeability (in cm^2) at saturation, p_b is the bubbling pressure (in dynes/cm^2) and the exponent n is an index of the pore-size distribution of the porous medium. Additional information on (6-32) and experimental verifications are given by Laliberte, Corey, and Brooks (1966). The various coefficients appearing in the above expressions are obtained by comparison with experimental data.

Bear (1972, p. 493) discusses the determination of unsaturated ·hydraulic conductivity in the laboratory, using specially designed permeameters.

6-4 THE CONTINUITY EQUATIONS

Assuming no sources or sinks of moisture within the unsaturated flow domain (e.g., due to uptake by roots), we may start from the mass conservation equation for the water phase

$$\partial(\rho_w n S_w)/\partial t + \text{div}(\rho_w \mathbf{q}) = 0; \quad \text{or} \quad \partial(\rho_w \theta_w)/\partial t + \text{div}(\rho_w \mathbf{q}) = 0 \quad (6\text{-}33)$$

where $\mathbf{q}$ is the water's specific discharge and we consider only the flow of water. If water is assumed incompressible in the range of pressures considered in unsaturated flow, and assuming also no changes in ρ_w due to dissolved solids, (6-33) becomes (see Eq. (A-21))

$$\partial(n S_w)/\partial t + \text{div}\,\mathbf{q} = 0; \quad \partial\theta_w/\partial t + \text{div}\,\mathbf{q} = 0 \quad (6\text{-}34)$$

For a nondeformable medium, $n = \text{const.}$, and the first equation of (6-34) reduces to

$$n\partial S_w/\partial t + \text{div}\,\mathbf{q} = 0 \quad (6\text{-}35)$$

In a deformable, or consolidating, medium, we must take into account variations in n, and also the fact that $\mathbf{q}$, as defined by the equations of motion (Sec. 6-2), is with respect to the moving grains. By introducing these two factors into the continuity equation, the partial differential equations describing flow and consolidation may be derived. In this section we neglect solid velocity.

By combining any of the continuity equations (6-33) through (6-35) with any of the equations of motion in Sec. 6-2, we obtain various forms of the continuity equation, each corresponding to a specific case, as defined by a set of assumptions. For example, from the first equation of (6-33) and (6-19) we obtain:

$$\frac{\partial}{\partial t}(\rho_w n S_w) - \frac{\partial}{\partial x_i}\left[\rho_w k_{ij}\frac{k_{rw}(S_w)}{\mu_w}\left(\frac{\partial p_w}{\partial x_j} + \rho_w g\frac{\partial z}{\partial x_j}\right)\right] = 0 \quad (6\text{-}36)$$

For an isotropic porous medium and a homogeneous fluid, $\rho_w = \text{const.}$, we obtain for $p_{\text{air}} = 0$

$$\partial\theta_w/\partial t - \text{div}[K(\theta_w)\,\text{grad}\,\phi] = 0; \quad \phi = z - \psi, \quad \theta_w = n S_w,$$

or

$$\partial\theta_w/\partial t + \text{grad}\,K(\theta_w)\cdot\text{grad}\,\psi + K(\theta_w)\nabla^2\psi - \partial K(\theta_w)/\partial z = 0 \quad \left.\right\} \quad (6\text{-}37)$$

Since $\theta_w = \theta_w(\psi)$, $K = K(\psi)$, $\partial\theta_w/\partial t = (\partial\psi/\partial t)(d\theta_w/d\psi) = -C(\theta_w)\,\partial\psi/\partial t$, and $\text{grad}\,K(\theta_w) = [dK(\theta_w)/d\psi]\,\text{grad}\,\psi$, and we assume that $\theta_w = \theta_w(\psi)$, $K = K(\theta_w)$

and $K = K(\psi)$ are known functions, we obtain

$$\frac{\partial \psi}{\partial t} \frac{d\theta_w}{d\psi} + \text{div}[K(\theta_w) \, \text{grad} \, \psi] - \frac{\partial K(\theta_w)}{\partial z} = 0 \qquad (6\text{-}38)$$

or

$$\partial \psi / \partial t = A(\psi)(\nabla \psi)^2 + D(\psi) \nabla^2 \psi - A(\psi) \partial \psi / \partial z \qquad (6\text{-}39)$$

where

$$D(\psi) = -\frac{K(\theta_w)}{d\theta_w/d\psi} = \frac{K(\theta_w)}{C(\theta_w)}; \qquad A(\psi) = -\frac{dK(\theta_w)/d\psi}{d\theta_w/d\psi}$$

$$\frac{A(\psi)}{D(\psi)} = \frac{dK(\theta_w)/d\psi}{K(\theta_w)} = \gamma_w \frac{dK(\theta_w)/dp_c}{K(\theta_w)} = \gamma_w d[\ln K(\theta_w)]/dp_c$$

$$\left. \right\} \qquad (6\text{-}40)$$

In terms of $p_c \, (= \gamma_w \psi)$, or $p \, (\equiv p_w = -p_c)$, and ψ, we obtain

(a) $$\partial p_c / \partial t = \frac{1}{\gamma_w} A(\psi)(\nabla p_c)^2 + D(\psi) \nabla^2 p_c - A(\psi) \partial p_c / \partial z$$

(b) $$\partial p_c / \partial t = -\frac{1}{\gamma_w} A(\psi) \nabla p_c \cdot \nabla(\gamma_w z - p_c) - D(\psi) \nabla^2 (\gamma_w z - p_c) \qquad (6\text{-}41)$$

(c) $$\partial p / \partial t = \frac{1}{\gamma_w} A(\psi)(\nabla p)^2 - D(\psi) \nabla^2 p + A(\psi) \partial p / \partial z$$

$$\left. \right\}$$

For one-dimensional flow in the vertical, z, direction, (6-38) reduces to

$$\frac{\partial \psi}{\partial t} \frac{d\theta_w}{d\psi} + \frac{\partial}{\partial z} \left[K(\theta_w) \frac{\partial \psi}{\partial z} - K(\theta_w) \right] = 0 \qquad (6\text{-}42)$$

known as the *Fokker–Planck equation*.

In the equations above, $D(\psi)$, or $D(\theta_w)$, is called *moisture diffusivity*, while $C(\theta_w) = -\gamma_w(d\theta_w/dp_c) \equiv -d\theta_w/d\psi$ has the meaning of water capacity, or *specific water capacity*, or *specific water storativity* of an unsaturated medium, in analogy to the thermal capacity appearing in the heat conduction equation. Because of hysteresis, the water capacity $C(\theta_w)$ is a single-valued function only if everywhere in the flow domain we have either drainage or imbibition. Otherwise, it depends upon the particular scanning curve (Fig. 6-11) that describes the process that the porous medium is undergoing. Describing the sequence of wetting and drying by $\theta_{w1}, \theta_{w2}, \theta_{w3}, \ldots$, we have

$$p_w = p_w(\theta, \theta_{w1}, \theta_{w2}, \theta_{w3}, \ldots)$$

so that for a process of m steps

$$dp_w = \frac{\partial p}{\partial \theta_w} d\theta_w + \sum_{\alpha=1}^{m} \frac{\partial p}{\partial \theta_{w\alpha}} d\theta_{w\alpha} \qquad (6\text{-}43)$$

One should note that in order to take into account the history of drying and wetting, (6-43) should be used to define all pressure gradients in the motion and continuity equations of unsaturated flow (Childs, 1967). When solving a problem by numerical methods using digital computers, the many-valuedness of $\theta_w = \theta_w(p_c)$ can be taken into account by first checking the type of process, and then picking up the value of water capacity from the appropriate curve.

An important advantage of using the continuity equation in the form (6-41c), in which the pressure $p = p(x, y, z, t)$ is the dependent variable, is that it is also valid for the saturated region below the water table where $K(\theta_w) = K_0$ and $\theta_w = n$. This enables one to treat a flow domain that is partly saturated (in the zone of saturation) and partly unsaturated (in the zone of aeration) as one continuous system having a single dependent variable. One should note that although it is assumed here that the water is an incompressible fluid, we may have $\partial p/\partial t \neq 0$ because of the initial and boundary conditions of the problem.

Another form of the continuity equation is based on (6-25) written for θ_w as the dependent variable:

$$\partial\theta_w/\partial t = \text{div}\left[D(\theta_w)\,\text{grad}\,\theta_w\right] + \partial K(\theta_w)/\partial z \qquad (6\text{-}44)$$

$$\partial\theta_w/\partial t = \left[dD(\theta_w)/d\theta_w\right](\nabla\theta_w)^2 + D(\theta_w)\nabla^2\theta_w + \left[dK(\theta_w)/d\theta_w\right]\partial\theta_w/\partial z \quad (6\text{-}45)$$

It is always assumed that $K(\theta_w)$, $C(\theta_w)$, $D(\theta_w)$, etc., appearing in the various forms of the continuity equation, are known functions of the water content. Even in homogeneous soils these functions vary also with position and time: $K = K[\theta_w(x, y, z, t)]$, $C = C[\theta_w(x, y, z, t)]$, etc.

All the partial differential equations presented above, as well as those presented in the remaining part of this section, which are various forms of the continuity equation for unsaturated flow, are nonlinear equations, as conductivity and diffusivity are functions of the dependent variable (θ_w, ψ, etc.). This fact dictates the methods of solution applicable to these equations. Sometimes the equations may be linearized by applying a certain transformation.

Whenever the gravity term, i.e., the last term on the right-hand side of the continuity equations given above, can be neglected, as in horizontal, or approximately horizontal, flow, the continuity equation has the mathematical form of the nonlinear heat conduction equation $\partial\theta/\partial t = \text{div}\left[D(\theta_w)\,\text{grad}\,\theta_w\right]$. Under such conditions, there is some advantage in using the form (6-44) without the gravity term, because of some simple expressions that have been suggested for $D(\theta_w)$. For example, Gardner and Mayhugh (1958) suggest for the intermediate range of water contents the exponential expression

$$D = D_0\exp\left[\beta(\theta_w - \theta_{w0})\right] \qquad (6\text{-}46)$$

where D_0 is the diffusivity corresponding to $\theta = \theta_{w0}$ and β and θ_{w0} are empirical parameters. Figure 6-19 shows data of $D(\theta_w)$ for three soils that seem to justify (6-46).

When the gravity term cannot be neglected, there is no advantage in using the form (6-44). On the other hand, there is some advantage in using the forms including $K(\theta_w)$, expecially where hysteresis is to be taken into account, as in general $K(\theta_w)$ is practically hysteresis free.

When the water density varies mainly because of dissolved solids (with practically no effect of compressibility) ρ_w cannot be eliminated from (6-33) and

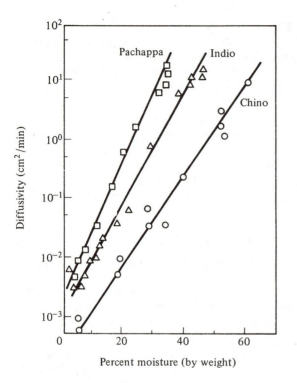

Figure 6-19 Diffusivity-water content data for three soils *(from Gardner, 1958)*.

hence from all the other continuity equations derived from it. In this case, combining (6-33) with (6-19) yields (6-36) for the water

$$\frac{\partial}{\partial t}(\rho_w n S_w) = \nabla \cdot \left[\frac{\rho_w k_w(S_w)}{\mu_w} \cdot (\nabla p_w + \rho_w g \mathbf{1z}) \right] \tag{6-47}$$

The left-hand side of (6-47) can be developed as follows. Using (5-22) and the condition of $d\sigma = 0$ leading to it, for compressible fluid and deformable solid matrix, we obtain

$$\frac{\partial(\rho_w n S_w)}{\partial t} = S_w \frac{\partial(\rho_w n)}{\partial t} + \rho_w n \frac{\partial S_w}{\partial t} = S_w \rho_w [\alpha(1-n) + \beta n] \frac{\partial p_w}{\partial t}$$

$$+ \rho_w n \frac{\partial S_w}{\partial p_w} \frac{\partial p_w}{\partial t}$$

$$= \rho_w n \left[S_w \frac{\alpha(1-n)}{n} + S_w \beta + \frac{\partial S_w}{\partial p_w} \right] \frac{\partial p_w}{\partial t} = \left(S_w S_{0p} + n \rho_w \frac{\partial S_w}{\partial p_w} \right) \frac{\partial p_w}{\partial t}$$

$$= n S_w \left(S_{0p}/n + \rho_w \frac{\partial \ln S_w}{\partial p_w} \right) \frac{\partial p_w}{\partial t} \tag{6-48}$$

where $S_{0p} = \rho_w [\alpha(1-n) + \beta n]$ is the specific storativity with respect to pressure and $\partial S_w/\partial p_w$ is a known function of S_w (or p_w).

Thus, in the square brackets of (6-48) we have the effects of solid matrix compressibility, of water compressibility and of the relationship $S_w(p_w)$. In the unsaturated zone, the effect of the first two is negligible with respect to the third (recalling that $d\sigma = 0$, that is, no changes in external loading of the soil which may cause subsidence). In the saturated zone, the third term vanishes.

Simultaneous Saturated–Unsaturated Flow

Sometimes we wish to treat the flow in the unsaturated zone together with that in the saturated one. One example is the movement of dissolved pollutants from the ground surface into and then through the aquifer, the pollutants being carried by the water. Equation (6-47), expressing mass conservation for $\rho_w \neq \text{const.}$ is obviously valid for both zones. The differences are

Saturated zone	Unsaturated zone

$$p_w > 0; \qquad S_w = 1,$$
$$k_w(S_w) = k_0 = \text{const.},$$

$$p_w < 0, \ S_w = S_w(p_w); \qquad S_{w0} \leq S_w < 1,$$

$$k_w \begin{cases} = k(S_w) & \text{for} \quad S_w > S_{w0}, \\ = 0 & \text{for} \quad 0 \leq S_w \leq S_{w0} \end{cases}$$

With (6-48), we may now write (6-47) in the form

$$\left(S_w S_{0p} + n\rho_w \frac{\partial S_w}{\partial p_w} \right) \frac{\partial p_w}{\partial t} = \nabla \cdot \left[\frac{\rho_w k(S_w)}{\mu_w} \cdot (\nabla p_w + \rho_w g \mathbf{1} z) \right] \qquad (6\text{-}49)$$

Assuming that $\mathbf{q} \cdot \nabla \rho_w \ll nS_w \partial \rho_w / \partial t$, so that (6-33) becomes:

$$\partial(\rho_w n S_w)/\partial t + \rho_w \nabla \cdot \mathbf{q} = 0 \qquad (6\text{-}50)$$

we obtain from (6-49)

$$\left\{ S_w[\alpha(1 - n) + \beta n] + n \frac{\partial S_w}{\partial p_w} \right\} \frac{\partial p_w}{\partial t} = \nabla \cdot \left[\frac{k(S_w)}{\mu_w} \cdot (\nabla p_w + \rho_w g \mathbf{1} z) \right] \qquad (6\text{-}51)$$

This approximation is valid for a homogeneous compressible fluid. We may then use (6-20) to express the flux in terms of ϕ or ψ, and to replace $(\partial S_w/\partial p_w) \partial p_w/\partial t$ by $(\partial S_w/\partial \psi)\partial \psi/\partial t$. If solid velocity is not neglected, $\alpha(1 - n)$ is replaced by α in (6-51).

When distributed sinks of magnitude $Q(x, y, z, t)$ exist within the flow domain, e.g., uptake of moisture by roots (in this case, Q itself may be a function of p_w, e.g., $Q = \alpha(p_w - \beta)$; α, β = coefficients), we have to add the term $+Q(x, y, z, t)$ on the left-hand side of the basic continuity equation (6-33) and therefore also of (6-36), (6-37), (6-44), (6-47), (6-49), and (6-51). For $Q(x, y, z, t)$ in the form of point sinks of strength Q_{wi} at point (x_i, y_i, z_i), we may write

$$\sum_{(i)} Q_{wi}(x_i, y_i, z_i, t) \, \delta(x - x_i, y - y_i, z - z_i)$$

where δ is the *Dirac delta function*.

Simultaneous Air and Water Flow

When solving for the flow of water only, neglecting any flow of air, we assume that the entire stationary air phase is at atmospheric pressure, taken as $p_{air} = 0$. However, it is possible (e.g., Noblanc and Morel-Seytoux, 1972; Morel-Seytoux, 1973) to consider the simultaneous flow of both the water and the air in the unsaturated zone. In this case, we have to solve the continuity equations for air (subscript a) and for water (subscript w) simultaneously

$$\left.\begin{array}{ll} \partial(\rho_w n S_w)/\partial t + \text{div}(\rho_w \mathbf{q}_w) = 0 & \text{or} \quad \partial(\rho_w \theta_w)/\partial t + \text{div}(\rho_w \mathbf{q}_w) = 0 \\ \partial(\rho_a n S_a)/\partial t + \text{div}(\rho_a \mathbf{q}_a) = 0 & \partial(\rho_a \theta_a)/\partial t + \text{div}(\rho_a \mathbf{q}_a) = 0 \\ \text{Also} \\ \qquad\qquad S_a + S_w = 1; & \theta_a + \theta_w = n \end{array}\right\} \quad (6\text{-}52)$$

If we assume: $n = $ const., $\rho_w = $ const., we obtain from (6-52)

$$n\partial S_w/\partial t + \text{div } \mathbf{q}_w = 0, \qquad n\partial(\rho_a S_a)/\partial t + \text{div}(\rho_a \mathbf{q}_a) = 0 \qquad (6\text{-}53)$$

Obviously, in this case of flow of both air and water, we have to provide appropriate initial and boundary conditions also for the air phase.

In studies of flow in large unsaturated flow domains in the field, the effect of air flow is, in general, neglected. However, as water infiltrates into a soil, air must escape through the boundaries by flow towards them. This flow of air may affect the flow of water (Morel-Seytoux, 1973). The fact that pressure in the air phase may differ significantly from atmospheric should not be overlooked (Vachaud et al., 1974). This is especially true when air cannot escape freely from the system.

6-5 MATHEMATICAL STATEMENT OF UNSATURATED FLOW

As with the solution of flow problems in the saturated zone, the solution of the partial differential equations of unsaturated flow requires the specification of initial and boundary conditions. The latter are mathematical statements of the space and time distributions of ψ (or p or θ_w, depending on the partial differential equation to be solved) or of the water fluxes on the external boundaries of the flow domain. However, unlike the case of saturated flow, the statement of water content distribution alone is not sufficient because $K(\theta_w)$ or $\psi(\theta_w)$ is subject to hysteresis. It is also necessary to state whether a drying or a wetting process is taking place along the boundary.

We assume that $K(\theta_w)$ (or $K(p_c)$), $p_c(\theta_w)$ (or $\psi(\theta_w)$) are known. If the flow reverses its direction during the period of study, we have to take hysteresis into account. For air flow, we have to know also $K(\theta_a)$.

The discussion on the nature of the macroscopic boundary presented in App. A8 is applicable here also.

Only water flow is being considered here. The extension to simultaneous air–water flow is obvious and requires no further discussion.

Initial conditions include the specification of θ_w or p_w (or $p_c = -p_w$) at every point inside the considered flow domain.

Boundary conditions may be of several types:

(a) *Prescribed water content* θ_w, *or head* ϕ_w, *or pressure* p_w ($= -p_c = \gamma_w \psi$) *at all points of the (macroscopic) boundary.* The latter situation occurs, for example, when we have ponded water on the soil surface, dictating there a certain water pressure. In a limit situation, we may have a thin sheet of water over the surface so that practically $p_w = 0$. Instead, we can always specify the θ_w at saturation, corresponding to $p_w = 0$. We shall do so when the partial differential equation is given in terms of θ_w. However, usually θ_w is not known.

This is a *boundary condition of the first type (Dirichlet boundary condition).*

(b) *Prescribed flux of water at the boundary.* This case occurs, for example, when water (rainfall or irrigation by sprinklers) reaches the ground surface, which serves as a boundary to the unsaturated flow domain, at a known rate. For a vertically downward accretion at a known rate $N(\mathbf{x}, t)$ we have

$$\mathbf{N} \cdot \mathbf{1n} = \mathbf{q} \cdot \mathbf{1n} \tag{6-54}$$

where $\mathbf{1n}$ is the unit vector normal to the boundary surface and $\mathbf{N} = -N\mathbf{1z}$ (see (5-45), with $\mathbf{V}_G = 0$). For evaporation, we replace $\mathbf{N}$ by $\mathbf{E} = E\nabla z$.

For $\mathbf{q}$ we can introduce any of the equations of motion given in Sec. 6-2. For example, using (6-23), we obtain

$$\mathbf{N} \cdot \mathbf{1n} = \left[K_w(S_w) \cdot \nabla \psi - K_w(S_w) \cdot \mathbf{1z} \right] \cdot \mathbf{1n} \tag{6-55}$$

and for a horizontal surface and an isotropic aquifer, $\mathbf{1n} \equiv \mathbf{1z}$, we obtain

$$K_w(\psi) \frac{\partial \psi}{\partial z} = K_w(\psi) - N \tag{6-56}$$

Or, using (6-25):

$$\mathbf{N} \cdot \mathbf{1n} = \mathbf{q} \cdot \mathbf{1n} = \left[-D(\theta_w) \nabla \theta_w - K_w(\theta_w) \mathbf{1z} \right] \cdot \mathbf{1n} \tag{6-57}$$

and for a horizontal surface

$$D(\theta_w) \frac{\partial \theta_w}{\partial z} = N - K_w(\theta) \tag{6-58}$$

If the rate of accretion, N, is such that not all of it can be taken up by the soil, we may have ponding. This happens when $N = K_0$ (i.e., K at saturation). At that time, θ_w reaches saturation at the surface, $p_w = 0$, $\partial p_w / \partial z = 0$, and the rate (= specific discharge) of downward flow is equal to K_0. If $N > K_0$, ponding (or surface runoff removing part of N) will take place.

The impervious boundary is a special kind of this type of boundary. On it $q_n = \mathbf{q} \cdot \mathbf{1n} = 0$.

The boundary condition of prescribed flux is thus either a *second type boundary condition,* or a *third type one.*

(c) *Semipervious boundary.* This situation occurs when a thin layer of reduced permeability is formed at the ground surface. This happens, for example, on the bottom of artificial recharge ponds. Assuming that the semipermeable membrane is saturated, the boundary condition is

$$\mathbf{q} \cdot \mathbf{ln} = \frac{\phi_0 - \phi}{\sigma'} \tag{6-59}$$

where σ' is the resistance ($= B'/K'$) of the semipermeable layer and $\mathbf{ln}$ is the outward normal to the boundary (see (5-40)). The specific discharge $\mathbf{q}$ can be expressed by any of the motion equations of Sec. 6-2. For example, for a horizontal boundary and an isotropic soil (Fig. 6-20)

$$- K_w(\psi) \frac{\partial \psi}{\partial z} + K_w(\psi) - \psi/\sigma' = (d + B')/\sigma' \tag{6-60}$$

This is a *boundary condition of the third type*, or a *Cauchy boundary condition.*

When the flow domain is made up of regions of different (homogeneous) porous media (e.g., a layered soil), we require that on points of the boundary between two media, both the flux and the pressure be continuous

$$p_{w1} = p_{w2} ; \qquad q_{n1} = q_{n2} \tag{6-61}$$

We have seen above (Sec. 6-1) that the requirement of pressure continuity means a discontinuity in water content.

If we wish to consider the flow of both air and water in the unsaturated zone, by solving (6-52), we have to state also boundary conditions, similar to those described above for the water phase, also for the air phase.

The complete mathematical statement of an unsaturated flow problem is similar to that of saturated flow, discussed in Sec. 5-4. We have to specify the flow domain, to determine which is the more appropriate dependent variable to be used (p_w, ψ, or θ_w), to state the appropriate partial differential equation to be solved, and to state the initial and boundary conditions according to the physics of the problem to be solved. A solution requires the knowledge of the functional relationships $K(\theta_w)$ or $K(\psi)$ and $\psi(\theta_w)$. Special attention should be given to the problem of hysteresis. Sometimes water and soil compressibility is also taken tnto account.

For the simultaneous flow of air and water, we also add :

$$S_a + S_w = 1 ; \qquad p_c = p_c(S_w) = p_a - p_w$$

where $p_c = p_c(S_w)$ is a known relationship.

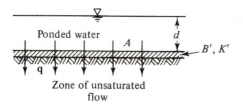

Ponded water

Zone of unsaturated flow

Figure 6-20 Nomenclature for semipermeable boundary.

6-6 METHODS OF SOLUTION

The general equation governing the flow of water in the unsaturated zone is (6-38) written (for a homogeneous medium) in the form

$$\left(\frac{d\theta_w}{d\psi} \right) \frac{\partial \psi}{\partial t} + \nabla \cdot (K\nabla\psi) - \left(\frac{dK}{d\psi} \right) \frac{\partial \psi}{\partial z} = 0 \qquad (6\text{-}62)$$

for ψ as a dependent variable, or (6-44)

$$\frac{\partial \theta_w}{\partial t} - \nabla \cdot (D\nabla\theta_w) - \left(\frac{dK}{d\theta_w} \right) \frac{\partial \theta_w}{\partial z} = 0 \qquad (6\text{-}63)$$

in terms of θ_w as dependent variable. In these (Fokker–Planck) equations, $d\theta_w(\psi)/d\psi$, $K(\psi)$, $dK(\psi)/d\psi$, $D(\theta_w) \equiv -K(\theta_w)/(d\theta_w/d\psi)$, $dK(\theta_w)/d\theta$ are known relationships. An equation in terms of p can also be written. One should keep in mind the set of assumptions leading to (6-62) and (6-63). The main difference between these equations and the continuity equations for saturated flow (e.g., (5-26)) is that the former are *non-linear*, since the coefficients K, D $dK/d\theta_w$ and $d\theta_w/d\psi$ strongly depend on θ_w (or ψ). This fact virtually precludes exact analytical solutions of unsaturated flow problems. Additional difficulties stem from: (a) the fact that the above coefficients are available only as experimental relationships, (b) they are subject to hysteresis (although $K(\theta_w)$ is less susceptible to it than $K(\psi)$), and (c) possible irregularity of boundaries and/or inhomogeneity of medium. Difficulties also result from the last terms on the left-hand side of (6-62) and (6-63), which express the effect of gravity.

Because of the difficulties of obtaining exact analytical solutions, which, whenever possible, are obviously preferable, researchers have been trying to derive semi-analytical, or approximate, ones, often involving numerical integrations. In most cases, they facilitate their solution by introducing analytical expressions, such as (6-30) through (6-32) and (6-46), as approximations of the relationships between the various coefficients and the dependent variables θ_w or ψ. Among many who present in the literature such solutions on unsaturated flow, we may mention Klute (1952), Gardner (1958), Philip (1955, 1957a–g, 1958a, b, 1960a, b, 1969, 1970), Youngs (1957, 1958, 1960), Irmay (1966), Rubin (1968), Brutsaert (1968a, b) and Brutsaert and Weisman (1970), Braester *et al.* (1971), Braester (1973), and Parlange and Babu (1976, 1977). Noblanc and Morel-Seytoux (1972), Morel-Seytoux (1973) and Sonu and Morel-Seytoux (1976) discuss cases where both air and water are flowing simultaneously. Some of these methods are reviewed by Bear (1972, pp. 503–513). Most of these works are related to one-dimensional (vertical or horizontal) flows.

Of all the models and analogs listed in Table 5-1, only the sand box can simulate unsaturated flow. However, because of both technical and scaling difficulties (Bear, 1972, p. 682) the sand box cannot be used as a tool for solving boundary value problems of practical interest. The column filled with a porous medium, however, is a very important laboratory research tool used for observing phenomena (also of solute transport with or without liquid–solid interaction) and verifying approximate solutions.

As in saturated flow, numerical techniques of various kinds are nowadays the main tool for solving problems of unsaturated flow encountered in the practice. The basic ideas of both the finite difference and the finite element techniques presented in Sec. 5-6 with respect to saturated flow are also applicable here. One should recall that parameters appearing in these equations are saturation dependent.

As an example, consider the application of the Galerkin finite element technique to (6-44) written for $\theta = \theta(x, z, t)$ in the flow domain Ω in the xz plane, in the form

$$\frac{\partial \theta}{\partial t} = \frac{\partial}{\partial x} \left[D(\theta) \frac{\partial \theta}{\partial x} \right] + \frac{\partial}{\partial z} \left[D(\theta) \frac{\partial \theta}{\partial z} + K(\theta) \right] \tag{6-64}$$

Let the boundary conditions be

$$\theta = \theta_s = \text{constant close to saturation along } \Gamma_1, \quad \text{for } 0 \leq t$$

$$q_n = [D(\theta) \nabla \theta + K(\theta) \mathbf{1z}] \cdot \mathbf{1n} = D(\theta) \frac{\partial \theta}{\partial x} l_x$$

$$+ \left[D(\theta) \frac{\partial \theta}{\partial z} + K(\theta) \right] l_z \quad \text{on } \Gamma_2, \quad \text{for } 0 \leq t \tag{6-65}$$

where Γ_1 and Γ_2 are segments of the total boundary, and l_x and l_z are direction cosines of the normal $\mathbf{1n}$.

Initial conditions are

$$\theta = \theta_0 \quad \text{at} \quad t = 0 \tag{6-67}$$

Following Bruch (1976), we start by approximating θ throughout the discretized flow domain at any time t by

$$\theta(x, z, t) = \sum_{i=1}^{n} N_i(x, z) \theta_i(t) \tag{6-68}$$

where N_i are the shape functions defined piecewise, element by element, which here depend on x and z only, $\theta_i(t)$ are the discrete nodal values and n is the number of nodes within the solution domain.

Using the weighted residual process in which the weighting function is equal to the shape function defining the approximation, the Galerkin representation of the flow problem described by (6-64) for point i is

$$\iint_{(\Omega)} N_i \left\{ \frac{\partial}{\partial x} \left[D(\theta) \frac{\partial \theta}{\partial x} \right] + \frac{\partial}{\partial z} \left[D(\theta) \frac{\partial \theta}{\partial z} \right] + \frac{\partial K(\theta)}{\partial z} - \frac{\partial \theta}{\partial t} \right\} d\Omega = 0 \tag{6-69}$$

We now apply Green's theorem to the first two terms of the integrand in order to avoid the continuity conditions imposed by the second derivative, and obtain

$$\iint_{(\Omega)} \left\{ D(\theta)\frac{\partial N_i}{\partial x}\frac{\partial \theta}{\partial x} + D(\theta)\frac{\partial N_i}{\partial z}\frac{\partial \theta}{\partial z} - N_i\frac{\partial K(\theta)}{\partial z} + N_i\frac{\partial \theta}{\partial t} \right\} d\Omega$$

$$- \int_{\Gamma_2} N_i\left[D(\theta)\frac{\partial \theta}{\partial x}l_x + D(\theta)\frac{\partial \theta}{\partial z}l_z \right]d\Gamma_2 = 0 \quad (6\text{-}70)$$

We still face the problem that (6-70) is non-linear. Bruch (1976) suggests the following iterative scheme. A value of water content θ is guessed at each node, $\{\theta^k\}$, where k denotes the number of the iteration; $k = 0$ is the initial guess. The values of $D(\theta)$ and $K(\theta)$ are approximated within the solution domain by

$$D(\theta) = \sum_{I=1}^{m} \bar{N}_I D_I; \qquad K(\theta) = \sum_{I=1}^{m} \bar{N}_I K_I \qquad (6\text{-}71)$$

respectively, where the shape functions $\bar{N}_I$ are not necessarily of the same form as N_i; N_I and K_I are evaluated using the guessed values of θ. Thus, within an element $D(\theta)$ and $K(\theta)$ are now functions of the space coordinates. Inserting the approximation (6-68) for θ and (6-71) for $D(\theta)$ and $K(\theta)$ in (6-70), we obtain

$$\sum_{(e)} \iint_{(\Omega^e)} \left\{ \sum \bar{N}_I D_I(\theta^k)\frac{\partial N_i}{\partial x}\frac{\partial \sum N_i\theta_i^{k+1}}{\partial x} + \sum \bar{N}_I D_I(\theta^k)\frac{\partial N_i}{\partial z}\frac{\partial \sum N_i\theta_i^{k+1}}{\partial z} \right.$$

$$\left. - N_i\frac{\partial \sum \bar{N}_I D_I(\theta^k)}{\partial z} + N_i\frac{\partial \sum N_i\theta_i^{k+1}}{\partial t} \right\} d\Omega^e$$

$$- \sum_{(e)} \int_{(\Gamma_2^e)} N_i\left\{ \sum \bar{N}_I K_I(\theta^k)\left[\frac{\partial \sum N_i\theta_i^{k+1}}{\partial x}l_x + \frac{\partial \sum N_i\theta_i^{k+1}}{\partial z}l_z \right] \right\} d\Gamma_2 = 0 \quad (6\text{-}72)$$

Equation (6-72) now gives a set of n linear ordinary differential equations for θ in Ω. These equations can be written in matrix form as

$$[K]\{\theta^{k+1}\} + [C]\frac{\partial \theta^{k+1}}{\partial t} + \{F\} = 0 \qquad (6\text{-}73)$$

where typical elements of the coefficient matrices are

$$K_{ij} = \sum_{(e)} \iint_{(\Omega^e)} \sum \bar{N}_I D_I(\theta^k)\left[\frac{\partial N_i}{\partial x}\frac{\partial N_j}{\partial x} + \frac{\partial N_i}{\partial z}\frac{\partial N_j}{\partial z} \right] d\Omega^e$$

$$C_{ij} = \sum_{(e)} \iint_{(\Omega^e)} N_i N_j d\Omega^e$$

$$F_i = -\sum_{(e)} \iint_{(\Omega^e)} N_i\frac{\partial \sum \bar{N}_I K_I(\theta^k)}{\partial z} d\Omega^e$$

$$- \sum_{(e)} \int_{(\Gamma_2^e)} N_i[q_n - \sum \bar{N}_I K_I(\theta^k)l_z] d\Gamma_2^e$$

where the summations are taken over the contributions of each element domain, Ω^e, and each pertinent element boundary, Γ_2^e.

Bruch (1976) then applies a Crank–Nicolson finite difference scheme (Sec. 5-6) to (6-73) and obtains a set of linear algebraic equations. These are solved for θ^{k+1}. An iteration technique with the criterion $|\theta^{k+1} - \theta^k| < \varepsilon$ is used. Bruch (1976) presents results of several examples.

Nelson (1962), Hanks and Bowers (1962), Reisenaur *et al.* (1963), Bhuijan *et al.* (1971), Neuman (1974), and Pinder and Gray (1977), among others, also present numerical solutions using various numerical techniques.

Numerical solutions are also presented in the literature to the problem of the saturated and unsaturated zones treated as a single flow domain, usually using pressure as the dependent variable (e.g., Freeze, 1971; Neuman, 1972, 1973, 1975; Narasimhan, 1975; Vauclin *et al.*, 1974; Pinder and Gray, 1977).

SEVEN

GROUNDWATER QUALITY PROBLEM (HYDRODYNAMIC DISPERSION)

So far, we have discussed only the movement and storage of water in aquifers of various types. We have overlooked a major problem which is of interest in any development and management of a water resources system, namely that of water *quality*. In fact, with the increased demand for water in most parts of the world, and with the intensification of water utilization, the quality problem becomes the limiting factor in the development of water resources in many parts of the world. Although in such regions, the quality of both surface and groundwater resources deteriorates as a result of pollution, special attention should be devoted to the pollution of groundwater in aquifers due to their very slow velocity. Hence, although it seems that groundwater is more protected than surface water against pollution, it is still subject to pollution, and when the latter occurs, the restoration to the original, non-polluted state, is more difficult.

The term "quality" usually refers either to energy—in the form of heat or nuclear radiation—or to materials contained in the water. Many materials dissolve in water, whereas others may be carried with the water in suspension. Given the very large number of possible constituents—and new materials are coming into the market every day—groundwater quality can be measured in terms of practically hundreds of parameters. The relevance of any of these materials depends on the use that is being considered. For example, salinity may be important if the water is intended for drinking, for irrigation, or for certain industries, but less important for recreation. Often standards are issued by health authorities with respect to the various constituents according to the origin of the water and the type of consumer. Standards have also been recommended by the World Health Organization.

When we speak of "water pollution" rather than of "water quality", we usually have in mind already a situation in which the quality of the water has been deteriorating towards the point of being hazardous to the consumer. However, even under undisturbed conditions, and without man's intervention, groundwater already contains a certain amount of dissolved matter, sometimes reaching levels which render the water unsuitable for certain usages. With this in mind, we shall use henceforth the term "pollutant" to denote dissolved matter carried with the water and accumulating in the aquifer, without inferring that concentrations have necessarily reached dangerous levels.

Groundwater pollution is usually traced back to four sources:

(a) *Environmental.* This type of pollution is due to the environment through which the flow of groundwater takes place. For example, in flow through carbonate rocks, water dissolves small, yet, sometimes, significant amounts of the rock. Sea water intrusion, or pollution of good quality aquifers by invading brackish groundwater from adjacent aquifers as a result of disturbing an equilibrium that existed between the two bodies of water, are also examples of environmental pollution.

(b) *Domestic.* Domestic pollution may be caused by accidental breaking of sewers, by percolation from septic tanks, by rain infiltrating through sanitary landfills, or by artificial recharge of aquifers by sewage water after being treated to different levels. Biological contaminants (e.g., bacteria and viruses) are also related to this source.

(c) *Industrial.* In many cases, a single sewage disposal system serves both industrial and residential areas. In this case, one cannot separate between industrial and domestic pollution, although their compositions—and hence the type of treatment they require and the pollution they cause—are completely different. Heavy metals, for example, constitute a major problem in industrial wastes. Industrial waste may also contain specially nondeteriorating toxic compounds and radioactive materials.

(d) *Agricultural.* This is due to irrigation water and rain water dissolving and carrying fertilizers, salts, herbicides, pesticides, etc., as they infiltrate through the ground surface and replenish the aquifer.

We are dealing with mass transport in porous media, where the considered "mass" is that of some solute (= pollutant) moving with the solvent (= water) in the interstices of a porous medium, both in the saturated and in the unsaturated zones.

The main mechanisms affecting the transport of a solute in a porous medium are: convection, mechanical dispersion, molecular diffusion, solid-solute interactions and various chemical reactions and decay phenomena, which may be regarded as source-sink phenomena for the solute.

Our objective in this section is to present and discuss the laws governing the movement and accumulation of pollutants in groundwater flow. As in the case of groundwater flow discussed in the previous chapters, our objective here is to describe the movement and accumulation of a solute in the form of a balance equation. This balance equation takes the form of a partial differential equation, the solution of which, subject to specified initial and boundary conditions, enables the engineer and planner to predict future pollutants' distributions in an aquifer.

We shall consider the general case of three-dimensional flow. However, the procedure of averaging along the vertical, employed in Chap. 5, will also be used here to derive equations for essentially horizontal flow in aquifers. Again, one should be careful to employ the aquifer flow concept only when justified. Whereas equipotentials in an aquifer are more or less vertical, even when an aquifer is stratified (i.e., consists of several layers of different hydraulic conductivities), velocities in the different strata may vary appreciably, resulting in a marked difference in the rates of advance and spreading of a pollutant in the different strata. A situation (e.g., with respect to a partially penetrating pumping well) may arise, where the average (along the vertical) concentration is meaningless and one should take into account the stratification in water quality.

Under certain conditions, the transition zone between two bodies of groundwater of different qualities may be approximated as an abrupt front. Water of one quality injected into an aquifer containing water of another quality may serve as an example. This approximate approach is treated in Sec. 7-10.

7-1 DISPERSION PHENOMENA

Consider saturated flow through a porous medium, and let a portion of the flow domain contain a certain mass of solute. This solute will be referred to as a *tracer*. The tracer, which is a labeled portion of the same liquid, may be identified by its density, color, electrical conductivity, etc. Experience shows that as flow takes place, the tracer gradually spreads and occupies an ever-increasing portion of the flow domain, beyond the region it is expected to occupy according to the average flow alone. This spreading phenomenon is called *hydrodynamic dispersion* (*dispersion, miscibile diplacement*) in a porous medium. It is a nonsteady, irreversible process (in the sense that the initial tracer distribution cannot be obtained by reversing the flow) in which the tracer mass mixes with the non-labeled portion of the liquid. If initially a tracer-labeled liquid occupies a separate region, with an abrupt interface separating it from an unlabeled one, this interface does not remain an abrupt one, the location of which may be determined by the average velocity as expressed by Darcy's law. Instead, an ever-widening transition zone is created, across which the tracer concentration, C, varies from that of the tracer liquid to that of the unmarked liquid (Fig. 7-1a).

Another example is an instantaneous point injection, say, through a well, into an aquifer where the average uniform velocity is V. Figure 7-1b shows how both longitudinal and transversal (to the direction of the flow) spreading takes place. The ellipsoidal contours of equal concentration can be obtained from a sufficient number of observation wells.

One of the earliest observations of these phenomena is reported by Slichter (1905), who used an electrolyte as a tracer in studying the movement of groundwater. Slichter observed that at an observation well downstream of a (continuous) injection point, the tracer's concentration increases gradually, and that even in a uniform (average) flow field the tracer advances in the direction of the flow in a pear-like shape that becomes longer and wider as it advances.

The dispersion phenomenon may be demonstrated also by a simple laboratory experiment. Consider steady flow in a cylindrical column of homogeneous sand,

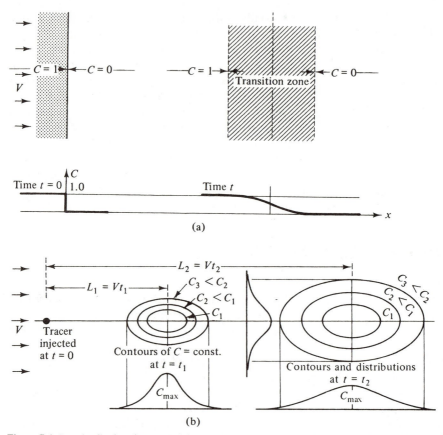

Figure 7-1 Longitudinal and transversal spreading of a tracer. (a) Longitudinal spreading of an initially sharp front. (b) Spreading of a point injection.

saturated with water. At a certain instant, $t = 0$, a tracer-marked particle of water (e.g., water with NaCl at a low concentration so that the effect of density variations on the flow pattern is negligible) starts to displace the original unlabeled water in the column. Let the tracer concentration, $C = C(t)$, at the end of the column be measured and presented in a graphic form, called a *breakthrough curve*, as a relationship between the relative tracer distribution ε and time t, or volume of effluent U

$$\varepsilon(U) = [C(U) - C_0]/(C_1 - C_0); \qquad \varepsilon(t) = [C(t) - C_0]/(C_1 - C_0)$$

where C_0 and C_1 are the tracer concentrations of the original and of the displacing water, respectively.

In the absence of dispersion, the breakthrough curve should have taken the form of the broken line in Fig. 7-2, where U_0 is the pore volume of the column, and Q is the constant discharge. Actually, owing to hydrodynamic dispersion, it will take the form of the S-shaped curve shown in full line in Fig. 7-2.

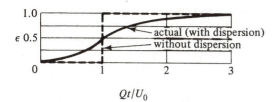

Figure 7-2 Breakthrough curves in one-dimensional flow in a sand column.

Mechanical Dispersion and Molecular Diffusion

We cannot explain all the above observations on the basis of the average, Darcy flow. We must refer to what happens at the microscopic level, namely, inside the pore. There we have velocity varying in both magnitude and direction across any pore cross section. We assume zero fluid velocity on the solid surface, with a maximum velocity at some internal point (compare with the parabolic velocity distribution in a straight capillary tube). The maximum velocity itself varies according to the size of the pore. Because of the shape of the interconnected pore space, the (microscopic) streamlines fluctuate in space with respect to the mean direction of flow (Fig. 7-3a and b). These phenomena cause the spreading of any initially close group of tracer particles; as flow continues, they will occupy an ever increasing volume of the flow domain. The spreading is in both the longitudinal direction, namely that of the average flow, and in the direction transversal to the average flow. We shall refer to this spreading, caused by the velocity variations at the microscopic level, as *mechanical dispersion*. It is sometimes called *convective diffusion*. It is spreading with respect to the average flow produced by velocity variation (in magnitude and direction) in the pore space.

The two basic elements in this kind of mixing are, therefore, *flow* and the *presence of a pore system* through which flow takes place. In addition to inhomogeneity on a microscopic scale (presence of pores and grains), we may also have inhomogeneity on a macroscopic scale due to variations in permeability from one portion of the flow domain to the next. This inhomogeneity also contributes to the mechanical dispersion of the tracer (see Sec. 7-8).

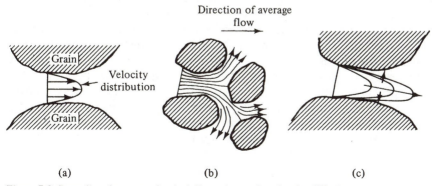

Figure 7-3 Spreading due to mechanical dispersion and molecular diffusion.

In general, we may have a convective mass transport in both a laminar flow regime, where the liquid moves along definite paths that may be averaged to yield streamlines, and a turbulent flow regime, where the turbulence may cause yet an additional mixing. In what follows, we shall focus our attention only on flow of the first type.

An additional mass transport phenomenon, which occurs *simultaneously* with mechanical dispersion, is that caused by molecular diffusion resulting from variations in tracer concentration within the liquid phase. Molecular diffusion produces an additional flux of tracer particles (at the microscopic level) from regions of higher tracer concentrations to those of lower ones. This means, for example, that as the tracer is spread *along* each microscopic streamtube as a result of mechanical dispersion, and a tracer concentration gradient is produced, molecular diffusion will tend to equalize the concentrations along the streamtube. At the same time, a tracer concentration gradient will also be produced between adjacent streamlines, causing lateral molecular diffusion across streamtubes (Fig. 7-3c).

Hydrodynamic Dispersion

We use the term hydrodynamic dispersion to denote the spreading (at the macroscopic level) resulting from both mechanical dispersion and molecular diffusion. Actually, the separation between the two processes is rather artificial as they are inseparable. However, molecular diffusion alone does take place also in the absence of motion (both in a porous medium and in a liquid continuum). Because molecular diffusion depends on time, its effects on the overall dispersion will be more significant at low flow velocities. It is molecular diffusion which makes the phenomenon of hydrodynamic dispersion in purely laminar flow irreversible.

In addition to convection (at the average velocity), mechanical dispersion and molecular diffusion, several other phenomena may affect the concentration distribution of a tracer as it moves through a porous medium. The tracer (= a solute) may interact with the solid surface of the porous matrix in the form of adsorption of tracer particles on the solid surface, deposition, solution of the solid matrix, ion exchange, etc. All these phenomena cause changes in the concentration of a tracer in a flowing liquid. Radioactive decay and chemical reactions within the liquid also cause tracer concentration changes.

In general, variations in tracer concentration cause changes in the liquid's density and viscosity. These, in turn, affect the flow regime (i.e., velocity distribution) that depends on these properties. We define an *ideal tracer* as one that is inert with respect to its liquid and solid surroundings and that does not affect the liquid's properties. At relatively low concentrations, the ideal tracer approximation is sufficient for most practical purposes. However, in certain cases, for example in the problem of sea water intrusion, the density may vary appreciably, and the ideal tracer approximation should not be used.

7-2 OCCURRENCE OF DISPERSION PHENOMENA

Hydrodynamic dispersion phenomena occur in many problems of groundwater flow, in chemical engineering, in oil reservoir engineering, etc. In groundwater flow, we encounter it in:

(*a*) The continuous variation of the concentration of some specific polluting constituents, or of total dissolved solids, as flow takes place in an aquifer,

(*b*) Groundwater pollution from some localized source, such as a faulty sewage installation. or waste dump.

(*c*) Groundwater pollution from a distributed source, such as fertilizers and pesticides applied to the area overlying an aquifer,

(*d*) Sea water intrusion into a coastal aquifer, producing a transition zone from fresh water to sea water,

(*e*) Encroachment of saline, or brackish water, into an aquifer as a result of changes in the hydrologic regime,

(*f*) Seepage of polluted surface water through pervious river beds or lakes.

(*g*) The movement of pollutants from the ground surface to the underlying water table, under unsaturated flow conditions, e.g., the movement of fertilizers or the leaching of salts from the soil in agriculture,

(*h*) Changes in the quality of water in an aquifer as a result of artificial recharge water spreading in it. Sometimes, reclaimed sewage is used to recharge an aquifer.

In all these cases, we are seeking some tool which should enable us to predict the concentration changes that will occur as a result of planned operations, superimposed on the natural flow regime in an aquifer.

7-3 COEFFICIENTS OF DISPERSION

A large number of articles have been published in the professional journals, mainly in the period 1950–1970, in which theories on dispersion have been developed. Detailed summaries are given, among others, by Bear (1969, Chap. 4 and 1972, Chap. 10) and Fried (1975). The main effort has been to express hydrodynamic dispersion macroscopically through a partial differential equation and to determine the nature of the coefficients which appear in this equation. Of special interest is the relationship between these coefficients and matrix and flow parameters.

Two approaches are commonly employed. In the first one, the porous medium is replaced by a fictitious, greatly simplified, model in which the spreading of a solute that occurs can be analyzed by exact mathematical methods. A single capillary tube, a bundle of capillaries, an array of mixing cells, are examples of such models. The second approach is to construct a statistical (conceptual) model of the microscopic motion of solute particles and to average these motions in order to obtain a macroscopic description of them (see, for example, de Josselin de Jong, 1958; Saffman, 1959, 1960; Wilson and Gelhar, 1974).

The partial differential equation itself can be obtained by averaging over a Representative Elementary Volume (REV) around a point (Sec. 7-4 and App. A4). However, in order to obtain an insight of the structure of the coefficients, models of the types mentioned above have to be employed. In what follows, a summary is presented of the coefficients and their structure. For the detailed development, the reader is referred to the literature or to the summaries mentioned above.

The average product $\overline{c_\alpha \mathbf{V}_\alpha}$ expresses the mass flux of a solute at a point in a porous medium. By the definition of the average, we have (App. A-3)

$$\overline{c_\alpha \mathbf{V}_\alpha} \equiv \overline{(\bar{c}_\alpha + \overset{\circ}{c}_\alpha)(\overline{\mathbf{V}}_\alpha + \overset{\circ}{\mathbf{V}}_\alpha)} = \bar{c}_\alpha \overline{\mathbf{V}}_\alpha + \overline{\overset{\circ}{c}_\alpha \overset{\circ}{\mathbf{V}}_\alpha} \tag{7-1}$$

This is a flux through the α phase only. In (7-1) c_α is mass of a solute, γ, per unit volume of α phase. The term $\bar{c}_\alpha \overline{\mathbf{V}}_\alpha$ expresses the *transport by convection*. The second term $\overline{\overset{\circ}{c}_\alpha \overset{\circ}{\mathbf{V}}_\alpha} \equiv \overline{c_\alpha \overset{\circ}{\mathbf{V}}_\alpha}$ is an additional flux at the macroscopic scale—the dispersive flux—introduced by the process of averaging. It is a flux of γ carried by the fluctuating velocity. This new phenomenon at the macroscopic scale (as it does not exist at the microscopic one) represents the loss of information by the passage from one scale of description to another, larger, one.

As a working hypothesis, we shall assume that the *dispersive flux* can be expressed as a Fickian type law

$$\overline{\overset{\circ}{c}_\alpha \overset{\circ}{\mathbf{V}}_\alpha} = -\boldsymbol{D} \cdot \operatorname{grad} \bar{c}_\alpha; \qquad \overline{\overset{\circ}{c}_\alpha \overset{\circ}{V}_{i\alpha}} = -D_{ij}\, \partial \bar{c}_\alpha / \partial x_j \tag{7-2}$$

or, for $\rho_\alpha = \text{const.}$, and a concentration $C = \bar{c}_\alpha / \rho_\alpha = $ mass of γ per unit mass of α phase

$$\overline{\overset{\circ}{C} \overset{\circ}{\mathbf{V}}} = -\boldsymbol{D} \cdot \operatorname{grad} C; \qquad \overline{\overset{\circ}{C} \overset{\circ}{V}_i} = -D_{ij}\, \partial C / \partial x_j \tag{7-3}$$

where $\boldsymbol{D}$, a second rank symmetric tensor, is the coefficient of mechanical (or convective) dispersion, and we have omitted the subscript α in $\mathbf{V}$.

Henceforth to the end of this section, we shall use $\overline{\mathbf{V}}$ to indicate average α phase velocity; in the following sections $\mathbf{V}$ will be used.

Bear and Bachmat (1967), who employed a model composed of a network of interconnected capillary tubes, derived the following expression for the relationship between the coefficient $\boldsymbol{D}$ and porous matrix geometry, flow velocity, and molecular diffusion

$$D_{ij} = a_{ijkm} \frac{\overline{V}_k \overline{V}_m}{\overline{V}} f(Pe, \delta) \tag{7-4}$$

where $\overline{V} = |\overline{\mathbf{V}}|$ is the average velocity, $Pe = $ Peclet number $= L\overline{V}/D_d$, L being some characteristic length of the pores, $D_d = $ coefficient of molecular diffusion of the solute in the considered liquid phase, $\delta = $ ratio of length characterizing an individual channel of a porous medium to its hydraulic radius, and $f(Pe, \delta)$ is a function which introduces the effect of transfer by molecular diffusion between adjacent streamlines at the microscopic level. This effect is coupled with mechanical dispersion. It is distinct from the macroscopic flux, $\bar{\mathbf{J}}$, due to molecular diffusion in a porous medium expressed by Fick's law for molecular diffusion in a porous medium which can be written as

$$\bar{\mathbf{J}} = -\mathbf{D}_d^* \cdot \text{grad } \bar{c}_\alpha, \qquad \text{or} \quad \bar{J}_i = -(D_d^*)_{ij} \, \partial \bar{c}_\alpha / \partial x_j \qquad (7\text{-}5)$$

where $\mathbf{D}_d^*$, a second rank tensor which reduces to a scalar in an isotropic porous medium, is the coefficient of molecular diffusion in a porous medium. The summation convention should be used in the second equation (7-5).

Thus, we have the effect of molecular diffusion in two different ways: one as a microscopic phenomenon in the fluid phase, which is averaged to yield a macroscopic phenomenon. It is expressed by $\bar{\mathbf{J}}$ in the dispersion equation and takes place even in the absence of bulk flow (i.e., $\bar{\mathbf{V}} = 0$). Secondly, we have to understand that "pure" mechanical dispersion cannot exist and that molecular diffusion acts between adjacent streamlines and along them, making the phenomenon of dispersion irreversible in laminar flow. When (conceptual) models are constructed for the movement of a solute, this molecular diffusion is taken into account at the elementary model unit (e.g., a capillary tube), thus introducing the effect of molecular diffusion through a Peclet number, Pe, in the expression for the coefficient of mechanical dispersion.

The coefficient of hydrodynamic dispersion, D_h, is defined as the sum of the coefficients of mechanical dispersion, D, and of molecular diffusion in a porous medium, $\mathbf{D}_d^*$

$$\mathbf{D}_h = \mathbf{D} + \mathbf{D}_d^* \qquad (7\text{-}5a)$$

The works of Saffman (1960) and Bear and Bachmat (1967) may be regarded as examples of investigations leading to the structure of D_h, D, and $\mathbf{D}_d^*$.

Saffman's (1960) work resulted in two dispersion coefficients. A *coefficient of longitudinal dispersion* D_{hL}

$$D_{hL} \cong \frac{1}{3} D_d + \frac{3}{80} \frac{R^2 \bar{V}^2}{D_d} + \frac{l^2 \bar{V}^2}{4} \int_0^1 (3\alpha^2 - 1)^2 \frac{M \coth M - 1}{D'' M^2} \, d\alpha$$

$$M = \frac{3}{2} \frac{\bar{V} l \alpha}{D''} ; \qquad D'' = D_d + \frac{3}{16} \alpha^2 \frac{R^2 \bar{V}^2}{D_d} ; \qquad \alpha = \cos\theta \qquad (7\text{-}6)$$

where R and l are the radius and length of a capillary tube, respectively, $\bar{V}$ is the average velocity in the porous medium model and θ is the angle that a capillary tube makes with the direction of $\bar{V}$, and a *coefficient of transverse dispersion* D_{hT}

$$D_{hT} = \frac{1}{3} D_d + \frac{1}{80} \frac{R^2 \bar{V}^2}{D_d} + \frac{9 l^2 \bar{V}^2}{8} \int_0^1 \alpha^2 (1 - \alpha^2) \frac{M \coth M - 1}{D'' M^2} \, d\alpha \qquad (7\text{-}7)$$

In both expressions, the first term on the right-hand side is the contribution of molecular diffusion (equivalent to D_d^* in (7-5)). Saffman, (1960) comments that laboratory experiments in granular material actually gave this contribution as $(2/3) D_d$ rather than $D_d/3$. He, therefore, suggested to use in the above expression some empirical coefficient m $(1/3 < m < 2/3)$ rather than $1/3$. For l he used the value of average grain size and for $R(\leq l/5)$ he used $R = (24k/n)^{1/2}$, where k and n are the medium's permeability and porosity, respectively.

For example

(a) $Pe = \overline{V}l/D_d \ll 1,$ $D_{hL}/D_d = m + Pe^2/15,$ $D_{hT}/D_d = m + Pe^2/40$

(b) $1 \leq Pe \leq 8(R/l)^2$

$$D_{hL}/D_d \cong (Pe/6)\ln(1.5Pe) - (17Pe/72) - (R/l)^2 \, Pe^2/48 + (m + 4/9)$$
$$+ \, 0(1/Pe)$$

$$D_{hT} \cong 3Pe/10 + (R/l)^2 \, Pe^2/40 + (m - 1/3) + 0(1/Pe)$$

In (7-6) and (7-7) we see the relationship between D_{hL}, D_{hT}, D_d^*, the velocity, $\overline{V}$ and the Peclet number.

Bear and Bachmat (1967) obtained (7-4) with

$$f(Pe, \delta) = Pe/(2 + Pe + 4\delta^2) \tag{7-8}$$

where δ is the ratio between the length of an elementary channel (of porous medium model) and some length characterizing its cross section.

The influence of the geometry of the void space is represented in (7-4) by the coefficient a_{ijkm} (dim. L), called the (*geometrical*) *dispersivity of the porous medium*, which in saturated flow is a property of the geometry of the solid matrix. It is a fourth rank tensor which has certain properties of symmetry (Scheidegger, 1961; Bear, 1961). Bear and Bachmat (1967), Bear (1972, p. 614) express a_{ijkm} by

$$a_{ijkm} = [\overline{\overset{\circ}{B}T_{il}^* \overset{\circ}{B}T_{jp}^*}/\overline{BT_{lk}^*} \, \overline{BT_{pm}^*}] \, \overline{L} \tag{7-9}$$

where B is the conductance of an elementary medium channel, BT_{ij}^* is an oriented conductance of a channel, $\overline{T}_{ij}^*$ is the *medium's tortuosity*, $n\overline{BT_{ij}^*} = k_{ij}$ is the medium's permeability and $\overline{L}$ is a characteristic length of the medium. Thus, the medium's dispersivity is related to the variance of BT_{ij}^* while its permeability is related to the average, $\overline{BT_{ij}^*}$, of BT_{ij}^*.

For an isotropic porous medium, the medium's dispersivity, a_{ijkm}, is related to two constants: $a_L = $ *longitudinal dispersivity of the porous medium* and $a_T = $ *transversal dispersivity of the porous medium*. With these parameters, the components of the medium's dispersivity may be expressed by

$$a_{ijkm} = a_T \delta_{ij} \delta_{km} + \frac{a_L - a_T}{2} (\delta_{ik}\delta_{jm} + \delta_{im}\delta_{jk}) \tag{7-10}$$

where δ_{ik} is the Kronecker delta. For an isotropic medium, the components of a_{ijkm} do not change with the rotation of the coordinate axes.

By combining (7-10) with (7-4), assuming $f(Pe, \delta) \equiv 1$, we obtain

$$D_{ij} = a_T \overline{V} \delta_{ij} + (a_L - a_T) \overline{V}_i \overline{V}_j/\overline{V} \tag{7-11}$$

For Cartesian coordinates and a velocity $\overline{V}$ (components $\overline{V}_x$, $\overline{V}_y$, $\overline{V}_z$), we obtain from (7-11)

$$D_{xx} = a_T \bar{V} + (a_L - a_T) \bar{V}_x^2 / \bar{V} = [a_T(\bar{V}_y^2 + \bar{V}_z^2) + a_L \bar{V}_x^2] / \bar{V}$$

$$D_{xy} = (a_L - a_T) \bar{V}_x \bar{V}_y / \bar{V} = D_{yx}$$

$$D_{xz} = (a_L - a_T) \bar{V}_x \bar{V}_z / \bar{V} = D_{zx}$$

$$D_{yy} = a_T \bar{V} + (a_L - a_T) \bar{V}_y^2 / \bar{V} = [a_T(\bar{V}_x^2 + \bar{V}_z^2) + a_L \bar{V}_y^2] / \bar{V}$$

$$D_{yz} = (a_L - a_T) \bar{V}_y \bar{V}_z / \bar{V} = D_{zy}$$

$$D_{zz} = a_T \bar{V} + (a_L - a_T) \bar{V}_z^2 / \bar{V} = [a_T(\bar{V}_x^2 + \bar{V}_y^2) + a_L \bar{V}_z^2] / \bar{V} \qquad (7\text{-}11a)$$

If in the xyz domain the flow is everywhere horizontal, that is, $\bar{V}_z = 0$, only $D_{xz}(= D_{zx})$ and $D_{yz}(= D_{zy})$ vanish. Lateral dispersion can still take place in the z direction.

If we choose a cartesian coordinate system at a point, such that one of its axes, say x_1, coincides with the direction of the average uniform velocity $\bar{V}$, then at that point (7-11) reduces to

$$D_{11} = a_L \bar{V}, \qquad D_{22} = a_T \bar{V}, \qquad D_{33} = a_T \bar{V}, \qquad D_{ij} = 0 \quad \text{for} \quad i \neq j \quad (7\text{-}12)$$

which can be written in the form of a matrix

$$[D_{ij}] = \begin{bmatrix} a_L \bar{V} & 0 & 0 \\ 0 & a_T \bar{V} & 0 \\ 0 & 0 & a_T \bar{V} \end{bmatrix} \qquad (7\text{-}13)$$

The axes of the coordinate system in which D_{ij} is expressed by (7-12)—namely, in the direction of the flow at a point and perpendicular to it—are called the *principal axes of the dispersion*. The coefficients D_{11}, D_{22}, and D_{33} are the principal values of the coefficient of mechanical dispersion. In this case, D_{11} is called the *coefficient of longitudinal dispersion* while D_{22} and D_{33} are called *coefficients of transversal dispersion*. Only in uniform flow (that is, $\bar{V}_1 = $ const., $\bar{V}_2 = \bar{V}_3 = 0$, everywhere) is (7-12) valid for the entire flow domain. In general, the principal axes of dispersion, being (in an isotropic medium) in the direction of the velocity and perpendicular to it, vary from point to point. In (7-11) through (7-13), we can always reintroduce the effect of $f(Pe, \delta)$ by replacing D by $D/f(Pe, \delta)$.

The coefficient of molecular diffusion in a porous medium, D_d^*, also a second rank symmetric tensor, is expressed by Bear and Bachmat (1967) in the form

$$(D_d^*)_{ij} = D_d \bar{T}_{ij}^* \qquad (7\text{-}14)$$

where $\bar{T}_{ij}^*$ is the *tortuosity* of the porous medium (Bear, 1972). For an isotropic medium, $\bar{T}_{ij}^* = \bar{T}^* g_{ij}$, where the tortuosity $\bar{T}^*$ is a scalar and g_{ij} is explained below.

Bachmat and Bear (1964) and Bear and Bachmat (1967) extend the above considerations to a general curvilinear coordinate system y^i. They obtain

$$D^{ij} = a_{km}^{ij} \frac{\bar{V}^k \bar{V}^m}{\bar{V}} f(Pe, \delta); \qquad (D_d^*)^{ij} = D_d \overline{T^{*ij}} \qquad (7\text{-}15)$$

$$a_{km}^{ij} = [\overline{B\mathring{T}^{*il} B\mathring{T}^{*jp}} / \overline{BT_{lk}^* BT_{pm}^*}] \bar{L} \qquad (7\text{-}16)$$

and for an isotropic medium

$$a_{km}^{ij} = \tfrac{1}{2}(a_L - a_T)(g_m^i g_k^j + g_k^i g_m^j) + a_T g^{ij} g_{km} \tag{7-17}$$

$$D^{ij} = \left[\frac{a_L - a_T}{2}(g_m^i g_k^j + g_k^i g_m^j) + a_T g^{ij} g_{km} \right] \frac{\overline{V}^k \overline{V}^m}{\overline{V}} f(Pe, \delta) \left. \right\} \tag{7-18}$$

$$(D_d^*)^{ij} = D_a \overline{T}^* g^{ij} = D_d^* g^{ij}$$

$$D^{ij} = \left[(a_L - a_T) \frac{\overline{V}^i \overline{V}^j}{\overline{V}} + a_T g^{ij} \overline{V} \right] f(Pe, \delta) \tag{7-19}$$

Also

$$D_{ij} = \left[\frac{a_L - a_T}{2}(g_{im} g_{jk} + g_{ik} g_{jm}) + a_T g_{ij} g_{km} \right] \frac{\overline{V}^k \overline{V}^m}{\overline{V}} f(Pe, \delta) \tag{7-20}$$

or

$$D_{ij} = a_T g_{ij} \overline{V} + (a_L - a_T) \overline{V}_i \overline{V}_j / \overline{V}$$

For an orthogonal curvilinear system, $g_{ij} = 0$ for $i \neq j$ and $g_{ii} = (h_i)^2$

$$D_{ij} = \left\{ a_T g_{ij} \overline{V} + \begin{array}{ll} (a_L - a_T) \overline{V}_i^2 / \overline{V}, & \text{for } i = j \\ (a_L - a_T) \overline{V}_i \overline{V}_j / \overline{V}, & \text{for } i \neq j \end{array} \right. \tag{7-21}$$

or

$$D_{ij} = \left\{ a_T (h_i)^2 \overline{V} + \begin{array}{ll} (a_L - a_T) \overline{V}_i^2 / \overline{V}, & \text{for } i = j \\ (a_L - a_T) \overline{V}_i \overline{V}_j / \overline{V}, & \text{for } i \neq j \end{array} \right. \tag{7-21a}$$

where superscripts and subscripts denote contravariant and covariant components, respectively, and g^{ij} is the fundamental tensor of the Riemannian space (or the metric tensor). Some of the main features of this tensor are

$$ds^2 = g_{ij} dy^i dy^j \, ; \qquad g^{pi} g_{iq} \equiv \delta_q^p = \begin{array}{ll} 1 & \text{for } p = q \\ 0 & \text{for } p \neq q \end{array} \, ; \qquad g = |g| \tag{7-21b}$$

where δ_j^i is the Kronecker delta.

$$g^{ij} = (\partial y^i / \partial x)(\partial y^j / \partial x) + (\partial y^i / \partial y)(\partial y^j / \partial y) + (\partial y^i / \partial z)(\partial y^j / \partial z) \tag{7-21c}$$

$$g_{ij} = (\partial x / \partial y^i)(\partial x / \partial y^j) + (\partial y / \partial y^i)(\partial y / \partial y^j) + (\partial z / \partial y^i)(\partial z / \partial y^j) \tag{7-21d}$$

$$\overline{V}^i = g^{ij} \overline{V}_j \, ; \qquad g_{ii} = h_i^2, \qquad h_i = \text{scale factor}; \qquad ds = h_i dy^i \tag{7-21e}$$

$$|\mathbf{V}|^2 = g_{km} \overline{V}^k \overline{V}^m \tag{7-21f}$$

where ds is length between two points in the considered space. For further information, the reader is referred to any text on curvilinear coordinates.

In the special case of an orthogonal coordinate system y^i, such that y^1 coincides at a point with the direction of the average velocity $\overline{V}(= \overline{V}^1)$, while $\overline{V}^2 = \overline{V}^3 = 0$, we have

$$[D^{ij}] = \begin{bmatrix} a_L (h^1)^2 \overline{V} & 0 & 0 \\ 0 & a_T (h^2)^2 \overline{V} & 0 \\ 0 & 0 & a_T (h^3)^2 \overline{V} \end{bmatrix} f(Pe, \delta) \tag{7-22}$$

at that point.

Practically no work has been carried out on the dispersivity of anisotropic porous media.

In (7-4), deleting the function $f(Pe, \delta)$, we have a linear relationship between the coefficient of mechanical dispersion D and the average velocity $\overline{V}$. However, $f(Pe, \delta)$ introduces a nonlinear effect of the velocity (as $Pe = \overline{V}l/D_d$). Many experiments and some analytical studies seem to indicate that the coefficient of dispersion is not exactly a linear function of the velocity. Often expressions of the form

$$D_{11} = a_L \overline{V}(Pe)^{m_1}, \qquad D_{22} = a_T \overline{V}(Pe)^{m_2} \qquad (7\text{-}23)$$

where m_1 and m_2 are constants, are suggested instead of the linear ones given by (7-12).

Figure 7-4 gives a schematic representation of results of a large number of experiments similar to those mentioned above. Practically all the experiments were conducted in unconsolidated porous media. In it, D_{hL} is the coefficient of longitudinal hydrodynamic dispersion as obtained from dispersion experiments in one-dimensional flow. The common experiment is one in which a fluid at constant concentration is introduced at one end of a porous medium packed column, saturated with another fluid of constant, but different, concentration. The concentration of the effluent is recorded in the form of a breakthrough curve (Fig. 7-2). The coefficient D_{hL} is obtained from the slope of this curve (see Sec. 7-9).

Figure 7-4 may be divided into several zones:

Zone I: In this zone, molecular diffusion predominates, as the average flow velocity is very small ($a_L \overline{V} \ll D_d \overline{T}^*$; $1/3 < \overline{T}^* < 2/3$).

Zone II: Corresponds approximately to Pe between 0.4 and 5. In this zone, the effects of mechanical dispersion and molecular diffusion are of the same order of magnitude.

Zone III: Here spreading is caused mainly by mechanical dispersion. In this zone

$$D_{hL}/D_d = \alpha(Pe)^m; \qquad \alpha \approx 0.5; \qquad 1 < m < 1.2 \qquad (7\text{-}24)$$

Zone IV: This is the region of dominant mechanical dispersion (as long as we stay in the range of validity of Darcy's law). The effect of molecular diffusion is negligible. In the diagram, we obtain a straight line at 45 degrees

$$D_{hL}/D_d = \beta Pe; \qquad \beta \approx 1.8 \qquad (7\text{-}25)$$

Zone V: This is another zone of pure mechanical dispersion, but beyond the range of Darcy's law, so that the effects of inertia and turbulence can no longer be neglected.

Much less information is available on transversal dispersion. Ratios of a_L/a_T between $5:1$ and $24:1$ have been mentioned in the literature. A relationship for D_{hT}/D_d similar to that given by (7-24) is often used but with different values for α and m (e.g., $\alpha = 1/40$ and $m = 1.1$ for Zone III).

In unsaturated flow, the components of a_{ijkm} e.g., a_L and a_T, as well as the various constants (e.g., m_1 and m_2 in (7-23)), are functions of the water content. Actually, the components of a_{ijkm} reflect the geometry of the portion of space

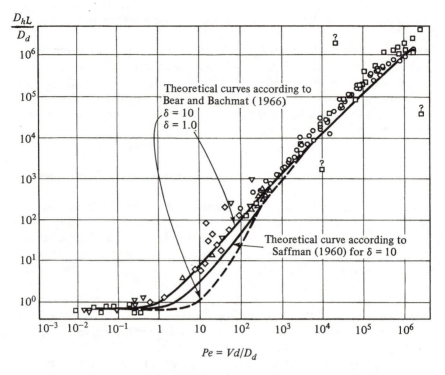

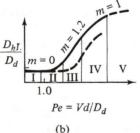

Figure 7-4 Relationship between molecular diffusion and hydrodynamic dispersion *(after Pfannkuch, 1963; Saffman, 1960)*.

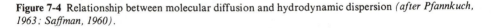

occupied by a considered α phase (water in unsaturated flow) which, in turn, is a function of the saturation. Similarly, the components $\overline{T_{ij}^*}$ of the tortuosity tensor are also a function of the saturation. Hence, when we consider unsaturated flow in an isotropic medium, we have to verify that the isotropy of $\overline{T}^*$ remains for all saturations.

Let us summarize the various dispersion coefficients:

$D_{hij} = D_{ij} + (D_d^*)_{ij} =$ coefficient of hydrodynamic dispersion,
$D_{ij} =$ coefficient of mechanical (or convective) dispersion,

$(D_d^*)_{ij} = D_d \overline{T}_{ij}^* =$ coefficient of molecular diffusion in a porous medium,

$a_{ijkm} =$ dispersivity of a porous medium,

$a_L, a_T =$ longitudinal and transversal dispersivities, respectively, of an isotropic porous medium,

$\overline{T}_{ij}^* =$ tortuosity of a porous medium.

In unsaturated flow, a_L, a_T and $\overline{T}_{ij}^*$ are functions of θ_w.

7-4 THE EQUATION OF HYDRODYNAMIC DISPERSION

In this section, the basic equations governing the movement of a solute γ ($=$ pollutant, tracer) in a porous medium are presented. These are second order partial differential equations which express the balance of the considered solute at a (macroscopic) point inside the flow domain. The coefficients which appear in these equations are discussed in the previous section.

The balance equation for a solute γ in an α-phase flowing through a porous medium is presented in App. A-1 through 7. Equation (A-25)

$$\frac{\partial(\theta_\alpha \bar{c}_\alpha)}{\partial t} = -\mathrm{div}\left[\bar{c}_\alpha \mathbf{q}_\alpha + \theta_\alpha \overline{\hat{c}_\alpha \mathring{\mathbf{V}}_\alpha} + \theta_\alpha \overline{\mathbf{J}(c_\alpha/\rho_\alpha)}\right] - <J_n(c_\alpha/\rho_\alpha)$$

$$+ c_\alpha(V_{n\alpha} - u_{n\alpha}) > \sigma'_\alpha + \theta_\alpha \rho_\alpha \overline{\Gamma(c_\alpha/\rho_\alpha)} \tag{7-26}$$

where $\bar{c}_\alpha$ is the mass of γ per unit volume of α phase, is called the equation of hydrodynamic dispersion.

Let us rewrite (7-26) in more common symbols, introducing macroscopic variables and fluxes, and

(a) assume that the dispersive flux is expressed by (7-2),

(b) use (7-5) to describe the diffusive flux $\overline{\mathbf{J}(c_\alpha/\rho_\alpha)}$, and

(c) assume that the source, or production, function can be expressed in the form: $\theta_\alpha \rho_\alpha \overline{\Gamma(c_\alpha/\rho_\alpha)} = \theta_\alpha \bar{\rho}_\alpha \overline{\Gamma}_{\alpha\gamma}$. Often the production function is such that $\overline{\Gamma} = \overline{\Gamma}(C)$. For example, if the production is due to radioactive decay (in this case negative production), we have

$$\overline{\Gamma}(C) = -\lambda C$$

where we have introduced the concentration $C = \overline{c_\alpha/\rho_\alpha} \approx \bar{c}_\alpha/\bar{\rho}_\alpha =$ mass of γ per unit mass of α phase and λ is the decay constant of the solute (i.e., the reciprocal of the solute's mean half life). The production function $\overline{\Gamma}$ may also result from chemical reactions among constituents in the α phase, leading to the production of γ.

(d) The term $<J_n(c_\alpha/\rho_\alpha) + c_\alpha(V_{n\alpha} - u_{n\alpha}) > \sigma'_\alpha$ expresses transfer of γ across the (possibly moving) boundaries of the α phase, e.g., by diffusion, solution of the solid, adsorption, precipitation, desorption, ion-exchange, etc. Let us denote this flux across the α phase boundaries by f.

With these assumptions and nomenclature, (7-26) becomes

$$\frac{\partial \theta_\alpha \bar{c}_\alpha}{\partial t} = -\mathrm{div}\left[\bar{c}_\alpha \mathbf{q}_\alpha - \theta_\alpha \mathbf{D} \cdot \mathrm{grad}\,\bar{c}_\alpha - \theta_\alpha \mathbf{D}_d^* \cdot \mathrm{grad}\,\bar{c}_\alpha\right] - f + \theta_\alpha \bar{\rho}_\alpha \overline{\Gamma}_\alpha \tag{7-27}$$

valid also for unsaturated flow ($\theta_\alpha = \theta_w$). In the square brackets we have the total flux.

Various expressions are presented in the literature for f, depending on the case (i.e., adsorption, solution, etc.) and on the assumptions underlying the investigations. Most expressions are already in a macroscopic form called *isotherm*.

A comprehensive review of adsorption models is given by Boast (1973). Some examples of *adsorption isotherms* are

1. Freundlich (in Adamson, 1960) suggested

$$F = \beta C_{eq}^{1/m} \tag{7-28}$$

where F ($=$ mass of solute per unit mass of solid) is the solute concentration on the solid phase; β and m are constants and C_{eq} is the concentration of the solute in the liquid phase under equilibrium. For $m = 1$, (7-28) is called the *linear equilibrium isotherm*.

2. A non-equilibrium isotherm for an irreversible system (Langmuir, in Adamson, 1960) is

$$\partial F / \partial t = kC \tag{7-29}$$

where C is the solute concentration in the liquid phase and k is a constant. This isotherm describes the solid phase as a sink for the solute, where the change in solute concentration on the solid phase is linearly related to its concentration in the liquid phase.

3. A non-equilibrium isotherm which describes a reversible system (e.g., Hougen and Marshall, 1947) is

$$\partial F / \partial t = k(C - mF) \tag{7-30}$$

where k and m are constants, $F = c_s / \rho_s$ is the mass of γ per unit mass of solid, and c_s is the mass of γ per unit volume of solid.

In both (7-29) and (7-30), the right-hand side expresses the rate of transfer, f_s, from the liquid phase onto the solid phase, through the interface area between the two (in terms of mass of γ per unit volume of system per unit time). We may express this in the general form

$$f_s = f_s(C, F) \tag{7-31}$$

Obviously, $f_s = -f$.

Since we have introduced here another dependent variable, namely F, we need another balance equation. We obtain it by writing (A-25) for the solid phase (denoted by the subscript s). Assuming $\mathbf{V}_s$ to be negligible, no diffusion in the solid phase, and $\theta_s \rho_s \overline{\Gamma(c_s / \rho_s)} = \theta_s \bar{\rho}_s \overline{\Gamma}_{s\gamma}$, we obtain

$$\frac{\partial (\theta_s \bar{c}_s)}{\partial t} = -f_s + \theta_s \bar{\rho}_s \overline{\Gamma}_{s\gamma} \tag{7-32}$$

where $\theta_s (= 1 - n)$ is the volumetric fraction of the solid phase.

If F also undergoes radioactive decay, $\overline{\Gamma}_{s\gamma} = -\lambda F$.

Obviously, sometimes (e.g., in unsaturated flow), the exchange between the α phase and the solid phase takes place only across part of the α phase boundary.

Introducing all these assumptions into (7-26), we obtain

$$\frac{\partial(\theta_\alpha\bar{\rho}_\alpha C)}{\partial t} = -\mathrm{div}\left[\bar{\rho}_\alpha C\mathbf{q}_\alpha - \theta_\alpha \mathbf{D}\cdot\mathrm{grad}(\bar{\rho}_\alpha C) - \theta_\alpha \mathbf{D}_d^*\cdot\mathrm{grad}(\bar{\rho}_\alpha C)\right]$$

$$-\frac{\partial(\theta_s\bar{\rho}_s F)}{\partial t} + \theta_s\bar{\rho}_s\bar{\Gamma}_{sy} + \theta_\alpha\bar{\rho}_\alpha\bar{\Gamma}_{\alpha y} \tag{7-33}$$

In (7-33) $\bar{\rho}C = \bar{c}_\alpha = $ mass of γ in the α phase per unit volume of the α phase. We also have

$$\frac{\partial(\theta_s\bar{\rho}_s F)}{\partial t} = -f_s(C, F) + \theta_s\bar{\rho}_s\bar{\Gamma}_{sy} \tag{7-34}$$

For the special case: $\bar{\rho}_\alpha = $ const., $\bar{\rho}_s = $ const., $\theta_s = $ const., $\bar{\Gamma}_{\alpha y} = -\lambda C$, $\bar{\Gamma}_{sy} = -\lambda F$, and $f_s(C, F)$ expressed by the right-hand side of (7-30), we obtain

$$\bar{\rho}_\alpha\frac{\partial(\theta_\alpha C)}{\partial t} + \bar{\rho}_s\theta_s\frac{\partial F}{\partial t} = -\mathrm{div}\left[\bar{\rho}_\alpha C\mathbf{q}_\alpha - \bar{\rho}_\alpha\theta_\alpha\mathbf{D}_h\cdot\mathrm{grad}\,C\right] - \bar{\rho}_s\theta_s\lambda F - \bar{\rho}_\alpha\theta_\alpha\lambda C \tag{7-35}$$

and

$$\frac{\partial F}{\partial t} = k'(C - mF) - \lambda F; \qquad k' = k/\theta_s\bar{\rho}_s \tag{7-36}$$

where $\mathbf{D}_h = \mathbf{D} + \mathbf{D}_d^*$ is the coefficient of hydrodynamic dispersion.

In the absence of adsorption ($k = 0$), (7-35) and (7-36) become independent. With $\theta_\alpha \equiv \theta$ and $\mathbf{q}_\alpha \equiv \mathbf{q}$, (7-35) reduces to

$$\frac{\partial(\theta C)}{\partial t} = -\mathrm{div}(C\mathbf{q} - \theta\mathbf{D}_h\cdot\mathrm{grad}\,C) - \theta\lambda C \tag{7-37}$$

In the absence of radioactive decay. we insert $\lambda = 0$ in (7-35) through (7-37).

From (A-21), it follows that for an incompressible fluid (div $\mathbf{V}_\alpha = 0$) and in the absence of volume transfer across α phase boundaries (that is, $V_{\alpha n} = u_{\alpha n}$), we obtain (see Sec. 6-4)

$$\partial\theta/\partial t = -\mathrm{div}\,\mathbf{q} \tag{7-38}$$

By combining (7-37) and (7-38) we obtain

$$\theta\frac{\partial C}{\partial t} = \mathrm{div}(\theta\mathbf{D}_h\cdot\mathrm{grad}\,C) - \mathbf{q}\cdot\mathrm{grad}\,C - \theta\lambda C \tag{7-39}$$

where, as in (7-35), $\bar{\rho}_\alpha = $ const.

If $\rho_\alpha \equiv \bar{\rho}_\alpha$ is not a constant, we have to use (7-33). Using the continuity equation

$$\frac{\partial(\theta\rho)}{\partial t} = -\mathrm{div}(\rho\mathbf{q}), \qquad \rho \equiv \rho_\alpha \tag{7-40}$$

(when applicable), (7-33), in the absence of adsorption, reduces to

$$\rho\theta\frac{\partial C}{\partial t} = \mathrm{div}(\theta D_h \cdot \mathrm{grad}(\rho C)) - \rho\mathbf{q}\cdot\mathrm{grad}\, C - \rho\theta\lambda C \qquad (7\text{-}41)$$

where C is the mass of the solute per unit mass of the solvent and ρC is the mass of the solute per unit volume of solvent.

It is often more convenient in the case of $\rho_\alpha \neq$ const. to use the definition of $\bar{c}_\alpha$. Then (7-33) becomes

$$\frac{\partial(\theta\bar{c}_\alpha)}{\partial t} = -\mathrm{div}\left[\bar{c}_\alpha\mathbf{q} - \theta D_h\cdot\mathrm{grad}\,\bar{c}_\alpha\right] - \theta\lambda\bar{c}_\alpha \qquad (7\text{-}42)$$

where $\bar{c}_\alpha$ is the mass of γ per unit volume of α. This can further be simplified when the continuity equation is $\partial\theta/\partial t + \mathrm{div}\,\mathbf{q} = 0$.

Obviously in all these equations we have to know $\mathbf{q} = \mathbf{q}(x, y, z, t)$ (see discussion in Sec. 7-6).

For the special case of saturated flow, $\theta = n =$ const., $\theta_s = 1 - n$, (7-35) becomes

$$\frac{\partial}{\partial t}(C^* + F^*) = -\mathrm{div}(C^*\mathbf{V} - D_h\cdot\mathrm{grad}\,C^*) - \lambda(C^* + F^*) \qquad (7\text{-}43)$$

where C^* and F^* denote mass of solute per unit volume of porous medium, in the liquid phase and in the solid phase, respectively, and

$$\frac{\partial F^*}{\partial t} = k(C^* - m''F^*) - \lambda F^*; \qquad m'' = m/\rho_s(1 - n) \qquad (7\text{-}44)$$

Similar equations can be written for the concentration of γ in the fluid phase in terms of mass of γ per unit volume of fluid, and on the solid phase in terms of mass of γ per unit volume of the solid phase.

For the equilibrium isotherm (7-28)

$$F^* = \beta C^*, \qquad \beta = \text{const.} \qquad (7\text{-}45)$$

and (7-43) can be written as

$$\frac{\partial C^*}{\partial t} = -\mathrm{div}\left(C^*\frac{\mathbf{V}}{R_d} - \frac{D_h}{R_d}\cdot\mathrm{grad}\,C^*\right) - \lambda C^* \qquad (7\text{-}46)$$

where $R_d = 1 + \beta$ can be called a *retardation factor*. From (7-46) it follows that the effect of adsorption is equivalent to that of reducing the average velocity (taking D to be linearly proportional to the velocity) and the coefficient of molecular diffusion in a porous medium (D_d^*).

A phenomenon often encountered in hydrodynamic dispersion is the presence of *dead-end pores* (Coats and Smith, 1964). These are pores, or more generally, stagnant (or practically so) static fluid bodies, e.g., in unsaturated flow, into and out of which a solute moves by molecular diffusion only. Thus, the behavior of this portion of the pore space is equivalent to that of sources or sinks for the solute.

This phenomenon may be represented by a model similar to that used above for describing adsorption. With θ_d and $\bar{c}_d$ denoting the fractional volume of the fluid filled dead-end pores and the mass of the solute γ per unit volume of fluid in these pores, respectively, we have

$$\partial(\theta_d\bar{c}_d)/\partial t = -f_d, \quad \text{or} \quad \partial(\bar{\rho}_\alpha\theta_d C_d)/\partial t = -f_d; \quad C_d = \bar{c}_d/\bar{\rho}_\alpha \quad (7\text{-}47)$$

where C_d represents the concentration of γ in the stagnant fluid, and $-f_d$ represents the net flow of γ by molecular diffusion from the moving fluid having the fractional volume $\theta_m = \theta_\alpha - \theta_d$ into the (practically) stagnant water in the dead-end pores (per unit time and unit volume of system). An expression for f_d could take the form $f_d = \beta\bar{\rho}_\alpha D_d^*(C_d - C)$, where C represents the concentration of γ in the moving fluid, and β is some coefficient representing the geometry of the region through which the transfer takes place. Equation (7-33), in the absence of sources and sinks, then becomes

$$\partial(\theta_m\bar{\rho}_\alpha C)/\partial t = -\operatorname{div}\{\theta_m\bar{\rho}_\alpha C\mathbf{V} - \theta_m\mathbf{D}_h \cdot \operatorname{grad}(\bar{\rho}_\alpha C)\} + f_d \quad (7\text{-}48)$$

Another interesting phenomenon is that of dispersion in a *fractured porous medium*. In such medium, most of the porosity is in the small pores of the porous medium and only a small part of it is in the fractures (or fissures). On the other hand, the permeability of the system of fractures is usually much larger than that of the porous medium. A solute (e.g., a pollutant) may advance and spread out rather fast in the fracture system, while penetrating the porous blocks between the fractures very slowly. The velocity in the porous medium is very slow and most of the solute transfer is by molecular diffusion from the fracture to the pore space of the porous medium. Similarly, when a pollutant is flushed by cleaner water, the fractures are flushed fast, but the process of pollution removal from the porous medium (mainly by molecular diffusion) is very slow. In addition, the porous medium with small pores has a very large surface area and stores pollutants adsorbed on it.

Other Forms of the Dispersion Equation

Let us consider some other forms of the hydrodynamic dispersion equation (7-39). Some of the coefficients appearing in these equations are discussed in Sec. 7-3.

So far, we have used the symbols div and grad to indicate that the equations are valid for any coordinate system. In cartesian coordinates, $\mathbf{x}$, we can write (7-39) for $\lambda = 0$ in the form

$$\theta\frac{\partial C}{\partial t} = \frac{\partial}{\partial x_i}\left(\theta D_{hij}\frac{\partial C}{\partial x_j}\right) - q_i\frac{\partial C}{\partial x_i} \quad (7\text{-}49)$$

which, when written in detail for the Cartesian xyz coordinate system, takes the form

$$\theta \frac{\partial C}{\partial t} = \frac{\partial}{\partial x}\left[\theta\left(D_{hxx}\frac{\partial C}{\partial x} + D_{hxy}\frac{\partial C}{\partial y} + D_{hxz}\frac{\partial C}{\partial z}\right)\right]$$

$$+ \frac{\partial}{\partial y}\left[\theta\left(D_{hyx}\frac{\partial C}{\partial x} + D_{hyy}\frac{\partial C}{\partial y} + D_{hyz}\frac{\partial C}{\partial z}\right)\right]$$

$$+ \frac{\partial}{\partial z}\left[\theta\left(D_{hzx}\frac{\partial C}{\partial x} + D_{hzy}\frac{\partial C}{\partial y} + D_{hzz}\frac{\partial C}{\partial z}\right)\right]$$

$$- q_x\frac{\partial C}{\partial x} - q_y\frac{\partial C}{\partial y} - q_z\frac{\partial C}{\partial z} \tag{7-49a}$$

where the coefficients of hydrodynamic dispersion D_{hxx}, D_{hxy}, etc., are obtained from the coefficients of mechanical dispersion D_{xx}, D_{xy}, etc., as defined by (7-11a) by adding (for an isotropic porous medium) D_d^* to D_{xx}, D_{yy}, and D_{zz}.

Equation (7-49a) demonstrates the summation convention used in writing (7-49). The dispersion equations which are presented below can also be written in detail in a similar way.

In generalized curvilinear coordinates, (7-39) for $\lambda = 0$ is written in the form

$$\theta \frac{\partial C}{\partial t} = \frac{\partial}{\partial y^i}\left[\theta D_h^{ij}\frac{\partial C}{\partial y^j}\right] - q^i\frac{\partial C}{\partial y^i} \tag{7-50}$$

where D_h^{ij} is properly expressed in the selected curvilinear coordinate system.

In the square brackets on the right-hand side of (7-50), we have the flux, q_d, due to both mechanical dispersion and molecular diffusion. In curvilinear coordinates, this vector is expressed in the form

$$(q_d)_i = -D_{hij}g^{jk}\,\partial C/\partial y^k \tag{7-50a}$$

where the metric tensor g is defined by (7-21b) through (7-21f). Since in the curvilinear coordinate system y^i, we have

$$\text{div}(q_d) \equiv \frac{\partial}{\partial y^i}(q_d)^i = \frac{1}{\sqrt{g}}\frac{\partial}{\partial y^i}(\sqrt{g}g^{ik}(q_d)_k) \tag{7-50b}$$

equation (7-50) can also be written as

$$\theta \frac{\partial C}{\partial t} = \frac{1}{\sqrt{g}}\frac{\partial}{\partial y^k}\left\{\theta\sqrt{g}D_h^{ik}\frac{\partial C}{\partial y^i}\right\} - g^{ik}q_k\frac{\partial C}{\partial y^i} \tag{7-51}$$

Note that the *physical components* $q(1)$, $q(2)$, $q(3)$ of the vector q are given by $q(i) = q^i(g_{ii})^{1/2} = (g_{ii})^{1/2}g^{ij}q_j$ (with no summation on i). In (7-51), we may replace $(D_h)^{ik}$ by: $(D_h)^{ik} = g^{ip}g^{kq}(D_h)_{pq}$ or: $D_h^{ik} = g^{km}(D_h)_m^i$. The relationships for D in Sec. 7-3 are always in terms of the physical components of the velocity.

In an orthogonal coordinate system, $g_{ij} = (h_i)^2$ for $i = j$, $g^{ij} = 0$ for $i \neq j$ and $g^{1/2} = h_1 h_2 h_3$. Recalling that: $D_h^{ij} = D^{ij} + g^{ij}D_d^*$ and $D_h^{ij} = g^{ip}g^{jq}(D_h)_{pq}$, we obtain for (7-51)

$$\theta \frac{\partial C}{\partial t} = \frac{1}{h_1 h_2 h_3} \frac{\partial}{\partial y^i} \left\{ \frac{h_1 h_2 h_3}{(h_i)^2} \theta g^{ik} (D_h)_{kj} \frac{\partial C}{\partial y^j} \right\} - \frac{q_i}{(h_i)^2} \frac{\partial C}{\partial y^i} \tag{7-52}$$

or in terms of a_L and a_T for an isotropic porous medium (assuming $f(Pe, \delta) = 1$)

$$\theta \frac{\partial C}{\partial t} = \frac{1}{h_1 h_2 h_3} \frac{\partial}{\partial y^i} \left\{ \frac{h_1 h_2 h_3}{(h_i)^2} \theta \left[(a_L - a_T) \frac{V_i V_k}{V} \frac{1}{(h_k)^2} \frac{\partial C}{\partial y^k} + a_T V \frac{\partial C}{\partial y^i} + D_d^* \frac{\partial C}{\partial y^i} \right] \right\}$$
$$- \frac{q_i}{(h_i)^2} \frac{\partial C}{\partial y^i} \tag{7-53}$$

with no summation in $(h_i)^2$ and $(h_k)^2$; $V = |\mathbf{V}|$.

When $V_1 = V \neq 0$, $V_2 = V_3 = 0$, (7-53) reduces to

$$\theta \frac{\partial C}{\partial t} = \frac{1}{h_1 h_2 h_3} \left\{ \frac{\partial}{\partial y^1} \left[\frac{h_2 h_3}{h_1} \theta (a_L V + D_d^*) \frac{\partial C}{\partial y^1} \right] + \frac{\partial}{\partial y^2} \left[\frac{h_3 h_1}{h_2} \theta (a_T V + D_d^*) \frac{\partial C}{\partial y^2} \right] \right.$$
$$\left. + \frac{\partial}{\partial y^3} \left[\frac{h_1 h_2}{h_3} \theta (a_T V + D_d^*) \frac{\partial C}{\partial y^3} \right] \right\} - \theta \frac{V}{h_1} \cos(\mathbf{V}, \mathbf{1y}^1) \frac{\partial C}{\partial y^1} \tag{7-54}$$

where: $V_i = V h_i \cos(\mathbf{V}, \mathbf{1y}^i)$.

If ds^i is an element of length along the coordinate dy^i, we have $ds^i = h_i dy^i$ and $(1/h_i) \partial C/\partial y^i = \partial C/\partial s^i$. Equation (7-54) becomes

$$\theta \frac{\partial C}{\partial t} = \frac{1}{h_2 h_3} \frac{\partial}{\partial s^1} \left[h_2 h_3 \theta (a_L V + D_d^*) \frac{\partial C}{\partial s^1} \right] + \frac{1}{h_1 h_3} \frac{\partial}{\partial s^2} \left[h_3 h_1 \theta (a_T V + D_d^*) \frac{\partial C}{\partial s^2} \right]$$
$$+ \frac{1}{h_1 h_2} \frac{\partial}{\partial s^3} \left[h_1 h_2 \theta (a_T V + D_d^*) \frac{\partial C}{\partial s^3} \right] - \theta V \cos(\mathbf{V}, \mathbf{1s}^1) \frac{\partial C}{\partial s^1} \tag{7-55}$$

In cartesian coordinates, the flow for which $V^1 = V \neq 0$, $V_2 = V_3 = 0$ is uniform flow. In general, (7-54) and (7-55) are written in orthogonal curvilinear coordinates in which the base vectors, which are tangent to the coordinate lines at each point, are the principal axes of dispersion in the neighborhood of that point.

An orthogonal curvilinear coordinate system which satisfies this condition is that in which one family of coordinate lines, say y^1, is the family of streamlines $\Psi = \text{const.}$ In two-dimensional plane saturated flow in a homogeneous isotropic porous medium, such a system is composed of the family of streamlines $\Psi = \text{const.}$ and that of equipotentials $\Phi = \text{const.}$ (Bear, 1972, p. 233), with $\Phi = \text{velocity potential} = -K\phi/n$, $\mathbf{V} = \text{grad}\,\Phi$, $\partial\Phi/\partial x = \partial\Psi/\partial y$, $\partial\Phi/\partial y = -\partial\Psi/\partial x$ (Cauchy–Riemann conditions). For this system

$$y^1 = \Phi(x, y); \qquad y^2 = \Psi(x, y), \qquad ds^i = h_i dy^i$$
$$h_1 \equiv h_\phi = |\text{grad}\,\Phi|^{-1}, \qquad h_2 \equiv h_\psi = |\text{grad}\,\Psi|^{-1}$$

Since $|\text{grad}\,\Phi| = |\text{grad}\,\Psi|$, $h_1 = h_2 = h$; $V = |\text{grad}\,\Phi|$, we obtain

$$\frac{\partial C}{\partial t} = V^2 \left\{ \frac{\partial}{\partial \Phi}\left[(a_L V + D_d^*)\frac{\partial C}{\partial \Phi} \right] + \frac{\partial}{\partial \Psi}\left[(a_T V + D_d^*)\frac{\partial C}{\partial \Psi} \right] \right\} - V^2 \cos(\mathbf{V}, \mathbf{1}\Phi)\frac{\partial C}{\partial \Phi}$$

$$(7\text{-}56)$$

As an example, consider the case of cylindrical coordinates (r, θ, z) and (7-51) written for saturated flow $\theta \equiv n = $ const., in the form

$$\frac{\partial C}{\partial t} = \frac{1}{\sqrt{g}}\frac{\partial}{\partial y^i}\left[\sqrt{g}\, g^{ip}g^{jq}D_{hpq}\frac{\partial C}{\partial y^j} \right] - g^{ik}V_k\frac{\partial C}{\partial y^i}$$

$$(7\text{-}57)$$

In this case: $g^{11} = 1$, $g^{22} = 1/r^2$, $g^{33} = 1$; $g^{ij} = 0$ for $i \neq j$ and $\det g^{ij} \equiv g = r$. We obtain

$$\begin{aligned}
\frac{\partial C}{\partial t} = \frac{1}{r}\Bigg\{ & \frac{\partial}{\partial r}\left[r\left(D_{hrr}\frac{\partial C}{\partial r} + \frac{1}{r^2}D_{hr\theta}\frac{\partial C}{\partial \theta} + D_{hrz}\frac{\partial C}{\partial z} \right) \right] \\
& + \frac{\partial}{\partial \theta}\left[r\left(\frac{1}{r^2}D_{h\theta r}\frac{\partial C}{\partial r} + \frac{1}{r^4}D_{h\theta\theta}\frac{\partial C}{\partial \theta} + \frac{1}{r^2}D_{h\theta z}\frac{\partial C}{\partial z} \right) \right] \\
& + \frac{\partial}{\partial z}\left[r\left(D_{hzr}\frac{\partial C}{\partial r} + \frac{1}{r^2}D_{hz\theta}\frac{\partial C}{\partial \theta} + D_{hzz}\frac{\partial C}{\partial z} \right) \right] \Bigg\} \\
& - V_r\frac{\partial C}{\partial r} - \frac{1}{r^2}V_\theta\frac{\partial C}{\partial \theta} - V_z\frac{\partial C}{\partial z}
\end{aligned}$$

$$(7\text{-}58)$$

For the special case of radial symmetry, that is $\partial(\ )/\partial \theta = 0$, we obtain

$$\begin{aligned}
\frac{\partial C}{\partial t} = \frac{1}{r}\frac{\partial}{\partial r}\left[r\left(D_{hrr}\frac{\partial C}{\partial r} + D_{hrz}\frac{\partial C}{\partial z} \right) \right] + \frac{\partial}{\partial z}\left(D_{hzr}\frac{\partial C}{\partial r} + D_{hzz}\frac{\partial C}{\partial z} \right) \\
- V_r\frac{\partial C}{\partial r} - V_z\frac{\partial C}{\partial z}
\end{aligned}$$

$$(7\text{-}59)$$

In plane radial flow, also, $\partial(\ )/\partial z = 0$, and we obtain

$$\frac{\partial C}{\partial t} = \frac{1}{r}\frac{\partial}{\partial r}\left[rD_{hrr}\frac{\partial C}{\partial r} \right] - V_r\frac{\partial C}{\partial r}$$

$$(7\text{-}60)$$

where we often assume $\partial(rV_r)/\partial r = 0$.

The coefficient of dispersion: D_{hpq} for an orthogonal system is given by

$$D_{hpq} = a_T g_{pq} V + (a_L - a_T)V_p V_q/V + D_d^* g_{pq}$$

$$(7\text{-}61)$$

where $g_{11} = 1$, $g_{22} = r^2$, $g_{33} = 1$, $g_{pq} = 0$ for $p \neq q$, and all V's have the non-physical values. Then

$$D_{hrr} = a_T V + (a_L - a_T)V_r^2/V + D_d^*; \qquad D_{hr\theta} = (a_L - a_T)V_r V_\theta/V = D_{h\theta r}$$

$$D_{hrz} = (a_L - a_T)V_r V_z/V = D_{hzr}; \qquad D_{h\theta\theta} = r^2 a_T V + (a_L - a_T)V_\theta^2/V + D_d^* r^2$$

$$D_{h\theta z} = (a_L - a_T)V_\theta V_z/V = D_{hz\theta}; \qquad D_{hzz} = a_T V + (a_L - a_T)V_z^2/V + D_d^*$$

$$(7\text{-}61a)$$

Recall that the physical components of $\mathbf{V}$ (from $|\mathbf{A}|^2 = g^{ij}A_iA_j = g_{ij}A^iA^j$, $A(j) = (g_{jj})^{1/2} A^j$; $A(j) = (g_{jj})^{1/2} g^{ij} A_i$) are: $V(r) = V_r = V^r$; $V(\theta) = rV^\theta = (1/r) V_\theta$, $V(z) = V_z = V^z$. Also $D(ij) = (g^{ii}/g_{jj})^{1/2} g^{im}D_{mj} = (g^{ii}/g_{jj})^{1/2} g_{jm}D^{im}$

Hence

$$D_{\theta\theta} = r^2\big(a_T V + (a_L - a_T) V^2(\theta)/V\big) = r^2 D(\theta\,\theta)$$

$$D_{rr} = D(r\,r); \qquad D_{rz} = D(r\,z); \qquad D_{zz} = D(z\,z)$$

$$D_{r\theta} = r(a_L - a_T) V(r) V(\theta)/V = rD(r\,\theta)$$

$$D_{\theta z} = r(a_L - a_T) V(\theta) V(z)/V = rD(\theta\,z) \tag{7-61b}$$

and

$$
\begin{aligned}
\frac{\partial C}{\partial t} = \frac{1}{r}\bigg\{ &\frac{\partial}{\partial r}\bigg[r\bigg(a_T V + (a_L - a_T)\frac{V^2(r)}{V} + D_d^* \bigg)\frac{\partial C}{\partial r} + (a_L - a_T)\frac{V(r) V(\theta)}{V}\frac{\partial C}{\partial \theta} \\
&+ r(a_L - a_T)\frac{V(r) V(z)}{V}\frac{\partial C}{\partial z} \bigg] + \frac{\partial}{\partial \theta}\bigg[(a_L - a_T)\frac{V(r) V(\theta)}{V}\frac{\partial C}{\partial r} \\
&+ \frac{1}{r}\bigg(a_T V + (a_L - a_T)\frac{V^2(\theta)}{V} + D_d^* \bigg)\frac{\partial C}{\partial \theta} + (a_L - a_T)\frac{V(\theta) V(z)}{V}\frac{\partial C}{\partial z} \bigg] \\
&+ \frac{\partial}{\partial z}\bigg[r(a_L - a_T)\frac{V(r) V(z)}{V}\frac{\partial C}{\partial r} + (a_L - a_T)\frac{V(\theta) V(z)}{V}\frac{\partial C}{\partial \theta} \\
&+ r\bigg(a_T V + (a_L - a_T)\frac{V^2(z)}{V} + D_d^* \bigg)\frac{\partial C}{\partial z} \bigg]\bigg\} - V(r)\frac{\partial C}{\partial r} - \frac{1}{r}V(\theta)\frac{\partial C}{\partial \theta} \\
&- V(z)\frac{\partial C}{\partial z}
\end{aligned}
\tag{7-62}
$$

For the case of radial symmetry

$$
\begin{aligned}
\frac{\partial C}{\partial t} = \frac{1}{r}\frac{\partial}{\partial r}\bigg[&r(a_T V + (a_L - a_T)\frac{V^2(r)}{V} + D_d^*)\frac{\partial C}{\partial r} + r(a_L - a_T)\frac{V(r) V(z)}{V}\frac{\partial C}{\partial z} \bigg] \\
+ \frac{\partial}{\partial z}\bigg[&(a_L - a_T)\frac{V(r) V(z)}{V}\frac{\partial C}{\partial r} + \bigg(a_T V + (a_L - a_T)\frac{V^2(z)}{V} + D_d^* \bigg)\frac{\partial C}{\partial z} \bigg] \\
&- V(r)\frac{\partial C}{\partial r} - V(z)\frac{\partial C}{\partial z}
\end{aligned}
\tag{7-63}
$$

For plane radial flow, $V(r) \equiv V$

$$\frac{\partial C}{\partial t} = \frac{1}{r}\frac{\partial}{\partial r}\bigg[r(a_L V + D_d^*)\frac{\partial C}{\partial r} \bigg] - V\frac{\partial C}{\partial r} \tag{7-64}$$

For $rV = $ const. and $D_d^* \ll a_L V$, this reduces to

$$\frac{\partial C}{\partial t} = a_L V\frac{\partial^2 C}{\partial r^2} - V\frac{\partial C}{\partial r} \tag{7-65}$$

For one-dimensional flow in the z-direction, $V(z) \equiv V = $ const.

$$\frac{\partial C}{\partial t} = (a_L V + D_d^*)\frac{\partial^2 C}{\partial z^2} - V\frac{\partial C}{\partial z} \tag{7-66}$$

Several other equations of practical interest are discussed by Bachmat and Bear (1964).

It is easy to insert in the above equations the effect of $f(Pe, \delta)$. There should also be no problem in adding the term corresponding to adsorption and/or radioactive decay when necessary.

7-5 INITIAL AND BOUNDARY CONDITIONS

The solution of any of the partial differential equations describing the transport of a solute in an aquifer, requires additional information in the form of initial and boundary conditions to be satisfied by the dependent variable—usually the concentration (say, C) of the solute. Sometimes, similar information is needed also with respect to the concentration, F, on the solid phase, if the latter is an additional dependent variable of the problem.

The boundary itself may be either some arbitrary (mathematical) surface within the porous medium (this is already a boundary in the macroscopic sense) or it may be one across which there exists an abrupt change ($=$ jump) in the volumetric fraction, θ_α, of any phase α, including the solid phase. External boundaries of the flow domain are of the latter type. However, it is of interest also to refer to the general case of a boundary between two media having a jump in θ_α because wherever abrupt jumps occur in the values of the parameters appearing in the transport equations, we have to divide the domain into subdomains along the surfaces of abrupt change and treat these surfaces as boundaries of their respective subdomains.

As initial conditions we must specify the concentration distribution at some initial time, $t = 0$, at all points of the flow domain

$$C(\mathbf{x}, 0) = f_1(\mathbf{x}), \qquad \text{or} \qquad c(\mathbf{x}, 0) = f_2(\mathbf{x}) \tag{7-67}$$

where the vector $\mathbf{x}$ indicates the position of a point (components, x, y, z in cartesian coordinates) and f_1, f_2 are known functions. We shall use c ($\equiv \bar{c}$) when $\bar{\rho}_\alpha \neq$ const., and C when $\bar{\rho}_\alpha =$ const.

The conditions along the boundaries of a flow domain throughout which the transport of a solute takes place depend on the type of medium and fluid present in the region just outside these boundaries. Essentially, however, *all* boundary conditions express the requirement that at any point on the boundary surface the mass flux of the solute (mass of solute per unit area of surface per unit time) normal to the boundary must be equal on both sides of a stationary boundary.

The detailed development of the general boundary condition is given in App. A-8 in the form of (A-29). Here we shall apply (A-29) to the case of $e_\alpha = \bar{c}_\alpha/\rho_\alpha$, where, following (7-2) and (7-5)

$$\mathbf{J}^*(e_\alpha) \equiv \hat{c}_\alpha \mathbf{\overset{\circ}{V}}_\alpha + \overline{\mathbf{J}}(e_\alpha) = -\mathbf{D} \cdot \operatorname{grad}(\bar{c}_\alpha) - \mathbf{D}_d^* \cdot \operatorname{grad}(\bar{c}_\alpha) = -\mathbf{D}_h \cdot \operatorname{grad}(\bar{c}_\alpha)$$

and $\mathbf{D}_h$ is the coefficient of hydrodynamic dispersion.

Let us consider two media, denoted by superscripts a and b, with a boundary between them moving at the macroscopic velocity u_n. On the basis of (A-29), omitting the overbar on c and $\mathbf{V}$, we can write

$$v_i [c_\alpha \theta_\alpha (V_{\alpha i} - u_i) - \theta_\alpha D_{hij} \partial c_\alpha / \partial x_j]_{a,b} = 0$$

or

$$v_i \{c_\alpha \theta_\alpha (V_{\alpha i} - u_i) - \theta_\alpha D_{hij} \partial c_\alpha / \partial x_j\}^a = v_i \{c_\alpha \theta_\alpha (V_{\alpha i} - u_i) - \theta_\alpha D_{hij} \partial c_\alpha / \partial x_j\}^b \quad (7\text{-}68)$$

where the v_i's are components of the unit vector $\mathbf{1}v$ in the direction of the outward normal to region a and the jump from side a to side b is denoted by $[\ \ ]_{a,b}$. It is also possible to add a term to represent an accumulation of solute on the surface itself. The coefficient of hydrodynamic dispersion can, in turn, be expressed in terms of a_L and a_T.

For $\bar{\rho}_\alpha = \text{const.} = \rho$, we can write

$$[\theta_\alpha C (V_{\alpha n} - u_n) - v_i \theta_\alpha D_{hij} \partial C / \partial x_j]_{a,b} = 0 \quad (7\text{-}69)$$

For a stationary boundary, $u_n = 0$ in (7-69).

Consider the following cases of particular interest, where, for the sake of simplicity, the flow is assumed one-dimensional in the $+x$ direction. $\bar{\rho}_\alpha = \text{const.}$ and $u_n \equiv 0$.

(i) Both a and b domains are porous media, but of different characteristics. The boundary condition (7-69) becomes

$$\{Cq_\alpha - \theta_\alpha D_h (\partial C / \partial x)\}^a = \{Cq_\alpha - \theta_\alpha D_h (\partial C / \partial x)\}^b \quad (7\text{-}70)$$

(ii) Region b, outside the porous medium domain a, is an α phase liquid continuum which is assumed to be continuously mixed so that it is maintained at a constant concentration C_2. On the boundary, we have also to maintain $q_\alpha^a = q_\alpha^b \equiv q_\alpha$. The boundary condition (7-70) reduces to

$$\{q_\alpha (C - C_2) - \theta_\alpha D_h \partial C / \partial x\}^a = 0 \quad (7\text{-}71)$$

as in the domain b we have no dispersion or diffusion. We could assume $C_2 \neq \text{const.}$ and molecular diffusion to take place in region b.

(iii) Assuming that in case (ii) above, we have, after a sufficiently long time, $C = C_2$, that is, the same concentration on both sides of the boundary, (7-71) reduces to

$$\{\partial C / \partial x\}^a = 0 \quad \text{or} \quad C(\mathbf{x}, t) = C_2 \text{ on the boundary.} \quad (7\text{-}72)$$

This boundary condition is also applicable to the case where the external domain b is a vacuum or a gas continuum.

It is important to note that if both regions are porous media of different properties, we may have $[\theta_\alpha]_{1,2} \neq 0$. This is, for example, the case of unsaturated flow in a layered domain.

(iv) Region b is completely impervious to flow of both water and solute. In this (one-dimensional) case, $q = 0$ and therefore the boundary condition is

$$\partial C / \partial x = 0 \tag{7-72a}$$

In the more general (three-dimensional) case, we have to use (7-68).

7-6 MATHEMATICAL STATEMENT OF A GROUNDWATER POLLUTION PROBLEM

We now have all the elements required for stating any problem of solute movement in groundwater flow.

An *ideal tracer problem* is one in which the density of the fluid remains constant in spite of variations in solute concentration C ($= c_\alpha/\rho_\alpha$). Otherwise, we have a *real tracer problem* where $\rho_\alpha \neq$ const. and we use c ($\equiv \bar{c}_\alpha$) as a variable. In saturated flow, we usually assume that the porosity $n \equiv \theta_\alpha =$ const. In unsaturated flow, θ_α is also a dependent variable. When adsorption (or other surface phenomena) takes place, F is another dependent variable.

In all problems, we have to know the velocity $\mathbf{V}_\alpha = \mathbf{q}_\alpha/\theta_\alpha$. This means that additional equations are required in order to solve also for the velocity distribution.

In each case, the statement of a solute ($=$ pollutant) transport problem consists of the following parts.

(a) Specifying the flow domain and its boundaries.
(b) Specifying the dependent variables $C(\mathbf{x}, t)$ or $c(\mathbf{x}, t)$ and additional ones, e.g., $\mathbf{V}(\mathbf{x}, t)$, or equivalently $\phi(\mathbf{x}, t)$ or $p(\mathbf{x}, t)$ from which $\mathbf{V}$ or $\mathbf{q}$ can be determined, $\rho_\alpha(\mathbf{x}, t)$, $\theta_\alpha(\mathbf{x}, t)$ when necessary. Certain relationships (e.g., between ρ and C) may exist among the dependent variables.
(c) Stating the partial differential equation for C and additional ones for the other dependent variables.
(d) Stating initial and boundary conditions for each of the partial differential equations.

For the ideal tracer case (that is, $\rho_\alpha =$ const. and no adsorption), we use (7-39). When $\rho_\alpha \neq$ const., we use (7-42). In both cases, θ_α is the volumetric fraction of the α phase; in saturated flow, it is the porosity n.

Our main objective is to solve for the concentration $C(\mathbf{x}, t)$ or $c(\mathbf{x}, t)$. However, only in the ideal tracer case, where density is unaffected by concentration changes, can we separate between the flow problem—i.e., determining the velocity distribution $\mathbf{V}_\alpha(\mathbf{x}, t)$ and the distribution of water content $\theta_\alpha(\mathbf{x}, t)$, in unsaturated flow—and the transport one. In this case, the flow problem is solved first and then $\mathbf{V}_\alpha(\mathbf{x}, t)$ and $\theta_\alpha(\mathbf{x}, t)$ are inserted in the transport equation. When $\rho_\alpha = \rho_\alpha(c)$, and since velocity depends on ρ_α, we have to solve for ρ_α, $\mathbf{V}_\alpha$, and c *simultaneously*.

As an example, consider the case of flow in a confined coastal aquifer, where an assumption of an abrupt interface is not justified and we have to solve the problem as one of mass transport with a variable density (actually also viscosity). We have saturated flow without adsorption.

The flow domain $ABCDE$ is shown in Fig. 7-5. We shall use c as the dependent variable (mass of salt per unit volume of water). The partial differential equation to be solved is (7-42)

$$\frac{\partial(nc)}{\partial t} = \nabla \cdot n\mathbf{D}_h \cdot \nabla c - \nabla \cdot \mathbf{q}c \qquad (7\text{-}73)$$

The mass balance for the fluid given by (5-24)

$$\partial(n\rho)/\partial t = -\nabla \cdot \rho\mathbf{q}, \quad \text{or} \quad \rho S_{0p}\,\partial p/\partial t + n\frac{\partial \rho}{\partial c}\frac{\partial c}{\partial t} = -\nabla \cdot \rho\mathbf{q} \quad (7\text{-}74)$$

and the motion equation by

$$\mathbf{q} = -\frac{k}{\mu}\cdot(\nabla p + \rho g\,\nabla z); \qquad \mathbf{q} = n\mathbf{V} \qquad (7\text{-}75)$$

where $\mathbf{D}_h$ is expressed in terms of a_L, a_T, D_d^* and $\mathbf{V}$ by (7-11), k is the permeability of the porous medium and S_{0p} is specific storativity related to pressure (Sec. 5-1). We need relationships $\rho = \rho(c)$ and $\mu = \mu(c)$. For example

$$\rho = \rho_0 + a(c - c_0); \qquad \mu = \mu_0 + b(c - c_0) \qquad (7\text{-}76)$$

where a, b, c_0 are known constants. For a compressible fluid, we need also a relationship for $\rho = \rho(p, c)$, however, usually (in sea water intrusion problems) we neglect the changes in ρ due to pressure changes, and in μ due to pressure and concentration changes. Segol *et al.* (1975) approximated (7-74) by $\nabla \cdot (\rho\mathbf{q}) = 0$.

As initial condition, we shall assume $c = 0$ everywhere.

The boundary conditions for c are as follows.

Assume that AB and DE are sufficiently removed from regions where changes in c take place. Hence

$$c = 0 \quad \text{along } DE; \qquad c = c_s \quad \text{along } AB$$

Along the impervious boundaries, the total flux perpendicular to the boundary

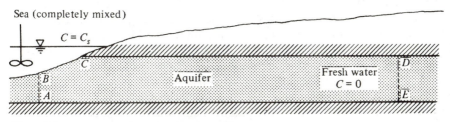

Figure 7-5 Cross section through a coastal aquifer.

is zero. Hence $q_n c - n D_{hij}(\partial c/\partial x_j)\nu_i = 0$, and since also $q_n = 0$, we have

$$\partial c/\partial n = 0 \quad \text{along } AE \text{ and } CD$$

Along the sea bottom, BC, we have continuity of salt fluxes and of water fluxes, while the sea is *assumed* to remain at a constant concentration, c_s. Hence

$$\nu_i q_i(c_s - c) = - n D_{hij}(\partial c/\partial x_j)\nu_i \quad \text{along } BC$$

Often, it is assumed that after a sufficiently long time $c = c_s$ along BC and then also $\partial c/\partial n = 0$ there.

The boundary conditions for p are as follows.

We assume a hydrostatic pressure distribution both along ED and CBA. Along the impervious boundaries, CD and AE, we have $q_n = 0$, or $\partial p/\partial z = -\rho g$.

This completes the mathematical statement of the problem.

In order to solve a forecasting problem for groundwater flow, information is required on aquifer parameters such as storativity and transmissivity. Similarly, in order to solve a forecasting problem for water quality as stated above, we have to know the various dispersion coefficients. These are derived by solving the *inverse problem*, or the *identification problem*, using as input field data with respect to solute concentration, water levels, etc.

7-7 AQUIFER DISPERSION EQUATION AND PARAMETERS

The equation of (three-dimensional) hydrodynamic dispersion at a point in an aquifer is discussed in Sec. 7-4. For example, for the case of saturated flow, $\theta_\alpha = n$, $\mathbf{q}_\alpha = \mathbf{q}$, and variable density, $\bar{\rho}_\alpha = \rho_{\text{water}} \neq$ const., we obtain from (7-27)

$$\frac{\partial(n\bar{c})}{\partial t} + \nabla \cdot \{\bar{c}\mathbf{q} + \mathbf{J}_c^*\} - \tau_c^* = 0 \tag{7-77}$$

where $\bar{c}$ denotes the mass of solute per unit volume of water, $\mathbf{J}_c^* = -n\mathbf{D}_h \cdot \nabla\bar{c}$ is the sum of dispersion and diffusive fluxes, $\mathbf{D}_h$ is the coefficient of hydrodynamic dispersion $(= \mathbf{D} + \mathbf{D}^*)$; and $\tau_c^* = -f + n\bar{\rho}_\alpha \bar{\Gamma}_\alpha$ denotes the total source/sink function due to both surface phenomena at the solid–water interface and chemical reactions and decay phenomena of the solute in the water.

In (7-77), we have $\bar{c} = \bar{c}(x, y, z, t)$ and $\mathbf{q} = \mathbf{q}(x, y, z, t)$. However, under certain conditions, the "hydraulic approach" of treating the dispersion problem as one of essentially two-dimensional flow in the horizontal xy plane is justified. Figure 7-6 shows, schematically, several cases of aquifer pollution. In Case 1 we have a continuous surface source, say leachate from a landfill, contaminating a phreatic aquifer with accretion. In the absence of dispersion, the pollutant will advance within a well-defined streamtube. Dispersion would cause both longitudinal and transversal spreading, the latter beyond the surface bounding the streamtube. At a sufficiently large distance from the source (say 10–15 times the thickness of the flow domain), the plume of contaminant will occupy most of the thickness of the aquifer. Beyond such distance, the two-dimensional approach

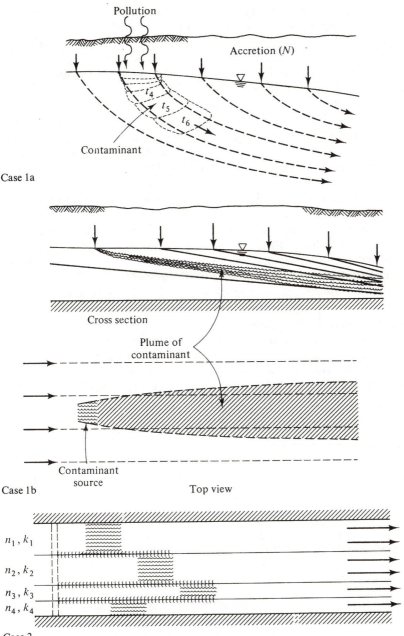

Figure 7-6 Some cases of aquifer pollution. *Case 1a:* Pollution of a homogeneous water table aquifer with accretion; local pollution source; short time and travel distance; phenomenon is 3-dimensional; some contamination beyond streamtube by lateral dispersion. *Case 1b:* Same as Case 1a, except that time is long and travel distance is large; may be considered 2-dimensional phenomenon (in horizontal plane). *Case 2:* Uniform flow in a stratified confined aquifer; fully penetrating pollution source; longitudinal spreading within layers and lateral dispersion between layers.

seems justified. In the absence of accretion, the contaminant will remain close to the surface, with lateral dispersion causing spreading in the downward direction. The lower layers may remain uncontaminated for a rather large distance.

Case 2 describes a fully (or partially) penetrating well injecting water of a different quality into a stratified (that is, $K = K(z)$) confined aquifer. Again, because of transversal dispersion, beyond some distance from the injection well, the average quality of the mixed water may be considered as depending on x and y only. One should remember, that a three-dimensional approach requires also measurements at points in space. On the other hand, a pumping well, even a partially penetrating one, performs an averaging, or mixing, of the water quality along the different elevations (or along different streamlines terminating in the well).

In many regional problems, either we have water which is well mixed (along the vertical) independent of the kind of source, or we can only measure values which are averaged ones along the vertical. This justifies the two-dimensional approach to be discussed here. However, one should avoid the use of this approach when the flow pattern indicates that we have a problem which is strictly three-dimensional in nature, or with clear quality stratification.

The averaged or integrated equation is obtained by integrating (7-77) along the vertical height of the saturated flow domain (App. B).

Before performing the integration, let us examine what happens at an impervious surface, a leaky surface (both in the absence of subsidence) and a phreatic surface which bound the saturated flow domain either from above or from below. Let the boundary elevations be denoted by $b(x, y, t)$, where the time dependence is applicable only to a phreatic surface. Accordingly, the equation of the bounding surface is

$$F = F(x, y, z, t) \equiv z - b(x, y, t) = 0 \qquad (7\text{-}78)$$

For this surface, we also have

$$\frac{dF}{dt} = \frac{\partial F}{\partial t} + \mathbf{u} \cdot \nabla F = 0, \quad \text{or} \quad \mathbf{u} \cdot \nabla F \begin{cases} = 0 \text{ for a stationary boundary} \\ = -\dfrac{\partial F}{\partial t} \text{ for a phreatic surface} \end{cases} \qquad (7\text{-}79)$$

where $\mathbf{u}$ is the velocity of the moving boundary ($=$ phreatic surface).

Assuming that there are no sinks or sources of solute on the boundary, the general continuity condition (A-29) on the boundary requires that

$$[\bar{c}(\mathbf{q} - n\mathbf{u}) + \mathbf{J}_c^*]_{1,2} \cdot \mathbf{1n} = 0 \qquad (7\text{-}80)$$

where $[A]_{1,2} = A|_1 - A|_2$ denotes a jump in A across the boundary and $\mathbf{1n}$ is the unit outward normal to the boundary.

Since $\mathbf{1n} = \nabla F / |\nabla F|$, we obtain from (7-79) and (7-80)

$$[\bar{c}(\mathbf{q} - n\mathbf{u}) + \mathbf{J}_c^*]_{1,2} \cdot \nabla F = 0$$

$$[\bar{c}\mathbf{q} + \mathbf{J}_c^*]_{1,2} \cdot \nabla F = [\bar{c}n]_{1,2}\,\mathbf{u} \cdot \nabla F \begin{cases} = 0 \quad \text{for a stationary boundary} \\ = -[\bar{c}n]_{1,2}\dfrac{\partial F}{\partial t} \text{ for a phreatic surface} \end{cases} \qquad (7\text{-}81)$$

or

$$(\bar{c}\mathbf{q} + \mathbf{J}_c^*)\bigg|_b \cdot \nabla F - (\bar{c}\mathbf{q} + \mathbf{J}_c^*)\bigg|_{ext} \cdot \nabla F \begin{cases} = 0 & \text{for a stationary boundary} \\[2ex] = -[\bar{c}n]_{b,\,ext}\dfrac{\partial F}{\partial t} & \text{for a phreatic surface} \end{cases}$$

$$(7\text{-}82)$$

where "ext" denotes the external side of the considered boundary.

From (7-82) it follows that for a stationary impervious boundary (to both flow and solute)

$$(\bar{c}\mathbf{q} + \mathbf{J}_c^*)\bigg|_b \cdot \nabla F = 0 \tag{7-83}$$

For a stationary leaky boundary

$$(\bar{c}\mathbf{q} + \mathbf{J}_c^*)\bigg|_b \cdot \nabla F = (\bar{c}\mathbf{q} + \mathbf{J}_c^*)\bigg|_{ext} \cdot \nabla F \tag{7-84}$$

For a moving phreatic surface

$$(\bar{c}\mathbf{q} + \mathbf{J}_c^*)\bigg|_b \cdot \nabla F = (\bar{c}\mathbf{q} + \mathbf{J}_c^*)\bigg|_{ext} \cdot \nabla F + [n\bar{c}]_{1,2}\frac{\partial F}{\partial t} \tag{7-85}$$

We can now perform the integration of (7-77) along the vertical thickness of the saturated domain of the considered aquifer, from the bottom at $z = b_1(x, y)$ to the top at $z = b_2(x, y)$. We shall assume that the bottom is always stationary and impervious while the top is stationary when impervious or leaky

$$\int_{b_1}^{b_2} \frac{\partial(n\bar{c})}{\partial t}\,dz + \int_{b_1}^{b_2} \nabla\cdot\{\bar{c}\mathbf{q} + \mathbf{J}_c^*\}\,dz - \int_{b_1}^{b_2} \tau_c^*\,dz = 0 \tag{7-86}$$

For a confined aquifer, with $b_2 = b_2(x, y)$, using the procedure described in App. B-2, we obtain from (7-86)

$$\frac{\partial}{\partial t}\int_{b_1}^{b_2} n\bar{c}\,dz + \nabla'\cdot\int_{b_1}^{b_2}\{\bar{c}\mathbf{q}' + \mathbf{J}_c^{*'}\}\,dz + \{\bar{c}\mathbf{q} + \mathbf{J}_c^*\}\bigg|_{F_2}\cdot\nabla F_2$$

$$- \{\bar{c}\mathbf{q} + \mathbf{J}_c^*\}\bigg|_{F_1}\cdot\nabla F_1 - \int_{b_1}^{b_2}\tau_c^*\,dz = 0 \quad (7\text{-}87)$$

where the primed symbols are defined in App. B-2.

In view of (7-83), and with an average as defined in App. B-2, we obtain from (7-87)

$$\frac{\partial}{\partial t}Bn\widetilde{\bar{c}} + \nabla'\cdot B\{\widetilde{\bar{c}\mathbf{q}'} + \widetilde{\mathbf{J}_c^{*'}}\} - B\widetilde{\tau_c^*} = 0 \tag{7-88}$$

For a leaky aquifer, in view of (7-83) applied to b_1 and (7-84) applied to b_2, we obtain

$$\frac{\partial}{\partial t}Bn\widetilde{\bar{c}} + \nabla'\cdot B\{\widetilde{\bar{c}\mathbf{q}'} + \widetilde{\mathbf{J}_c^{*'}}\} + \{\bar{c}\mathbf{q} + \mathbf{J}_c^*\}\bigg|_{ext}\cdot\nabla(z - b_2) - B\widetilde{\tau_c^*} = 0 \tag{7-89}$$

In this equation, the third term on the left-hand side expresses the leakage of solute by convection, dispersion, and molecular diffusion through the surface $F_2 = 0$.

Finally, for a phreatic aquifer with accretion at a rate $\mathbf{N}$ $(= -N\mathbf{1z} \equiv \mathbf{q}|_{\text{ext}})$ with solute concentration $\bar{c}_N$, we obtain from (7-85) and (7-87)

$$\frac{\partial}{\partial t} Bn\widetilde{\bar{c}} + \nabla' \cdot B\{\widetilde{\bar{c}\mathbf{q}'} + \widetilde{\mathbf{J}_c^{*\prime}}\} - (n\bar{c}_N)\bigg|_{\text{on } h} \frac{\partial h}{\partial t} - \bar{c}_N N - B\widetilde{\tau_c^*} = 0 \qquad (7\text{-}90)$$

or

$$B\frac{\partial}{\partial t} n\widetilde{\bar{c}} + \nabla' \cdot B\{\widetilde{\bar{c}\mathbf{q}'} + \widetilde{\mathbf{J}_c^{*\prime}}\} + \left\{ \widetilde{n\bar{c}} - (\theta_{w0}\bar{c}_N)\bigg|_{\text{on } h} \right\} \frac{\partial h}{\partial t} - \bar{c}_N N - B\widetilde{\tau_c^*} = 0 \qquad (7\text{-}91)$$

where $b_2 \equiv h$ and we have assumed that $\mathbf{J}_c^*|_{\text{ext}} = 0$. The development can easily be extended to the case of a consolidating medium, where both b_1 and b_2 are time dependent and $\mathbf{u} \neq 0$, by using the equation $\mathbf{u} \cdot \nabla F = \partial F/\partial t$ and $[\bar{c}n]_{1,2} \mathbf{u} \cdot \nabla F = -[\bar{c}n]_{1,2} \partial F/\partial t$. When the upper boundary is a phreatic surface, $b_2 = b_2(x, y, t) = h(x, y, t)$.

It is of interest to note that, similar to the integrated continuity equations discussed in Sec. 5-4, the integrated solute balance equations (7-88) through (7-91) can also be derived by assuming only that we have essentially horizontal flow in the aquifer. It is also of interest to note how by the integration, the conditions on the upper and lower boundaries of the aquifer are incorporated in the partial differential balance equation itself. In this equation:

$B n\widetilde{\bar{c}}$ = total mass of solute per unit (horizontal) area of aquifer,
$B\widetilde{\bar{c}\mathbf{q}'}$ = total horizontal flux of solute (by convection) through the entire thickness of the aquifer,
$B\widetilde{\mathbf{J}_c^{*\prime}}$ = total horizontal flux of solute (by hydrodynamic dispersion) through the entire thickness of the aquifer,
$B\widetilde{\tau_c^*}$ = total source/sink function for the entire thickness of the aquifer (per unit area and unit time).

All these variables are now *functions of x and y only.*

Let us now discuss the two fluxes $B\widetilde{\bar{c}\mathbf{q}'}$ and $B\widetilde{\mathbf{J}_c^{*\prime}}$ mentioned above. For the first, $\widetilde{\bar{c}\mathbf{q}'} \equiv \widetilde{(n\bar{c})}\mathbf{\bar{V}'}$, we can write

$$\mathbf{q}' = \widetilde{\mathbf{q}}' + \hat{\mathbf{q}}'; \qquad \bar{c} = \widetilde{\bar{c}} + \hat{c}; \qquad \widetilde{\bar{c}\mathbf{q}'} = \widetilde{\bar{c}}\widetilde{\mathbf{q}}' + \widetilde{\hat{\bar{c}}\hat{\mathbf{q}}'} \qquad (7\text{-}92)$$

or, since $\mathbf{q}' = n\mathbf{\bar{V}'}$

$$\widetilde{n\bar{c}\mathbf{\bar{V}'}} = \widetilde{n\bar{c}}\widetilde{\mathbf{\bar{V}'}} + \widetilde{(\widehat{n\bar{c}})\hat{\mathbf{\bar{V}'}}}; \qquad \widetilde{(n\bar{c})}\,\widetilde{\mathbf{\bar{V}'}} = \widetilde{(n\bar{c})}\,\widetilde{\mathbf{\bar{V}'}} \qquad (7\text{-}93)$$

where $n\bar{c}$ is the mass of solute per unit volume of porous medium and the hat (^) symbol indicates deviation from the average over the vertical.

As in the passage from the microscopic level to the macroscopic one (e.g., (7-1)), the total solute flux here is also made up of two parts, a convective part, with the solute carried by the averaged specific discharge $\widetilde{\mathbf{q}}'$, where $B\widetilde{\mathbf{q}}' = \mathbf{Q}'$ as defined, say, by (4-28) or (4-59), and a *dispersive flux*, expressed by $\widetilde{\hat{\bar{c}}\hat{\mathbf{q}}'}$, which results from the fluctuations in $\mathbf{q}'$ (or in $\overline{\mathbf{V}}'$) along the vertical. An appropriate name for this flux would be *macrodispersion*. This dispersive flux is distinct from $\widehat{\overset{\circ}{c}_\alpha \mathbf{V}_\alpha}$ in (7-1) and (7-2) which results from fluctuations in the microscopic velocity $\mathbf{V}_\alpha$ (of an α phase). However, in principle, one should regard both dispersive fluxes as resulting from basically the same mechanism, except that each involves velocity fluctuations and averaging at a different level and scale, respectively. In mechanical dispersion we have $\overset{\circ}{\mathbf{V}}$ at the microscopic level and the averaging is over an REV the size of which is related to the microscopic inhomogeneity (i.e., presence of solids and voids). In macrodispersion, the fluctuations $\hat{\mathbf{q}}'$ (or $\hat{\mathbf{V}}'$) are in $\mathbf{q}'$ (or $\overline{\mathbf{V}}'$) and the averaging is performed over the entire thickness of the aquifer.

In a stratified aquifer, the hydraulic conductivity varies only with z and the flow is essentially horizontal, i.e., vertical equipotentials. Hence:

$$\mathbf{q}'(z) = -K'(z) \cdot \nabla'\overline{\phi}; \qquad K' = \tilde{K}' + \hat{K}'; \qquad \mathbf{q}' = \tilde{\mathbf{q}}' + \hat{\mathbf{q}}' = -\tilde{K}' \cdot \nabla\overline{\phi} - \hat{K}' \cdot \nabla'\overline{\phi} \tag{7-94}$$

The macrodispersive flux defined in (7-92) is therefore expressed by

$$\widehat{\overset{\approx}{c}\tilde{\mathbf{q}}'} = -\widehat{\overset{\approx}{c}\tilde{K}'} \cdot \nabla'\overline{\phi} \tag{7-95}$$

Thus macrodispersion is related to the variability in hydraulic conductivity.

We may continue the analogy between dispersion and macrodispersion and define, as a *working hypothesis*, a coefficient of macrodispersion $\tilde{D}_{ij}$ $(i, j = 1, 2)$ and a *macrodispersivity* A_{ijkm} $(i. j, k, m = 1, 2)$ related to longitudinal macrodispersivity, A_L, and transversal macrodispersivity, A_T, of the aquifer. With these, the macrodispersive flux will be given by

$$\left.\begin{array}{c} \widehat{\overset{\approx}{\tilde{c}}\tilde{\mathbf{q}}'} = -\tilde{D} \cdot \nabla\tilde{\tilde{c}}; \qquad \tilde{D}_{ij} = \dfrac{\widehat{\tilde{q}_i\tilde{q}_j}}{\tilde{q}'}\,\tilde{L} = A_{ijkm}\,\dfrac{\tilde{q}'_k\tilde{q}'_m}{\tilde{q}'}\,f(Pe^*, \delta^*); \\[1.5em] A_{ijkm} = \dfrac{\widehat{\hat{K}'_{in}\hat{K}'_{jl}}}{\hat{K}'_{nk}\hat{K}'_{lm}}\,\tilde{L}; \qquad i, j, k, m = 1, 2 \ (\equiv x, y) \end{array}\right\} \tag{7-96}$$

(or equivalent expressions in terms of $\overset{\approx}{\tilde{\mathbf{V}}}$ and $\widetilde{n\tilde{c}}$; compare with (7-4) and (7-9); see also Bear, 1972, p. 614), where Pe^* is a Peclet number related to the dispersivity of the medium, and especially to transversal dispersivity of the medium, δ^* is some dimensionless parameter describing the thickness of the layers relative to the thickness of the aquifer, and $\tilde{L}$ is a length characterizing the inhomogeneity of the aquifer due to the stratification. Here the transversal dispersivity plays the same role as that played by molecular diffusion inside an individual channel of a porous medium in producing mixing between adjacent streamlines and making the reference to the average concentration $\bar{c}$ meaningful (see, for example, Saffman, 1960; Bear and Bachmat, 1967).

Gelhar (1976) analyzed the dependence of macrodispersion on permeability variations. For horizontal flow in a confined aquifer he suggested

$$A_L = \frac{1}{3}\frac{L_1^2\sigma_{\ln k}^2}{a_T} \tag{7-97}$$

where L_1 is a correlation distance (= distance along which permeabilities are still correlated) and $\sigma_{\ln k}$ is the standard deviation of $\ln k$ (= natural logarithm of the permeability k).

Following the same line of thinking, one may extend the above ideas to inhomogeneity, say in permeability, in a general three-dimensional flow domain. For example, we may have the flow domain shown in Fig. 2-7, or a similar flow domain, except that the lenses may have a higher permeability than that of the surrounding porous material. This kind of inhomogeneity will produce a dispersion, or spreading, phenomenon with respect to the average flow, similar to that occurring in a layered aquifer, where under the same head gradient flow is faster in the layer with larger K/n value.

Accordingly, (7-96) may intuitively be extended to three-dimensional flow with $i, j, k, m = 1, 2, 3$ (or: x, y, z) and $\mathbf{q}$ replacing $\mathbf{q}'$.

Let us now go back to (7-88) through (7-91) and examine the averaged dispersive flux $\widetilde{\mathbf{J}}_c^{*\prime}$

$$\widetilde{\mathbf{J}}_c^{*\prime} = \frac{1}{B}\int_{b_1}^{b_2} \mathbf{J}_c^{*\prime}\, dz = -\frac{1}{B}\int_{b_1}^{b_2} n\mathbf{D}_h' \cdot \nabla \bar{c}\, dz = -n\widetilde{\mathbf{D}}_h' \cdot \widetilde{\nabla \bar{c}} - \widetilde{(n\mathbf{D}_n')\cdot(\nabla'\bar{c})} \quad (7\text{-}98)$$

Neglecting the second term on the right-hand side of (7-98), we obtain

$$\widetilde{\mathbf{J}}_c^{*\prime} = -n\widetilde{\mathbf{D}}_h' \cdot \widetilde{\nabla \bar{c}} = -n\widetilde{\mathbf{D}}_h' \cdot \frac{1}{B}\int_{b_1}^{b_2} \nabla'\bar{c}\, dz$$

$$= -n\widetilde{\mathbf{D}}_h' \cdot \frac{1}{B}\left[\nabla' \int_{b_1}^{b_2} \bar{c}\, dz - \bar{c}\Big|_{b_2} \nabla b_2 + \bar{c}\Big|_{b_1} \nabla b_1 \right] \quad (7\text{-}99)$$

$$= -n\widetilde{\mathbf{D}}_h' \cdot \left[\nabla'\tilde{\bar{c}} + \frac{1}{B}\left\{ \tilde{\bar{c}}\nabla B - \bar{c}\Big|_{b_2} \nabla b_2 + \bar{c}\Big|_{b_1} \nabla b_1 \right\} \right] \approx -n\widetilde{\mathbf{D}}_h' \cdot \nabla'\tilde{\bar{c}}$$

where we have assumed $\tilde{\bar{c}} \approx \bar{c}|_{b_2} \approx \bar{c}|_{b_1}$. From the discussion above, it seems reasonable to assume that $\widetilde{D} \gg \widetilde{nD_n'}$. This explains why the dispersivity as derived from field experiments of solute spreading is several orders of magnitude that of the dispersivity related to size of individual grains or openings as obtained in laboratory experiments with a homogeneous medium.

In conclusion, the movement and accumulation of a contaminant in a confined aquifer in which we can justify an averaged approach (i.e., the hydraulic approach with respect to concentration), are governed by

$$\frac{\partial}{\partial t} Bn\tilde{\bar{c}} + \nabla' \cdot B\{\tilde{\bar{c}}\tilde{\mathbf{q}}' - \widetilde{D} \cdot \nabla'\tilde{\bar{c}}\} - B\tilde{\tau}_c^* = 0 \quad (7\text{-}100)$$

obtained from (7-88). The solute is carried by convection and macrodispersion and we have neglected averaged hydrodynamic dispersion. Similar integrated equations can be written for a leaky aquifer and a phreatic one.

The extension to macrodispersion in three-dimensional flow requires the definition of a new, larger REV and averaging over it.

Marle *et al.* (1967) considered dispersion in a stratified aquifer and proposed for it an expression for an equivalent coefficient of aquifer hydrodynamic dispersion.

7-8 METHODS OF SOLUTION

As in the case of unsaturated flow, the only model which, in principle, can simulate a regional solute transport problem is the sand box model (Bear, 1972, p. 683). However, because of both scaling and technical difficulties, the application of this tool to the solution of problems of practical interest should be ruled out. Similarly, analytic methods of solution are applicable mainly to simple one-dimensional problems. This leaves numerical methods, mainly of the finite difference and finite element types, as the main tools for solving quality fore-casting problems of practical interest. The *method of characteristics* is also used occasionally (Gardner et al., 1964; Fried, 1975, p. 256).

Although, by now, the application of numerical techniques, especially the Galerkin finite element method, to the solution of solute transport problems in aquifers is well established and documented in the literature, certain difficulties and errors, some of which are discussed below, make the actual solution a rather difficult task to the nonexpert. At the present stage, the solution of a solute transport problem should not be considered a matter of routine, even when a program can be found in the open literature. The reader is referred to Pinder and Gray (1977) for a detailed discussion on numerical methods.

Method of Finite Differences

The basic ideas underlying this method and the various numerical schemes are discussed in Sec. 5-6. We may demonstrate the method by analyzing the case of one-dimensional saturated flow, where the solute transport is governed by (7-66)

$$\frac{\partial C}{\partial t} = D_h \frac{\partial^2 C}{\partial x^2} - V \frac{\partial C}{\partial x} \tag{7-101}$$

The centered finite difference approximation used for both spatial derivatives and then weighted and summed together at two different time levels, yields

$$\frac{C_i^{k+1} - C_i^k}{\Delta t} = D_h \left\{ \varepsilon \frac{C_{i-1}^{k+1} - 2C_i^{k+1} + C_{i+1}^{k+1}}{(\Delta x)^2} + (1 - \varepsilon) \frac{C_{i-1}^k - 2C_i^k + C_{i+1}^k}{(\Delta x)^2} \right\}$$

$$- V \left\{ \varepsilon \frac{C_{i+1}^{k+1} - C_{i-1}^{k+1}}{2(\Delta x)} + (1 - \varepsilon) \frac{C_{i+1}^k - C_{i-1}^k}{2(\Delta x)} \right\}, \quad 0 \le \varepsilon \le 1 \tag{7-102}$$

For $\varepsilon = 1.0$, 0.5, and 0.0, (7-102) becomes the implicit, or backward, difference formulation, the centered Crank–Nicolson formulation and the explicit, or forward, difference formulation, respectively. Knowing the values of C^k at time $k\,\Delta t$, we solve for C^{k+1}, that is, at time $(k + 1)\,\Delta t$. The explicit scheme is stable only when the time step, Δt, satisfies

$$\Delta t < (\Delta x)/(2D_h/\Delta x + V) \tag{7-103}$$

The other two schemes are unconditionally stable.

When treating problems in the xy plane, ADI schemes (Sec. 5-6) may also be employed.

Method of Finite Elements

The method itself is presented in Sec. 5-6. Here, let us demonstrate it by applying it to the two-dimensional dispersion equation (7-100) written as

$$L(c) \equiv \frac{\partial}{\partial t}(Bnc) - \frac{\partial}{\partial x}\left(\tilde{D}_{xx} \frac{\partial c}{\partial x} \right) - \frac{\partial}{\partial x}\left(\tilde{D}_{xy} \frac{\partial c}{\partial y} \right) - \frac{\partial}{\partial y}\left(\tilde{D}_{yx} \frac{\partial c}{\partial x} \right)$$

$$- \frac{\partial}{\partial y}\left(\tilde{D}_{yy} \frac{\partial c}{\partial y} \right) + \frac{\partial}{\partial x}(cQ'_x) + \frac{\partial}{\partial y}(cQ'_y) - \tau^* = 0 \qquad (7\text{-}104)$$

where $n\bar{c} \equiv nc$, $\tilde{\bar{c}} \equiv c$, $\tilde{D}_{xx} \equiv B\tilde{D}_{xx}$, etc., $B\tilde{\mathbf{q}}' \equiv \mathbf{Q}'$, $B\tilde{\tau}^* = \tau^*$. It is assumed that $\tilde{D}$ can be expressed by (7-96).

Following the Galerkin formulation presented in Sec. 5-6 and Pinder (1974), we approximate the unknown concentration, c, and also the components of the spatially dependent parameters in terms of basis functions $w_i(x, y)$, by

$$c \cong \hat{c} = \sum_{j=1}^{N} C_j(t)\, w_j(x, y) \qquad (7\text{-}105)$$

where here the hat (^) symbol is not a deviation as in Sec. 7-7, and $\tilde{D}_{xx} \approx \hat{D}_{xx} = \sum_{j=1}^{K} \tilde{D}_{xxj} w_j(x, y)$, and similar expressions for $\tilde{D}_{xy}$, $\tilde{D}_{yx}$, and $\tilde{D}_{yy}$. Also

$$Q'_x \cong \hat{Q}_x = \sum_{j=1}^{N} Q'_{xj} w_j(x, y); \qquad Q'_y \cong \hat{Q}_y = \sum_{j=1}^{N} Q'_{yj} w_j(x, y)$$

where the flux $\mathbf{Q}'$ is obtained from an appropriate Galerkin formulation of the flow problem.

One should note that although the coefficient of dispersion, $\tilde{D}$, plays a role similar to transmissivity, T, in the flux expression, the Galerkin formulation of the dispersion tensor is considerably more complex. Recalling that $\tilde{D}$ depends on the velocity distribution, there does not appear to be a unique form of $\tilde{D}$ which will accurately represent all field situations (Pinder, 1974). When the flow and solute transport problems are separable (Sec. 7-4), the velocity distribution can be introduced as a known input to determine the components of $\tilde{D}$ at every point. Otherwise, we have to regard the head, ϕ, as another unknown and solve the aquifer continuity equation and (7-104), simultaneously for ϕ and c.

The approximating integral equations are obtained from Galerkin's scheme by making the residual, generated by introducing (7-105) into (7-104), orthogonal to each of the N basis functions w_j

$$\iint_{(A)} L(\hat{c})\, w_j\, dA = \iint_{(A)} L\left(\sum_{j=1}^{N} C_j w_j \right) w_i\, dA = 0 \qquad (7\text{-}106)$$

$i = 1, 2, \dots, N$.

Equation (7-106) may be expanded and reformulated in matrix form as (Pinder, 1974)

$$[E]\{C\} + [M]\left\{\frac{dC}{dt}\right\} + \{F\} = 0 \qquad (7\text{-}107)$$

where typical elements of E, M, and F are

$$E_{km} = \iint_{(A)} \left(\hat{D}_{xx} \frac{\partial w_m}{\partial x} \frac{\partial w_k}{\partial x} + 2\hat{D}_{xy} \frac{\partial w_m}{\partial y} \frac{\partial w_k}{\partial x} + \hat{D}_{yy} \frac{\partial w_m}{\partial y} \frac{\partial w_k}{\partial y} + \hat{Q}'_x w_k \frac{\partial w_m}{\partial x} \right.$$

$$\left. + \hat{Q}'_x w_k \frac{\partial w_m}{\partial y} + w_k w_m \frac{\partial \hat{Q}'_x}{\partial x} + w_k w_m \frac{\partial \hat{Q}'_y}{\partial y} \right) dA$$

$$M_{km} = \iint_{(A)} Bn w_k w_m \, dA; \qquad F_k = \iint_{(A)} w_k \tau^* \, dA - \int_{(S)} (w_k \tilde{D} \cdot \nabla c) \cdot \mathbf{1n} \, dS \qquad (7\text{-}108)$$

where $\mathbf{1n}$ is the unit normal vector to the boundary S of A. The time derivative is approximated by using differences.

When solving simultaneously for ϕ and c, we end up with two sets of algebraic equations which have to be solved simultaneously

$$[E]\{C\} + [M]\left\{\frac{dC}{dt}\right\} + \{F\} = 0$$

$$[P]\{H\} + [R]\left\{\frac{dH}{dt}\right\} + \{U\} = 0 \qquad (7\text{-}109)$$

where we have represented $\phi(x, y, t)$ by: $\phi \cong \hat{\phi} = \sum_{j=1}^{N} H_j(t) u_j(x, y)$ in the continuity equation, say: $L(\phi) \equiv S \, \partial \phi / \partial t - \nabla \cdot (\mathbf{T} \cdot \nabla \phi) + Q = 0$, H_j is an undetermined coefficient and u_j is a basis function. The solution is obtained by first solving for ϕ, then calculating $\mathbf{q}$ and finally solving for c.

Galerkin finite element solutions of the solute transport problem are presented, among others, by Guymon *et al.* (1970), Pinder (1973), Gupta *et al.* (1975), Segol (1976), van Genuchten (1977), and Pinder and Gray (1977).

Numerical Dispersion and Overshooting

To illustrate what is meant by numerical dispersion, consider the following finite difference representation of the convective term only in a problem of one-dimensional solute transport in saturated flow (Fig. 7-7). The flux of solute, q_c, from the x_{i-1} cell across its boundary at x_i is given by

$$q_c \Big|_i = q_i C_{i-1/2}^t \qquad (7\text{-}110)$$

If an initially abrupt front starts at time t at some point between x_{i-1} and x_i, say, $x_i + \varepsilon \Delta x$, where $\varepsilon < 1$ (Fig. 7-7), then

$$C_{i-1/2}^t = (C_{i-1}^t + C_i^t)/2 = C_0/2 \qquad (7\text{-}111)$$

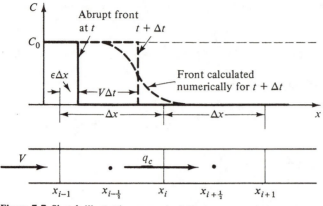

Figure 7-7 Sketch illustrating numerical dispersion.

Hence, the flux at x_i at time t is given by

$$q_c^t \bigg|_i = \tfrac{1}{2} q_i C_0 \qquad (7\text{-}112)$$

and after any time interval, Δt, no matter how small, solute will have already entered the next cell, centered at $x_{i+1/2}$. However, in reality, this is not possible as it takes a period $(1 - \varepsilon) \Delta x / V$, where $V \equiv q/n$, for the solute front to reach x_i by convection. Because of the finite difference representation, solute seems to advance faster than it actually does. This behavior is equivalent to that of dispersion. Thus, although the equation for pure convection is $\partial C/\partial t = -V \partial C/\partial x$, we seem to have solved an equation which is similar to the dispersion equation, namely $\partial C/\partial t = D_h \partial^2 C/\partial x^2 - V \partial C/\partial x$, where D_h is a coefficient of numerical dispersion. It has no physical meaning, but is related to the selection of Δx and Δt with respect to a given flow velocity V.

Mathematically speaking, numerical dispersion is a truncation error resulting from the fact that in approximating the first derivative, $\partial C/\partial x$, we have neglected the term proportional to the second derivative

$$V \frac{\partial C}{\partial x} \cong V \frac{C_i - C_{i-1}}{\Delta x} - V \frac{\Delta x}{2} \frac{\partial^2 C}{\partial x^2} \qquad (7\text{-}113)$$

(see (5-96)). A numerical dispersion coefficient $D_h = V \Delta x / 2$ is thus introduced.

Overshooting and undershooting (Fig. 7-8) are oscillations of a solution as an abrupt front is approached from its upstream and downstream sides. respectively. In the first case we obtain values of concentration that are above the maximum possible ones (say, C_0 in Fig. 7-7). In the latter case, we obtain negative values. Like numerical dispersion, this is another error which is inherent in most numerical solutions of the solute transport equation. The two kinds of error are interrelated.

Bender *et al.* (1975) and van Genuchten (1976) analyze these errors in different numerical schemes and conclude that finite element schemes generally provide more accurate solutions than finite difference ones.

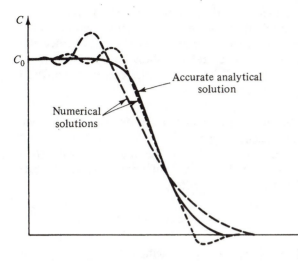

Figure 7-8 Schematic illustrations of oscillations in numerical solutions of an advancing front *(after Bender et al., 1975 and van Genuchten, 1976)*.

Pinder and Gray (1977) comment that when a scheme is developed to minimize numerical dispersion, overshoot is encountered, and when overshoot is controlled it is at the expense of numerical dispersion.

Solute flux is due to both convection and dispersion. As the former becomes more predominant (say, when $VL \gg D_h$, where L is some characteristic length of the flow domain) fronts tend to be sharper and errors due to numerical dispersion become larger.

Several correction methods are proposed in the literature (e.g., van Genuchten, 1976) to reduce (yet not totally eliminate) the effect of numerical dispersion and overshoot. A common technique to reduce numerical dispersion is to use $D_h - V\Delta x/2$ as a dispersion coefficient.

For large scale dispersion problems, numerical dispersion effects may be negligible if mesh sizes are properly selected. Also, errors are large when relatively sharp fronts are involved. In a regional problem where concentrations vary gradually in the flow domain, and concentration gradients are small, the errors introduced by a numerical solution are relatively small.

As in the case of flow of water, we may conclude that we have now practical tools in the form of numerical methods of different kinds to forecast the movement and accumulation of pollutants in aquifers.

7-9 SOME SIMPLE ANALYTIC SOLUTIONS

Analytic solutions of the dispersion equation in cases of practical interest are not feasible, due to the irregularity of the boundaries, inhomogeneity of aquifer dispersivity, etc., and numerical techniques (Sec. 7-8) must be employed. Only for a rather limited number of relatively simple, usually one-dimensional, problems, can an analytical solution be derived. Let us review some of these cases in order to gain some insight into the dispersion pattern that is produced. In all cases we shall assume that the tracer is an *ideal* one (i.e., of constant density and viscosity, in spite of changes in concentration), that the fluid (= water) is incompressible, and that the medium is rigid, homogeneous and isotropic. Only saturated flow is considered. For unsaturated flow, we need relationships between dispersivity components and the degree of saturation. Because of the constant density, C is mass of solute per unit volume of water.

Case 1. Movement of a tracer in an infinite column Consider one-dimensional flow along the x axis. The specific discharge q satisfies $\partial q/\partial x = 0$, $q = q(t)$, that is, specific discharge is constant along the column, but may vary with time; $q > 0$ describes flow in the $+x$ direction. The partial differential equation which governs the solute (= tracer) distribution, reduces to

$$\frac{\partial C}{\partial t} = D_h \frac{\partial^2 C}{\partial x^2} - \frac{q}{n}\frac{\partial C}{\partial x}, \qquad -\infty < x < +\infty \qquad (7\text{-}114)$$

where n = porosity = const., $D_h = a_L|q|/n + D_d^*$ = coefficient of hydrodynamic dispersion.

The initial and boundary conditions are

$$t \leq 0, \quad -\infty < x < 0, \quad C = C_1, \quad t > 0, \quad x = \pm\infty, \quad \partial C/\partial x = 0,$$
$$0 \leq x < +\infty, \quad C = C_0, \qquad\qquad\qquad x = -\infty, \quad C = C_1,$$
$$x = +\infty, \quad C = C_0$$
$$(7\text{-}115)$$

Bear (1960) solves this problem by applying the Laplace transform to (7-114) and (7-115). The solution is

$$\frac{C(x,t) - C_0}{C_1 - C_0} = \frac{1}{2}\,\mathrm{erfc}\left\{ -\frac{x - \int_0^t [q(t)/n]\,dt}{2[\int_0^t (a_L|q|/n + D_d^*)\,dt]^{1/2}} \right\} \qquad (7\text{-}116)$$

For $q = nV$ = const., and neglecting molecular diffusion (that is, $D_d^* \ll a_L V$), (7-116) reduces to

$$\varepsilon(x,t) \equiv \frac{C(x,t) - C_0}{C_1 - C_0} = \frac{1}{2}\,\mathrm{erfc}\left\{ -\frac{x - Vt}{2[Dt]^{1/2}} \right\} = \frac{1}{2}\,\mathrm{erfc}\left\{ -\frac{x - Vt}{\sqrt{2}\sigma} \right\} \quad (7\text{-}117)$$

where $\mathrm{erf}\,x = (2/\sqrt{\pi})\int_0^x e^{-\alpha^2}\,d\alpha$, $\mathrm{erfc}\,x = 1 - \mathrm{erf}\,x = (2\sqrt{\pi})\int_x^\infty e^{-\alpha^2}\,d\alpha$, $\sigma^2 = 2Dt = 2a_L Vt = 2a_L L$ is the variance of the distribution, and $L = Vt$ is the (average) distance travelled by the water during t. Figure 7-9 shows (7-117) in a graphical form for the case $C_0 > C_1$. From (7-117) and Fig. 7-9, it follows that for such a case the point $\varepsilon(x,t) = 0.5$ travels with the mean flow and that the variance of the concentration distribution is proportional to the length (L) travelled. The corresponding conclusions from (7-116), where $q = q(t)$, are that the point $\varepsilon = 0.5$ travels again with the mean flow (that is, with velocity V) and that σ^2 is *proportional to the total path travelled*.

Figure 7-10 shows a case with fluctuating flow, $q = q(t)$ and $C_0 < C_1$.

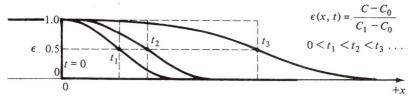

$$\epsilon(x, t) = \frac{C - C_0}{C_1 - C_0}$$

$$0 < t_1 < t_2 < t_3 \cdots$$

Figure 7-9 Tracer concentration distribution in one-directional flow in an infinite column.

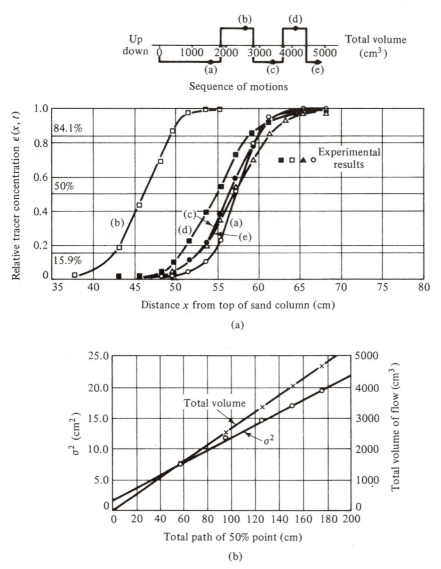

(a)

(b)

Figure 7-10 Dispersion in one-dimensional fluctuating flow *(Bear, 1961a)*. (a) Relative tracer concentration. (b) Relations of total volume of flow and variance to total path of 50% concentration.

Case 2. Infinite column with adsorption For this case, with adsorption described by an equilibrium isotherm, we obtain the partial differential equation from (7-45) and (7-46)

$$\frac{\partial C}{\partial t} = \frac{D_h}{R_d}\frac{\partial^2 C}{\partial x^2} - \frac{q}{nR_d}\frac{\partial C}{\partial x} \tag{7-118}$$

where $C\,(\equiv C^*/n)$ is the mass of solute per unit volume of water, $F\,(\equiv F^*)$ is the mass of solute on the solid matrix per unit volume of porous medium and $R_d(=1+\beta)$ is the *retardation factor*. For steady flow, $\partial C/\partial t = 0$.

It can be seen that (7-118) can be obtained from (7-114) by replacing q/n by q/nR_d and D_h by D_h/R_d. Hence, for the initial and boundary conditions given by (7-115) the solution is

$$\frac{C(x,t) - C_0}{C_1 - C_0} = \frac{1}{2}\mathrm{erfc}\left\{ -\frac{R_d x - (q/n)\,t}{2[R_d D_h t]^{1/2}} \right\} \tag{7-119}$$

which is obtained from (7-117) also by replacing q/n by q/nR_d and D_h by D_h/R_d.

Case 3. Injection of a tracer slug in an infinite column At $t = 0$, a very thin slug of tracer marked particles is injected into a column at $x = 0$. As the slug moves downstream with the specific discharge q in the $+x$ direction, it spreads out, occupying an ever increasing portion of the column. The entire column is initially at $C = 0$. The tracer concentration distribution $C(x,t)$ is governed by (7-114).

For an observer moving with the (average) flow, the governing equation becomes

$$\partial C/\partial t = D_h \partial^2 C/\partial x'^2, \qquad x' = x - (q/n)\,t \tag{7-120}$$

The initial conditions, $C = C(x,0)$, are in the form of a *Dirac delta function*, $\delta(x)$

$$C(x,0) = \frac{M}{n}\delta(x); \quad \delta(x) = \lim_{m\to 0}\delta_m(x) = \begin{cases} (1/m) & \text{for} \quad 0 < x < m, \quad m > 0, \\ 0 & \text{elsewhere} \end{cases}$$

$$M = \int_{-\infty}^{+\infty} nC(x',t)\,dx' \tag{7-121}$$

The boundary conditions, specified for $C(x',t)$ are

$$\lim C(x',t) = 0, \qquad |x'| \to \infty \tag{7-122}$$

The solution is (Crank, 1956)

$$C(x,t) = \frac{M/n}{(4\pi D_h t)^{1/2}}\exp\left\{ -\frac{x'^2}{4D_h t} \right\} = \frac{M/n}{(4\pi D_h t)^{1/2}}\exp\left\{ -\frac{[x - (q/n)\,t]^2}{4D_h t} \right\} \tag{7-123}$$

Figure 7-11 shows the shape of the curves $C = C(x',t)$.

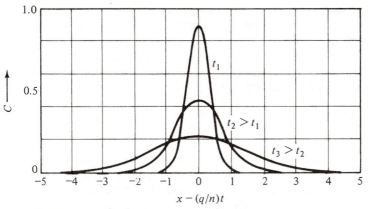

Figure 7-11 Progress of a slug.

Case 4. A continuous injection at $x = 0$ of a radioactive tracer into an infinite column The partial differential equation in this case is

$$\frac{\partial C}{\partial t} = D_h \frac{\partial^2 C}{\partial x^2} - \frac{q}{n}\frac{\partial C}{\partial x} - \lambda C \qquad (7\text{-}124)$$

The elementary solution is

$$C(x, t, t') = \frac{dM}{[4\pi D_h(t - t')]^{1/2}} \exp\left\{ -\frac{[x - (q/n)(t - t')]^2}{4D_h(t - t')} - \lambda(t - t') \right\} \qquad (7\text{-}125)$$

where $dM = C_0(\dot{q}/n)\, dt'$ is the tracer mass injection during dt' at $x = 0$ and $t = t'$; its concentration is C_0. For a continuous injection at a constant rate, we obtain, by integration

$$C(x, t) = \frac{C_0 q/n}{(4\pi D_h)^{1/2}} \exp\left\{ \frac{qx}{2nD_h} \right\} \int_0^{\tau = t} \frac{1}{\sqrt{\tau}} \exp\left\{ -\frac{a}{\tau} - b\tau \right\} d\tau \qquad (7\text{-}126)$$

where $a = x^2/4D_h$, $b = q^2/4D_h n^2 + \lambda$.
As $t \to \infty$, we obtain

$$C(x, \infty) = \frac{C_0}{[1 + 4\lambda\, D_h n^2/q^2]^{1/2}} \exp\left\{ \frac{qx}{2D_h n}(1 - [1 + 4\lambda D_h n^2/q^2]^{1/2}) \right\} \qquad (7\text{-}127)$$

For $x = 0$, (7-126) can be integrated to yield

$$C(0, t) = \frac{C_0}{[1 + 4\lambda D_h n^2/q^2]^{1/2}} \operatorname{erf}\left\{ \frac{q^2 t}{4D_h n^2} + \lambda t \right\} \qquad (7\text{-}128)$$

which shows that $C(0, t) \neq C_0$. As $t \to \infty$, we obtain

$$C(0, \infty) = C_0/[1 + 4\lambda D_h n^2/q^2]^{1/2} \neq C_0 \qquad (7\text{-}129)$$

Case 5. Movement of a radioactive tracer in a semi-infinite column This is the case where the column $(x > 0)$, initially at tracer concentration $C = 0$, is connected to a reservoir containing a tracer solution of constant concentration C_0. The flow in the column is maintained at a constant specific discharge q in the $+x$ direction. The tracer continuously undergoes radioactive decay.

The partial differential equation here is

$$\frac{\partial C}{\partial t} = D_h \frac{\partial^2 C}{\partial x^2} - \frac{q}{n} \frac{\partial C}{\partial x} - \lambda C \tag{7-130}$$

where $\lambda = \ln2/T$ and T is the half-life of the radioactive tracer.

We *assume*, that at $x = 0$ the concentration immediately reaches its ultimate level C_0 upon commencement of flow. This is equivalent to an assumption that at $x = 0$, $\lim_{t \to 0} \partial C / \partial x = 0$. We refer to this condition as an assumption since from (7-71) it follows that at $x = 0$ the boundary condition should actually be

$$C_0 q = Cq - nD_h \, \partial C / \partial x$$

or
$$\tag{7-131}$$

$$q(C_0 - C) = -nD_h \, \partial C / \partial x$$

Accordingly, the initial and boundary conditions are

$$t \leq 0, \qquad x \geq 0, \qquad C = 0$$

$$t > 0, \qquad x = 0, \qquad C = C_0 \tag{7-132}$$

$$x = \infty, \qquad C = 0$$

By applying the Laplace transform to (7-130) and (7-132), we obtain the solution (Bear, 1972, p. 630)

$$C(x, t) = \frac{C_0}{2} \exp\left\{ \frac{qx}{2nD_h} \right\} \cdot \left[\exp(-x\beta) \cdot \text{erfc} \frac{x - [(q/n)^2 + 4\lambda D_h]^{1/2} t}{2[D_h t]^{1/2}} \right.$$

$$\left. + \exp(\beta x) \cdot \text{erfc} \frac{x + [(q/n)^2 + 4\lambda D_h]^{1/2} t}{2[D_h t]^{1/2}} \right] \tag{7-133}$$

where $\beta^2 = q^2/4n^2 D_h^2 + \lambda/D_h$.

For $\lambda = 0$ (i.e., without radioactive decay), (7-133) reduces to

$$C(x, t) = \frac{C_0}{2} \left\{ \text{erfc} \frac{x - (q/n) t}{2[D_h t]^{1/2}} + \exp\left(\frac{qx}{nD_h} \right) \cdot \text{erfc} \frac{x + (q/n) t}{2[D_h t]^{1/2}} \right\} \tag{7-134}$$

If we now introduce adsorption (see Case 2), (7-134) becomes

$$C(x, t) = \frac{C_0}{2} \left\{ \text{erfc} \frac{R_d x - (q/n) t}{2[R_d D_h t]^{1/2}} + \exp\left(\frac{qx}{nD_h} \right) \text{erfc} \frac{R_d x + (q/n) t}{2[R_d D_h t]^{1/2}} \right\}. \tag{7-135}$$

Curves describing (7-134) are shown in Fig. 7-12. According to Ogata and Banks (1961), who also obtained (7-134), the second term in (7-134) may be neglected

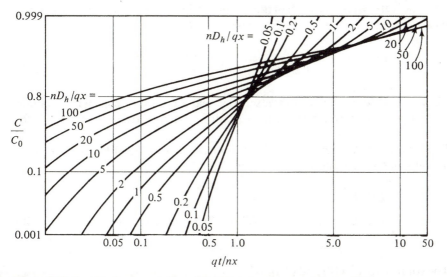

Figure 7-12 Graphical representation of (7-134) *(after Ogata and Banks, 1961)*.

when x/a_L is sufficiently large, a condition usually satisfied in the practice (e.g., an error of less than 3 percent when $x/a_L > 500$). The approximate form of (7-134) is then

$$C(x, t) = \frac{C_0}{2} \operatorname{erfc} \left\{ \frac{x - (q/n) t}{2[D_h t]^{1/2}} \right\}$$ (7-136)

similar to (7-117).

Equation (7-134) is also an approximation of (7-133) when $4\lambda D_h n^2/q^2 \ll 1$. When molecular diffusion is neglected (justifie$d$ when $D_d^* \ll a_L q/n$), D_h reduces to $a_L q/n$. Then the parameter $4\lambda a_L n/q$ is a criterion for the relative importance of the radioactive decay.

When the second term of (7-133) may be neglected, we obtain

$$C(x, t) = \frac{C_0}{2} \exp \left\{ \frac{qx}{2nD_h} \left[1 - \left(1 + \frac{4\lambda D_h n^2}{q^2} \right)^{1/2} \right] \right.$$
$$\left. \times \operatorname{erfc} \left\{ \frac{x - (q/n) [1 + 4\lambda D_h n^2/q^2]^{1/2} t}{2[D_h t]^{1/2}} \right\} \right.$$ (7-137)

All these results can always be modified to include also adsorption, described by an equilibrium isotherm, by replacing D (in the absence of molecular diffusion) by D/R_d and q/n by q/nR_d.

As $t \to \infty$, (7-137) reduces to

$$C(x) = C_0 \exp \left\{ \frac{qx}{2nD_h} [1 - (1 + 4\lambda D_h n^2/q^2)^{1/2}] \right\}$$ (7-138)

which is also the steady state solution of the dispersion equation.

For the case with both adsorption and radioactive decay, and neglecting molecular diffusion, as $t \to \infty$, (7-137) is modified to yield

$$C(x) = C_0 \exp\left\{\frac{qx}{2nD}\left[1 - (1 + 4\lambda D R_d n^2/q^2)^{1/2}\right]\right\} \qquad (7\text{-}139)$$

Case 6. A semi-infinite column with adsorption and a third type boundary condition at $x = 0$ Again, the partial differential equation, obtained from (7-45) and (7-46) is (7-118).

Following Bastian and Lapidus (1956), Gershon and Nir (1969) solve (7-118) for the following initial and boundary conditions

$$
\begin{aligned}
t \leq 0, &\qquad x > 0 &\qquad C = 0, \\
t > 0, &\qquad x = 0^+ &\qquad (C_0 - C)\,q = -nD\,\partial C/\partial x, &\qquad (7\text{-}140) \\
&\qquad x \to \infty &\qquad C = 0.
\end{aligned}
$$

The solution of (7-118) and (7-140) is

$$
\begin{aligned}
\frac{C(x,t)}{C_0} = {} & \frac{1}{2}\,\mathrm{erfc}\left(\frac{R_d x + qt/n}{2[R_d D_h \to D_h t]^{1/2}}\right) - \frac{1}{2}\exp\left\{\frac{qx}{nD_h \to D_h}\right\} \cdot \mathrm{erfc}\left(\frac{R_d x - qt/n}{2[R_d D_h \to D_h t]^{1/2}}\right) \\
& \times \left(1 + \frac{R_d x + qt/n}{R_d n D_h \to D_h/q}\right) + \left(\frac{q^2 t}{\pi R_d n^2 D_h \to D_h}\right)^{1/2}\exp\left\{\frac{qx}{nD_h \to D_h} - \frac{(R_d x - qt/n)^2}{4 R_d D_h t \to D_h t}\right\}
\end{aligned}
$$
$$(7\text{-}141)$$

Gershon and Nir (1969) also present a steady state solution for the same type of boundary conditions, but with radioactive decay (otherwise steady flow cannot occur). The partial differential equation in this case is

$$D_h\frac{\partial^2 C}{\partial x^2} - \frac{q}{n}\frac{\partial C}{\partial x} - \lambda R_d C = 0 \qquad (7\text{-}142)$$

and the solution is

$$\frac{C(x,t)}{C_0} = \left[\frac{1}{2} + \left(\frac{1}{4} + \frac{D_h R_d \lambda n^2}{q^2}\right)^{1/2}\right]^{-1/2}\exp\left\{\frac{qx}{2D_h n}\left(1 - \left[1 + \frac{4R_d \lambda D_h n^2}{q^2}\right]^{1/2}\right)\right\} \qquad (7\text{-}143)$$

Case 7. Movement of a tracer with adsorption in a finite column of length L Bastian and Lapidus (1956) solve this case which is governed by the partial differential equation (7-118). The boundary and initial conditions are

$$
\begin{aligned}
t \leq 0, &\qquad x > 0, &\qquad C = 0, \\
t > 0, &\qquad x = 0, &\qquad (C_0 - C)\,q = -nD\,\partial C/\partial x &\qquad (7\text{-}144) \\
&\qquad x = L, &\qquad \partial C/\partial x = 0
\end{aligned}
$$

Their solution for the effluent concentration at the end of the column, $x = L$, is

$$\frac{C(L, t)}{C_0} = 1 - 2 \sum_{n=1}^{\infty} \frac{\alpha_n L \sin \alpha_n L}{(\alpha_n L)^2 + \frac{q^2 L^2}{4n^2 D^2} + \frac{qL}{nD}} \exp \left\{ \frac{qL}{2nD} - \left(\frac{q^2}{4R_d n^2 D} + \frac{D\alpha_n^2}{R_d} \right) t \right\}$$

(7-145)

where the α_n's are roots of $\alpha_n L \cot \alpha_n L + (qL/4nD) = (\alpha_n L)^2/(qL/nD)$.

They also present the solution for $C = C(x, t)$.

Gershon and Nir (1969) present several other cases with radioactive decay and adsorption for the different types of boundary and initial conditions encountered in Cases 1–4 above. They compare the results for $C = C(x, t)$ for different boundary conditions, e.g., those appearing in Cases 3 and 4 above. They show that in most steady state cases of practical interest, the results are influenced by less than 0.5 percent by the difference in boundary conditions. In cases of nonsteady state the results are influenced by up to 5 percent in the region of $C/C_0 \cong 0.5$.

The results of the one-dimensional cases described above, serve as the basis for the laboratory determination of the coefficient of dispersion, D_h, and from it of the longitudinal dispersivity $a_L = D_h/V$. A typical experiment is, one in which a column is initially filled with water at a certain tracer concentration. During the experiment, the water in the column is displaced by water of another (constant) concentration, introduced at one end of the column ($x = 0$) from $t = 0$ onward. The effluent concentration is recorded at $x = L$, for $t \geq 0$. Although the column is of a finite length L, (7-117) is usually employed to describe the effluent concentration at $x = L$ and hence also for the derivation of D_h. Actually, (7-117) describes the tracer distribution in an infinite column (Case 1). It is shown to be a good approximation for a semi-infinite column (Case 3). It is also considered a good approximation of (7-145).

The relationship (7-117), written for the column outlet at $x = L$, takes the form

$$C(L, t) = \frac{C_0}{2} \operatorname{erfc} \left\{ \frac{L - Vt}{2 [D_h t]^{1/2}} \right\}$$

(7-146)

With $Q = qA = nVA = VU_p/L$, $U_p = ALn$ = pore volume of the column, $U = Qt$ = volume of effluent and A = cross-sectional area of the column, (7-146) can be rewritten in the form

$$C(L, t) = \frac{C_0}{2} \operatorname{erfc} \frac{1 - U/U_p}{2 \left[\dfrac{D_h}{LV} \right]^{1/2} \left[\dfrac{U}{U_p} \right]^{1/2}}$$

(7-147)

During the experiment, C is recorded as a function of effluent volume, U. Figure 7-13 shows results of a typical experiment. At a time corresponding to the displacement of one pore volume, that is, $U = U_p$, the slope of the curve is given by

$$\frac{d(C/C_0)}{d(U/U_p)}\bigg|_{U/U_p=1} = \frac{1}{2\sqrt{\pi}}\left\{\frac{LV}{D_h}\right\}^{1/2}, \qquad V = \frac{QL}{U_p} \qquad (7\text{-}148)$$

from which D_h can be obtained.

Another method. using the same results shown in Fig. 7-13, is based on a certain property of the normal distribution. Equation (7-147) is the normal distribution of $(1 - U/U_p)/(2D_h/LV)^{1/2}(U/U_p)^{1/2}$ with mean at $U/U_p = 1$ and standard deviation $\sigma = (2D_h/LV)^{1/2}(U/U_p)^{1/2}$. Referring to Fig. 7-13, a known property of the normal distribution at $U = U_p$ is

$$\frac{U_{0.841} - U_{0.159}}{U_p} = 2\sigma = 2\left[\frac{2D_h}{LV}\right]^{1/2} \qquad (7\text{-}149)$$

Hence

$$D_h = \frac{\sigma^2}{2}LV = \frac{1}{2}\left(\frac{U_{0.841} - U_{0.159}}{2U_p}\right)^2 LV \qquad (7\text{-}150)$$

which is used for determining D_h from the experimental results plotted in the form of Fig. 7-13. Actually, the graph in Fig. 7-13 is only approximately a normal distribution (with respect to $U(t)$).

Case 8. Dispersion of an initially sharp front in uniform flow in a plane The flow at $q = $ const. is in the $+x$ direction. Initially, an abrupt form along $y = x\tan\alpha$ separates the two zones with concentrations $C = C_0$ and $C = 0$ (Fig. 7-14).

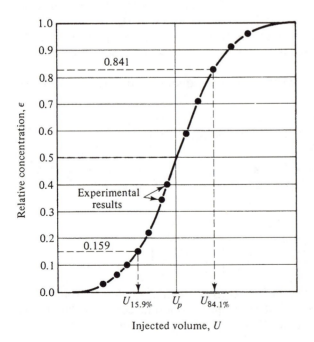

Figure 7-13 Computing D_h from a column displacement experiment.

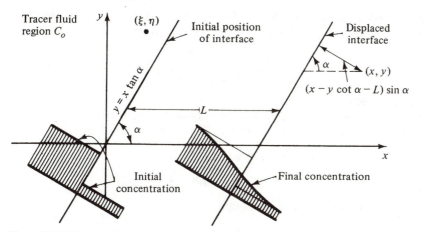

Figure 7-14 The transition zone at a moving interface in uniform planar flow.

The equation governing the changes in $C(x, y, t)$ for this case is

$$\frac{\partial C}{\partial t} = D_{hL} \frac{\partial^2 C}{\partial x^2} + D_{hT} \frac{\partial^2 C}{\partial y^2} - \frac{q}{n} \frac{\partial C}{\partial x}$$

$$D_{hL} = a_L q/n + D_d^*, \qquad D_{hT} = a_T q/n + D_d^* \qquad (7\text{-}151)$$

where a_L and a_T are the longitudinal and transversal dispersivities, of the porous medium, respectively.

The sought distribution is obtained by integrating an elementary solution, similar to (7-123) over the entire region occupied by tracer labeled water. If M is the mass of tracer injected at a point source, (ξ, η), at time $t = 0$, the concentration areal distribution at any later time is given by

$$C(x, y, t) = \frac{M/n}{4\pi [D_{hL} D_{hT}]^{1/2} t} \exp \left\{ -\frac{(x - \xi - qt/n)^2}{4D_{hL}t} - \frac{(y - \eta)^2}{4D_{hT}t} \right\}$$

$$M/n = \int_{-\infty}^{+\infty} \int_{-\infty}^{+\infty} C(x', y', t) \, dx' \, dy' \qquad (7\text{-}152)$$

Curves of $C = \text{const.}$, described by (7-152), have the shape of ellipses centered at $(\xi + (q/n) t, \eta)$.

To obtain the concentration distribution resulting from a moving front, we integrate the effect of an infinite number of small point sources, each with $M = C_0 n d\xi d\eta$, initially located behind the front. The result is

$$C(x, y, t) = \frac{C_0}{2} \operatorname{erfc} \left\{ \frac{[x - (q/n) t] \sin \alpha - y \cos \alpha}{[4(D_{hL} \sin^2\alpha + D_{hT} \cos^2\alpha) t]^{1/2}} \right\} \qquad (7\text{-}153)$$

which describes a normal distribution perpendicular to the displaced front (Fig. 7-14).

When the water motion is parallel to the initial front, $\alpha = 0$, and initially $C = 0$ for $y \leq 0$ and $C = C_0$ for $y > 0$, then (7-153) reduces to

$$C(y, t) = \frac{C_0}{2} \operatorname{erfc} \left\{ - \frac{y}{[4D_{hT}t]^{1/2}} \right\} \tag{7-154}$$

which is independent of x and D_{hL} (and also of V!).

Verruijt (1971) studies the case of a steady state distribution in the semi-infinite plane $x > 0$. Initially, the tracer distribution is described by

$$x = 0, \qquad 0 < y < \infty, \qquad C = 0$$
$$x = 0, \qquad -\infty < y < 0, \qquad C = C_0$$

For sufficiently large values of x, he obtains

$$C = \frac{C_0}{2} \operatorname{erfc} \left\{ \frac{y}{2[a_T x]^{1/2}} \right\} \tag{7-155}$$

For large values of x, (7-155) describes a normal distribution in the y direction, with the width of the transition zone being proportional to $x^{1/2}$. Lines of constant concentration have the form of parabolas, at least for large x. It is of interest to note that from (7-155) it follows that $C(y, t)$ is independent of the velocity $V (= q/n)$.

Verruijt (1971) also studies the case where the velocity (still uniform in the $+x$ direction) in the two zones is different. Specifically, $V \neq 0$ for $y > 0$ and $V = 0$ for $y < 0$.

Case 9. A continuous injection at the origin into a uniform steady flow in an infinite plane As in Case 4 the resulting tracer distribution is obtained by integrating the elementary solution of the partial differential equation (7-151). The effect of an instantaneous slug of tracer mass $dM = C_0 Q \, dt$ is

$$dC(x, y, t) = \frac{dM}{2\pi [2D_{hL}t]^{1/2} [2D_{hT}t]^{1/2}} \exp \left\{ - \frac{(x - qt/n)^2}{4D_{hL}t} - \frac{y^2}{4D_{hT}t} \right\} \tag{7-156}$$

For a continuous injection, we obtain

$$\frac{C}{C_0} = \frac{Q}{4\pi [D_{hL}D_{hT}]^{1/2}} \int_{\tau=0}^{t} \frac{1}{t - \tau} \exp \left\{ - \frac{1}{t - \tau} \left(\frac{x^2}{4D_{hL}} + \frac{y^2}{4D_{hT}} \right) \right.$$
$$\left. + \frac{2xq/n}{4D_{hL}} - \frac{(t - \tau) q^2/n^2}{4D_{hL}} \right\} d\tau \tag{7-157}$$

For the steady concentration distribution, we obtain, by inserting $t = \infty$ in (7-157)

$$\frac{C}{C_0} = \frac{Q}{2\pi [D_{hL}D_{hT}]^{1/2}} \exp \left(\frac{xq/n}{2D_{hL}} \right) \times K_0 \left[\left\{ \frac{q^2/n^2}{4D_{hL}} \left(\frac{x^2}{D_{hL}} + \frac{y^2}{D_{hT}} \right) \right\}^{1/2} \right] \tag{7-158}$$

where K_0 is the modified Bessel function of second kind and zero order.

Case 10. Dispersion in radial flow; molecular diffusion is neglected Consider the case of a fully penetrating well in a confined aquifer, recharging water at a constant rate Q_w. The concentrations of the indigenous water in the aquifer and of the recharge water are 0 and C_0, respectively. Let the governing equation for dispersion in this case be (7-65)

$$\frac{\partial C}{\partial t} = a_L V \frac{\partial^2 C}{\partial r^2} - V \frac{\partial C}{\partial r}; \qquad Q_w = 2\pi r B n V \qquad (7\text{-}159)$$

Because $V = V(r) = (Q_w/2\pi Bn)/r$, (7-159) is nonlinear and an analytic solution is most difficult.

Ogata (1958; see Bear, 1972, p. 636) gives an analytic solution for the case

$$t \leq 0, \qquad r > r_w, \qquad C = 0; \qquad t > 0, \qquad r = r_w, \qquad C = C_0$$

$$r \to \infty, \qquad C = 0 \qquad (7\text{-}160)$$

His solution is

$$\frac{C}{C_0} = 1 + \frac{2}{\pi} \exp\left\{ \frac{r - r_w}{2a_L} \right\} \int_0^\infty \frac{\exp(-v^2 t)}{v} \left(\frac{v^2 r - G/4a_L}{v^2 r_w - G/4a_L} \right) M(v)\, dv \qquad (7\text{-}161)$$

where

$$M(v) = \frac{J_{1/3}(\sigma)\, Y_{1/3}(\sigma') - Y_{1/3}(\sigma)\, J_{1/3}(\sigma')}{J^2_{1/3}(\sigma') + Y^2_{1/3}(\sigma')}; \qquad G = Q_w/2\pi Bn = Vr$$

$$\sigma = \frac{2}{3\sqrt{a_L G}} \frac{(v^2 r - G/4a_L)^{3/2}}{v^2}; \qquad \sigma' = \frac{2}{3\sqrt{a_L G}} \frac{(v^2 r_w - G/4a_L)^{3/2}}{v^2}$$

and $J_{1/3}$, $Y_{1/3}$ are Bessel functions of order 1/3, of the first and second kinds, respectively.

Bondarev and Nikolaevskii (1962) give for $Q_w = Q_w(t)$ the solution

$$\frac{C}{C_0} = 1 - \frac{1 - \exp(r/a_L) + (r/a_L)\exp\sqrt{6\tau}}{1 - \exp\sqrt{6\tau} + \sqrt{6\tau}\exp\sqrt{6\tau}}; \qquad \tau = \frac{1}{a_L^2} \int_0^t \frac{Q_w(t)}{2\pi Bnr}\, dt \qquad (7\text{-}162)$$

valid except during the first period of injection.

Because of the difficulties involved in trying to derive an exact analytical solution, attempts have been made to obtain an approximate one. For example, de Josselin de Jong (in Lau *et al.*, 1959) suggests an approximate solution for dispersion in radially diverging flow from a well. His solution is based on two assumptions: that the tracer distribution is sufficiently near a normal one, and that it is produced as a linear sum of two effects—one due to longitudinal dispersion and the other due to the divergence of streamlines. The corresponding solution is

$$\frac{C(r, t)}{C_0} = \frac{1}{2} \operatorname{erfc}\left\{ \frac{r - \bar{r}}{[\frac{4}{3}a_L \bar{r}]^{1/2}} \right\}, \qquad \sigma_r^2 = \frac{2}{3} a_L \left(\bar{r} - \frac{r_w^3}{\bar{r}^2} \right) \qquad (7\text{-}163)$$

where the total volume recharged during time t through a well of radius r_w is

$$U_r = \int_0^t Q_w(t)\,dt = \pi \bar{r}^2 Bn; \qquad r_w \ll \bar{r}$$

Accordingly, $C/C_0 = 0.5$ when $r = \bar{r}$.

Mercado and Bear (1965) extend de Josselin de Jong's approach also to the pumping stage, which follows a certain period of recharge. For the variance of the distribution after a volume U_p has been pumped from the aquifer, they obtain

$$\sigma_r^2 = \frac{2}{3} a_L \left(\frac{U_r}{\pi B n} \right) \left[\frac{2}{1 - \alpha_0} - (1 - \alpha_0)^{1/2} \right]; \qquad \alpha_0 = \frac{U_p}{U_r} \qquad (7\text{-}164)$$

and for the entire distribution

$$\frac{C}{C_0} = \frac{1}{2} \operatorname{erfc} \left[\frac{r - \bar{r}}{[(4/3)\, a_L R \{2(R/\bar{r})^2 - (r/R)\}]^{1/2}} \right]; \qquad \bar{r}^2 = \frac{U_r - U_p}{\pi B n}$$

$$R^2 = \frac{U_r}{\pi B n} \qquad (7\text{-}165)$$

where U_r is the volume of injected water.

Equation (7-165) can also be written as $C = C(U_p)$, so that it can be used for determining a_L by using data from laboratory or field experiments (e.g., Mercado and Bear, 1965).

Numerical solutions for the radial case are presented, among others, by Ogata (1958). Shamir and Harleman (1966), Hoopes and Harleman (1967), and Prakash (1976).

7-10 MOVEMENT OF WATER BODIES INJECTED INTO AQUIFERS

So far in this chapter, we have been dealing with a single inhomogeneous fluid, water, where the inhomogeneity is due to variations in the concentration of dissolved matter. Sometimes, this concentration affects the density and the viscosity of the water. Our objective was to predict the areal and temporal variations of concentration of any dissolved species as it is carried with the water and spreads in the aquifer. This objective is achieved by solving the appropriate equations of hydrodynamic dispersion for $c(x, y, z, t)$, given initial and boundary conditions, as well as inputs and outputs of both water and the considered species. We have seen that when water characterized by the concentration of some dissolved species is introduced into an aquifer with indigenous water at a different concentration, through part of its boundary (e.g., sea water intrusion; Chap. 9), or through wells (artificial recharge; Sec. 3-4), a transition zone is created (e.g., Fig. 7-1) across which the concentration varies. The width of the transition zone increases with the length of flow of the advancing front.

Sharp Front Approximation

However, often the transition between the two zones is narrow, relative to the length dimensions of the individual areas (or volumes) occupied by each of the two kinds of water, and can be neglected. We then consider an *approximation* in which the transition zone is replaced by an *assumed abrupt* front, which, while moving, continuously separates the two zones, each occupied by water at a different concentration. This is a practical approach which enables the planner to arrive at first estimates of contamination phenomena. In the following paragraphs, we shall apply this approach to several cases of practical interest, especially in connection with artificial recharge. Nevertheless, the planner should always remember that a transition zone does exist in reality; in certain cases, a correction due to its presence should be introduced in results obtained by assuming the presence of an abrupt front only.

For the sake of simplicity, we shall assume that the two kinds of water, the indigenous water in the aquifer and the water injected into the aquifer through wells, are distinguishable, say, by their salinity which serves as a label. Unless otherwise specified, the density and viscosity of the two kinds of water are identical. The two liquids (or zones) are separated by an abrupt front. The aquifer is assumed homogeneous and isotropic.

The problem of two kinds of water separated by an assumed abrupt interface is treated also in Chap. 9, in connection with the fresh water–sea water interface in a coastal aquifer. There the density effect is predominant and therefore the elevation of the interface is part of the sought solution. Here, however, we shall treat the flow in the aquifer as being essentially horizontal, the abrupt front is vertical so that only its trace in the horizontal plane is considered.

We shall also assume that aquifer storativity can be neglected, so that we have everywhere steady flow, although the front is moving.

Mathematical Statement of Front Movement

Let $F(x, y, z, t) = 0$ represent the surface of a considered abrupt front in space. Since as it moves, the same fluid particles continuously remain on it, we have

$$\frac{\partial F}{\partial t} + \mathbf{V} \cdot \nabla F = 0 \tag{7-166}$$

(see Sec. 5-3), where $\mathbf{V}$ is the velocity of points on the front.

The abrupt front divides the flow domain, D, into two subdomains (D_1 and D_2), each occupied by a different fluid. Let us define two piezometric heads: ϕ_1 for D_1 and ϕ_2 for D_2. Figure 7-15 shows this situation.

Approaching the front, F, from either side, (7-166) can be written as

$$\frac{\partial F}{\partial t} + \mathbf{V_i} \cdot \nabla F \equiv \frac{\partial F}{\partial t} - \left(\frac{K_i}{n}\right) \nabla \phi_i \cdot \nabla F = 0, \qquad i = 1, 2 \tag{7-167}$$

where $K_i = k\gamma_i/\mu_i$. We often make the assumption that $K_1 = K_2 = K = $ const. The problem can now be stated as follows.

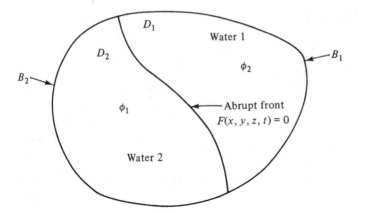

Figure 7-15 An abrupt front in a flow domain.

Determine ϕ_1 in D_1 and ϕ_2 in D_2 such that

(a) $\qquad\qquad\qquad\qquad \phi_1$ satisfies $\nabla^2\phi_1 = 0$ in D_1,

(b) $\qquad\qquad\qquad\qquad \phi_2$ satisfies $\nabla^2\phi_2 = 0$ in D_2,

or other appropriate partial differential equations for other sets of assumptions with respect to storativity, permeability, etc.,

(c) $\qquad\qquad\qquad\qquad\qquad p_1 = p_2$ on F,

or $\phi_1 = \phi_2$ when density effects are neglected,

(d) $\qquad\qquad V_{n1} = V_{n2}$,   or $\mathbf{V}_1 \cdot \mathbf{1n} = \mathbf{V}_2 \cdot \mathbf{1n}$ on F,

where $\mathbf{1n} = \nabla F/|\nabla F|$ is the normal to F at a point on the latter, and

$$\mathbf{V}_i = -\frac{K_i}{n}\nabla\phi_i, \qquad i = 1, 2,$$

(e) appropriate boundary conditions for ϕ_1 on B_1 and ϕ_2 on B_2 (and initial conditions for ϕ_1 in D_1 and for ϕ_2 in D_2),

(f) the expression for F satisfies (7-167).

The solution of the problem as stated in this way is a rather difficult task, except for very simple cases (Muskat, 1937, pp. 459–466; Bear, 1972, pp. 527–528), as the shape of the separating front F must be found simultaneously with the potential distributions. Muskat (1937, p. 466) presents an approach for cases in which viscosity differences are neglected, a situation which corresponds to most groundwater problems.

As an objective Muskat (1937) sets out to trace the history of a line of fluid particles originally lying on a given initial front, as the latter moves in the field of flow. The entire flow domain is characterized by a single potential ϕ. Equipotential surfaces can be drawn in the flow domain. At each point on an equipotential surface, a streamline perpendicular to the surface, can be drawn.

Muskat (1937) introduces a set of orthogonal curvilinear coordinates

$$\xi_1 = \xi_1(x, y, z), \qquad \xi_2 = \xi_2(x, y, z), \qquad \xi_3 = \xi_3(x, y, z),$$
$$\nabla\xi_1 \cdot \nabla\xi_2 = \nabla\xi_2 \cdot \nabla\xi_3 = \nabla\xi_3 \cdot \nabla\xi_1 = 0,$$

with

$$\nabla\phi \cdot \nabla F = \frac{\partial\phi}{\partial\xi_1}\frac{\partial F}{\partial\xi_1}|\nabla\xi_1|^2 + \frac{\partial\phi}{\partial\xi_2}\frac{\partial F}{\partial\xi_2}|\nabla\xi_2|^2 + \frac{\partial\phi}{\partial\xi_3}\frac{\partial F}{\partial\xi_3}|\nabla\xi_3|^2 \qquad (7\text{-}168)$$

Selecting the surfaces $\xi_1 = $ const. to coincide with the equipotential surfaces $\phi = $ const., (7-168) reduces to

$$\nabla\phi \cdot \nabla F = \frac{\partial F}{\partial\xi_1}|\nabla\xi_1|^2 \qquad (7\text{-}169)$$

In the $\xi_1\,\xi_2\,\xi_3$ system of coordinates, and assuming a homogeneous fluid, (7-167), describing F, takes the form

$$\frac{\partial F}{\partial t} - \frac{K}{n}\frac{\partial}{\partial\xi_1}|\nabla\xi_1|^2 = 0 \qquad (7\text{-}170)$$

where $|\nabla\xi_1|^2$ is to be expressed in terms of ξ_1, ξ_2, and ξ_3.

By integrating (7-170), we obtain

$$F = t + \frac{n}{K}\int\frac{d\xi_1}{|\nabla\xi_1|^2} + f(\xi_2, \xi_3) = \text{const.} \qquad (7\text{-}171)$$

where $f(\xi_2, \xi_3)$ is an arbitrary function to be adjusted so that F assumes its known initial shape at $t = 0$.

For two-dimensional flow in the horizontal plane, $\xi_1 = $ const. coincides with the equipotential curves $\phi = $ const., and $\xi_2 = $ const. are the streamlines $\psi = $ const. Equation (7-171) becomes

$$F(\phi, \psi, t) = t + \frac{n}{K}\int\frac{d\phi}{|\nabla\phi|^2} + f(\psi) = \text{const.} \qquad (7\text{-}172)$$

in which the integral is to be evaluated along a streamline $\psi = $ const.

Examples of Muskat's Approach

As an example, Muskat (1937) considers the steady flow pattern produced by a single pumping well at $(0, d)$ near an equipotential boundary $\phi = \phi_0$ (Fig. 7-16) along the x axis. One may visualize the situation as a flow domain in the semi-infinite plane $(y > 0)$, bounded along $y = 0$ by a reservoir (e.g., lake) with water of a different quality. This boundary also coincides with the front at $t = 0$.

For this case (see Sec. 8-10)

$$\phi_0 - \phi = -\frac{Q_w}{4\pi T}\ln\frac{x^2 + (y - d)^2}{x^2 + (y + d)^2}; \qquad \psi = \frac{Q_w}{2\pi T}\tan^{-1}\frac{2dx}{x^2 + y^2 - d^2} \qquad (7\text{-}173)$$

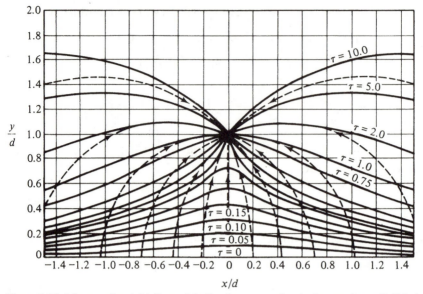

Figure 7-16 Advance of an initially straight line front toward a single pumping well *(Muskat, 1937)*.

Hence

$$|\nabla\phi|^2 = \frac{Q_w^2}{4\pi^2 d^2 T^2}(\cosh\alpha + \cos\beta); \qquad \alpha = \frac{2\pi T(\phi_0 - \phi)}{Q_w}; \qquad \beta = -\frac{2\pi T\psi}{Q_w}$$

$$(7\text{-}174)$$

$$F = t - \frac{n}{K}\cdot\frac{2\pi d^2 T}{Q_w}\int\frac{d\alpha}{(\cosh\alpha + \cos\beta)} + f(\beta) = C = \text{const.} \quad (7\text{-}175)$$

The initial conditions require that $f(\beta) = C$. Hence

$$t = \frac{2d^2 nB}{Q_w \sin^2\beta}\left[\frac{\sinh\alpha}{\cosh\alpha + \cos\beta} - 2\cot g\,\beta\,\tan^{-1}\left(\tanh\frac{\alpha}{2}\tan\frac{\beta}{2}\right)\right] \quad (7\text{-}176)$$

Figure (7-17) shows (7-176) with some streamlines. The advancing front will reach the pumping well at

$$t = 2\pi n d^2 B/3Q_w$$

where B is the thickness of the aquifer. The total area swept by the advancing front is $2\pi d^2/3$, or two-thirds of the area of a circle of radius d.

Muskat (1937) also determines the shape of the front advancing from a recharging well at $(0, -d)$ to a pumping one at $(0, d)$; both wells are of equal strength Q_w. For this case ϕ and ψ are also given by (7-173).

The front, composed of fluid particles leaving the injection well at $t = 0$ is advancing from the injection well to the pumping one. The front position at

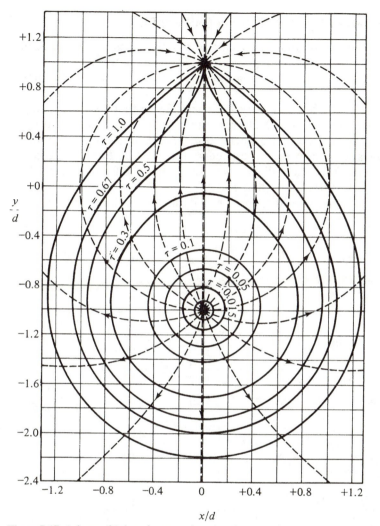

$$\frac{x}{d}$$

Figure 7-17 A front of injected water advancing from an injection well to a pumping one *(Muskat, 1937).*

different times is given by

$$t = \frac{2\pi n d^2 B}{Q_w \sin^2\beta} \left[\frac{\sinh\alpha}{\cosh\alpha + \cos\beta} - 2\cotg\beta \tan^{-1}\left(\tanh\frac{\alpha}{2}\tan\frac{\beta}{2} \right) \right.$$

$$\left. - \frac{\sinh\alpha_0}{\cosh\alpha_0 + \cos\beta} + 2\cotg\beta \tan^{-1}\left(\tan\frac{\alpha_0}{2}\tan\frac{\beta}{2} \right) \right] \qquad (7\text{-}177)$$

where $\alpha = \alpha_0$ is the equipotential defining the injection well at $(0, -d)$. The advancing front is shown in Fig. 7-17. In this case, the injected water will reach the pumping well at $t = 4\pi n d^2 B/3Q_w$.

Movement of a Body of Injected Water in Uniform Flow

Bear and Jacobs (1965) also study the shape of the advancing front separating the indigenous water of a confined aquifer from a body of water injected into it, through a well at the origin, by following the movement of water particles along streamlines.

Differences in density and viscosity are neglected so that we have a single homogeneous fluid, a portion of which (say, the injected water) is labeled by a tracer. A uniform flow at a constant specific discharge q_0 in the $+x$ direction is taking place in the aquifer of constant thickness, B (Fig. 7-18). The front separating the injected water from the indigenous water of the aquifer is composed of all particles leaving the injection well at $t = 0$.

For this case, the potential distribution and the stream function are given by

$$\phi = -\frac{q_0}{K}x - \frac{Q_w}{4\pi T}\ln(x^2 + y^2); \qquad \psi = -\frac{q_0}{K}y - \frac{Q_w}{2\pi T}\tan^{-1}\frac{y}{x} \qquad (7\text{-}178)$$

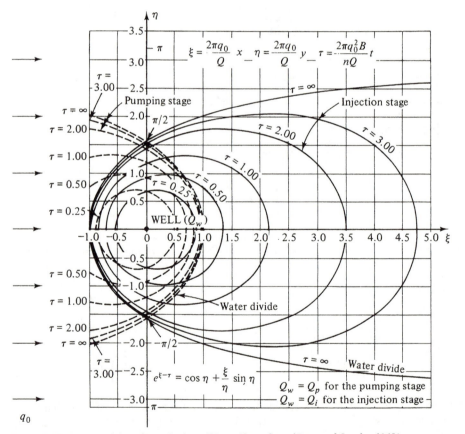

Figure 7-18 Front positions for a single well in uniform flow *(Bear and Jacobs, 1965)*.

where: along $y = 0$, $\tan^{-1}(y/x) = 0$, $\psi = 0$; $y = 0$, $x < 0$, $\psi = -Q_w/2T$; $y = 0$, $x > 0$, $\psi = -Q_w/T$, and $Q_w > 0$ for injection.

The velocity components V_x, V_y in the $+x$ and $+y$ directions, respectively, are given by

$$V_x = \frac{dx}{dt} = \frac{q_0}{n} + \frac{Q_w x}{2\pi n B(x^2 + y^2)}; \qquad V_y = \frac{dy}{dt} = \frac{Q_w y}{2\pi n B(x^2 + y^2)} \qquad (7\text{-}179)$$

From (7-178), it follows that along a streamline ($=$ pathline in steady flow) $\psi = $ const., we have

$$x = y \cot g[2\pi(K\psi + q_0 y) B/Q_w] \qquad (7\text{-}180)$$

which, when inserted into (7-179), yields

$$dt = -\frac{2\pi n B y}{Q_w} \text{cosec}^2[-2\pi(K\psi + q_0 y) B/Q_w] \, dy \qquad (7\text{-}181)$$

By integrating (7-181) for particles leaving the origin at $t = 0$, and introducing the dimensionless variables

$$\xi = \frac{2\pi q_0 B}{Q_w} x; \qquad \eta = \frac{2\pi q_0 B}{Q_w} y, \qquad \tau = \frac{2\pi q_0^2 B}{n Q_w} t$$

we obtain

$$\tau = \xi + \ln[\sin\theta/\sin(\eta + \theta)], \qquad \theta = \tan^{-1}(y/x) \qquad (7\text{-}182)$$

which defines the time, τ, required for the front to reach point (ξ, η). The equation of the front at $\tau = \tau_0$ is

$$\left[\cos\eta + \frac{\xi}{\eta}\sin\eta\right]\exp(-\xi) = \exp(-\tau_0) = \text{const.} \qquad (7\text{-}183)$$

Figure 7-18 shows the shapes of the advancing fronts. As $\tau \to \infty$, we have $\xi = -\eta/\tan\eta$, which defines the water divide which is the limiting front position. The front does not advance beyond $\eta = \pm\pi$. The stagnation point along the x axis ($V_x = 0$, $V_y = 0$) is at $\eta = 0$, $\xi = -1$. Along the ξ axis, the front moves according to

$$\tau = \xi - \ln(1 + \xi) \qquad (7\text{-}184)$$

Often, following a period of recharge, pumping takes place in the same well. As we shall see below, the tracer concentration of the pumped water varies between that of the indigenous water of the aquifer and that of the recharge water. By alternating recharge and pumping operations, even in a single well, it is possible to control the salinity of the pumped water between the above limits. Needless to add that a further mixing actually takes place due to hydrodynamic dispersion which is not taken into account in the present discussion (e.g., due to hydrodynamic dispersion, the first trace will arrive at the pumping well earlier than the time computed by (7-176) or (7-177)).

Let $Q_p(> 0)$ denote the pumping rate in the same well and under the same conditions as those considered above for the period of recharge. One may then determine the shapes of the curves representing the loci of all water particles arriving at the pumping well at the same time under the combined effects of the natural uniform flow and of the converging radial flow. The shapes of the curves, also called fronts in the following discussion, may be obtained by taking the mirror image of the previous curves with respect to the η axis and with Q_p replacing Q_w in the expressions developed above.

The stagnation point is now downstream of the pumping well at $\xi = +1$, $\eta = 0$. The limiting front

$$\xi = \eta/\tan\eta$$

represents the water divide; all water within the area bounded by this curve will eventually reach the pumping well.

Consider the case (Fig. 7-19a) of a well which begins pumping immediately after a recharge period, τ_i, during which a certain volume of water has been injected at a rate Q_i. At first (e.g., t_{p1}), the well pumps only from water recharged during τ_i. After a certain time, the front representing all particles having the same time of arrival will become tangent at $t_{pc'}$, and then intersect the ultimate front of the injected water body (say, at t_{p2}).

If pumping continues beyond $t_{pc'}$, the well starts pumping water both from the injected water body and from the native water of the aquifer. For example, during $t_{p2} < \tau < t_{p3}$, pumped water will be a mixture of native water from the region $ABCDEF$ and recharged water, $AFDC$. At a later period $t = t_{pc''}$, the downstream edge of the pumping front will become tangent to the ultimate injected front. At this instant the last drop of injected water reaches the pumping well. For $t > t_{pc''}$, only native water is pumped.

The above description corresponds to the case in which the ultimate injected front does not penetrate the water divide of the pumping stage. If, on the other hand, the injection period continues to such an extent that the injected water body extends beyond the water divide of the pumping stage, part of the injected water (dashed area in Fig. 7-19b) can never be recovered by the pumping. In this case $t_{pc''}$ does not exist; some injected water will always reach the pumping well, although at an ever decreasing rate.

The knowledge of the recovery ratio, λ, and the ultimate recovery ratio, λ_∞, is most important in all injection-pumping operations for the purpose of water quality control

$$\lambda = \frac{\text{volume of pumped recharge water}}{\text{total volume of recharge water}} \; ; \qquad \lambda_\infty = \lim_{t_p \to \infty} \lambda \qquad (7\text{-}185)$$

As has already been explained above, a certain volume of recharge water, namely, that portion of the injected water body extending beyond the water divide for pumping, can never be recovered (Fig. 7-19b). The value of λ depends on t_p, reaching a maximum value λ_∞ as $t_p \to \infty$. The value of λ_∞ itself depends on the ratio $\beta = Q_p/Q_i$ and on t_i. For small values of t_i, λ_∞ reaches the value 1.0 already at finite times.

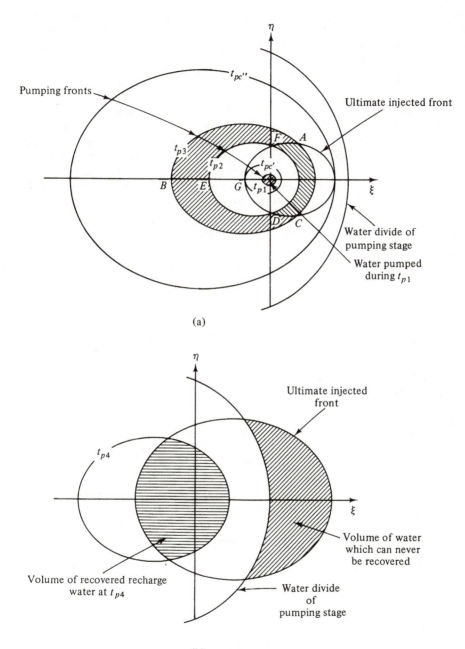

(a)

(b)

Figure 7-19 Pumping in the same well after a recharge period *(Bear and Jacobs, 1965)*.

In order to obtain the relationship $\lambda = \lambda(\tau_p, \tau_i, \beta)$, the curves of Fig. 7-18 are used. For $\beta = 1$, one determines the ratio of the common area of the injected and the pumped water bodies for various values of τ_p to the total area of the injected water body for various values of τ_i. The results are shown on Fig. 7-20. The curve for $\tau_i = 0.31$ (not shown in Fig. 7-20) is asymptotic to the line $\lambda = 1$; all the injected water is recovered only at $t = \infty$. This value of τ_i is obtained from $\tau_i = \xi_i - \ln(1 + \xi_i)$, with $\xi_i = 1$.

Bear and Jacobs (1965) also consider the cases of $\beta = \frac{1}{2}$ and $\beta = 2$.

The proportion of injected water in the water pumped at each moment is important in recharge-pumping operations, especially when recharge water and native water have different salinities. The instantaneous relative concentration is $\varepsilon = (C - C_0)/(C_i - C_0)$, where C is salinity concentration in the pumped water, C_i is the concentration in the injected water and C_0 is the concentration of the native water; it may be derived from the recovery curves of Fig. 7-20. The relative concentration, ε, is also the rate of pumpage of recharge water divided by the total rate of pumpage. Its relation to λ may be derived as follows.

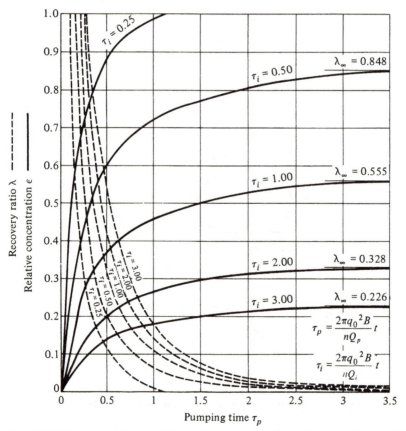

Figure 7-20 Recovery ratio and relative concentration for $\beta = 1$ *(Bear and Jacobs, 1965)*.

Let Q_p be decomposed into $Q_p^{(1)}$ and $Q_p^{(2)}$, where $Q_p^{(1)}$ and $Q_p^{(2)}$ are the instantaneous rates of pumping of previously injected and of native water, respectively

$$Q_p = Q_p^{(1)} + Q_p^{(2)}$$

Then

$$\lambda = \int_0^{t_p} \frac{Q_p^{(1)} \, dt}{Q_i t_i} \tag{7-186}$$

Differentiating (7-186) and inserting into the definition for ε yields

$$\varepsilon = \frac{Q_p^{(1)}}{Q_p} = \frac{t_i}{\beta dt_p} \frac{d\lambda}{dt_p} = \frac{\tau_i}{\beta^2} \frac{d\lambda}{d\tau_p} \tag{7-187}$$

Curves $\varepsilon = \varepsilon(\tau_p)$ are derived by (7-187) and by differentiating graphically the curves $\lambda = \lambda(\tau_p, \tau_i)$. One should notice that since at $\tau_p = 0$, $\varepsilon = 1.0$, all the curves $\lambda = \lambda(\tau_p)$ make at the origin an angle α with the τ_p axis such that

$$\tan \alpha = \frac{\beta^2}{\tau_i} \tag{7-188}$$

according to the corresponding values of τ_i and β. Curves of $\varepsilon = \varepsilon(\tau_p)$ are shown in dashed lines on Fig. 7-20.

If a certain delay time elapses between the end of the recharge stage and the beginning of the pumping one, the injected water body will move during this interval together with the uniform flow in the aquifer. The front of the injected water is translated without changing its shape. The shape of the injected water-body at the end of the recharge period t_{i0} is given by

$$\exp(\xi - \tau_{i0}) = \cos \eta + \frac{\xi}{\eta} \sin \eta \tag{7-189}$$

where ξ, η, and τ_{i0} are defined with respect to Q_i and the well is at the origin. After a delay time $\tau_d = \tau - \tau_{i0}$, the equation defining the shape of this body will be

$$\exp(\xi - \tau) = \cos \eta + \frac{\xi - (\tau - \tau_{i0})}{\eta} \sin \eta \tag{7-190}$$

where $\tau \, (> \tau_{i0})$ is the running time with respect to Q_i.

Along the ξ axis, where $\eta = 0$, (7-190) becomes

$$\exp(\xi - \tau) = 1 + \xi - (\tau - \tau_{i0})$$

or

$$\tau = \xi - \ln(1 + \xi - \tau + \tau_{i0}) \tag{7-191}$$

During the pumping period which follows the delay stage, water is pumped only from within a certain region bounded by the limiting streamline, independent of the length of the pumping period. The stagnation point is at the intersection of this line with the x axis. Depending on its size, on the delay time, and on the pumping rate, this injected water body may completely have passed beyond the stagnation point of the pumping stage. No portion of it can then be recovered by pumping. In general, by having this delay time, the period of pumping only injected water ($\varepsilon = 1.0$) is shortened. The stagnation point for pumping is at $\xi = \beta$, where $\beta = |Q_p|/|Q_i|$ and ξ is defined with respect to Q_i. Hence, if the injected water body has just passed beyond $\xi = \beta$, the relationship

$$\tau_{io} = (\beta - \tau_d) - \ln[1 + (\beta - \tau_d)] \tag{7-192}$$

holds. Here τ_d is the dimensionless delay time. Given τ_{io} and β (i.e., Q_p), we solve (7-192) for τ_d; the larger of the two solutions for τ_d is the one we are seeking.

Bear and Jacobs (1965) discuss also the case of unsteady flow from a well in an infinite aquifer with uniform flow.

Bear and Zaslavsky (1962; also Harpaz and Bear, 1963) discuss several additional simple cases of recharging and pumping wells.

In the case of steady flow from a single well of radius r_w injecting a constant rate Q_w into an infinite confined aquifer (porosity n and thickness B), the circular front moves in time such that

$$t = \frac{n\pi r^2 B}{Q_w} \tag{7-193}$$

where $r(t)$ is the radius of the circular injected water body.

If a pair of injection $(-d, 0)$ and pumping $(+d, 0)$ wells, a distance $2d$ apart, are operating simultaneously (Fig. 7-21a), one can determine the position of the advancing front nearest to the pumping well and farthest from it (that is, along the x axis), by integrating the proper expression for the velocity

$$V = \frac{dx}{dt} = \frac{Q_w}{2\pi n B}\left(\frac{1}{d - x} + \frac{1}{d + x}\right) \tag{7-194}$$

Integration of (7-194), with $r_w \ll d$ and $\xi = x/d$; $\tau = t/T$; $T = \pi n B d^2/Q_w$, yields

$$\tau = \frac{2}{3} + \xi - \frac{\xi^3}{3} \tag{7-195}$$

Equation (7-195) holds for points A and B in Case A of Fig. 7-21a. Breakthrough time $\tau_b = \frac{4}{3}$ is obtained when $x = d$ or $\xi = 1$.

It is interesting to note in Fig. 7-21a how the rate of front movement varies. It is slow at the initial stages and becomes faster as the pumping well is approached.

The case of an injection well and a pumping well which do not have the same strength can be handled in a similar manner. However. if no change in storage takes place, one has to assume the existence of sources or sinks in the aquifer at infinity in order to satisfy continuity conditions.

In a similar manner, one may consider the case of an injection well at $(0, 0)$ recharging at a rate Q_w and two pumping wells located at $(d, 0)$ and $(-d, 0)$, each pumping at a rate $Q_w/2$ (Fig. 7-21b). Here, with $r_w \ll d$, the velocity of a particle at $(x, 0)$ is expressed by

$$\frac{dx}{dt} = \frac{1}{2\pi n B}\left(\frac{Q_w/2}{d-x} + \frac{Q_w}{x} - \frac{Q_w/2}{d+x}\right) \qquad (7\text{-}196)$$

By integrating (7-196) along the x axis, we obtain

for point A

$$\tau = \xi^2 - \frac{1}{2}\xi^4; \qquad \xi = \frac{x}{d}$$

for point C

$$\tau = \eta^2 + \frac{1}{2}\eta^4; \qquad \eta = \frac{y}{d} \qquad (7\text{-}197)$$

$$\tau = \frac{t}{T}; \qquad T = \frac{n\pi d^2 B}{Q_w}$$

Breakthrough is at $\tau_b = \frac{1}{2}$ for which $\xi_b = 1.0$, $\eta_b = 0.644$.

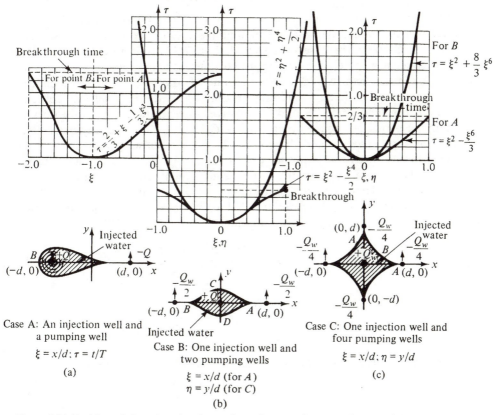

Figure 7-21 Position of the advancing front along the x- and y-axes *(Bear and Zaslavsky, 1962; Harpaz and Bear, 1963)*.

For the symmetrical case shown in Fig. 7-21c, where one injection well, Q_w, is located at the origin and four wells at $(d, 0)$, $(-d, 0)$, $(0, d)$, and $(0, -d)$ are pumping, each $Q_w/4$, one obtains

for point A

$$\tau = \xi^2 - \frac{\xi^6}{3} \quad \text{and} \quad \tau = \eta^2 - \frac{\eta^6}{3} \tag{7-198}$$

for point B

$$\tau = \xi^2 + \frac{8}{3}\xi^6 \quad \text{and} \quad \tau = \eta^2 + \frac{8}{3}\eta^6 \tag{7-199}$$

Breakthrough is at $\tau_b = \frac{2}{3}$ when $\xi_B = \eta_B = 0.548$.

In all the above cases, one can add the effect of the velocity of the natural flow in the aquifer to that produced by the recharging and pumping wells.

Although it is rather easy to determine the movement of the front along the axes, it is not always possible to determine the detailed shape of the advancing front analytically. Numerical techniques (see below) are usually employed. In the past, analog methods (especially the horizontal Hele–Shaw analog; Harpaz and Bear, 1963) were also used.

An estimate of the mixing in a pumping well (without the effect of hydro-dynamic dispersion) can be obtained in a simple way for some special cases. As an example, consider the case of a well A (Fig. 7-22a) injecting a volume U_1 of water (concentration C_1) into a confined aquifer as above, while no other well is operating in the vicinity. When well A is shut off, another well, B, starts pumping, withdrawing a volume U_2 (concentration C). If the original salt content in the aquifer is denoted by C_0, and the relative concentration at any time during the pumping period is ε defined by $\varepsilon = (C - C_0)/(C_1 - C_0)$, one may determine the relationship between ε and the volume U_2 in the following way.

The flow towards the pumping well is radial. All particles on the circumference of the cylinder with radius r and height B will reach the pumping well at the same time. The concentration of the pumped water depends, therefore, on the ratio between the length of the arc CDE (concentration C_1) and that of the arc EFC (concentration C_0). With the nomenclature of Fig. 7-22b

$$\varepsilon = \frac{1}{180} \cos^{-1}\left[\frac{r^2 + d^2 - R^2}{2rd}\right] \tag{7-200}$$

or

$$\varepsilon = \frac{1}{180} \cos^{-1}\left[\frac{U_2/U_0 + 1 - U_1/U_0}{2(U_2/U_0)^{1/2}}\right] \tag{7-201}$$

The maximum values of ε for various values of U_1/U_0 are determined from (7-201)

$$\varepsilon_{max} = \frac{1}{180} \cos^{-1}\left(1 - \frac{U_1}{U_0}\right)^{1/2} \tag{7-202}$$

since at $\alpha = \alpha_{max}$, $U_2 + U_1 = U_0$ (see dashed line on Fig. 7-22).

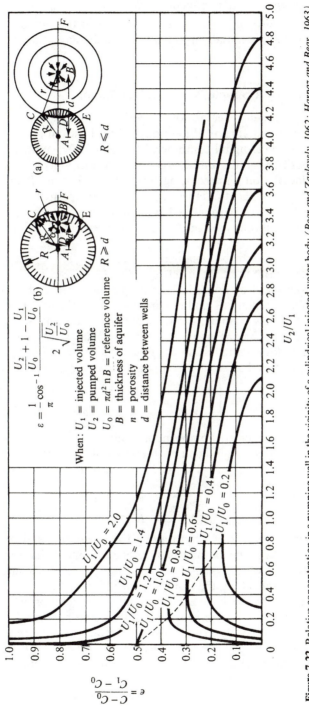

Figure 7-22 Relative concentration in a pumping well in the vicinity of a cylindrical injected water body (*Bear and Zaslavsly, 1962; Harpaz and Bear, 1963*).

291

Equations (7-200) through (7-202) remain valid also when $R > d$ (Fig. 7-22b). Figure 7-22 shows also the relative concentration in the pumping well for this case. Other shapes of the initially injected water body may be handled in a similar manner.

Front Movement by a Simple Numerical Technique

As has already been mentioned above, numerical techniques may yield the shape of a moving front when analytical methods fail to do so. In general, the solution itself is carried out on a digital computer. The basis for the numerical determination of front positions is the knowledge of the velocity at every instant for all fluid particles located on an advancing front at that instant. If this velocity can be determined for a sufficient number of fluid particles (points) on a front, the new position of the latter can be obtained by considering the displacement of each particle during a small time interval Δt. The displacement of each particle is given by $\Delta s = \mathbf{V} \Delta t$.

The velocity itself can be obtained in several ways. For example, in steady flow (of the water, although the front is moving), we can use potential theory (or any other analytical or numerical technique) in order to determine the flow pattern, i.e., the equipotentials and streamlines. Once the streamlines have been established for the flow domain, the movement of the front is obtained by displacing fluid particles belonging to it along *streamlines*. Thus, if ds denotes an element of length along a streamline, the time required for a fluid particle belonging to the front to travel a length $s(t)$ along a streamline, is given by

$$t = \int_0^{s(t)} \frac{ds}{V_s(s)} = -\frac{n}{K} \int_0^{s(t)} \frac{ds}{\partial \phi / \partial s} \tag{7-203}$$

where the aquifer is assumed homogeneous and isotropic and V_s is the velocity component in the direction tangent to the streamline. When the potential distribution, ϕ, is obtained numerically, the computation of t by (7-203) is also carried out numerically

$$t = \sum_{(i)} (\Delta t)_i = \frac{n}{K} \sum_{(i)} \frac{(\Delta s)_i^2}{(\Delta \phi)_i} \tag{7-204}$$

where $(\Delta s)_i$ is an increment of length along the streamline and $(\Delta \phi)_i$ is the potential drop along $(\Delta s)_i$. The variations of velocity during Δt have been neglected.

In unsteady flow, we can determine the velocity (magnitude and direction) at every instant for points along the front, and displace the latter during Δt accordingly. At the end of Δt, we have a new potential distribution which, in turn, is used to determine new velocities. However, since we do not have here streamlines (as also in the case of steady flow, if streamlines are not determined), an accumulating error is introduced due to the variations in velocity (in magnitude and direction) during the time interval Δt. The accuracy of the front shape and position is increased by reducing Δt.

Of special interest is the case of an aquifer with a number of injection and pumping wells, each having its own operation schedule, $Q(t)$. We assume that the effect of aquifer storativity can be neglected (see Bear and Jacobs, 1965). The flow pattern produced at every instant by these wells is one of steady flow. The velocity at time t of any fluid particle i located at point (x_i, y_i) which is at a distance r_{ij} from well j located at (x_j, y_j), in which the rate of flow is $Q_{wj}(t)$, is given by $V_i(t) = Q_{wj}(t)/2\pi r_{ij}Bn$. Together, all the wells produce at point i, a velocity which is the *vector sum* of the individual velocities induced by the wells operating at that time. Accordingly, the position $(x_i|_{t+\Delta t}, y_i|_{t+\Delta t})$, of the particle at $t + \Delta t$ is given by

$$
\left.\begin{aligned}
x_i\Big|_{t+\Delta t} &= x_i\Big|_t + (\Delta t)\sum_{(j)} \frac{Q_{wj}(t)}{2\pi Bn}\cdot\frac{x_i|_t - x_j}{r_{ij}^2}, \\
y_i\Big|_{t+\Delta t} &= y_i\Big|_t + (\Delta t)\sum_{(j)} \frac{Q_{wj}(t)}{2\pi Bn}\cdot\frac{y_i|_t - y_j}{r_{ij}^2}, \\
r_{ij}^2 &= \left(x_i\Big|_t - x_j\right)^2 - \left(y_i\Big|_t - y_j\right)^2
\end{aligned}\right\}
\tag{7-205}
$$

obtained from $\Delta x = V_x\Delta t$, $\Delta y = V_y\Delta t$. When a natural flow exists in the aquifer, with velocity components V_{x0}, V_{y0}, we have to add the terms $V_{0x}\Delta t$ and $V_{0y}\Delta t$ to the displacements Δx and Δy, respectively. In writing (7-205) we have neglected aquifer storativity and changes in velocity during Δt. This introduces an accumulating error in the process as each particle gradually deviates from the streamline on which it was initially located. Smaller time increments will reduce this error.

Thus, given a front at some initial instant of time, we start by designating a number of fluid particles along the front, sufficient to represent the entire front. We then move these particles, by repeating the procedure outlined above (Fig. 7-23) for successive time intervals to obtain the history of the front movement. The whole procedure is usually performed by a digital computer. In the calculations, it is convenient to assume that at $t = \Delta t$, the front in the vicinity of each

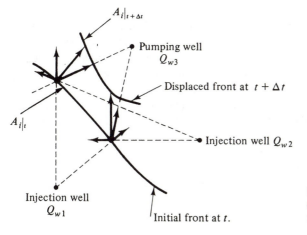

Figure 7-23 Graphical method for determining successive front positions.

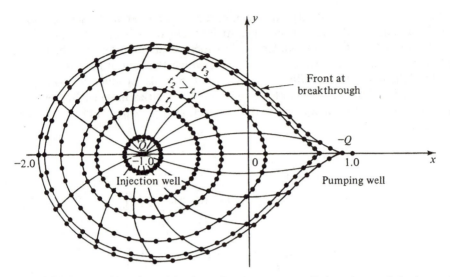

Figure 7-24 Front positions for an injection well and a pumping well of equal strength, by the graphical method.

injection well has the shape of a circle of radius r_0 determined from $Q_{wj}(t) \Delta t = \pi r_0^2 Bn$.

The advantage of this approach (when applicable) is that one need not solve for the ϕ (or ϕ, ψ) distribution in the entire field at the end of each time interval. Figure 7-24 shows the application of this approach to an injection well and a pumping well of equal strength. Figure 7-25 (Harpaz and Bear, 1963) shows typical front positions for a field of pumping and injection wells.

The concentration of injected water in the water pumped from a well is obtained from the angle of the boundary of injected water body at that well.

Arrival Distributions

Nelson (1977), in a series of papers on the evaluation of environmental consequences of groundwater contamination, introduces the concept of *arrival distributions*. These quantities define how much of a contaminant reaches a specific point in a flow domain at a specific time. This information serves as basic input to the solution of many quality management problems.

Essentially, the technique presented by Nelson (1977) is based on determining travel times along path lines by one of the methods described above. Let us demonstrate his approach by the case of a homogeneous isotropic aquifer in which, superimposed on the natural flow, we introduce a recharge pond, an injection well, and a pumping well. The situation is shown in Fig. 7-26a. The flow domain is a two-dimensional one in the horizontal xy plane.

The potential distribution can be obtained by superposition of elementary flows: a uniform flow in the $+x$ direction, a flow produced by a source (the pond),

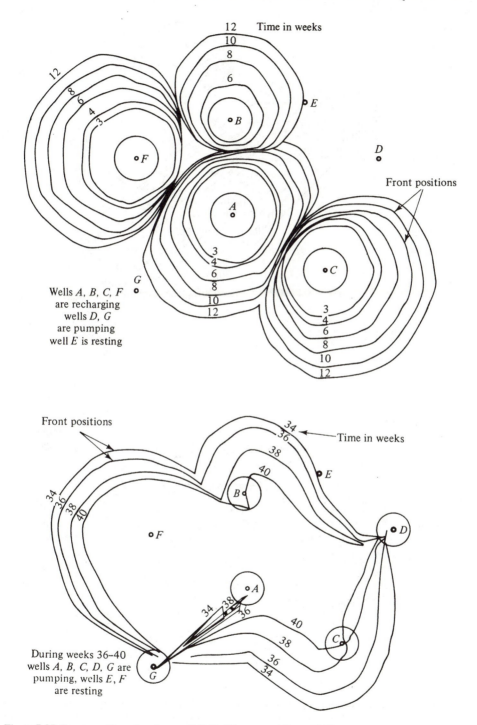

Figure 7-25 Front positions for given well field (*Harpaz and Bear, 1963*).

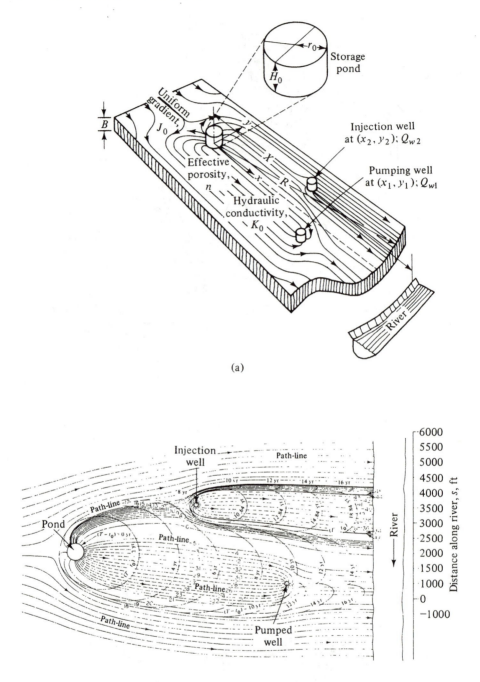

(a)

(b)

Figure 7-26 Determination of arrival distributions *(Nelson, 1977)*. (a) Layout and parameters. (b) Flow pattern. (c) The location/arrival-time distribution along the river for the pond leakage. (d) The location/outflow-quantity distributions at the river.

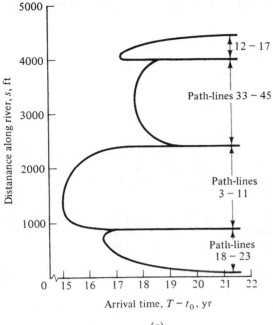

(c)

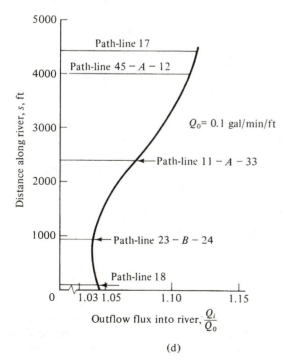

(d)

Figure 7-26 (contd.)

and a flow produced by the doublet of pumping and injection wells of equal strength. The resulting potential is

$$\phi = xJ_0\left(\frac{r_0^2}{x^2 + y^2} - 1\right) + H'H_0\left[1 - \frac{\ln\left[(x^2 + y^2)/r_0^2\right]}{\ln(R^2/r_0^2)}\right]$$

$$- \sum_{j=1}^{N} \frac{Q_{wj}}{4\pi T} \ln \frac{(x - x_j)^2 + (y - y_j)^2}{x_j^2 + y_j^2} \qquad (7\text{-}206)$$

where $N = 2$, H' is the transient head in the pond (here $H' = 1$ for steady flow), J_0 is the constant gradient of the uniform flow, T is aquifer transmissivity and ϕ is defined for the domain $x^2 + y^2 > r_0^2$, except at the points x_j, y_j. The potential ϕ satisfies the Laplace equation and the boundary conditions: $\phi = H_0 \pm \varepsilon$ when $x^2 + y^2 = r_0^2$, $\phi = 0 \pm \varepsilon'$ when $x^2 + y^2 = R^2$; $\varepsilon, \varepsilon'$ are the approximation errors introduced by the wells in the head at the pond and at the remote boundary, R is a large radial distance to the remote outer boundary, and r_0 is the radius of the pond. Nelson (1977) suggests that ε and ε' are negligible for $r_0^2/(x_j^2 + y_j^2) <$ 0.01 and $(x_j^2 + y_j^2)/R^2 < 0.7$.

Once the potential distribution is obtained (as shown above or by any other method, say, a numerical solution), one can determine path-lines for the individual fluid particles through the aquifer to the points of emergence at outflow boundaries.

The locations and the arrival times of fluid particles at outflow boundaries are determined by a *path function*. By solving

$$\left. \begin{array}{l} V_x = \dfrac{dx}{dt} = -\dfrac{K_0}{n}\dfrac{\partial\phi}{\partial x}, \qquad V_y = \dfrac{dy}{dt} = -\dfrac{K_0}{n}\dfrac{\partial\phi}{\partial y}, \\[3mm] x\bigg|_{t_0} = x_0, \qquad y\bigg|_{t_0} = y_0; \qquad x_0^2 + y_0^2 = r_0^2 \end{array} \right\} \qquad (7\text{-}207)$$

simultaneously, usually numerically, we obtain path-lines and travel times to the pumped well, or to the remote river boundary. Figure 7-26b shows path-lines and travel times.

The flux, or fluid flow rate, is obtained from the *stream function*, which, in turn, is obtained by superposition of the simple basic flows.

$$\frac{Q_i}{Q_0} = \frac{J_0 \ln(R/r_0)}{2\pi H_0} y + \frac{J_0 r_0^2 \ln(R/r_0)}{2\pi H_0}\cdot\frac{y}{x^2 + y^2} + \frac{H'}{2\pi}\tan^{-1}\frac{y}{x}$$

$$+ \sum_{j=1}^{N} \frac{Q_{wj}\ln\dfrac{R}{(r_0)}}{4\pi^2 T H_0}\left[\tan^{-1}\left(\frac{y - y_j}{x - x_j}\right) - \tan^{-1}\left(\frac{-y_j}{r_0 - x_j}\right)\right] \qquad (7\text{-}208)$$

where $Q_0 = 2\pi T H_0/\ln(R/r_0)$; $H' = 1$, $N = 2$, and Q_i is the cumulative flow rate of contaminated fluid. This equation is used to determine the cumulative flow rate for the two outflow boundaries of interest in this example: the periphery of the pumping well and the river. The major flow pattern results are plotted in

Fig. 7-26b. In particular the figure illustrates the flow paths and fronts (isochronal lines) of the contaminated fluid advancing from the pond towards the river. Between pathlines 24 and 2, the contaminated pond outflow is intercepted by the pumping well.

From this information, Nelson (1977) obtains the location/arrival time distribution along the river for the pond leakage (Figs 7-26c and d) and the location/ outflow quantity distribution at the river. Nelson (1977) shows how the various distributions lead to the determination of the concentration of contaminants which reach the aquifer, that will, with passing time, appear at the aquifer outlets (lakes, rivers, springs, and pumping wells). Once this information is made available to the planner, he can control the movement of the contaminants in the aquifer by controlling (say, via pumping and recharge operations) the flow pattern in the aquifer and/or initiate corrective measures to protect the environment endangered by the outflow of contaminated groundwater.

EIGHT

HYDRAULICS OF PUMPING AND RECHARGING WELLS

Most of this book is concerned with the regional approach in which we seek the elevations of the water table (or piezometric surface) resulting from inputs and outputs such as natural replenishment, artificial recharge, and pumping. We are interested in the general flow pattern in an aquifer, rather than in the details of what happens in the vicinity of an individual well (= point sink or source). Wells are considered only as inputs or outputs which affect the overall flow pattern in the aquifer. In the present chapter, we shall focus our attention on the vicinity of an individual well and determine the *drawdown* (or *build-up*) that will be produced in the well itself and in its vicinity.

Only slight modifications have to be introduced in the relationships derived for pumping wells in order to make them applicable to recharging wells. These are considered in Sec. 8-13.

The actual structure of wells, as well as drilling and completion techniques, are not considered. The reader is referred to specialized texts (e.g., Johnson Inc. 1966, and many articles in the Johnson Driller's Journal; Campbell and Lehr, 1973).

Various cases of hydraulics of wells in confined, leaky, and phreatic aquifers are treated in numerous publications in the literature. Reviews on this subject are presented by Bruin and Hudson (1955), Ferris *et al.* (1962), Hantush (1964), Bear *et al.* (1968), Huisman (1972). Huyakorn and Dudgeon (1972) presented an extensive annotated bibliography on well hydraulics. The following sections contain a summary of various selected cases of interest to the hydrologist and planning engineer. Only vertical wells are discussed here. Hantush and Papadopulos (1962) and Hantush (1964) discuss also flow to *horizontal wells* (generally known as *collector wells*) which are sometimes used for extracting groundwater from aquifers. When located in aquifers close to streams and lakes (which are hydraulically connected to the aquifer), they often yield large quantities of water. The advantage lies in the filtration of the water on its way through the aquifer from the stream to the well. In general, the drawdown in collector wells is small: one may visualize them as wells of very large diameter.

8-1 INTRODUCTION

Pumping from a phreatic aquifer removes water from the void space leaving there a certain quantity of water which is held against gravity. As a result, the water table at each point is lowered with respect to its initial position by a vertical distance called *drawdown* ($s(x, y, t)$ in Fig. 8-1a). The *drawdown surface* in the vicinity of a pumping well (or the *cone of depression*) shows the variation of drawdown with the distance from the well. Taking the product of the specific yield (Sec. 5-1) and the drawdown, and integrating it over the entire area (horizontal projection) of the cone of depression gives the volume of water removed by pumping. This statement assumes that (i) water is instantaneously removed from the aquifer upon a decline of head (see Sec. 5-4), and (ii) the expanding cone of depression has not reached any source boundary. Here and below, in unsteady flow, drawdown is with respect to water table without pumping.

In a confined aquifer, the entire discharge of the well is provided by the release of water stored in the aquifer due to its compressibility and the compressibility of the water. In this case, the piezometric surface is lowered by the pumpage. The drawdown, $s(x, y, t)$, is the vertical distance between the initial piezometric surface, and the piezometric surface at some later time t at the same point (Fig. 8-1b). The total volume of water removed by pumping from a confined aquifer is equal to the product of the aquifer storativity and the drawdown, integrated over the (horizontal) area of the cone of depression.

Thus, in both confined and phreatic aquifers, *steady flow cannot exist in an extensive aquifer*. The flow is *unsteady* until a source, or a region of replenishment, is intercepted by the growing cone of depression. In a confined aquifer, the cone of depression expands theoretically with the speed of sound, though a measurable drawdown is observed at points sufficiently removed from the pumping well only after a long period of pumping. From the practical point of view, the flow approaches a quasi-steady state in which no significant additional drawdown is observed. The distance from a pumping well to points at which practically no drawdown can be observed is called the *radius of influence* of the well (see below).

When the flow in a confined aquifer is everywhere horizontal, or assumed practically so, the equipotential surfaces are vertical, and the piezometric surface obtained from a number of observation wells (or piezometers) is unique; it describes the piezometric head distribution in the aquifer. The drawdown is a function of x, y and t, but not of z, and the depth at which the opening of an observation well is located is immaterial. However, when the flow is not horizontal (e.g., close to a partially penetrating well, Sec. 8-8), the equipotentials are no longer vertical, and the location of the opening of an observation well becomes important (Fig. 8-2). In this case the drawdown $s = \phi(x, y, z, 0) - \phi(x, y, z, t)$, where ϕ is the piezometric head at a point.

In a phreatic aquifer, the flow is never *exactly* horizontal. The water level in

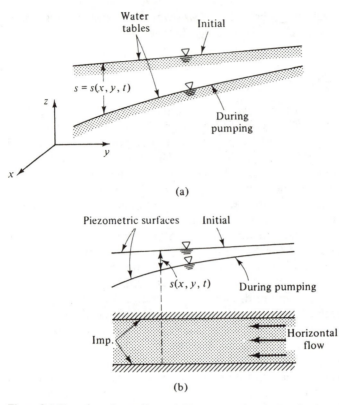

Figure 8-1 Drawdown in aquifers. (a) Phreatic aquifer. (b) Confined aquifer.

an observation well indicates the elevation of the phreatic surface at the location of the well only approximately (depending on the slope of the phreatic surface); it gives exactly the piezometric head at the point where water enters the observation well. The discrepancy is shown in Fig. 8-2b. In the practice, however, because the slope of the phreatic surface is small (Sec. 4-5), the error is small. It is mentioned here to emphasize the error which might be introduced by locating observation wells too close to pumping wells, and by making their screens too long.

In the vicinity of a recharging (= injection) well, the piezometric surface (or the water table in the case of a phreatic aquifer) will rise. The vertical distance between the initial and the instantaneous piezometric surface at a point is called *build-up*. Because of clogging which takes place in the vicinity of a recharging well (Sec. 8-13), for the same rate we have a higher build-up than drawdown.

The wells considered here may be drilled wells of a relatively small diameter (say, 3–24 in), or larger dug wells or shafts, several feet in diameter. In the latter case, flow through the well's bottom cannot be neglected.

It is difficult to define the *radius of a well*. Depending on the type of material comprising the aquifer formation (consolidated or unconsolidated, fine or coarse material), the well may be left as an uncased hole, it may be equipped with a screen of various types, or even with a gravel-pack (say, 2 in thick) outside the

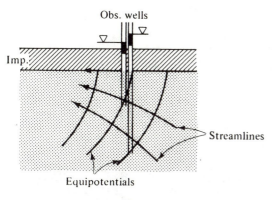

Obs. wells

Imp.

Streamlines

Equipotentials

(a)

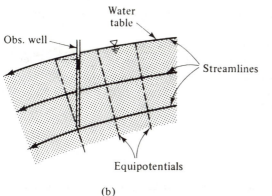

Water
table

Obs. well

Streamlines

Equipotentials

(b)

Figure 8-2 Observation wells in non-horizontal flow.

screen. Also, when a well is developed upon completion, fine material is removed from the interstices in the vicinity of the well, thus increasing the permeability in this region. The increased permeability (by the gravel pack and/or the development procedure) in the vicinity of a well reduces the drawdown in and close to the well. This is shown in Fig. 8-3 (after Jacob, 1950). In addition, we have entrance head losses (depending on the quality of the well's screen) and head losses caused by the upward flow through the well's casing (Sec. 8-12). This means that in a confined aquifer the piezometric head just outside the screen is at a higher elevation than that of the water level inside the casing of a pumping well.

For an uncased well, or for a well equipped with a screen, but without a gravel pack or an increased permeability zone in its vicinity, the nominal radius of the hole (or screen) is usually used as radius of well, r_w. For a developed, or gravel-packed well (Fig. 8-3c) we define an *effective well radius* equal to the distance from the axis of the well at which the theoretical drawdown in steady flow (i.e., a logarithmic distribution of drawdown) equals the actual drawdown just outside the screen. The dashed curve in Fig. 8-3c gives the drawdown distribution that would occur if the formation were left undisturbed with uniform permeability

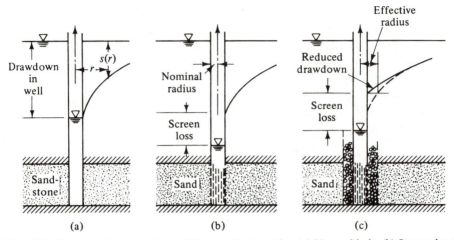

Figure 8-3 Drawdown in a pumping well in a confined aquifer. (a) Uncased hole. (b) Screened well. (c) Gravel-packed well.

(Jacob, 1950); it is the same drawdown curve as in Figs. 8-3a and b. Even with this definition, it is difficult to estimate r_w in practical cases, as the drawdown curve in steady flow close to the well is usually unavailable. It is recommended to avoid the use of r_w whenever possible.

The following assumptions underlie the discussion in the present section, unless specified differently:

(i)   The flow in the aquifer obeys Darcy's law.
(ii) Water is instantaneously removed from storage upon a decline of head.
(iii) The aquifer is homogeneous, isotropic, and of infinite areal extent.
(iv) The aquifer's bottom is horizontal. A confined aquifer has a constant thickness.
(v)   The water table, or piezometric surface, without pumping is horizontal.
(vi) Storage is ignored in aquitard.

The effect of anisotropy $(K_x \neq K_y)$ can be introduced by using the transformations discussed in Sec. 5-9. Near partially penetrating wells, we have also vertical flow components; the effect of $K_z \neq K_x = K_y$ may also be introduced by the same procedure.

In the practice, before applying the results derived in this chapter, actual field conditions should carefully be checked against the assumptions listed here and against additional ones introduced in the course of the discussion. In general, in spite of some of the simplifying assumptions, a good agreement between theory and field observations is obtained.

8-2 STEADY FLOW TO A WELL IN A CONFINED AQUIFER

Figure 8-4 shows the radially converging flow to a well fully penetrating a homogeneous confined aquifer (hydraulic conductivity K and constant thickness B)

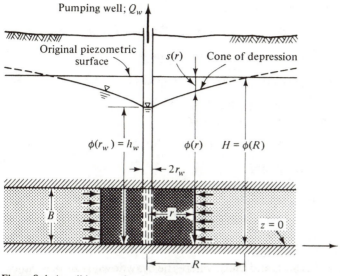

Figure 8-4 A well in a confined aquifer.

of infinite areal extent. The well's constant pumping rate is Q_w. Initially the piezo-metric surface is at $\phi(r) = H = $ const. When pumping takes place, a cone of depression is formed with a drawdown $s = s(r) = H - \phi(r)$ in the vicinity of the pumping well. Since we have here radially symmetric, horizontal, steady flow, $\phi(r)$ can be obtained by solving the Laplace equation (5-61), which in polar coordinates takes the form

$$\partial^2\phi/\partial r^2 + (1/r)\,\partial\phi/\partial r = 0 \qquad (8\text{-}1)$$

with the boundary condition at the well, $r = r_w$, $\phi = h_w$, and at some distance $r, \phi = \phi(r)$. We may also integrate between $r = r_w$, $\phi = h_w$ and $r = R$, $\phi = H$.

The potential distribution $\phi(r)$ can also be derived directly by considering continuity for the portion of the aquifer between two concentric cylinders of radii r_w and r. Equating inflow, Q, to outflow, Q_w, we obtain

$$Q = A \times q_r = 2\pi r B \times K\partial\phi/\partial r = Q_w = \text{const.} \qquad (8\text{-}2)$$

where q_r is the specific discharge in the radial direction. By integrating, between $r = r_w$ where $\phi(r_w) = h_w$, and r, we obtain

$$\phi(r) - h_w = (Q_w/2\pi T)\ln(r/r_w); \qquad T = KB \qquad (8\text{-}3)$$

Equation (8-3) describes the drawdown curve $\phi(r)$ shown in Fig. 8-4.

From (8-3), it follows that $\phi(r)$ will increase indefinitely as r increases. As the piezometric surface cannot rise above H, one must conclude that *steady flow in an infinite aquifer is impossible* and it is meaningless to consider such flow. This means that (8-3) is valid only in the close proximity of a well where steady flow has been established.

At some distance $r = R$, we have $\phi(r) = H$, and the drawdown $s(r) = 0$.

By integrating (8-1) from r_w to R, we obtain

$$s_w = H - h_w = \phi(R) - \phi(r_w) = (Q_w/2\pi T)\ln(R/r_w) \tag{8-4}$$

Between any two distances r_1 and $r_2 (> r_1)$, we obtain

$$\phi(r_2) - \phi(r_1) = s(r_1) - s(r_2) = (Q_w/2\pi T)\ln(r_2/r_1) \tag{8-5}$$

Equation (8-5) is called the Thiem equation (Thiem, 1906).

Between any two distances r and R, we obtain

$$s(r) = \phi(R) - \phi(r) = (Q_w/2\pi T)\ln(R/r) \tag{8-6}$$

By dividing (8-4) by (8-6), we obtain

$$\phi(r) - h_w = (H - h_w)\frac{\ln(r/r_w)}{\ln(R/r_w)} \tag{8-7}$$

showing that the shape of the curve $\phi = \phi(r)$, given h_w and H at r_w and R, respectively, is independent of Q_w and T.

The distance R in (8-4), (8-6), and (8-7), where the drawdown is zero, is called the *radius of influence of the well*. Since we have established above that steady flow cannot prevail in an infinite aquifer, the distance R should be interpreted as a parameter which indicates the distance beyond which the drawdown is negligible, or unobservable. In general, this parameter has to be estimated from past experience. Fortunately, R appears in (8-6) in the form of $\ln R$ so that even a large error in estimating R does not appreciably affect the drawdown determined by (8-6). The same observation is true also for another parameter—the radius of the well r_w (Sec. 8-1).

Various attempts have been made to relate the radius of influence, R, to well, aquifer, and flow parameters in both steady and unsteady flow in confined and phreatic aquifers. Some relationships are purely empirical, others are semi-empirical. For example (Bear, Zaslavsky, and Irmay, 1968).

Semi-empirical formulas are

Lembke (1886, 1887):	$R = H(K/2N)^{1/2},$	(8-8)
Weber (Schultze, 1924):	$R = 2.45(HKt/n_e)^{1/2},$	(8-9)
Kusakin (Aravin and Numerov, 1953):	$R = 1.9(HKt/n_e)^{1/2}$	(8-10)

Empirical formulas are

Siechardt (Chertousov, 1962):	$R = 3000 s_w K^{1/2},$	(8-11)
Kusakin (Chertousov, 1949):	$R = 575 s_w (HK)^{1/2}$	(8-12)

where R, s_w (= drawdown in pumping well), and H are in meters and K in meters per second.

In phreatic aquifers (Sec. 8-3) N, H, and n_e represent accretion from precipitation, the initial thickness of the saturated layer, and the specific yield (or effective porosity) of the aquifer, respectively. In confined aquifers, H and n_e have to be

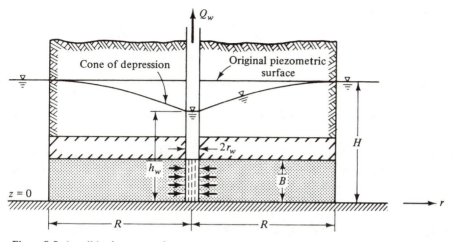

Figure 8-5 A well in the center of a circular island.

replaced by the aquifer's thickness, B, and its storativity, S. A further discussion on R is given in Sec. 8-5.

Although in an infinite aquifer R should be considered as a coefficient, one can visualize a case of an aquifer of finite areal extent in the form of a circle, where along the boundary, $r = R$, the head, H, is maintained constant. An example of such a case is shown in Fig. 8-5.

When along $r = R$, ϕ varies with the angle θ, we have to solve the continuity equation (5-61), written in polar coordinates, for $\phi = \phi(r, \theta)$

$$\partial^2 \phi / \partial r^2 + (1/r)\, \partial \phi / \partial r + (1/r^2)\, \partial^2 \phi / \partial \theta^2 = 0 \qquad (8\text{-}13)$$

Muskat (1937) presents a solution for $\phi = \phi(r, \theta)$, where the boundary conditions are defined by $\phi = \phi(r_w, \theta)$ and $\phi = \phi(R, \theta)$. It may be of interest to note that the well's discharge Q_w in this case can be obtained from (8-4) with $\phi(R)$ and $\phi(r_w)$ replaced by their average values $[\phi(R)]_{av}$ and $[\phi(r_w)]_{av}$ defined by

$$[\phi(r_w)]_{av} = \frac{1}{2\pi} \int_0^{2\pi} \phi(r_w, \theta)\, d\theta; \qquad [\phi(R)]_{av} = \frac{1}{2\pi} \int_0^{2\pi} \phi(R, \theta)\, d\theta \qquad (8\text{-}14)$$

Nonlaminar Flow Regime

Let us use this opportunity of dealing with the relatively simple flow pattern around a well in a confined aquifer to examine a *nonlaminar flow regime*. As above, we start from the continuity statement for the portion of the aquifer between r and R

$$Q = -Aq_r = Q_w = \text{const.}; \qquad A = 2\pi r B \qquad (8\text{-}15)$$

where the minus sign is introduced because the radial specific discharge, q_r, is opposite to the direction $+r$. We relate q_r to the hydraulic gradient $-\partial \phi / \partial r$ by

$$-\partial \phi / \partial r = W q_r + b q_r^2; \qquad W = 1/K \qquad (8\text{-}16)$$

By combining (8-15) and (8-16), we obtain

$$-\frac{\partial \phi}{\partial r} = -W\frac{Q_w}{A} + b\frac{Q_w^2}{A^2} = -\frac{WQ_w}{2\pi B}\frac{1}{r} + \frac{bQ_w^2}{4\pi^2 B^2}\frac{1}{r^2} \qquad (8\text{-}17)$$

Integrating from $r = r_w$, $\phi = \phi_w$ to any distance r, we obtain

$$\phi(r) - \phi_w = \frac{Q_w}{2\pi T}\ln\frac{r}{r_w} + \frac{bQ_w^2}{4\pi^2 B^2}\left(\frac{1}{r_w} - \frac{1}{r}\right) \qquad (8\text{-}18)$$

Without the second term on its right-hand side, caused by the quadratic term in (8-16), (8-18) is the same as (8-4) based on Darcy's law. One should note, however, that equipotentials are still circles centered at the well.

8-3 STEADY FLOW TO A WELL IN A PHREATIC AQUIFER

Figure 8-6 shows the cone of depression in the vicinity of a well pumping at a rate Q_w from an isotropic phreatic aquifer. The flow is radially symmetric between circular equipotential boundaries at $r = R$ and $r = r_w$. Hence, the potential distribution $\phi = \phi(r, z)$ satisfies the continuity equation

$$\partial^2\phi/\partial r^2 + (1/r)\partial\phi/\partial r + \partial^2\phi/\partial z^2 = 0 \qquad (8\text{-}19)$$

which is the Laplace equation (5-61) for radially symmetric flow. The boundary conditions, assuming no well losses, are

$$\phi(R, z) = H_0, \qquad 0 \le z \le H_0, \qquad \text{(equipotential)}$$

$$\phi(r_w, z) = h_w, \qquad 0 \le z \le h_w, \qquad \text{(equipotential)}$$

$$\phi(r_w, z) = z, \qquad h_w \le z \le h_s, \qquad \text{(seepage face)} \qquad (8\text{-}20)$$

$$\partial\phi/\partial z = 0, \qquad z = 0; r_w \le r \le R, \qquad \text{(impervious bottom)}$$

$$\left.\begin{array}{l}\phi(r, h) = h, \\ \partial\phi/\partial n = 0,\end{array}\right\} \quad \begin{array}{l}r_w \le r \le R \\ z = h\end{array} \right\} \qquad \text{(phreatic surface)}$$

where n is distance measured in the direction of the normal to the phreatic surface. A seepage face (Sec. 5-3) is always present when a phreatic surface approaches a downstream body of liquid continuum (here in the well). The situation is different when the well is cased (=impervious), with a screened (or perforated) section as its lower portion.

Another possible boundary condition at the well is that of constant discharge $Q_w = $ const. Then the second condition is replaced by

$$K\int_0^{h_w} 2\pi r[\partial\phi(r, z)/\partial r]\, dz = Q_w$$

Like other unconfined flow cases, the problem is nonlinear and, in general, cannot be solved analytically. Kirkham (1964) presents an exact solution for the

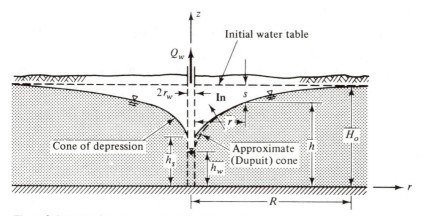

Figure 8-6 Radial flow to a well in a phreatic aquifer.

height of the phreatic surface, h, in the form of an equation which is solvable by iteration. His potential function is obtained by assuming that a certain fictitious flow exists in the region above the phreatic surface and below the horizontal plane at $z = H_0$, such that the boundary conditions on the phreatic surface are satisfied also by the potential of this flow.

Numerical methods have also been often applied to the solution of the problem as stated by (8-19) and (8-20).

By using the *Dupuit assumptions*, an easily integrable linear continuity equation can be derived. The results are accurate enough for distances $r > 1.5h$ from a well. In this approach, the seepage face is neglected. Hansen (1949) gives graphs of Q/Kr_w^2 as a function of h_s/r_w and h_w/r_w (Fig. 8-7). Boulton (1951) suggests

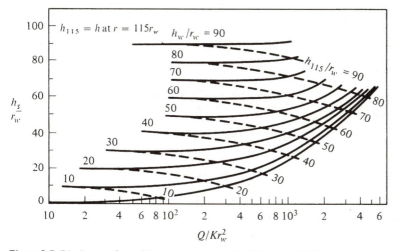

Figure 8-7 Discharge of a well in a phreatic aquifer *(Hansen, 1949)*.

the relationship

$$h_s - h_w \approx (H_0 - h_w) - 3.75 Q_w / 2\pi K H_0 \qquad (8\text{-}21)$$

where 3.75 is replaced by 3.5 if r_w/H_0 is of the order 0.25.

Consider a cylinder of radius r around the well. For the considered steady flow, the Dupuit assumptions lead to

$$Q_w = 2\pi r h q_r = 2\pi r h K dh/dr = 2\pi r K \partial (h^2/2)/\partial r \qquad (8\text{-}22)$$

where q_r is the specific discharge in the radial direction. Integrating between $h = h_w$ at $r = r_w$ and $h = H_0$ at $r = R$, we obtain

$$H_0^2 - h_w^2 = \frac{Q_w}{\pi K} \ln (R/r_w) \qquad (8\text{-}23)$$

In this integration, we have completely neglected the seepage face and made h_s identical to h_w. By integrating from some distance r to the external boundary at R, we obtain

$$H_0^2 - h^2 = \frac{Q_w}{\pi K} \ln (R/r) \qquad (8\text{-}24)$$

Dividing (8-24) by (8-23) gives

$$H_0^2 - h^2 = (H_0^2 - h_w^2) \frac{\ln (R/r)}{\ln (R/r_w)} \qquad (8\text{-}25)$$

The dashed curve in Fig. 8-6 gives the phreatic surface elevations, $h = h(r)$, as expressed by (8-25). It is interesting to note that neither Q_w nor K appear in (8-25). From (8-24), it follows that as $r \to \infty$, $h \to \infty$, which is obviously impossible. This means that *steady flow is impossible in an infinite aquifer.* The equation is, therefore, valid only in the vicinity of the well.

Equation (8-23) is known as the *Dupuit–Forchheimer well discharge formula.* It is an exact solution of the continuity equation (in polar coordinates) based on the Dupuit assumptions

$$\partial Q/\partial r = 0 = \partial (2\pi r h K \partial h/\partial r)/\partial r = \partial (\pi K r \partial h^2/\partial r)/\partial r \qquad (8\text{-}26)$$

or

$$\partial^2 (h^2)/\partial r^2 + (1/r) \, \partial (h^2)/\partial r = 0 \qquad (8\text{-}27)$$

which is linear in h^2.

Equation (8-24) may also be written as

$$H_0 - h = \frac{1}{(H_0 + h)} \frac{Q_w}{\pi K} \ln (R/r) \qquad (8\text{-}28)$$

For a thick aquifer and small drawdown, $(H_0 - h) \ll H_0$, $H_0 + h \approx 2H_0$, and (8-24) may be approximated by

$$s = \frac{Q_w}{\pi K (H_0 + h)} \ln \frac{R}{r} \quad \text{or} \quad s = \frac{Q_w}{2\pi T} \ln \frac{R}{r} \qquad (8\text{-}29)$$

where $s = H_0 - h$ is the drawdown and $T = K(H_0 + h) \approx KH_0$ is the average aquifer transmissivity. Equation (8-29) is identical to (8-6) describing steady flow in a confined aquifer. This means that *for small drawdowns* (*relative to H or h*), a *phreatic aquifer may be treated as a confined one.*

Between any two points r_1 with drawdown s_1 and r_2 with drawdown s_2, (8-24) becomes

or

$$\left. \begin{array}{c} h_2^2 - h_1^2 = \dfrac{Q_w}{\pi K} \ln \dfrac{r_2}{r_1} : \qquad (H_0 - s_2)^2 - (H_0 - s_1)^2 = \dfrac{Q_w}{\pi K} \ln \dfrac{r_2}{r_1} \\[2em] (s_1 - s_1^2/2H_0) - (s_2 - s_2^2/2H_0) = \dfrac{Q_w}{2\pi K H_0} \ln \dfrac{r_2}{r_1} \\[2em] s_1' - s_2' = \dfrac{Q_w}{2\pi K H_0} \ln \dfrac{r_2}{r_1} \end{array} \right\} \quad (8\text{-}30)$$

where $s' = s - s^2/2H_0$ is called the *corrected drawdown*, i.e., the drawdown that would occur in an equivalent confined aquifer.

Although (8-23) is an approximate expression for $h(r)$ based on the Dupuit assumption, Charni (1951) and Polubarinova-Kochina (1952, 1962) showed that it gives *exactly* the discharge Q_w.

The exact expression for the well's discharge can be written as

$$Q_w = 2\pi K r \int_0^{h(r)} [\partial\phi(r, z)/\partial r] \, dz = 2\pi r K \partial\phi'/\partial r$$

$$\phi' = \int_0^{h(r)} \phi(r, z) \, dz - h^2(r)/2$$

(8-31)

In the Dupuit approximation (8-22) for Q_w, we have replaced ϕ' by $h^2/2$. It can be shown (Bear, 1972, p. 363) that in an anisotropic aquifer ($K_r \neq K_z$), we have

$$0 < \frac{h^2/2 - \phi'(r)}{h^2/2} < \frac{(K_r/K_z) \, i^2}{1 + (K_r/K_z) \, i^2} \quad (8\text{-}32)$$

where $i = dh/dr$. Thus since $i \ll 1$, the error in replacing ϕ' by $h^2/2$ is small.

Figure 8-8 shows steady axisymmetric flow to a well in a phreatic aquifer. The well is fed by a constant rate of accretion (N) reaching the water table. Again, by assuming that the flow in the aquifer is essentially horizontal (i.e., using the Dupuit assumptions), we obtain

$$Q_w = \pi(r^2 - r_w^2) N + 2\pi r K h \partial h/\partial r \quad (8\text{-}33)$$

Integrating between $r = r_w$, $h = h_w$ (i.e., neglecting the seepage face) and a point at some distance r, yields

$$h^2 - h_w^2 = \frac{Q_w}{\pi K} \ln \frac{r}{r_w} - \frac{N}{2K}(r^2 - r_w^2) + \frac{r_w^2 N}{K} \ln \frac{r}{r_w} \quad (8\text{-}34)$$

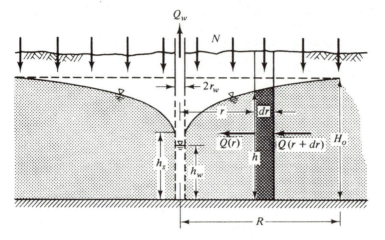

Figure 8-8 Radial flow to a well in a phreatic aquifer with accretion.

8-4 STEADY FLOW TO A WELL IN A LEAKY CONFINED AQUIFER

The leaky aquifer is defined in Sec. 2-3. When pumping takes place in such an aquifer, the resulting drawdown produces leakage into or out of the aquifer through the leaky (or semiconfining) layers (Fig. 8-9). De Glee (1930), Steggewentz (1939), Jacob (1946), Hantush and Jacob (1954, 1955a, b), Hantush (1956, 1957, 1959, 1960, 1964, 1967a, b), DeWiest (1961, 1963), Polubarinova-Kochina (1952, 1962), and many others, studied steady and unsteady flow in leaky aquifers (see discussion on unsteady flow in Sec. 8-7).

When $K' \ll K$, $B' \ll B$ (i.e., a thin semipervious layer; Fig. 8-9), an approximate solution may be derived by assuming that the flow in the less permeable layer ascends or descends vertically, depending on the direction of the hydraulic gradient within this layer. From (5-54) and Fig. 8-9 it then follows that under such conditions, the flow in the aquifer is practically horizontal, although it is augmented or diminished by the leakage through the semipervious beds. With this approximation, the problem is reduced to one of axisymmetrical horizontal flow with superimposed leakage.

After a certain period of pumping at a constant rate Q_w, a steady state of flow is reached, with a steady potential distribution in the aquifer. The flow in the aquifer (Fig. 8-9) is sustained almost entirely by the leakage. We assume that the supply of water to the upper phreatic aquifer (say, by infiltration) is sufficient to maintain ϕ_0 constant. Since the flow in the aquifer is assumed horizontal, equipotentials are vertical.

The discharge $Q(r)$ into a cylinder of radius r and height B (centered at the well; Fig. 8-9), is given by

$$Q(r) = 2\pi r B K \partial \phi / \partial r = 2\pi r T \partial \phi / \partial r \qquad (8\text{-}35)$$

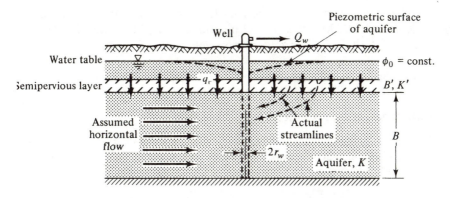

(a)

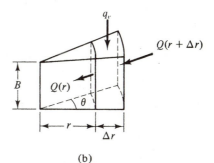

(b)

Figure 8-9 Schematic representation of approximate flow to a well in a confined leaky aquifer with ponded water on top of semipervious bed.

Continuity considerations for the portion of aquifer between two cylinders of radii r and $(r + \Delta r)$ lead to

$$Q(r + \Delta r) - Q(r) + (2\pi r \Delta r)\, q_v = 0 \qquad (8\text{-}36)$$

where q_v is shown in Fig. 8-9. In the limit, as $\Delta r \to 0$, this yields

$$\partial Q / \partial r + 2\pi r q_v = 0; \qquad q_v = K' \frac{\phi_0 - \phi}{B'} = \frac{\phi_0 - \phi}{\sigma'} \qquad (8\text{-}37)$$

$$\frac{1}{r} \frac{\partial}{\partial r}\left(r \frac{\partial \phi}{\partial r} \right) + \frac{\phi_0 - \phi}{\lambda^2} = 0; \qquad \lambda^2 = \sigma' T = \frac{B' B K}{K'} \qquad (8\text{-}38)$$

where λ is a characteristic length of the leaky aquifer called *leakage factor*. Equation (8-38) could be obtained from (5-70) with $\lambda^{(2)} = \infty$, $\phi_1 = \phi_0$, and $\partial \phi / \partial t = 0$.

Equation (8-38) is a Modified Bessel equation of order zero. Its general solution is

$$\phi_0 - \phi(r) = \alpha I_0(r/\lambda) + \beta K_0(r/\lambda) \qquad (8\text{-}39)$$

where α and β are constants to be derived from the boundary conditions, and $I_0(x)$ and $K_0(x)$ are the Modified Bessel functions of the first kind of order zero and of the second kind of order zero, respectively (Table 8-1).

Consider the following cases.

Drawdown in an Infinite Aquifer

From the boundary condition $\phi = \phi_0$ at $r = \infty$, it follows that $\alpha = 0$. Hence $\phi_0 - \phi(r) = \beta K_0(r/\lambda)$. As a second condition we use the well's constant discharge. This leads to

$$Q_w = 2\pi r_w BK \partial\phi/\partial r \bigg|_{r=r_w} = 2\pi r_w BK\beta K_1(r_w/\lambda)/\lambda;$$

$$\beta = Q_w/[2\pi T (r_w/\lambda) K_1(r_w/\lambda)]$$

and

$$s(r) = \phi_0 - \phi(r) = \frac{Q_w}{2\pi T} \frac{K_0(r/\lambda)}{(r_w/\lambda) K_1(r_w/\lambda)} \tag{8-40}$$

Table 8-1 Modified Bessel functions

x	$K_0(x)$	$K_1(x)$	$I_0(x)$	$I_1(x)$
0.010	4.7212	99.9739	1.0000	.0050
0.020	4.0285	49.9547	1.0001	.0100
0.030	3.6235	33.2715	1.0002	.0150
0.040	3.3365	24.9233	1.0004	.0200
0.050	3.1142	19.9097	1.0006	.0250
0.060	2.9329	16.5637	1.0009	.0300
0.070	2.7798	14.1710	1.0012	.0350
0.080	2.6475	12.3742	1.0016	.0400
0.090	2.5310	10.9749	1.0020	.0451
0.1	2.4271	9.8538	1.0025	.0501
0.2	1.7527	4.7760	1.0100	.1005
0.3	1.3725	3.0560	1.0226	.1517
0.4	1.1145	2.1843	1.0404	.2040
0.5	0.9244	1.6564	1.0635	.2579
0.6	0.7775	1.0283	1.0921	.3137
0.7	0.6605	1.0503	1.1263	.3719
0.8	0.5663	.8618	1.1665	.4327
0.9	0.4867	.7165	1.2130	.4971
1.0	0.4210	.6019	1.2661	.5652
1.5	0.2138	.2774	1.6467	.9817
2.0	0.1139	.1399	2.2796	1.5906
2.5	0.0624	.0739	3.2898	3.5167
3.0	0.0347	.0402	4.8808	3.9534
3.5	0.0196	.0222	7.3782	6.2058
4.0	0.0112	.0125	11.3019	9.7595
4.5	0.0064	.0071	17.4812	15.3892
5.0	0.0037	.0040	27.2399	24.3356

where K_1 is the Modified Bessel function of the second kind and first order (Table 8-1).

In the practice $r_w/\lambda \ll 1$. Since for $x \ll 1$, $xK_1(x) \approx 1$ with an error of less than one percent for $x < 0.02$, we may approximate (8-40) by

$$s(r) = \frac{Q_w}{2\pi T} K_0(r/\lambda) \tag{8-41}$$

Under the conditions leading to (8-41), $s(r)$ is independent of r_w.

In the vicinity of the pumping well, $r/\lambda \ll 1$. For $x \ll 1$, $K_0(x) \approx \ln(1.123/x)$. Equation (8-41) then becomes

$$s(r) = \frac{Q_w}{2\pi T} \ln \frac{1.123\lambda}{r} \tag{8-42}$$

with an error of less than five percent for $r/\lambda < 0.35$, and less than one percent for $r/\lambda < 0.18$.

Comparison of (8-29) and (8-42) shows that λ (or 1.123λ) expresses the *radius of influence of a leaky aquifer*. This can also be shown by deriving the ratio $Q(r)/Q_w$ which for every distance r indicates the portion of the well's discharge flowing through the aquifer; the remaining part $Q_w - Q(r)$ enters the aquifer through the semipervious cover. We obtain

$$Q(r)/Q_w = (r/\lambda) K_1(r/\lambda) \tag{8-43}$$

Figure 8-10 gives a schematic representation of (8-43). For example, for $r = 4\lambda$, $Q(r)/Q_w = 0.05$, which means that 95% of Q_w enters the cylinder of radius $r = 4\lambda$ through the semipervious layer.

In a similar manner, we may also treat cases where the potential on top of the semipervious layer, ϕ_0, or $\phi(r, B + B')$, varies, say as a result of pumping in the upper phreatic aquifer. We then have to introduce another equation which describes the variations of this potential.

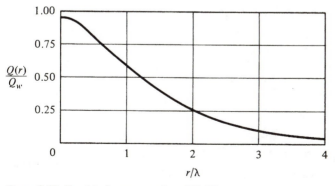

Figure 8-10 Graphical representation of (8-43).

Drawdown in a Finite Aquifer

The boundary conditions for a leaky aquifer of finite areal extent are $\phi = \phi_0$ at $r = R$ and $Q(r_w) = Q_w = $ const. For $r_w \ll \lambda$, we use the approximations for $x \ll 1$: $xK_1(x) \approx 1$, $xI_1(x) \approx 0$. These lead to the approximate solution

$$s(r) = \phi_0 - \phi(r) = \frac{Q_w}{2\pi T}\left[K_0(r/\lambda) - I_0(r/\lambda)\frac{K_0(R/\lambda)}{I_0(R/\lambda)}\right] \qquad (8\text{-}44)$$

where I_0 is the Modified Bessel function of the first kind and zero order (Table 8-1). For $R \ll \lambda$, except when r is nearly R, (8-44) may be approximated by (8-41). Again, $s(r)$ in (8-44) is independent of r_w as long as $r_w \ll \lambda$. For this case, the ratio of the portion of pumped water derived from the circumference of the cylinder of radius R to the total pumped discharge is approximated for $r_w/\lambda \ll 1$ by

$$Q(R)/Q_w = 1/I_0(R/\lambda) \qquad (8\text{-}45)$$

Figure 8-11 gives a graphical representation of (8-45). If $R \gg \lambda$ most of the pumped water enters through the semipervious layer.

For $R \ll \lambda$, (8-44) can also be approximated by

$$s(r) = \phi_0 - \phi(r) = \frac{Q_w}{2\pi T}\left[\left(1 + \frac{r^2}{4\lambda^2} + \cdots\right)\ln\frac{R}{r} - \frac{R^2 - r^2}{\lambda^2} + \cdots\right] \qquad (8\text{-}46)$$

For high values of λ, this reduces to (8-29).

For an impervious boundary condition at $r = R$, the solution is

$$s(r) = \frac{Q_w}{2\pi T}\left[K_0(r/\lambda) + I_0(r/\lambda)\frac{K_1(R/\lambda)}{I_1(R/\lambda)}\right] \qquad (8\text{-}47)$$

where I_1 is the Modified Bessel function of the first kind and of the first order.

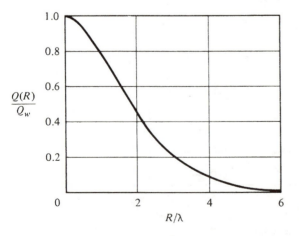

Figure 8-11 Graphical representation of (8-45).

Drawdown in a Multilayered Aquifer

Consider several aquifers separated by semipervious leaky beds as shown in Fig. 8-12. With the nomenclature of the figure, we write

$$Q_1 = 2\pi r K_1 B_1\, \partial\phi_1/\partial r; \qquad Q_2 = 2\pi r K_2 B_2\, \partial\phi_2/\partial r$$

$$\frac{\partial Q_1}{\partial r} = 2\pi r \left(\frac{\phi_1}{\sigma'_1} + \frac{\phi_1 - \phi_2}{\sigma'_2} \right); \qquad \frac{\partial Q_2}{\partial r} = 2\pi r\, \frac{\phi_2 - \phi_1}{\sigma'_2} \qquad (8\text{-}48)$$

where $\sigma'_1 = B'_1/K'_1$, $\sigma'_2 = B'_2/K'_2$. Inserting Q_1 and Q_2 into (8-48) yields

$$\frac{\partial^2 \phi_1}{\partial r^2} + \frac{1}{r}\frac{\partial \phi_1}{\partial r} = \frac{\phi_1}{\lambda_{11}^2} + \frac{\phi_1 - \phi_2}{\lambda_{12}^2}, \qquad \frac{\partial^2 \phi_2}{\partial r^2} + \frac{1}{r}\frac{\partial \phi_2}{\partial r} = \frac{\phi_2 - \phi_1}{\lambda_{22}^2} \qquad (8\text{-}49)$$

where $\lambda_{ij}^2 = K_i B_i \sigma'_j = T_i \sigma'_j$. These are two Bessel equations which have to be solved simultaneously, subject to the appropriate boundary conditions

$$\lim_{r \to \infty} \phi_i(r) = 0, \quad \text{for } i = 1, 2; \qquad Q_w = 2\pi T_1 \left(r\frac{\partial\phi_1}{\partial r} \right)\bigg|_{r=r_w \approx 0},$$

$$Q_w = 0 = 2\pi T_2 \left(r\frac{\partial\phi_2}{\partial r} \right)\bigg|_{r=r_w \approx 0} \qquad (x)$$

The solution is

$$\phi_1(r) = \frac{Q_w}{2\pi T_1} \frac{1}{\chi_1^2 - \chi_2^2} \{ (\chi_1^2 - \alpha_{22}^2)\, \mathrm{K}_0(\chi_1 r) + (\alpha_{22}^2 - \chi_2^2)\, \mathrm{K}_0(\chi_2 r) \},$$

$$\phi_2(r) = \frac{Q_w}{2\pi T_1} \frac{\alpha_{22}^2}{\chi_1^2 - \chi_2^2} \{ -\mathrm{K}_0(\chi_1 r) + \mathrm{K}_0(\chi_2 r) \} \qquad (8\text{-}50)$$

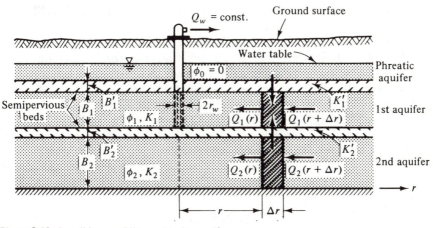

Figure 8-12 A well in a multilayered leaky aquifer.

where

$$\left.\begin{array}{r}\chi_1^2 \\ \chi_2^2\end{array}\right\} = \tfrac{1}{2}\{(\alpha_{11}^2 + \alpha_{22}^2 + \alpha_{12}^2) \pm [(\alpha_{11}^2 + \alpha_{22}^2 + \alpha_{12}^2)^2 - 4\alpha_{11}^2\alpha_{22}^2]^{1/2}\}$$

$$\alpha_{ij}^2 = 1/\lambda_{ij}^2$$

For the case of pumping from the lower layer only, the same procedure leads to

$$\phi_1(r) = \frac{Q_w}{2\pi T_2} \frac{\alpha_{12}^2}{\chi_1^2 - \chi_2^2} [-K_0(\chi_1 r) + K_0(\chi_2 r)]$$

$$\phi_2(r) = \frac{Q_w}{2\pi T_2} \frac{1}{\chi_1^2 - \chi_2^2} [(\alpha_{22}^2 - \chi_2^2) K_0(\chi_1 r) + (\chi_1^2 - \alpha_{22}^2) K_0(\chi_2 r)]$$

(8-51)

Huisman (1972) discusses several cases of a well in a two layered aquifer.

Estimate of Error

The main idea in the procedure presented in this section is that the flow in an aquifer is practically horizontal, whereas in a semipervious layer it is vertical. Obviously, with this assumption, the effect of a lower vertical permeability in the case of an anisotropic aquifer ($K_x > K_z$) cannot be introduced. An estimate of the error introduced by this assumption is (Bear, 1972, p. 218)

$$0 < \frac{\tilde{\phi} - \phi(r, B)}{\phi(r, B) - \phi_0} < \frac{B}{2\sigma' K_z} = \frac{1}{2}\left(\frac{B}{B'}\right)\left(\frac{K_z'}{K_z}\right) = \frac{1}{2}\left(\frac{K_r}{K_z}\right)\left(\frac{B^2}{\lambda}\right) \ll 1 \quad (8\text{-}52)$$

In general, $K_r > K_z$ and $B \ll \lambda$; $\tilde{\phi}$ is the average (along the vertical) value of ϕ which satisfies (8-38).

8-5 UNSTEADY FLOW TO A WELL IN A CONFINED AQUIFER

Figure 8-13 shows a fully penetrating well pumping at a constant rate, Q_w, from a confined aquifer. The aquifer is homogeneous, isotropic, and of constant thickness B. It is assumed that the aquifer's storativity, S (Sec. 5-1), resulting from the elastic properties of both the water and the aquifer matrix, is constant (in place and time), and that water is immediately released from storage in the aquifer upon a decline of head. Theis (1935) and Jacob (1940) pioneered the work on this topic.

Consider an aquifer domain between two concentric cylinders of radii r and $r + \Delta r$ centered at a well. The excess of water volume leaving this domain, over the volume of water entering it, during a time interval Δt is drawn from storage within this domain, causing a decline of head $\Delta\phi = \phi(t) - \phi(t + \Delta t)$. This balance can be written in the form

$$\Delta t[Q(r) - Q(r + \Delta r)] = S \times 2\pi r \, \Delta r [\phi(t) - \phi(t + \Delta t)]; Q(r) = 2\pi r T \, \partial\phi/\partial r$$

(8-53)

Piezometric surfaces (drawdown curve)

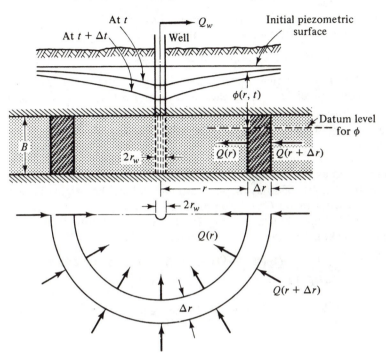

Figure 8-13 Unsteady flow to a well in a confined aquifer.

Dividing by $\Delta r\Delta t$ and passing to the limit as $\Delta r \to 0$ and $\Delta t \to 0$, leads to the partial differential equation

$$\partial^2\phi/\partial r^2 + (1/r)\,\partial\phi/\partial r = (S/T)\,\partial\phi/\partial t \tag{8-54}$$

which describes unsteady (converging or diverging) flow in a confined aquifer. It is the same as (5-60) written in radial coordinates, with $N = 0$.

Consider now the following cases.

Drawdown in an Infinite Aquifer; $Q_w = $ const.

To solve (8-54) for an aquifer of infinite areal extent, we assume that the well is *infinitesimally narrow* (which is practically true for distances $r \gg r_w$). The boundary and initial conditions are

$\phi(r,0) = \phi_0, \quad r_w \le r \le \infty,$ (initially uniform head)

$\phi(\infty, t) = \phi_0, \quad t \ge 0,$ (no influence at infinity) (8-55)

$\lim\limits_{r=r_w \to 0} 2\pi r T \partial\phi/\partial r = Q_w = $ const., $\quad t > 0,$ (Q_w is the well's discharge)

     Equation (8-54) may also be written in terms of the drawdown $s(r,t) = \phi_0 - \phi(r,t)$

$$\partial^2 s/\partial r^2 + (1/r)\,\partial s/\partial r = (S/T)\,\partial s/\partial t \tag{8-56}$$

with boundary conditions: $s(r,0) = 0$, $s(\infty,t) = 0$, $\lim_{r=r_w \to 0}(2\pi r T \partial s/\partial r) = -Q_w$.
To derive the solution of (8-56), we note that the solution

$$s = (A/t)\exp(-u), \qquad A = \text{const.}, \qquad u = Sr^2/4Tt \tag{8-57}$$

satisfies (8-56), and at $t = 0$ vanishes everywhere except at the origin where it becomes infinite. Also, for any $t > 0$, by the definition of S (Sec. 5-1), the total volume of water withdrawn from the aquifer is $U_0 = \int_0^\infty s \times S \times 2\pi r\,dr = 4\pi TA$. Hence, $A = U_0/4\pi T$. Equation (8-57) may be rewritten as

$$s = (U_0/4\pi Tt)\exp(-u) \tag{8-58}$$

where U_0 is interpreted as the volume of water instantaneously withdrawn from the aquifer at $t = 0$ (strength of sink) through a *point sink* at the origin. This is related to the rate of withdrawal Q_w (=pumpage) by: $U_0 = Q_w\,dt$. If the rate of withdrawal varies with time, that is $Q_w = Q_w(t)$, the corresponding elementary solution becomes (Muskat, 1937; Carslaw and Jaeger, 1946)

$$s = \frac{1}{4\pi T}\int_0^t \frac{Q_w(\tau)}{t-\tau}\exp\left\{-\frac{r^2 S}{4T(t-\tau)}\right\}d\tau \tag{8-59}$$

Abu-Zied and Scott (1963) solve the case of $Q(t) = Q_0(a + be^{-ct})$, where a, b and c are constants. In the present case we have $Q(t) = Q_w = \text{const.}$, which reduces (8-59) to

$$s = \frac{Q_w}{4\pi T}\int_{x=u}^{\infty}\frac{e^{-x}}{x}dx = -\frac{Q_w}{4\pi T}Ei(-u) \tag{8-60}$$

Table 8-2 Well function $W(u)$ for

N \ u	$N \times 10^{-15}$	$N \times 10^{-14}$	$N \times 10^{-13}$	$N \times 10^{-12}$	$N \times 10^{-11}$	$N \times 10^{-10}$	$N \times 10^{-9}$	$N \times 10^{-8}$
1.0	33.9616	31.6590	29.3564	27.0538	24.7512	22.4486	20.1460	17.8435
1.5	33.5561	31.2535	28.9509	26.6483	24.3458	22.0432	19.7406	17.4380
2.0	33.2684	30.9658	28.6632	26.3607	24.0581	21.7555	19.4529	17.1503
2.5	33.0453	30.7427	28.4401	26.1375	23.8349	21.5323	19.2298	16.9272
3.0	32.8629	30.5604	28.2578	25.9552	23.6526	21.3500	19.0474	16.7449
3.5	32.7088	30.4062	28.1036	25.8010	23.4985	21.1959	18.8933	16.5907
4.0	32.5753	30.2727	27.9701	25.6675	23.3649	21.0623	18.7598	16.4572
4.5	32.4575	30.1549	27.8523	25.5497	23.2471	20.9446	18.6420	16.3394
5.0	32.3521	30.0495	27.7470	25.4444	23.1418	20.8392	18.5366	16.2340
5.5	32.2568	29.9542	27.6516	25.3491	23.0465	20.7439	18.4413	16.1387
6.0	32.1698	29.8672	27.5646	25.2620	22.9595	20.6569	18.3543	16.0517
6.5	32.0898	29.7872	27.4846	25.1820	22.8794	20.5768	18.2742	15.9717
7.0	32.0156	29.7131	27.4105	25.1079	22.8053	20.5027	18.2001	15.8976
7.5	31.9467	29.6441	27.3415	25.0389	22.7363	20.4337	18.1311	15.8280
8.0	31.8821	29.5795	27.2769	24.9744	22.6718	20.3692	18.0666	15.7640
8.5	31.8215	29.5189	27.2163	24.9137	22.6112	20.3086	18.0060	15.7034
9.0	31.7643	29.4618	27.1592	24.8566	22.5540	20.2514	17.9488	15.6462
9.5	31.7103	29.4077	27.1051	24.802	22.4999	20.1973	17.8948	15.5922

This is also the solution given by Theis (1935) in the form

$$s(r, t) = \phi_0 - \phi(r, t) = (Q_w/4\pi T)\, W(u);$$
$$W(u) = -\, Ei(-u) = \int_{x=u}^{\infty} (e^{-x}/x)\, dx \qquad \left.\right\} \qquad (8\text{-}61)$$

where $W(u)$ is the *well function* of $u = Sr^2/4Tt$ for a *confined aquifer* (Jacob, 1940). The integral $-Ei(-u)$ is the *exponential integral* (Jahnke and Emde, 1945). Table 8-2 and Fig. 8-14a give values of $W(u)$. Theis (1935) obtained (8-61) by analogy to heat flow.

Equation (5-190) gives the drawdown $s = s(r, t)$ for an anisotropic aquifer.

Figure 8-14b shows the drawdown $s(r, t)$. An *inflection point* occurs at $u = 1$ (that is, $t = Sr^2/4T$). Thereafter, the *rate of drawdown* $\partial s/\partial t$ ($= -\partial\phi/\partial t$) decreases, but theoretically *never vanishes*.

The well function (or the exponential integral) is obtained from the series

$$W(u) = -\,0.5772 - \ln u + u - u^2/2 \times 2! + u^3/3 \times 3! - u^4/4 \times 4! + \cdots$$

For small values of u, say, $u < 0.01$ (i.e., for a large time at a given distance), this series may be approximated by its first two terms (Cooper and Jacob, 1946; Jacob, 1950)

$$s(r, t) \cong \frac{Q_w}{4\pi T}\left(-0.5772 - \ln\frac{r^2 S}{4Tt}\right) = \frac{Q_w}{4\pi T}\ln\frac{2.25\, Tt}{r^2 S} \qquad (8\text{-}62)$$

With this approximation, plotting $s = s(\ln t)$, $s = s(\ln r)$ and $s = s[\ln(r^2/t)]$ gives straight lines (see Sec. 11-1).

a confined aquifer (after Wenzel, 1942)

$N \times 10^{-7}$	$N \times 10^{-6}$	$N \times 10^{-5}$	$N \times 10^{-4}$	$N \times 10^{-3}$	$N \times 10^{-2}$	$N \times 10^{-1}$	N
15.5409	13.2383	10.9357	8.6332	6.3315	4.0379	1.8229	0.2194
15.1354	12.8328	10.5303	8.2278	5.9266	3.6374	1.4645	0.1000
14.8477	12.5451	10.2426	7.9402	5.6394	3.3547	1.2227	0.04890
14.6246	12.3220	10.0194	7.7172	5.4167	3.1365	1.0443	0.02491
14.4423	12.1397	9.8371	7.5348	5.2349	2.9591	0.9057	0.01305
14.2881	11.9855	9.6830	7.3807	5.0813	2.8099	0.7942	0.006970
14.1546	11.8520	9.5495	7.2472	4.9482	2.6813	0.7024	0.003779
14.0368	11.7342	9.4317	7.1295	4.8310	2.5684	0.6253	0.002073
13.9314	11.6280	9.3263	7.0242	4.7261	2.4679	0.5598	0.001148
13.8361	11.5330	9.2310	6.9289	4.6313	2.3775	0.5034	0.0006409
13.7491	11.4465	9.1440	6.8420	4.5448	2.2953	0.4544	0.0003601
13.6691	11.3665	9.0640	6.7620	4.4652	2.2201	0.4115	0.0002034
13.5950	11.2924	8.9899	6.6879	4.3916	2.1508	0.3738	0.0001155
13.5260	11.2234	8.9209	6.6190	4.3231	2.0867	0.3403	0.0000658
13.4614	11.1589	8.8563	6.5545	4.2591	2.0269	0.3106	0.0000376
13.4008	11.0982	8.7957	6.4939	4.1990	1.9711	0.2840	0.0000216
13.3437	11.0411	8.7386	6.4368	4.1423	1.9187	0.2602	0.0000124
13.2896	10.9870	8.6845	6.3828	4.0887	1.8695	0.2387	0.0000071

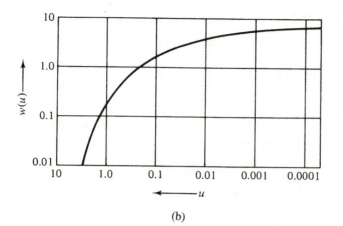

(b)

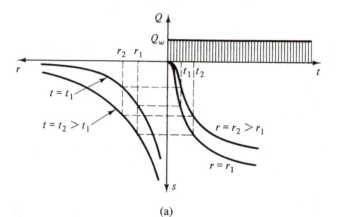

(a)

Figure 8-14 (a) Drawdown $s(r, t)$ in an infinite confined aquifer. (b) Type curve for a confined aquifer.

From (8-61), or (8-62), it follows that equipotentials ($s = $ const.) are circles, centered at the well, described by

$$x^2 + y^2 = \frac{2.25Tt}{S} \exp\left(-\frac{4\pi Ts}{Q_w}\right) \qquad (8\text{-}63)$$

Equation (8-62) may be rewritten in the form

$$s = \frac{Q_w}{2\pi T} \ln \frac{1.5(tT/S)^{1/2}}{r} \qquad (8\text{-}64)$$

By comparing (8-64) with (8-6) we may define a *radius of influence R* (that is, a distance where $s = 0$) by

$$R = 1.5(tT/S)^{1/2} = (2.25T/S)^{1/2} t^{1/2} \qquad (8\text{-}65)$$

which in this case is *time dependent*. This result is somewhat different from (8-9) or from (8-10). Coefficients in the range 1.5–4.3 (instead of 1.5) are mentioned in the literature (Van Poolen, 1964).

For very large times, it follows from (8-62) that between two points at distances r_1 and r_2, we have

$$s_1 - s_2 = \frac{Q_w}{2\pi T} \ln \frac{r_2}{r_1} \tag{8-66}$$

which is the steady state equation (8-5). For $r_1 = r_w$ and $r_2 = r$, we obtain (8-3).

The rate of propagation of the cone of depression can be derived from (8-61) by observing that for $s = $ const., we have $u = $ const., or $r^2/t = $ const. Also, when (8-64) is valid, the rate of drawdown which decreases with time becomes $\partial s/\partial t = Q_w/4\pi Tt$.

Drawdown in an Infinite Aquifer; $Q_w = Q_w(t)$

Equation (8-54) and the boundary conditions (8-55) are *linear*, so that the *principle of superposition* (Sec. 5-7) is applicable. Accordingly, (8-59) gives the drawdown for $Q_w = Q_w(t)$ by superposition of (8-58). For the case of Q_w varying in steps (Fig. 8-15c), the resulting drawdown may be obtained by summation of increments Δs (positive or negative) resulting from increments (positive or negative) of ΔQ_w. Each reduction in pumping rate, whether partial or complete, produced a *recovery* of heads around the well. If a well pumps at a rate Q_w^1 from $t = 0$ up to $t = t_1$ and then the rate is changed to Q_w^2, the drawdown at every instant is given by

$$s(r, t) = \frac{Q_w^1}{4\pi T} W \left(\frac{Sr^2}{4Tt} \right); \qquad t \leq t_1,$$

$$s(r, t) = \frac{Q_w^1}{4\pi T} W \left(\frac{Sr^2}{4Tt} \right) + \frac{Q_w^2 - Q_w^1}{4\pi T} W \left(\frac{Sr^2}{4T(t - t_1)} \right); \qquad t > t_1 \tag{8-67}$$

When a well pumps Q_w^1 during t_1 and is then shut off, we obtain the residual drawdown, $s(r, t)$ for $t > t_1$, by setting $Q_w^2 = 0$ in (8-67)

$$s(r, t) = \frac{Q_w^1}{4\pi T} \left[W \left(\frac{Sr^2}{4Tt} \right) - W \left(\frac{Sr^2}{4T(t - t_1)} \right) \right]; \qquad t > t_1 \tag{8-67a}$$

Figure 8-15a shows $s(r, t)$ for the case of $Q_w^{(2)} = 2Q_w^{(1)}$. Figure 8-15b shows $s(r, t)$ for the case of shutoff at t_1, that is, $Q_w^{(2)} = 0$. Since (8-61) is a solution for a well pumping at a constant rate indefinitely, the case of $Q_w^{(2)} = 2Q_w^{(1)}$ is obtained by a superposition of two imaginary wells at the same location: the first starting at $t = 0$ and pumping $Q_w^{(1)}$ and the second starting at t_1 and pumping $Q_w^{(1)}$. The case of well shutoff is obtained in a similar way, except that the second (imaginary) well is a recharge well, i.e., injecting $Q_w^{(1)}$, starting at time t_1. Note that in each case, the incremental drawdown, whether positive or negative, is with respect to that produced by the first well's continued operation.

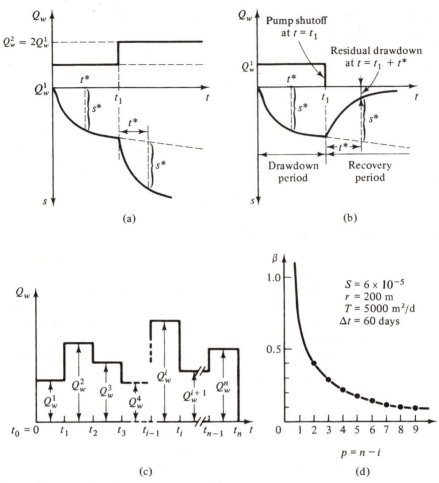

Figure 8-15 Drawdown for a varying pumping rate.

The assumption made here that the aquifer is of infinite areal extent, introduces no significant error in practical application, although such an aquifer does not exist. This statement is valid as long as the cone of depression has not reached any aquifer boundary. When an aquifer boundary is reached, one should use equations developed for wells near boundaries (Sec. 8-10).

When a well pumps Q_w^1 from $t = t_0 = 0$ to $t = t_1$, then Q_w^2 from $t = t_1$ to $t = t_2$ and in general Q_w^i from t_{i-1} to $t_i (i = 1, 2, \ldots, n)$, Fig. 8-15(c), the drawdown at any time $t \, (\equiv t_n)$ is given by

$$s(r, t) = \frac{1}{4\pi T} \sum_{i=1}^{n} (Q_w^i - Q_w^{i-1}) \, W \left\{ \frac{Sr^2}{4T(t - t_{i-1})} \right\}; \qquad t_{n-1} < t \le t_n \qquad (8\text{-}68)$$

For $t = t_n$, since $W(\infty) = 0$, (8-68) can be written as

$$s(r, t_n) = \frac{1}{4\pi T} \sum_{i=1}^{n} Q_w^i \left[W \left\{ \frac{Sr^2}{4T(t - t_{i-1})} \right\} - W \left\{ \frac{Sr^2}{4T(t - t_i)} \right\} \right] \qquad (8\text{-}68a)$$

If $t_i - t_{i-1} = \Delta t = $ const., independent of i, then (8-68a) reduces, for $t = t_n$, to

$$s(r, t_n) = \frac{1}{4\pi T} \sum_{i=1}^{n} Q_w^i \left[W \left\{ \frac{Sr^2}{4T(n - i + 1)\, \Delta t} \right\} - W \left\{ \frac{Sr^2}{4T(n - i)\, \Delta t} \right\} \right]$$

$$= \sum_{i=1}^{n} Q_w^i \beta(n - i) \qquad (8\text{-}68\text{b})$$

where $n\,\Delta t = t_n$, $\beta(p) = [W\{Sr^2/4T(p + 1)\,\Delta t\} - W\{Sr^2/4Tp\,\Delta t\}]/4\pi T$, and $p = 0$, $u = Sr^2/4Tp\,\Delta t = \infty$, $W(\infty) = 0$. Figure 8-15d gives an example of $\beta(p)$. The function $\beta(p)$ shows how the contribution of the ith period decreases rapidly with the time elapsed from t_i to t_n.

When $u < 0.01$ such that (8-62) is valid, the residual drawdown after a complete shutoff (recovery curve, Fig. 8-15b) is

$$s(r, t) = \frac{Q_w^1}{4\pi T} \left[\ln \frac{2.25Tt}{Sr^2} - \ln \frac{2.25T(t - t_1)}{Sr^2} \right] = \frac{Q_w^1}{4\pi T} \ln \frac{t}{t - t_1} \qquad (8\text{-}69)$$

with $s(r, t) \to 0$ as $t \to \infty$. From (8-69) it follows that the effect of pumpage may persist a very long time. However, in most practical cases, this time is relatively short. Note that Fig. 8-15b is drawn from (8-68).

If initially the drawdown is not zero everywhere, but a function of place, e.g., $s(x, y, 0) = g(x, y)$, the solution (8-59) becomes, using the principle of superposition (Sec. 5-7; Muskat, 1937)

$$s = \frac{1}{4\pi T} \left[\frac{S}{t} \int_{-\infty}^{+\infty} d\xi \int_{-\infty}^{+\infty} g(\xi, \eta) \exp[-\{(x - \xi)^2 + (y - \eta)^2\}\, S/4Tt]\, d\eta \right.$$

$$\left. + \int_{0}^{t} \frac{Q_w(\tau)}{t - \tau} \exp\left\{ -\frac{r^2 S}{4T(t - \tau)} \right\} d\tau \right] \qquad (8\text{-}70)$$

Drawdown in a Finite Aquifer; $Q_w = Q_w(t)$

Several cases of a single well located at the center of a circular aquifer of external radius r_e are treated here, following Muskat (1937) and Carslaw and Jaeger (1946). Equation (8-54) still describes the potential distribution $\phi(r, t)$ in the aquifer, assuming radially symmetric convergent or divergent flow.

The initial and boundary condition may be stated in general terms as

$$\left.\begin{array}{lll} t = 0, & r_w \le r \le r_e, & \phi = \phi(r, 0) = g(r) \\[4pt] t \ge 0, & r = r_w, & \phi = f_1(t) \\[4pt] & r = r_e, & \phi = f_2(t) \end{array}\right\} \qquad (8\text{-}71)$$

To derive a solution, we employ the principle of superposition (Sec. 5-7). Let us assume that we can derive solutions ϕ_0, ϕ_1, ϕ_2 to the following three subproblems

A solution $\phi_0 = \phi_0(r, t)$, for $g(r) \neq 0$, $f_1(t) \equiv 0$, $f_2(t) \equiv 0$

A solution $\phi_1 = \phi_1(r, t)$, for $g(r) \equiv 0$, $f_1(t) = 1$, $f_2(t) \equiv 0$ (8-72)

A solution $\phi_2 = \phi_2(r, t)$, for $g(r) \equiv 0$, $f_1(t) \equiv 0$, $f_2(t) = 1$

Once these solutions are derived, the solution to the original problem, where $\phi(r, t)$ satisfies (8-71), is given by

$$\phi(r, t) = \phi_0(r, t) + \int_0^t \left[f_1(\tau) \frac{\partial \phi_1(r, t - \tau)}{\partial t} + f_2(\tau) \frac{\partial \phi_2(r, t - \tau)}{\partial t} \right] d\tau \quad (8\text{-}73)$$

Muskat (1937) gives solutions for ϕ_0, ϕ_1 and ϕ_2 in terms of infinite series of Bessel functions for various cases of $r_w \neq 0$ and for $r_w \to 0$.

A Flowing Well in an Infinite Aquifer

Jacob and Lohman (1952) solve the case of a flowing (artesian) well discharging naturally from a homogeneous, infinite, confined aquifer. Here the drawdown, s_w, at the well is maintained constant whereas its discharge, Q_w, varies with time. The flow is governed by (8-54) with boundary and initial conditions

$$\phi(r, 0) = \phi_0, \qquad\qquad r_w \leq r \leq \infty$$

$$\phi(\infty, t) = \phi_0, \qquad\qquad t > 0$$

$$\phi(r_w, t) = \phi_0 - s_w = \text{const.} \qquad t > 0$$

The solution is

$$s_w = \frac{Q_w}{2\pi T} \frac{1}{G(1/4u_w)}; \qquad Q_w(t) = 2\pi T s_w G(1/4u_w) \qquad (8\text{-}74)$$

where

$$u_w = S r_w^2 / 4Tt; \quad G(1/4u_w) = \frac{1}{\pi u_w} \int_0^\infty x e^{-x^2/4u_w} \left\{ \frac{\pi}{2} + \tan^{-1}[Y_0(x)/J_0(x)] \right\} dx$$

J_0, Y_0 are Bessel functions of order zero of the first and second kinds, respectively. Table 8-3 (Jacob and Lohman, 1952) gives values of $G(1/4u_w)$. For large values of t, G approaches $2/W(u_w)$ so that we have

$$s_w = \frac{Q_w}{4\pi T} W\left(\frac{S r_w^2}{4Tt} \right) \qquad (8\text{-}75)$$

This means that we obtain the same *specific well capacity* (Sec. 8-12) Q_w/s_w as in the case of a well with constant discharge Q_w. For small values of u_w, we may also approximate G by $2/\ln(2.25\, Tt/Sr_w^2)$.

Effect of Storage in a Well of Finite Diameter

In the discussion so far, except for one case, we have assumed that the well's radius was infinitesimally small. When a well of finite radius was considered, however, the storage capacity in the well itself was not taken into account. It was assumed that the well's discharge is $Q_w = 2\pi r T \partial\phi/\partial r$, evaluated at $r = r_w$. If we wish to take storage in a well into account, as when a well has a relatively large diameter (e.g., a dug well), the boundary condition at the well, $r = r_w$, has to be changed. With the nomenclature of Fig. 8-16 (Papadopulos and Cooper, 1967), we have the condition

$$t > 0, \qquad 2\pi r_w T \left.\frac{\partial s(r, t)}{\partial r}\right|_{r=r_w} - \pi r_c^2 \frac{\partial s_w(t)}{\partial t} = -Q_w \qquad (8\text{-}76)$$

where r_c is the radius of the well casing in the interval over which s_w occurs, and r_w is the effective radius of the well screen or open hole. If losses in the well itself are neglected, we have

$$s(r_w, t) = s_w(t) \qquad (8\text{-}77)$$

Otherwise, $\phi(r_w, t) - \phi(r, t) = s_w(t) - s(r_w, t) = $ losses in pipe.

Papadopulos and Cooper (1967) solve the problem of flow to a well of large diameter. The partial differential equation is (8-56) with boundary conditions (8-77) and $s(\infty, t) = 0$, $s(r, 0) = 0$ for $r \geq r_w$, and $s_w(0) = 0$. They obtain

$$s = \frac{2Q_w\alpha}{\pi^2 T} \int_0^\infty (1 - e^{-\beta^2/4u_w}) \{J_0(\beta r/r_w)[\beta Y_0(\beta) - 2\alpha Y_1(\beta)]$$

$$- Y_0(\beta r/r_w)[\beta J_0(\beta) - 2\alpha J_1(\beta)]\} [1/\beta^2 \Delta(\beta)] \, d\beta \quad (8\text{-}78)$$

where $\alpha = r_w^2 S/r_c^2$, $u_w = r_w^2 S/4Tt$, $\Delta(\beta) = [\beta J_0(\beta) - 2\alpha J_1(\beta)]^2 + [\beta Y_0(\beta) - 2\alpha Y_1(\beta)]^2$, J_i and Y_i, are Bessel functions of ith order of the first and second kinds, respectively. The drawdown s_w inside the well is

$$s_w = \frac{Q_w}{4\pi T} F(u_w, \alpha) \qquad (8\text{-}79)$$

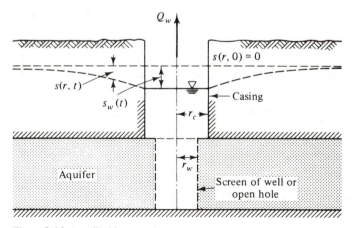

Figure 8-16 A well of large diameter *(after Papadopulos and Cooper, 1967)*.

Table 8-3 Values of the well function $G(\alpha)$

α	$\times 10^{-4}$	10^{-3}	10^{-2}	10^{-1}	1	10	10^2	10^3
1	56.9	18.34	6.13	2.249	0.985	0.534	0.346	0.251
2	40.4	13.11	4.47	1.716	0.803	0.461	0.311	0.232
3	33.1	10.79	3.74	1.477	0.719	0.427	0.294	0.222
4	28.7	9.41	3.30	1.333	0.667	0.405	0.283	0.215
5	25.7	8.47	3.00	1.234	0.630	0.389	0.274	0.210
6	23.5	7.77	2.78	1.160	0.602	0.377	0.268	0.206
7	21.8	7.23	2.60	1.103	0.580	0.367	0.263	0.203
8	20.4	6.79	2.46	1.057	0.562	0.359	0.258	0.200
9	19.3	6.43	2.35	1.018	0.547	0.352	0.254	0.198
10	18.3	6.13	2.25	0.985	0.534	0.346	0.251	0.196

where $F(u_w, \alpha) = (32\alpha^2/\pi^2) \int_0^\infty [(1 - e^{-\beta^2/4u_w})/\beta^3 \, \Delta(\beta)] \, d\beta$. Papadopulos and Cooper (1967) give Table 8-4 for $F(u_w, \alpha)$. They show that for $(u_w/\alpha) < 10^{-3}$, that is $t > 250 \, r_c^2/T$, $F(u_w, \alpha)$ may be closely approximated by $W(u_w)$. For large values of u_w we obtain

$$s_w = \frac{Q_w}{4\pi T} \frac{\alpha}{u_w} = \frac{Q_w t}{\pi r_c^2} \qquad (8\text{-}80)$$

i.e., all the pumped water is derived from storage within the well.

As another example of the influence of the storage in a well, consider (8-58) which gives the drawdown, s, in the vicinity of a well (of zero radius) in a confined aquifer as a result of an instantaneous withdrawal of a volume U_0 from it. The storage capacity in the well's casing is neglected. At the well itself, we have, by inserting $r = r_w \to 0$ in (8-58)

$$s_w = U_0/4\pi Tt \qquad (8\text{-}81)$$

Cooper et al. (1967) start by considering the build-up of piezometric heads caused by the injection of an instantaneous volume U_0 rather than a withdrawal of this volume from a well. The volume U_0 may be expressed by $U_0 = \pi r_c^2 s_0$, where (Fig. 8-16) $s_0 = s_w$ at $t = 0$ is the initial build-up in the well's casing, and r_c is the radius of the casing in the interval over which the build-up varies between $s_w(t) = s_0$ and $s_w(t) = 0$. With these definitions, (8-58) and (8-81) become

$$s/s_0 = (r_c^2/4Tt) \exp(-r^2 S/4Tt) \quad \text{and} \quad s_w(t)/s_0 = r_c^2/4Tt \qquad (8\text{-}82)$$

They then solve (8-56) subject to appropriate initial and boundary conditions for an instantaneous withdrawal of a volume U_0, followed by a recovery $s_w(t)$ at the well and $s(r, t)$ around it

$$s(r, t) = \frac{2s_0}{\pi} \int_0^\infty e^{-\beta^2/4u_w} \{J_0(\beta r/r_w) [\beta Y_0(\beta) - 2\alpha Y_1(\beta)] - Y_0(\beta r/r_w)$$

$$\times [\beta J_0(\beta) - 2\alpha J_1(\beta)]\} [1/\Delta(\beta)] \, d\beta \qquad (8\text{-}83)$$

for a flowing well in a confined aquifer

10^4	10^5	10^6	10^8	10^8	10^9	10^{10}	10^{11}	α
0.1964	0.1608	0.1360	0.1177	0.1037	0.0927	0.0838	0.0764	1
0.1841	0.1524	0.1299	0.1131	0.1002	0.0899	0.0814	0.0744	2
0.1777	0.1479	0.1266	0.1106	0.0982	0.0882	0.0801	0.0733	3
0.1733	0.1449	0.1244	0.1089	0.0968	0.0872	0.0792	0.0726	4
0.1701	0.1426	0.1227	0.1076	0.0958	0.0864	0.0785	0.0720	5
0.1675	0.1408	0.1213	0.1066	0.0950	0.0857	0.0779	0.0716	6
0.1654	0.1393	0.1202	0.1057	0.0943	0.0851	0.0774	0.0712	7
0.1636	0.1380	0.1192	0.1049	0.0937	0.0846	0.0770	0.0709	8
0.1621	0.1369	0.1184	0.1043	0.0932	0.0842	0.0767	0.0706	9
0.1608	0.1360	0.1177	0.1037	0.0927	0.0838	0.0764	0.0704	10

After Jacob and Lohman, 1952.

Table 8-4 Well function $F(u_w, \alpha)$ for a well of large diameter

u_w	$\alpha = 10^{-1}$	$\alpha = 10^{-2}$	$\alpha = 10^{-3}$	$\alpha = 10^{-4}$	$\alpha = 10^{-5}$
10	9.755×10^{-3}	9.976×10^{-4}	9.998×10^{-5}	1.000×10^{-5}	1.000×10^{-6}
1	9.192×10^{-2}	9.914×10^{-3}	9.991×10^{-4}	1.000×10^{-4}	1.000×10^{-5}
5×10^{-1}	1.767×10^{-1}	1.974×10^{-2}	1.997×10^{-3}	2.000	2.000
2	4.062	4.890	4.989	4.999	5.000
1	7.336	9.665	9.966	9.997	1.000×10^{-4}
5×10^{-2}	1.260×10^{0}	1.896×10^{-1}	1.989×10^{-2}	1.999×10^{-3}	2.000
2	2.303	4.529	4.949	4.995	5.000
1	3.276	8.520	9.834	9.984	1.000×10^{-3}
5×10^{-3}	4.255	1.540×10^{0}	1.945×10^{-1}	1.994×10^{-2}	2.000
2	5.420	3.043	4.725	4.972	4.998
1	6.212	4.545	9.069	9.901	9.992
5×10^{-4}	6.960	6.031	1.688×10^{0}	1.965×10^{-1}	1.997×10^{-2}
2	7.866	7.557	3.523	4.814	4.982
1	8.572	8.443	5.526	9.340	9.932
5×10^{-5}	9.318	9.229	7.631	1.768×10^{0}	1.975×10^{-1}
2	1.024×10^{1}	1.020×10^{1}	9.676	3.828	4.861
1	1.093	1.087	1.068×10^{1}	6.245	9.493
5×10^{-6}	1.163	1.162	1.150	8.991	1.817×10^{0}
2	1.255	1.254	1.249	1.174×10^{1}	4.033
1	1.324	1.324	1.321	1.291	6.779
5×10^{-7}	1.393	1.393	1.392	1.378	1.013×10^{1}
2	1.485	1.485	1.484	1.479	1.371
1	1.554	1.554	1.554	1.551	1.513
5×10^{-8}	1.623	1.623	1.623	1.622	1.605
2	1.705	1.705	1.705	1.714	1.708
1	1.784	1.784	1.784	1.784	1.781
5×10^{-9}	1.854	1.854	1.854	1.854	1.851
2	1.945	1.945	1.945	1.945	1.940
1	2.015	2.015	2.015	2.015	2.015

After Papadopulos and Cooper, 1967

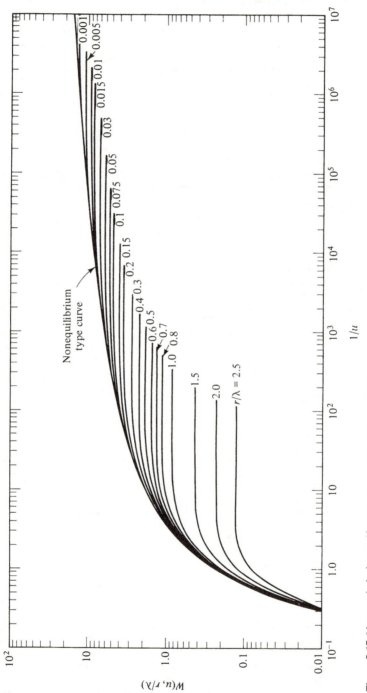

Figure 8-17 Nonsteady leaky aquifer type curve *(after Walton, 1960)*.

$$s_w(t) = s(r_w, t) = \frac{8s_0\alpha}{\pi^2} \int_0^\infty e^{-\beta^2/4u_w} d\beta / [\beta \, \Delta(\beta)] \qquad (8\text{-}84)$$

Cooper et al. (1967) show that for $Tt/r_c^2 > 100$ (or $s/s_0 < 0.0025$), the relationships (8-82) are a close approximation of (8-84).

8-6 UNSTEADY FLOW TO A WELL IN A PHREATIC AQUIFER

In principle, the (radially converging) flow to a fully penetrating well is described by

$$\partial^2\phi/\partial r^2 + (1/r)\,\partial\phi/\partial r + \partial^2\phi/\partial z^2 = (S_0/K)\,\partial\phi/\partial t \qquad (8\text{-}85)$$

which is the continuity equation (5-27) written in cylindrical coordinates. In (8-85), S_0 is specific aquifer storativity due to the elastic properties of the solid matrix and the water. Boundary conditions are: (5-48) with $N = 0$, written in cylindrical coordinates on the phreatic surface, $\phi = \phi_0$ as $r \to \infty$, $\partial\phi/\partial z = 0$ on the impervious bottom of the aquifer, $Q_w = \text{const.} = \int_0^{h_w(t)} 2\pi r_w K (\partial\phi/\partial r)\, dz$ on $r = r_w$ $0 \leq z \leq h_w$ and $\phi = z$ on $r = r_w$, $h_w \leq z \leq h_s$ (seepage face). If we neglect the elastic storage (as we should in a phreatic aquifer, since $n_e \gg BS_0$), the right-hand side of (8-85) becomes zero.

Because of the nonlinear boundary condition along the phreatic surface, and because the shape and position of the phreatic surface are unknown (and in fact they constitute part of the sought solution), an exact analytical solution of the problem is not possible.

By employing the Dupuit assumptions (Sec. 4-5), (8-85) is replaced by

$$\frac{\partial^2 h^2}{\partial r^2} + \frac{1}{r}\frac{\partial h^2}{\partial r} = \frac{2S}{K}\frac{\partial h}{\partial t} \qquad (8\text{-}86)$$

which is (5-76) written in radial coordinates, with $N = 0$. Here S is the phreatic aquifer storativity (or specific yield, or effective porosity, n_e).

Although (8-86) is still nonlinear in h, the advantage gained by using it stems from the fact that the nonlinear phreatic boundary condition (5-48) is now incorporated in (8-86) and no longer appears as a boundary condition of the problem. Also, the variable now is $h = h(r, t)$ instead of $\phi = \phi(r, z, t)$, as in the original statement of the problem. All these points have already been presented in Sec. 5-2.

It is important to recall that in developing (8-86) it is assumed that (1) S is a constant, (2) water is released instantaneously from storage upon any decline of the phreatic surface, and (3) the releases are linearly proportional to the changes in storage. Actually, in unsteady flow with a declining water table, the dewatering of the interstices is not instantaneous, but lags behind the drawdown; the unsaturated region above the water table continues to supply water to the declining water table. Lenses of silt and clay considerably increase the lag. From these considerations it follows that S (or n_e) increases at a diminishing rate with duration of pumpage (Bear, 1972, p. 487). If pumpage continues for a sufficiently long time, an asymptotic stage is reached in which a practically constant value of storativity is obtained. Neuman (1972) considers delayed response of the water table.

If we substitute $h = H_0 - s$ in (8-86) (Fig. 8-6), we obtain (Jacob, 1944)

$$\frac{\partial^2 s'}{\partial r^2} + \frac{1}{r}\frac{\partial s'}{\partial r} = \frac{H_0}{(H_0 - s)}\frac{S}{KH_0}\frac{\partial s'}{\partial t}; \qquad s' = s - s^2/2H_0$$

$$\frac{\partial s'}{\partial t} = \frac{H_0 - s}{H_0}\frac{\partial s}{\partial t} \qquad (8\text{-}87)$$

Denoting $T = KH_0 = $ *initial transmissivity* and $S' = SH_0/(H_0 - s) = $ *apparent storativity*, and assuming S' and T to be constants, (8-87) becomes

$$\frac{\partial^2 s'}{\partial r^2} + \frac{1}{r}\frac{\partial s'}{\partial r} = \frac{S'}{T}\frac{\partial s'}{\partial t} \qquad (8\text{-}88)$$

which is linear in s'. This assumption is a good approximation as long as $s \ll H_0$.

The solution of (8-88) is (8-61) in which s and S are replaced by s' and S', respectively

$$s' = \frac{Q_w}{4\pi T} W\left(\frac{S'r^2}{4Tt}\right) \qquad (8\text{-}89)$$

However, when drawdowns are sufficiently small compared to the average depth of flow, (say $s \ll 0.02H$; Jacob, 1950), we obtain from (8-88)

$$\frac{\partial^2 h}{\partial r^2} + \frac{1}{r}\frac{\partial h}{\partial r} = \frac{S}{T}\frac{\partial h}{\partial t} \qquad \text{or}: \qquad \frac{\partial^2 s}{\partial r^2} + \frac{1}{r}\frac{\partial s}{\partial r} = \frac{S}{T}\frac{\partial s}{\partial t} \qquad (8\text{-}90)$$

with $T = KH$, where H is the average depth of flow. The solution for $s = s(r, t)$ is then given by (8-61). If in (8-86) we assume that water table gradients are small, so that $(\partial h/\partial r) \ll (h/r)$ or that $(\partial h/\partial r)^2 \ll h\partial^2 h/\partial r^2$, we obtain (8-90) with $T = Kh$.

Boulton (1954; see also Schoeller, 1959; Hantush, 1964) gives an approximate drawdown solution for unsteady flow in a phreatic aquifer with $Q_w = $ const. He suggests that when drawdowns are small (e.g., $s_{\max} = H_0 - h_w < 0.5 H_0$; Fig. 8-6) the nonlinear phreatic surface boundary condition (5-48) be approximated for $N = 0$ by

$$-K\frac{\partial\phi(r, h, t)}{\partial z} = n_e\frac{\partial\phi(r, h, t)}{\partial t} \qquad (8\text{-}91)$$

that is, neglecting the quadratic terms. Equation (8-91) may be interpreted as meaning that lowering the water table is caused only by vertical flow; radial flow components are neglected.

Boulton (1954) also replaces the exact expression for the well's constant discharge

$$Q_w = \int_\zeta^{h_w(t)} 2\pi r K (\partial\phi/\partial r)\bigg|_{r = r_w} dz$$

by

$$Q_w = 2\pi K H_0 \lim_{r \to 0}\left(r\frac{\partial\phi}{\partial r}\right) \qquad (8\text{-}91a)$$

that is, an infinitesimally narrow well with a discharge, Q_w, entering uniformly along the original depth of saturation, H_0, although the actual length of saturation at the face of the well varies during pumpage. Finally, he assumes that the contribution to the flow by water and aquifer compression may be neglected, except during the very early period of pumping, i.e., the right-hand side of (8-85) may be assumed zero. One should notice that he nevertheless solves for $\phi = \phi(r, z, t)$, because of the nonsteady boundary condition (8-91). Under these conditions Boulton's solution is

$$H_0 - \phi(r, z, t) = \frac{Q_w}{2\pi T} \int_0^\infty \frac{J_0(\chi r/H_0)}{\chi} \left\{ 1 - \frac{\cosh \chi z/H_0}{\cosh \chi} \exp\left[-\tau\chi \tanh \chi\right] \right\} d\chi$$

(8-91b)

and

$$s(r, t) = H_0 - h(r, t) = \frac{Q_w}{2\pi T} V(\rho, \tau)$$

$$\equiv \frac{Q_w}{2\pi T} \int_0^\infty \frac{J_0(\chi\rho)}{\chi} \left\{ 1 - \exp(-\tau\chi \tanh \chi) \right\} d\chi \quad \text{(8-91c)}$$

where $\rho = r/H_0$, $\tau = Kt/n_e H_0$, $T = KH_0$. To minimize the errors involved in obtaining (8-91c), a correction factor C_f is applied to the latter

$$s = \frac{Q_w}{2\pi T} (1 + C_f) V(\rho, \tau) \qquad \text{(8-91d)}$$

where C_f depends on ρ, τ, r_w/H_0 and Q_w/H_0^2. Boulton (1954), Schoeller (1959) and Hantush (1964) give values of V and C_f (in the form of graphs and tables) which range from about -0.30 to about 0.16. The values of C_f are given separately for $\tau > 5$ and for $\tau < 0.05$. In the range $0.05 < \tau < 5$, C_f may be assumed zero with an error not exceeding six percent.

The values of the *gravity well function for phreatic aquifers* $V(\rho, \tau)$ may also be approximated as follows
For $\tau < 0.05$

$$V(\rho, \tau) \approx \sinh^{-1}(1/\rho) + \sinh^{-1}(\tau/\rho) - \sinh^{-1}\left[(1 + \tau)/\rho\right]$$

For $\tau > 5$

$$V(\rho, \tau) \approx \tfrac{1}{2} W(\rho^2/4\tau) = \tfrac{1}{2} W(u); \qquad u = n_e r^2/4Tt$$

where $W(u)$ is the *well function of a confined aquifer*.
For $\tau < 0.01$

$$V(\rho, \tau) \approx \sinh^{-1}(\tau/\rho) - \tau/\sqrt{1 + \rho^2}$$

For $\tau < 0.01$ and $\tau/\rho > 10$

$$V(\rho, \tau) \approx \ln(2\tau/\rho)$$

For the drawdown in the pumped well, Boulton (1954) suggests for $\tau < 0.05$, to compute the drawdown by (8-91d), with $\rho \to \rho_w = r_w/H_0$ and $h \to h_w$; for $0.05 < \tau < 5$, the drawdown may be computed from

$$s_w = H_0 - h_w = \frac{Q_w}{2\pi T}(m - \ln \rho_w) \tag{8-92}$$

where

τ	0.05	0.2	1.00	5.00
m	-0.043	0.087	0.512	1.288

and intermediate values may be obtained by interpolation.

For $\tau > 5$, we have

$$H_0^2 - h_w^2 = \frac{Q_w}{\pi K}\ln\left(\frac{1.5\sqrt{\tau}}{\rho_w}\right) = \frac{Q_w}{2\pi K}\ln\frac{2.25Tt}{n_e r_w^2} \tag{8-93}$$

to be compared with (8-24) and (8-92).

The drawdown for an anisotropic aquifer may be obtained by replacing τ by $\tau' = K_z t/n_e H_0$ and ρ by $\rho' = (r/H_0)(K_z/K_r)^{1/2}$.

Hantush (1964) discusses the drawdown in observation wells at some distance from a pumping well. In the region $r > 1.5H$, equipotentials are practically vertical and the average drawdown in observation wells is practically equal to the drawdown of the water table given by (8-91d).

In the region $r < 1.5H_0$, Hantush (1964) suggests for $(H_0 - h_w) < 0.5H_0$ and $t > 30\, r^2 n_e/T$

$$H_0^2 - \tilde{h}^2 = (Q_w/2\pi K)\, W(u); \qquad u = n_e r^2/4Tt; \qquad T \approx KH_0 \tag{8-94}$$

where $\tilde{h}$ is the average head defined by $\tilde{h} = (1/h)\int_0^h \phi(r,t)\,dz$ ($\tilde{h} < h$). For $u < 0.05$, or $t > 5\, r^2 n_e/T$

$$H_0^2 - \tilde{h}^2 = (Q/2\pi K)\ln\frac{2.25\, Tt}{n_e r^2} \tag{8-95}$$

Closer to a pumping well, (8-94) gives approximately the depth $\tilde{h}$ of water in a piezometer open at the impervious horizontal base of an aquifer (smaller than the actual water level in an observation well located above that point).

One should note, that in (8-94) and (8-95)

$$H_0^2 - \tilde{h}^2 = (H_0 - \tilde{h})(H_0 + \tilde{h}) = s(H_0 + H_0 - s) = 2H_0 s(1 - s/2H_0)$$

which for $s \ll 0.02H_0$ may be approximated by $2H_0 s$ so that the Theis formula (8-61) and Jacob's straight line approximation (8-62) are applicable.

Boulton (1965) solves also the case of a well pumping from an aquifer of infinite areal extent with a constant drawdown at the pumping well. Here also the exact non-linear phreatic surface boundary condition is replaced by the approximate one (8-87), so that the boundary and initial conditions may be summarized by (Fig. 8-6)

$$
\left.
\begin{aligned}
&t = 0: && r \ge r_w, && 0 \le z \le H_0 && \phi = H_0, \\
&t > 0: && r = r_w, && 0 \le z \le h_w, && \phi = h_w, \\
& && r \to \infty, && 0 \le z \le H_0, && \phi \to H_0, \\
& && r \ge r_w, && z = 0, && \partial\phi/\partial z = 0, \\
& && r = r_w, && h_w \le z \le h_s, && \phi = z, \\
& && r \ge r_w, && z = h, && n_e\,\partial\phi/\partial t + K\,\partial\phi/\partial z = 0 \\
\end{aligned}
\right\} \quad (8\text{-}96)
$$

$$
Q_w = 2\pi r_w K \int_0^{h_w} (\partial\phi/\partial z)\ \Big|_{r=r_w} dz
$$

The equation to be solved is again (8-85) with the right-hand side assumed zero, so that the time dependence of ϕ is introduced only through the phreatic surface boundary condition.

Boulton's solution takes the form

$$
Q_w = \frac{8}{\pi} KH_0^2 \int_0^\infty \frac{\cosh\chi - \cosh(\chi h_w/H_0)}{\chi^3 \cosh\chi} \times \frac{\exp(-\tau\chi\tanh\chi)}{J_0^2(\rho_w\chi) + Y_0^2(\rho_w\chi)}\, d\chi \quad (8\text{-}97)
$$

where $\rho = r/H_0$; $\tau = Kt/n_e H_0$, and J_0 and Y_0 are zero order Bessel functions of the first and of the second kind, respectively. Equation (8-97) may also be written as

$$
Q_w = \pi KH_0^2(1 - (h_w/H_0)^2)\,G', \qquad \text{or} \qquad H_0^2 - h_w^2 = \frac{Q_w}{\pi K}\frac{1}{G'} \quad (8\text{-}98)
$$

where G' is defined by (8-97). For $\tau \gg 1$, $1/G' = \frac{1}{2}\ln(2.25KH_0 t/n_e r_w^2)$. For large t, G' is thus identical with G of (8-74). For sufficiently small drawdowns (say $s_w/H_0 \ll 1$), the discharge is given by: $Q_w = 2\pi T s_w G'$, $T = KH_0$, which is identical to (8-74).

For large times, we obtain

$$
H_0^2 - h_w^2 = \frac{Q_w}{\pi K}\ln\left(\frac{R}{r_w}\right), \qquad R = 1.5H(Kt/n_e H)^{1/2} \quad (8\text{-}99)
$$

which is identical with the Dupuit–Forchheimer formula (8-23), but with a *time-dependent* radius of influence. The same conclusion may also be drawn from (8-95).

In the discussion above, the difficulty inherent in solving a phreatic flow problem was circumvented mainly by assuming, in one form or another, that the flow is essentially horizontal and that vertical flow components are negligible. Boulton (1954) and Hantush (1964) suggest that the vertical flow components significantly affect the drawdown when $t > 5H_0 n_e/K$ (or $t > 5H_0 n_e/K_z$ in an significantly affect the drawdown when $t > 5H_0 n_e/K$ (or $t > 5H_0 n_e/K_z$ in an an-isotropic aquifer) in the region $0 < r < 0.2H$. Stallman (1965) showed vertical flow components to be significant in the region: $Tt/n_e r^2 < 1$ or $(r/H_0)(K_z/K_r)^{1/2} > 3$.

Table 8-5 The well function $W(u_{xy}, r/D_t)$ for a phreatic aquifer with delayed storage

$$1/u_a = N_n \times 10^n$$

$r/D_t = 0.01$			$r/D_t = 0.1$			$r/D_t = 0.2$			$r/D_t = 0.316$			$r/D_t = 0.4$			$r/D_t = 0.6$		
N	n	$W(u_a, r/D_t)$	N	n	$W(u_a, r/D_t)$	N	n	$W(u_a, r/D_t)$	N	n	$W(u_a, r/D_t)$	N	n	$W(u_a, r/D_t)$	N	n	$W(u_a, r/D_t)$
1	1	1.82	1	1	1.80	5	0	1.19	1	0	0.216	1	0	0.213	1	0	0.206
1	2	4.04	5	1	3.24	1	1	1.75	2	0	0.544	2	0	0.534	2	0	0.504
1	3	6.31	1	2	3.81	5	1	2.95	5	0	1.153	5	0	1.114	5	0	0.996
5	3	7.82	2	2	4.30	1	2	3.29	1	1	1.655	1	1	1.564	1	1	1.311
1	4	8.40	5	2	4.71	5	2	3.50	5	1	2.504	5	1	2.181	2	1	1.493
1	5	9.42	1	3	4.83	1	3	3.51	1	2	2.623	1	2	2.225	5	1	1.553
1	6	9.44	1	4	4.85				1	3	2.648	1	3	2.229	1	2	1.555

$r/D_t = 0.8$			$r/D_t = 1.0$			$r/D_t = 1.5$			$r/D_t = 2.0$			$r/D_t = 2.5$			$r/D_t = 3.0$		
N	n	$W(u_a, r/D_t)$	N	n	$W(u_a, r/D_t)$	N	n	$W(u_a, r/D_t)$	N	n	$W(u_a, r/D_t)$	N	n	$W(u_a, r/D_t)$	N	n	$W(u_a, r/D_t)$
5	-1	0.046	5	-1	0.0444	5	-1	0.0394	3.33	-1	0.0100	5	-1	0.0271	5	-1	0.0210
1	0	0.197	1	0	0.1855	1	0	0.1509	5	-1	0.0335	1	0	0.0803	1	0	0.0534
2	0	0.466	2	0	0.421	1.25	0	0.199	1	0	0.114	1.25	0	0.0961	1.25	0	0.0607
5	0	0.857	5	0	0.715	2	0	0.301	1.25	0	0.144	2	0	0.1174	2	0	0.0681
1	1	1.050	1	1	0.819	5	0	0.413	2	0	0.194	5	0	0.1247	5	0	0.0695
2	1	1.121	2	1	0.841	1	1	0.427	5	0	0.227	1	1	0.1247	1	1	0.0695
5	1	1.131	5	1	0.842	2	1	0.428	1	1	0.228						

$$1/u_v = N_n \times 10^n$$

$r/D_t = 0.01$

N	n	$W(u_y, r/D_t)$
4	2	9.45
4	3	9.54
4	4	10.23
4	5	12.31
4	6	14.61

$r/D_t = 0.1$

N	n	$W(u_y, r/D_t)$
4	0	4.86
4	1	4.95
2	2	5.64
4	3	7.72
1.5	4	10.01

$r/D_t = 0.2$

N	n	$W(u_y, r/D_t)$
4	−1	3.51
4	0	3.54
2	1	3.69
4	1	3.85
1.5	2	4.55
4	2	5.42

$r/D_t = 0.316$

N	n	$W(u_y, r/D_t)$
4	−1	2.66
4	0	2.74
4	1	3.38
4	2	5.42
4	3	7.72

$r/D_t = 0.4$

N	n	$W(u_y, r/D_t)$
1	−1	2.23
1	0	2.26
5	0	2.40
1	1	2.55
3.75	1	3.20
1	2	4.05

$r/D_t = 0.6$

N	n	$W(u_y, r/D_t)$
4.44	−1	1.586
2.22	0	1.707
4.44	0	1.844
1.67	1	2.448
4.44	1	3.255

$r/D_t = 0.8$

N	n	$W(u_y, r/D_t)$
2.5	−2	1.133
2.5	−1	1.158
1.25	0	1.264
2.5	0	1.387
9.37	0	1.938
2.5	1	2.704

$r/D_t = 1.0$

N	n	$W(u_y, r/D_t)$
4	−2	0.844
4	−1	0.901
4	0	1.356
4	1	3.140

$r/D_t = 1.5$

N	n	$W(u_y, r/D_t)$
7.11	−2	0.444
3.55	−1	0.509
7.11	−1	0.587
2.67	0	0.963
7.11	0	1.569

$r/D_t = 2.0$

N	n	$W(u_y, r/D_t)$
4	−2	0.239
2	−1	0.283
4	−1	0.337
1.5	0	0.614
4	0	1.111

$r/D_t = 2.5$

N	n	$W(u_y, r/D_t)$
2.56	−2	0.1321
1.28	−1	0.1617
2.56	−1	0.1988
9.6	−1	0.3990
2.56	0	0.7977

$r/D_t = 3.0$

N	n	$W(u_y, r/D_t)$
1.78	−2	0.0743
8.89	−2	0.0939
1.78	−1	0.1189
6.67	−1	0.2618
1.78	0	0.5771

From Boulton, 1963.

Another difficulty results from the basic assumption that water is released from storage immediately upon a decline of the water table, thus disregarding the phenomenon of *delayed storage*. In later works, Boulton (1954a and 1963) suggests a solution assuming that the delayed yield from storage is expressed as an exponential function of time.

Boulton (1954a and 1963) assumed that the amount of water derived from storage in a phreatic aquifer (per unit horizontal area) due to an increment of drawdown Δs during an interval of time from τ to $\tau + \Delta\tau$ consists of two parts: (i) a volume of water, $S \Delta s$, instantaneously released from storage, and (ii) a delayed yield from storage at any time $t > \tau$ from the beginning of pumpage, given by $\alpha \Delta s \times S' \exp[-\alpha(t - \tau)]$, where α is an empirical parameter ($=$ reciprocal of the *delay index*; dim. T^{-1}) and S' is the delayed storativity defined as the total volume of delayed yield from storage per unit drawdown and per unit horizontal area (commonly referred to as *specific yield*, see Sec. 6-1).

Accordingly, the partial differential equation governing the flow is

$$\frac{\partial^2 s}{\partial r^2} + \frac{1}{r}\frac{\partial s}{\partial r} = \frac{S}{T}\frac{\partial s}{\partial t} + \alpha\frac{S'}{T}\int_0^t \frac{\partial s}{\partial t}\exp[-\alpha(t - \tau)]\,d\tau \qquad (8\text{-}100)$$

where τ is the elapsed time measured from the beginning of the delayed storage. Boulton (1963) solved (8-100) and obtained expressions for drawdown which take delayed storage into account. His solution may be written as

$$s = \frac{Q_w}{4\pi T} W\left(u_{\alpha y}, \frac{r}{D_t}\right) \qquad (8\text{-}101)$$

where $W(u_{\alpha y}, r/D_t)$ is the *well function for phreatic aquifer with fully penetrating well*

$$u_{\alpha y} = \begin{cases} u_\alpha = Sr^2/4Tt \text{ (applicable for small values of } t) \\ u_y = S'r^2/4Tt \text{ (applicable for large values of } t) \end{cases}$$

$$r/D_t = r/\sqrt{T/\alpha S'}$$

$$\alpha = (r/D_t)^2/4u_y t$$

$$u_y = u_\alpha/(N - 1)$$

$$N = (S + S')/S$$

Boulton (1963; see also Prickett, 1965; Walton, 1970) gives values of $W(u_{\alpha y}, r/D_t)$. Some values for the practical range of u_α, u_y, and r/D_t are given in Table 8-5.

In (8-100) we have used the symbol S for storativity rather than n_e to indicate that actually n_e is some function of both S and S', whereas S is only the effective, early time, storativity. Boulton (1963) gives type curves which facilitate the determination of drawdown under various conditions. Schoeller (1959) and Prickett (1965) discuss Boulton's solutions. Some additional comments on delayed storage are given by Prickett (1965).

The problem of unsteady flow to wells in phreatic aquifers is also discussed by Kriz *et al.* (1966) and Streltsova (1972, 1973).

Numerical solutions were obtained, among others, by Prickett (1965), Streltsova and Rushton (1973) and Rushton and Chan (1977).

8-7 UNSTEADY FLOW TO A WELL IN A LEAKY CONFINED AQUIFER

All the assumptions and the nomenclature of Fig. 8-9 leading to the flow pattern discussed in Sec. 8-4 (fully penetrating well, etc.) are applicable also here. References to publications on unsteady flow are given at the beginning of Sec. 8-4.

If a constant head ϕ_0 is maintained on top of the semipervious stratum, and the rate of leakage is assumed to be proportional to the drawdown in the aquifer, continuity yields the following partial differential equation for $\phi = \phi(r, t)$

$$\frac{\partial^2 \phi}{\partial r^2} + \frac{1}{r}\frac{\partial \phi}{\partial r} + \frac{\phi_0 - \phi}{\lambda^2} = \frac{S}{T}\frac{\partial \phi}{\partial t} \qquad (8\text{-}102)$$

where S and T are the storativity and the transmissivity of the aquifer, respectively and $\lambda = (B'T/K')^{1/2}$ is the *leakage factor*. In deriving (8-102) it is assumed that the storativity of the semipervious layer may be neglected. Sometimes ϕ_0 is taken as a reference head, $\phi_0 = 0$. Equation (8-102) is the same as (5-70) written in radial coordinates, with $\lambda^{(2)} = \infty$ and $\phi_1 = \phi_0$. It is also assumed that the leakage through the semipervious layer is given by $K'(\phi_0 - \phi)/B'$ or by $(\phi_0 - \phi)/\sigma'$. Jacob (1946) solves (8-102) for $\phi(R, t) = \phi_0 = \text{const.}$ on an outer circular boundary $r = R$. The solution takes the form of the sum of a steady-state drawdown, s, given by (8-44), and a transient drawdown, s', given by

$$s' = -\frac{Q_w}{\pi T}\sum_{n=1}^{\infty}\{J_0(r\alpha_n)\exp[-(a\alpha_n^2 + c)t]/R^2\alpha_n^2\,J_1^2(R\alpha_n) \times (1 + c/a\alpha_n^2)]\} \qquad (8\text{-}103)$$

where $a = T/S$; $c = K'/B'S$; $a/c = \lambda^2$; J_0 and J_1 are Bessel functions of the first kind of orders zero and one, respectively. The values of $(\alpha_n R)$ are roots of the equation $J_0(\alpha R) = 0$. For an impervious outer circular boundary, s is the sum of (8-47) and

$$s' = -\frac{Q_w}{\pi T}\sum_{n=1}^{\infty}\{J_0(\alpha_n r)\exp[-(a\alpha_n^2 + c)t]/[R^2\alpha_n^2 J_0^2(\alpha_n R) \times (1 + c/a\alpha_n^2)]\} \qquad (8\text{-}104)$$

where $(\alpha_n R)$ is the nth root of the equation $J_0'(\alpha R) = -J_1(\alpha R) = 0$.

For an infinite aquifer with leakage proportional to drawdown, Hantush and Jacob (1955b) find

$$s(r, t) = \frac{Q_w}{4\pi T}\int_{y=u}^{\infty}\frac{1}{y}\exp\left(-y - \frac{r^2}{4\lambda^2 y}\right)dy = \frac{Q_w}{4\pi T}W(u, r/\lambda) \qquad (8\text{-}105)$$

$$s(r, t) = \frac{Q_w}{4\pi T} \left\{ 2K_0(r/\lambda) - \int_{y=a}^{\infty} \frac{1}{y} \exp(-y - r^2/4\lambda^2 y) \, dy \right\}$$

$$= \frac{Q_w}{4\pi T} [2K_0(r/\lambda) - W(a, r/\lambda)] \qquad (8\text{-}106)$$

where $u = r^2 S/4Tt$, $a = r^2/4\lambda^2 u = Tt/S\lambda^2$. Hantush (1956, 1961, 1964) gives tables of *the well function for a leaky aquifer* $W(u, r/\lambda)$. Table 8-6 gives some selected values taken from these tables. Figure 8-17 gives the nonsteady leaky aquifer type curves. Of special interest are the following relationships (Hantush, 1961)

$$W(0, r/\lambda) = 2K_0(r/\lambda); \qquad W(u, 0) = W(u)$$

$$W(u, r/\lambda) = 2K_0(r/\lambda) - W(r^2/4\lambda^2 u, r/\lambda) = 2K_0(r/\lambda) - W(Tt/S\lambda^2, r/\lambda)$$

$$W(u, r/\lambda) \approx W(u) \quad \text{for} \quad u > 2r/\lambda$$

$$\approx W(u) \quad \text{for} \quad u > 5r^2/\lambda^2 \quad \text{if} \quad r/\lambda < 0.1$$

$$W(u, r/\lambda) \approx 2K_0(r/\lambda) - I_0(r/\lambda) W(Tt/S\lambda^2) \quad \text{for} \quad u < r^2/20\lambda^2 \quad \text{if} \quad u < 1$$

From the first of these relationships it follows that as $t \to \infty, u \to 0, W(0, r/\lambda) \to 2K_0(r/\lambda)$ and 8-105) reduces to (8-41).

For variable discharge, $Q_w = Q_w(t)$, (8-105) becomes

$$s(r, t) = \frac{1}{4\pi T} \int_0^t \frac{Q_w(\tau)}{t - \tau} \exp \left\{ - \frac{Sr^2}{4T(t - \tau)} - \frac{(t - \tau) T}{S\lambda^2} \right\} d\tau \qquad (8\text{-}106a)$$

which is obtained in a way similar to that leading to (8-59) for a confined aquifer.

Hantush (1959, 1964) determines also the potential distribution and the variation of discharge with time for a flowing (artesian) well discharging by natural flow from a homogeneous aquifer into or out of which there is leakage in proportion to the drawdown. He considers three cases of a homogeneous isotropic aquifer of constant thickness.

I A flowing well in a leaky aquifer of infinite areal extent.
II A flowing well at the center of a circular aquifer with zero drawdown on its outer boundary.
III A flowing well at the center of a circular aquifer with zero flux through its outer (impervious) boundary.

In all cases, a constant piezometric head is maintained in the aquifer above the semipervious layer. Once the drawdown distribution $s = s(r, t)$ or $\phi = \phi(r, t)$ in the vicinity of the flowing well is found, the well's discharge is determined by: $Q_w = 2\pi T r_w \, \partial s(r_w, t)/\partial r$.

Table 8-6 Well function for a leaky aquifer $W(u, r/\lambda)$ (after Hantush, 1956)

u \ r/λ	0	0.002	0.004	0.005	0.007	0.01	0.02	0.04	0.06	0.08	0.10
0		12.6611	11.2748	10.8286	10.1557	9.4425	8.0569	6.6731	5.8658	5.2950	4.8541
1×10^{-6}	13.2383	12.4417	11.2711	10.8283	10.1557						
2×10^{-6}	12.5451	12.1013	11.2259	10.8174	10.1554						
5×10^{-6}	11.6289	11.4384	10.9642	10.6822	10.1290	9.4425					
8×10^{-6}	11.1589	11.0377	10.7151	10.5027	10.0602	9.4313					
1×10^{-5}	10.9357	10.8382	10.5725	10.3963	10.0034	9.4176	8.0569				
2×10^{-5}	10.2426	10.1932	10.0522	9.9530	9.7126	9.2961	8.0558				
5×10^{-5}	9.3263	9.3064	9.2480	9.2052	9.0957	8.8827	8.0080	6.6730			
7×10^{-5}	8.9899	8.9756	8.9336	8.9027	8.8224	8.6625	7.9456	6.6726			
1×10^{-4}	8.6332	8.6233	8.5937	8.5717	8.5145	8.3983	7.8375	6.6693	5.8658	5.2950	
2×10^{-4}	7.9402	7.9352	7.9203	7.8958	7.8800	7.8192	7.4472	6.6242	5.8637	5.2949	4.8541
5×10^{-4}	7.0242	7.0222	7.0163	7.0118	6.9999	6.9750	6.8346	6.3626	5.8011	5.2848	4.8530
7×10^{-4}	6.6879	6.6865	6.6823	6.6790	6.6706	6.6527	6.5508	6.1917	5.7274	5.2618	4.8478
1×10^{-3}	6.3315	6.3305	6.3276	6.3253	6.3194	6.3069	6.2347	5.9711	5.6058	5.2087	4.8292
2×10^{-3}	5.6394	5.6389	5.6374	5.6363	5.6334	5.6271	5.5907	5.4516	5.2411	4.9848	4.7079
5×10^{-3}	4.7261	4.7259	4.7253	4.7249	4.7237	4.7212	4.7068	4.6499	4.5590	4.4389	4.2990
7×10^{-3}	4.3916	4.3915	4.3910	4.3908	4.3899	4.3882	4.3779	4.3374	4.2719	4.1839	4.0771
0.01	4.0379	4.0379	4.0375	4.0373	4.0368	4.0351	4.0285	4.0003	3.9544	3.8920	3.8190
0.02	3.3547	3.3547	3.3545	3.3544	3.3542	3.3536	3.3502	3.3365	3.3141	3.2832	3.2442
0.05	2.4679	2.4679	2.4678	2.4678	2.4677	2.4675	2.4662	2.4613	2.4531	2.4416	2.4271
0.07	2.1508	2.1508	2.1508	2.1508	2.1507	2.1506	2.1497	2.1464	2.1408	2.1331	2.1232
0.10	1.8229	1.8229	1.8229	1.8229	1.8228	1.8227	1.8222	1.8200	1.8164	1.8114	1.8050
0.20	1.2227	1.2226	1.2226	1.2226	1.2226	1.2226	1.2224	1.2215	1.2201	1.2181	1.2155
0.50	0.5598	0.5598	0.5598	0.5598	0.5598	0.5598	0.5597	0.5595	0.5592	0.5587	0.5581
0.70	0.3738	0.3738	0.3738	0.3738	0.3738	0.3738	0.3737	0.3736	0.3734	0.3732	0.3729
1.00	0.2194	0.2194	0.2194	0.2194	0.2194	0.2194	0.2194	0.2193	0.2192	0.2191	0.2190
2.00	0.0489	0.0489	0.0489	0.0489	0.0489	0.0489	0.0489	0.0489	0.0489	0.0489	0.0488
5.00	0.0011	0.0011	0.0011	0.0011	0.0011	0.0011	0.0011	0.0011	0.0011	0.0011	0.0011
7.00	0.0001	0.0001	0.0001	0.0001	0.0001	0.0001	0.0001	0.0001	0.0001	0.0001	0.0001
8.00	0.0000	0.0000	0.0000	0.0000	0.0000	0.0000	0.0000	0.0000	0.0000	0.0000	0.0000

The following solutions were obtained for Case I (Hantush, 1959)

$$\frac{s}{s_w} = K_0\left(\frac{r}{\lambda}\right) \Big/ K_0\left(\frac{r_w}{\lambda}\right) + \frac{2}{\pi} \exp\left(-\alpha r_w^2/\lambda^2\right) \int_0^\infty \frac{\exp\left(-\alpha\mu^2\right)}{\mu^2 + (r_w/\lambda)^2}$$

$$\times \frac{J_0\left(\frac{\mu r}{r_w}\right) Y_0(\mu) - Y_0\left(\frac{\mu r}{r_w}\right) J_0(\mu)}{J_0^2(\mu) + Y_0^2(\mu)} \mu d\mu = Z(\alpha, \rho, \beta)$$

$$\alpha = Tt/sr_w^2, \qquad \rho = r/r_w, \qquad \beta = r_w/\lambda \qquad (8\text{-}107)$$

where the infinite integral can be evaluated numerically.

As $t \to \infty$, $\alpha \to \infty$ and (8-108) reduces to the steady state solution (8-41), or (8-42) for $r/u < 0.05$.

The well's discharge corresponding to the drawdown described by (8-107) is

$$Q_w = -2\pi T r_w \, \partial s(r_w, t)/\partial r \Big|_{r=r_w} = 2\pi T s_w G(\alpha, r_w/\lambda)$$

$$G(\alpha, r_w/\lambda) = (r_w/\lambda) K_1(r_w/\lambda)/K_0(r_w/\lambda)$$

$$+ (4/\pi^2) \exp\left[-\alpha(r_w/\lambda)^2\right] \times \int_0^\infty \frac{\mu \exp(-\alpha\mu^2)}{J_0^2(\mu) + Y_0^2(\mu)} \times \frac{d\mu}{\mu^2 + (r_w/\lambda)^2} \qquad (8\text{-}108)$$

where K_1 is the first order Modified Bessel function of the second kind. Hantush (1959) tabulates the function $G(\alpha, r_w/\lambda)$ which he calls the *flowing well discharge function for leaky aquifers*. As α or t becomes very large, the steady state discharge, or the minimum yield of the well under the assumed conditions, is obtained. Then

$$Q_w = 2\pi T s_w(r_w/\lambda) K_1(r_w/\lambda)/K_0(r_w/\lambda) \qquad (8\text{-}109)$$

In the practice $r_w < \lambda < 0.01$. Then the steady state solutions (8-41) and (8-42) are applicable for Q_w.

Hantush (1959) also discusses asymptotic solutions for small values of time and for large values of time. In the latter case

$$s \approx \left[s_w/2K_0(s_w/\lambda)\right] W(u, r/\lambda) \qquad (8\text{-}110)$$

where $W(u, r/\lambda)$ is the well function for a leaky aquifer given in Table 8-6. The corresponding well's discharge is

$$Q_w \approx 4\pi T s_w/W(1/4\alpha, r_w/\lambda) \qquad (8\text{-}111)$$

which is the same as that giving the drawdown (Hantush, 1956) in a well of radius r_w discharging at a steady state for a period of time t. In other words, the ratios of discharge to drawdown in two isolated wells — one with constant discharge and the other with constant drawdown — become equal at a sufficiently large value of time. The same observation was made in the case of similar wells in a confined aquifer (Jacob and Lohman, 1952).

Hantush also gives solutions for Cases II and III.

DeWiest (1961, 1963) solved unsteady flow in a finite leaky aquifer, with time-dependent boundary condition at the contour of influence and with initially different heads in the main aquifer and in the semiconfining overlying layer. He considered two cases: (a) a well pumped at a constant discharge; (b) a flowing well at constant drawdown.

Numerical solutions for a well in a leaky aquifer were derived by Neuman and Witherspoon (1969, 1969a).

In the considerations leading to (8-102), it was assumed that the rate of leakage through the semiconfining layer is proportional to the decline of head in the main aquifer. However, in unsteady flow, such proportionality is possible only if the hydraulic gradient through the semiconfining layer adjusts instantaneously to the decline in head in the aquifer. This condition can be approached if (i) B' is sufficiently small, (ii) the rate of decline in head in the aquifer is sufficiently slow, or (iii) $K'B'/S'$ (*diffusivity of the confining bed*) is small. In general, in addition to leakage, water entering the main aquifer is derived also from storage in the semiconfining stratum. Hantush (1960, 1964) presents a theory in which the semipervious layer, although of very low hydraulic conductivity, may yield significant amounts of water from storage. Again the flow is assumed to be practically vertical in the semiconfining bed and horizontal in the main aquifer because $K' \ll K$. The problem is also discussed by Neuman and Witherspoon (1969, 1969a) who use analytical solutions. (See discussion in Sec. 5-4.)

Defining a piezometric head $\phi' = \phi'(r, z, t)$ and a corresponding drawdown $s' = s'(r, z, t)$ in the semipervious formation, the problem of flow in a horizontal homogeneous leaky aquifer of constant thickness B (Fig. 8-9) is stated by the following equations simultaneously

$$\frac{\partial^2 s}{\partial r^2} + \frac{1}{r}\frac{\partial s}{\partial r} + \frac{K'}{T}\frac{\partial s'}{\partial z}\bigg|_{z=B} = \frac{S}{T}\frac{\partial s}{\partial t} \qquad (8\text{-}112)$$

$$K'\frac{\partial^2 s'}{\partial z^2} = S'_s\frac{\partial s'}{\partial t} \qquad (8\text{-}113)$$

where $S'_s(= S'/B')$ is the *specific storativity* of the semipervious layer (compare with the discussion on the effect of storage in a semipermeable layer in Sec. 5-4).

Boundary and initial conditions are

$$\left.\begin{array}{ll} s(r, 0) = 0, & s'(r, z, 0) = 0 \\[6pt] s(\infty, t) = 0, & s'(r, B + B', t) = 0 \\[6pt] \text{for} \quad Q_w = \text{const.,} \ \lim\limits_{r \to 0}(r\,\partial s/\partial r) = Q_w/2\pi T, & s'(r, B, t) = s(r, t) \\[6pt] s_w = s(r_w, t) = \text{const. (i.e., an artesian, or flowing, well)} \end{array}\right\} \qquad (8\text{-}114)$$

Hantush (1960, 1964) presents various asymptotic solutions of this problem. For example:

For long times, that is, t larger than both $2B'S'/K'$ and $30\delta_1 r_w^2/v[1 - (10 r_w/\lambda^2)]$, with

$$r_w/\lambda < 0.1, \qquad \delta_1 = 1 + S'/3S, \qquad v = T/S$$

and a well of constant discharge Q_w, the drawdown is given by

$$s(r, t) = \frac{Q_w}{4\pi T} W(\delta_1 u, r/\lambda) \tag{8-115}$$

where $W(u, r/\lambda)$ is *the well function of a leaky aquifer* (Table 8-6).

For an artesian well and $t > 2B'S'/K'$, he obtains

$$s(r, t) = s_w Z(\tau/\delta_1, \rho, \beta) \tag{8-116}$$

where the function Z is defined by (8-107) and $\tau = vt/r_w^2$, $\rho = r/r_w$.

For short times, $t \leq B'S'/10K'$, he obtains

$$s = \frac{Q_w}{4\pi T} H(u, \beta') \tag{8-117}$$

where

$$u = Sr^2/4Tt, \qquad \beta' = \frac{r}{4\lambda} \sqrt{\frac{S'}{S}} = r \sqrt{\frac{K'S'_s}{16TS}}$$

and

$$H(u, \beta') = \int_0^\infty \frac{e^{-y}}{y} \operatorname{erfc} \frac{\beta' \sqrt{u}}{\sqrt{y(y - u)}} \, dy$$

Tables of $H(u, \beta')$ are given by Hantush (1964). He also presents solutions for the case where an impervious confining layer overlies the semipervious formation (Hantush, 1964). Various other cases are also given in his extensive treatise of Hydraulics of Wells (Hantush, 1964).

8-8 PARTIALLY PENETRATING WELLS

A *partially penetrating well* is one whose length of screen (or water entry portion) is less than the saturated thickness of the phreatic or confined aquifer from which it withdraws water. This situation often occurs in the practice. Flow to a partially penetrating well is three-dimensional in nature, often with radial symmetry.

Several examples of partially penetrating wells are shown in Fig. 8-18. When compared with a fully penetrating well, an additional head loss is observed in the vicinity of a partially penetrating well, due to the convergence of streamlines and their extended length. Hence, for a given pumping rate Q_w, the drawdown in a partially penetrating well is larger than in a fully penetrating one. Also, because a different drawdown curve corresponds to each streamline, the location of observation wells (distance from pumping well and depth of well's opening) becomes important (Fig. 8-18c). However, as will be shown below, beyond a distance from the well of $1.5B–2B$, where B is the thickness of the aquifer, the effect of partial penetration becomes negligible and the streamlines are practically horizontal. In anisotropic aquifers, with $K_x > K_z$, the distance becomes (Hantush, 1964; 1966) $1.5(K_x/K_z)^{1/2} B$ to $2(K_x/K_z)^{1/2} B$.

De Glee (1930), Kozeny (1933), Muskat (1937), Jacob (1945), Polubarinova Kochina (1951, 1952, 1962), Narhgang (1954), Boreli (1955), Hantush (1957,

1961a, 1964), Kirkham (1959), Kipps (1973), among many others, present analytical solutions and corrections for the drawdown in the vicinity of a partially penetrating well. Several cases of interest may be considered.

(a) Steady flow to a well of zero penetration When a well just penetrates the impervious ceiling of a very thick confined aquifer which approximates a semi-infinite domain bounded from above by an impervious boundary, it may be considered as a point sink with spherically converging flow towards it (Fig. 8-18d). With

$$Q_w = 2\pi r^2 K \, \partial\phi/\partial r; \qquad \phi = \phi_w \text{ at } r = r_w \text{ and } \phi = \phi_R \text{ at } r = R$$

(i.e., assuming that the well's bottom has the shape of a hemisphere of radius r_w), we obtain by integration

$$\phi_R - \phi_w = \frac{Q_w}{2\pi K}\left(\frac{1}{r_w} - \frac{1}{R}\right); \qquad \phi - \phi_w = \frac{Q_w}{2\pi K}\left(\frac{1}{r_w} - \frac{1}{r}\right)$$

$$\phi_R - \phi_w = \frac{Q_w}{2\pi K}\left(\frac{1}{r_w} - \frac{1}{R}\right)$$

$$(8\text{-}118)$$

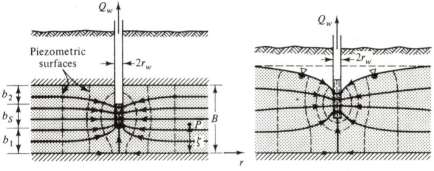

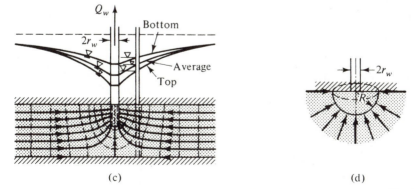

Figure 8-18 Partially penetrating wells. (a) In a confined aquifer. (b) In a phreatic aquifer. (c) Drawdown curves along streamlines. (d) Zero penetration in a thick aquifer.

In general we may assume $R \gg r_w$, so that (8-118) becomes

$$\phi_R - \phi_w = \frac{Q_w}{2\pi K r_w} \tag{8-118a}$$

If $\phi_R = 0$, then as $r_w \to 0$, $\phi_w \to -\infty$. Equation (8-118a), when compared with (8-4) for a fully penetrating well yields

$$\frac{Q_w \text{ of zero penetrating well}}{Q_w \text{ of fully penetrating well}} = \frac{r_w}{B} \ln \frac{R}{r_w} \tag{8-118b}$$

which, since $r_w \ll B$, shows that a zero-penetrating well with spherically converging flow is highly inefficient.

(b) Steady flow to a well of finite length in an infinite domain Consider a well with a screen of finite length $2L$ partially penetrating a very thick homogeneous isotropic aquifer. We may approximate the well as a finite line sink (i.e., assuming $r_w \to 0$) in an infinite (in all directions) porous medium domain (Fig. 8-19). The potential drop $d(\Delta\phi)$ produced at point $P(r, z)$ in the vicinity of the well by a point sink of strength dQ_w located at point $M(0, \zeta)$ on the line sink is

$$d(\Delta\phi) = \frac{dQ_w}{4\pi K \rho} = \frac{dQ_w}{4\pi K \left[(z - \zeta)^2 + r^2 \right]^{1/2}} \tag{8-119}$$

By dividing the length of the screen $(-L \le z \le +L)$ into a large number of such point sinks and integrating the potential drops $d\phi$ caused by them at P, we obtain

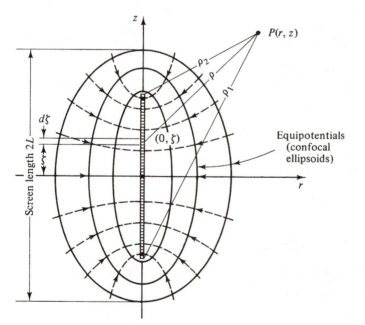

Figure 8-19 A partially penetrating well in a very thick aquifer (axially symmetric flow).

the total drop caused by the entire screen. As we shall see below, $dQ_w/d\zeta \neq$ const. However, as an approximation, let us *assume* that the well's discharge Q_w is uniformly distributed along the screen, so that the strength of an elementary line sink of length $d\zeta$ is $dQ_w = (Q_w/2L)\, d\zeta$. Then

$$d(\Delta\phi) = \frac{Q_w}{2L} \frac{d\zeta}{4\pi K [(z - \zeta)^2 + r^2]^{1/2}}$$

By integrating along $-L < \zeta < +L$, the effect of the entire line sink ($=$ screen) is found to be:

$$\Delta\phi(r, z) = \frac{Q_w}{8\pi KL} \int_{-L}^{+L} \frac{d\zeta}{[(z - \zeta)^2 + r^2]^{1/2}} = \frac{Q_w}{8\pi KL} \ln \frac{L - z + [(L - z)^2 + r^2]^{1/2}}{-L - z + [(L + z)^2 + r^2]^{1/2}}$$

(8-119a)

$$\Delta\phi(r, z) = \frac{Q_w}{8\pi KL} \ln \frac{\rho_1 + \rho_2 + 2L}{\rho_1 + \rho_2 - 2L}$$

(8-119b)

where $\rho_1 = [(z + L)^2 + r^2]^{1/2}$; $\quad \rho_2 = [(z - L)^2 + r^2]^{1/2}$; $\quad \rho_1^2 - \rho_2^2 = 4zL$.

For very large values of ρ_1 and ρ_2 ($\gg L$), we have $\phi = 0$. We may introduce $\phi = H =$ const. as a reference potential at infinity so that (8-119b) becomes

$$H - \phi = \frac{Q_w}{8\pi KL} \ln \frac{\rho_1 + \rho_2 + 2L}{\rho_1 + \rho_2 - 2L}$$

(8-119c)

As $(\rho_1 + \rho_2) \to 2L$, we have $(H - \phi) \to \infty$.

From (8-119c) it follows that equipotential surfaces $\phi =$ const. are confocal ellipsoids, $z^2/a^2 + r^2/b^2 = 1$, on which $\rho_1 + \rho_2 =$ const. $= 2a$ (with $a > L$). The foci of these ellipsoids are at points $(0, \pm L)$.

When $a \to L$ and $b \to 0$, the ellipsoid degenerates in the limit to the line sink, with $\phi = -\infty$ there. When $(z^2 + r^2)^{1/2} = \rho \gg L$, $[(z \pm L)^2 + r^2]^{1/2} \simeq (z^2 + r^2)^{1/2} \pm zL/(z^2 + r^2)^{1/2}$ so that (8-119c) becomes

$$H - \phi = \frac{Q_w}{8\pi KL} \ln \frac{1 + L/(z^2 + r^2)^{1/2}}{1 - L/(z^2 + r^2)^{1/2}} \simeq \frac{Q_w}{8\pi K(z^2 + r^2)^{1/2}} = \frac{Q_w}{8\pi K\rho}$$

(8-119d)

which describes equipotential surfaces which are concentric spheres. This means that at a sufficiently large distance from the finite line sink the potential distribution is the same as that of a point sink having the same total discharge.

For a point $P(r, z)$, where $r \ll L$ and z not too close to $\pm L$, an expansion of (8-119a) leads to

$$H - \phi \simeq \frac{Q_w}{2\pi K(2L)} \ln \frac{2L}{r}$$

(8-119e)

(compare with 8-6).

De Glee (1930), derived the additional steady state drawdown Δs_w due to the partial penetration of a well (Fig. 8-18a), for $10r_w \leq b_s \leq 0.8B$, in the form

$$\Delta s_w = \frac{Q_w}{2\pi KB} \frac{1 - p}{p} \ln \frac{(1.2 - p) b_s}{\beta r_w}$$

(8-120)

where, with the nomenclature of Fig. 8-18a, $p = b_s/B$, $\beta = 1$ for a top screen ($b_2 = 0$), or a bottom screen ($b_1 = 0$), and $\beta = 2$ for a screen in the middle. The drawdown is the sum of the drawdown Δs_w and that corresponding to a fully penetrating well.

Muskat (1932, 1937) also determined the potential distribution for steady flow to the partially penetrating well shown in Fig. 8-18c (or Fig. 8-18a with $b_2 = 0$). He integrates point sinks along the well's screen, and employs the method of images (Sec. 8-10) to introduce the effect of the impervious bottom and ceiling of a confined aquifer. The assumption of uniform distribution of inflow along the screen leads to a result in which the well's screen is not an equipotential surface.

Actually, in addition to the ordinary flux contribution of a radial character entering each unit length along the well, the lower parts receive most of the flux coming from that part of the aquifer which is not penetrated by the well. This additional flux is not uniformly distributed along the well; it is more concentrated near its extremities. To reduce the error introduced by this nonuniform distribution, Muskat introduces correction factors. Under certain conditions, he obtains, for example

$$H - \phi_w = \frac{Q_w}{4\pi KB} \frac{b}{b_s} \left[2\ln(4B/r_w) - G(b_s/B) - 2(b_s/B)\ln(4B/R) \right] \quad (8\text{-}121)$$

$$G(b_s/B) = \frac{\Gamma(0.875\, b_s/B)\, \Gamma(0.125\, b_s/B)}{\Gamma(1-0.875\, b_s/B)\Gamma(1-0.125\, b_s/B)}$$

Formulae for other cases were also presented by Muskat (1937). Another case treated by Muskat (1937) is that of a partially penetrating well in a stratified aquifer.

Kozeny (1933) summarized Muskat's analysis by a somewhat simpler empirical expression which is sufficiently accurate for most practical cases

$$s_w = \frac{1}{C} \times \frac{Q_w}{2\pi KB} \ln \frac{R}{r_w}; \qquad C = \left(\frac{b_s}{B} \right) \left[1 + 7(r_w/2b_s)^{1/2} \cos(\pi b_s/2B) \right] \quad (8\text{-}122)$$

where s_w is the drawdown in the pumped well. From (8-122) and (8-4) for a fully penetrating well it follows that for the same drawdown s_w, C is the ratio between Q_w of a partially penetrating well and that of a fully penetrating one. The correction C was also given by Wenzel (1942).

Based on Muskat's results, Jacob (1945 and 1963) presents correction (with respect to a fully penetrating well) curves for the drawdowns $s(r,0)$ and $s(r,t)$ along the top and the bottom of an aquifer, respectively (again for $b_2 = 0$ or $b_1 = 0$ in Fig. 8-18a). These curves give the deviation Δs (expressed as $2\pi T\Delta s/Q_w$) of the drawdown from the purely logarithmic distribution resulting from a fully penetrating well

$$\frac{\Delta s(r,B)}{\Delta s(r,0)} = \frac{Q_w}{\pi b_s} \sum_{n=1}^{\infty} \frac{(+1)^n}{(-1)^n} K_0\left(\frac{n\pi r}{B} \right) \sin\left(\frac{n\pi b_s}{B} \right) \quad (8\text{-}123)$$

where K_0 is the Modified Bessel function of the second kind of order zero. For the analysis of a pumping test Jacob (1963) presents a method of successive approximations based on the curves for Δs.

Kirkham (1959) derived formulas describing steady flow towards a partially penetrating well by exactly solving the problem shown in Fig. 8-18c. His boundary condition at the well was: $\lim\limits_{r \to 0} \int_{z=b_1}^{b_1+b_s} r(\partial\phi/\partial r)\,dz = Q_w/2\pi K$.

(c) Unsteady flow to a partially penetrating well Hantush (1957, 1961, 1961a, 1964) derived solutions and developed pumping test techniques for unsteady flow towards partially penetrating wells in confined and in leaky aquifers, neglecting storativity in the partial differential equation for three-dimensional flow in a confined aquifer. Introducing the leakage as a boundary condition at $z = B$, he suggests that the leakage actually crossing the boundary at $z = B$ be hypothetically assumed to be generated within the main aquifer whose confining beds are considered as completely impervious. In this way, a solution amenable to relatively easy calculation, yet sufficiently accurate for practical purposes, is obtained. The rate of hypothetically generated leakage per unit volume at each point in the aquifer is taken as $(K'/B')\,s(r, t)/B$, that is, $q_v(r)$ of (8-37) divided by B. Hantush claims that this idealization of the flow system does not materially affect the flow pattern of the actual flow. By introducing this source function, the continuity equation to be solved becomes

$$\frac{\partial^2 s}{\partial r^2} + \frac{1}{r}\frac{\partial s}{\partial r} + \frac{\partial^2 s}{\partial z^2} - \frac{s}{\lambda^2} = \frac{S}{T}\frac{\partial s}{\partial t} \tag{8-124}$$

The boundary and initial conditions are (Fig. 8-18a)

$$s(r, z, 0) = s(\infty, z, t) = 0; \qquad \partial s(r, 0, t)/\partial z = \partial s(r, B, t)/\partial z = 0$$

$$\lim_{r \to 0} b_s r \partial s/\partial r = \begin{cases} 0 & \text{for} \quad 0 < z < b_1 \\ -Q_w/2\pi K & \text{for} \quad b_1 < z < b_1 + b_s \\ 0 & \text{for} \quad b_1 + b_s < z < B \end{cases} \tag{8-125}$$

Some of the solutions presented by Hantush for a top penetration $b_2 = 0$, are

(i) Nonsteady flow in a leaky aquifer

$$s = \frac{Q_w}{4\pi T}\left[W(u, r/\lambda) + \frac{2B}{\pi b_s}\sum_{n=1}^{\infty}\frac{1}{n}\cos\frac{n\pi z}{B}\sin\frac{n\pi b_s}{B} W_n\left(u, \left\{\left(\frac{r}{\lambda}\right)^2 + \left(\frac{n\pi r}{B}\right)^2\right\}^{1/2}\right) \right]$$

$$W_n(u, x) = \int_u^\infty \exp\left(-y - \frac{x^2}{4y}\right)\frac{dy}{y} \tag{8-126}$$

(ii) Steady flow in a leaky aquifer

$$s = \frac{Q_w}{2\pi T}\left[K_0(r/\lambda) + \frac{2B}{\pi b_s}\sum_{n=1}^{\infty}\frac{1}{n}\cos\frac{n\pi z}{B}\sin\frac{n\pi b_s}{B} K_0\left[\left\{\left(\frac{r}{\lambda}\right)^2 + \left(\frac{n\pi r}{B}\right)^2\right\}^{1/2}\right]\right]$$

$$\tag{8-127}$$

(iii) Nonsteady flow in a confined aquifer

$$s = \frac{Q_w}{4\pi T}\left[W(u) + \frac{2B}{\pi b_s}\sum_{n=1}^{\infty}\frac{1}{n}\cos\frac{n\pi z}{B}\sin\frac{n\pi b_s}{B}W_n\left(u, \frac{n\pi r}{B}\right)\right] \quad (8\text{-}128)$$

A detailed analysis of the various cases is given by Hantush (1964). The most important conclusion reached by him is that for all practical purposes, for $r \geq 1.5B$, we may employ all the results derived for flow toward a fully penetrating well. For an anisotropic aquifer, this range is given by $r \geq 1.5B\sqrt{K_z/K_r}$. Moreover, in the latter case the flow will be as if the aquifer were isotropic with a conductivity equal to K_r.

8-9 MULTIPLE WELL SYSTEMS

When wells are spaced at distances smaller than their radius of influence R (Sec. 8-2), they affect each other's drawdown and discharge rate. As the equations for flow in a confined aquifer and a semiconfined one (with the assumption of horizontal flow in the latter) are linear in $\phi(x, y)$, superposition (Sec. 5-7) is applicable.

In steady flow in a phreatic aquifer, employing the Dupuit assumptions, the partial differential equation (5-76) with $\partial h/\partial t = 0$ is *linear* in h^2, hence superposition is applicable to h^2.

In unsteady flow in phreatic aquifers, (5-76) is *non-linear* in h and therefore superposition is not applicable. However, if we use the linearized form, (5-81), which is linear in h, or (5-82) which is linear in h^2, we may apply superposition. One should note that (5-81) is practically the same as (5-60), with h and ϕ measured from the same datum level.

All the linear (or linearized) aquifer equations, can be rewritten in terms of the drawdown $s = \phi_0 - \phi(t)$, or $s = h_0 - h(t)$, where $\phi_0 = \text{const.}$, or $h_0 = \text{const.}$, represents initial conditions of a horizontal water table. If initially $\phi_0 = f(x, y) \neq \text{const.}$, we have to decompose the problem according to one of the cases described in Sec. 5-7. We have also to watch for the conditions along the boundary of the flow domain. Superposition of drawdowns is directly applicable when we have homogeneous boundary conditions (e.g., $\phi = 0$, $s = 0$, $\partial\phi/\partial n = 0$). Otherwise, we have first to use decomposition as described in Sec. 5-7 in order to obtain homogeneous boundary conditions, and then apply superposition.

Steady Flow

Consider steady flow in an infinite aquifer (or in a bounded one, with homogeneous boundary and initial conditions) in which N wells are operating at constant pumping (or recharge) rates. Let Q_j denote the pumping rate ($Q_j < 0$ denotes recharge) at point (x_j, y_j), s_i be the total drawdown at an observation well located at (x_i, y_i) and s_{ij} be the drawdown caused at observation point (x_i, y_i) by the pumping well at (x_j, y_j), as if that well was operating alone in the considered field. Then

$$s_i = \sum_{j=1}^{N} s_{ij} \quad (8\text{-}129)$$

In a confined or a leaky aquifer (Secs 8-2 and 8-4), the drawdown s_i is given by an expression which has the general form $s_i = (Q_j/2\pi T) F(x_i, y_i; x_j, y_j)$, or: $s_i = (Q_j/2\pi T) F(r_{ij})$, where $r_{ij}^2 = (x_i - x_j)^2 + (y_i - y_j)^2$; the function F depends on the type of aquifer, on its parameters and on the shape of its boundaries. Hence, (8-129) can be written as

$$s_i = \sum_{j=1}^{N} s_{ij} = \sum_{j=1}^{N} (Q_j/2\pi T) F(r_{ij}) \tag{8-130}$$

For example, for steady flow in an infinite confined aquifer, $F(r_{ij}) = \ln(R_j/r_{ij})$, and (8-130) becomes

$$s_i = \sum_{j=1}^{N} (Q_j/2\pi T) \ln(R_j/r_{ij}) \tag{8-131}$$

assuming a different R_j for every pumping well.

For the special case $Q_1 = Q_2 = \ldots = Q = $ const. and $R_1 = R_2 = \ldots = R = $ const., (8-131) becomes

$$s_i = \sum_{j=1}^{N} (Q/2\pi T) \ln(R/r_{ij}) = \frac{NQ}{2\pi T} \ln \frac{R}{r_i^*} \tag{8-132}$$

where $(r_i^*)^N = r_{i1}r_{i2}r_{i3} \ldots r_{iN}$. This means that the drawdown s_i is equal to that produced by a single well pumping at a rate NQ located at an equivalent distance r_i^* from (x_j, y_j).

In a similar way, for steady flow to wells in a leaky aquifer, we have by (8-41), or (8-42)

$$s_i = \sum_{j=1}^{N} \frac{Q_j}{2\pi T} K_0\left(\frac{r_{ij}}{\lambda}\right); \qquad s_i = \sum_{j=1}^{N} \frac{Q_j}{2\pi T} \ln \frac{1.123\,\lambda}{r_{ij}} \tag{8-133}$$

Rewriting (8-131) for $i = 1, 2, \ldots, M$, we obtain a set of linear equations relating the Q_j's to s_i's. Where observations are made at the $M = N$ pumping wells, and $r_{ii} = r_{jj} = (r_w)_j$, given N values of Q_j we may solve for the values of s_i. Or, given the values of s_i (say, in a planning problem) we may solve for the corresponding Q_j's.

In the latter case so far, we have assumed that the radii of the wells are very small compared to distances between wells, so that the face of each well may be considered as practically an equipotential boundary (whereas, in principle, the influence of neighboring wells is to produce unequal drawdown along the face of each well). Recall that boundary condition at the wells, in most cases, is one of constant discharge, Q_w.

The Dutch Committee for Hydrological Research T.N.O. (Hydrologisch Colloquium, 1964) suggests the following formula when the effect of neighboring wells cannot be neglected

$$\phi(P) = [1 - \phi_2(P_1)/\phi_1^w]\phi_1(P) - [1 - \phi_1(P_2)/\phi_2^w]\phi_2(P) \tag{8-134}$$

where ϕ_1^w and ϕ_2^w are the potentials, respectively, on the faces of two wells at points P_1 and P_2; $\phi_1(P), \phi_2(P), \phi(P)$ are the potentials, respectively, at an observation well at P when only the well at P_1 is present, when only the well at P_2 is present and when both wells are operating, and we assume $\phi_1(P_2) \ll \phi_2^w$ and $\phi_2(P_1) \ll \phi_1^w$. The same procedure can be applied to a larger number of wells.

Muskat (1937) discusses several arrangements of wells in a confined aquifer, and determines in each case the drawdown at the various wells, using (8-131) with $R_j = R = \text{const.}$ For example: for two wells of equal drawdown $s_1 = s_2 = s_w$ at a distance L apart, we have

$$Q_1 = Q_2 = \frac{2\pi T s_w}{\ln(R^2/r_w L)} \tag{8-135}$$

The total discharge $(Q_1 + Q_2)$ is that of a single well of radius $(r_w L)^{1/2} \gg r_w$. This shows that a multiple well system is more efficient than a single large well having the same total discharge.

For three wells of equal drawdown forming an equilateral triangle of side L

$$Q_1 = Q_2 = Q_3 = \frac{2\pi T s_w}{\ln(R^3/r_w L^2)} \tag{8-136}$$

In order that a single well of discharge $(Q_1 + Q_2 + Q_3)$ will give the same drawdown s_w, its radius has to be $(r_w L^2)^{1/3}$.

For an infinite *array of wells* at $P_k(ka, 0)$, $k = \ldots, -2, -1, 0, 1, 2, \ldots$ in a confined aquifer in the xy plane, with $Q_k = \text{const.} = Q_w$, and $\phi(x, \pm R) = \text{const.} = H$, R being an equivalent distance of influence, Muskat (1937) gives

$$s(x, y) \equiv H - \phi(x, y) = \frac{Q_w}{4\pi T} \ln \frac{\cosh 2\pi(y - R)/a - \cos 2\pi x/a}{\cosh 2\pi(y + R)/a - \cos 2\pi x/a} \tag{8-137}$$

Figure 8-20 shows streamlines and equipotentials. At a distance of the order of the mutual spacing, $y > a$, the equipotentials become parallel to the array, as if the latter had been replaced by a continuous line sink.

For a line of three equally spaced wells a distance L apart, all having the same drawdown s_w, the outer wells discharge at

$$Q_1 = Q_3 = \frac{2\pi T s_w \ln(L/r_w)}{2 \ln(R/L) \ln(L/R) + \ln(R^2/2r_w L) \ln R/r_w} \tag{8-138}$$

while the middle well discharges at

$$Q_2 = \frac{2\pi T s_w \ln(L/2r_w)}{2 \ln(R/L) \ln(L/R) + \ln(R^2/2r_w L) \ln(R/r_w)} \tag{8-139}$$

Figure 8-21 shows the individual and composite drawdown curves for the three wells for $Q_1 = Q_2 = Q_3, s_1 = s_3 \neq s_2$.

The discharge of each of four wells forming a square of side L, all having the same drawdown s_w, is

$$Q_1 = Q_2 = Q_3 = Q_4 = \frac{2\pi T s_w}{\ln(R^4/\sqrt{2} r_w L^3)} \tag{8-140}$$

When N wells are pumping in a phreatic aquifer with a horizontal bottom, and the Dupuit approximation is used to determine the drawdown (steady flow!) in an

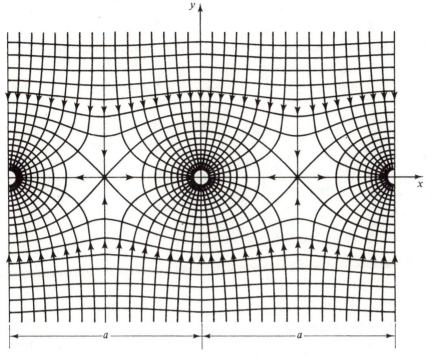

Figure 8-20 Streamlines and equipotentials about an infinite array of wells.

observation well, the principle of superposition (here with respect to h^2) leads to

$$H_0^2 - h_i^2 = \sum_{j=1}^{N} \frac{Q_j}{\pi K} \ln \frac{R_j}{r_{ij}} \qquad (8\text{-}141)$$

where $H_0 = $ const. is the initial (undisturbed) height of the water table above the impervious bottom, h_i is the height of the water table above the impervious bottom at the observation well (x_i, y_i), and the R_j's are the radii of influence of the pumping wells, (assuming that they are sufficiently large so that drawdown is produced at the observation well). When all R_j's are the same and all Q_j's are equal to Q/N, we obtain from (8-141)

$$H^2 - h_i^2 = \frac{Q}{\pi K} \ln(R/r^*); \qquad r^* = (r_{i1} r_{i2} r_{i3} \dots r_{iN})^{1/N} \qquad (8\text{-}142)$$

One should note here that because we have initially the nonhomogeneous conditions $h = H_0$ (and not $h = 0$), the superposition is actually not with respect to h^2. Instead, because initially $H_0^2 - h^2 = 0$ everywhere, the superposition is with respect to the difference $H_0^2 - h^2$. In a similar way, if in a confined aquifer we have initially $\phi_0 \neq 0$, the superposition is with respect to $\phi_0 - \phi$ (i.e., with respect to s!).

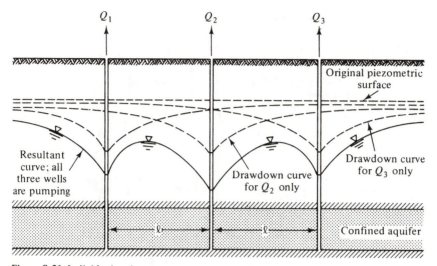

Figure 8-21 Individual and composite drawdown curves for three wells on a line.

Unsteady Flow

The method of superposition is applicable also to unsteady flow.

When N wells of discharge Q_j at points (x_j, y_j) are steadily discharging from a confined, or a semiconfined, aquifer, the drawdown s_i at an observation well (x_i, y_i) is given by

$$s_i = \sum_{j=1}^{N} Q_j Z_{ij}(r_{ij}, t_j) \qquad (8\text{-}143)$$

where $Q_j Z_{ij}$ is the drawdown produced by a single pumping well located at a distance r_{ij} from the observation well, and t_j is the time from the commencement of pumping in the jth well until the time of observation. When observations are made in a pumping well, we have $r_{ii} = r_{jj} = (r_w)_j$. For a confined aquifer $Z = (1/4\pi T) W(u)$. When all observations are only in pumping wells, (8-143) is a set of N linear equations in the $2N$ variables s_j and Q_j. Given N of these values, one can determine the remaining N.

Consider the following cases.

(a) For N wells starting to pump simultaneously in a homogeneous isotropic confined aquifer of infinite areal extent, the total drawdown at point $P_i(x_i, y_i)$ is

$$s(P_i, t) = (1/4\pi T) \sum_{j=1}^{N} Q_j W(u_j); \qquad u_j = S r_{ij}^2 / 4Tt \qquad (8\text{-}144)$$

(b) For wells of variable discharge, using (8-68), we obtain

$$s(P_i, t) = (1/4\pi T) \sum_{j=1}^{N} \sum_{k=1}^{n} [Q_j^{(k)} - Q_j^{(k-1)}] W(r_{ij}, t - t_{k-1}); \qquad t_{n-1} < t \le t_n$$

$$(8\text{-}145)$$

(c) For two wells a distance $L \gg r_w$ apart in an infinite confined aquifer, with the same drawdown s_w at both wells, we obtain from (8-144)

$$Q_1 = Q_2 = 4\pi T s_w / [W(Sr_w^2/4Tt) + W(SL^2/4Tt)] \qquad (8\text{-}146)$$

(d) For three wells forming an equilateral triangle, we obtain

$$Q_1 = Q_2 = Q_3 = 4\pi T s_w / [W(Sr_w^2/4Tt) + 2W(SL^2/4Tt)] \qquad (8\text{-}147)$$

If in the above examples the time is such that $SL^2/4Tt$ is sufficiently small (say, less than 0.01), then (8-62) may be used. Equation (8-146) then becomes

$$Q_1 = Q_2 = 2\pi T s_w / \ln(R_t^2/Lr_w); \qquad R_t = 1.5\sqrt{Tt/S} \qquad (8\text{-}148)$$

and (8-147) becomes

$$Q_1 = Q_2 = Q_3 = 2\pi T s_w / \ln(R_t^3/L^2 r_w) \qquad (8\text{-}149)$$

Sternberg and Scott (1967) discuss some problems related to interference among wells.

Sometimes a large number of wells is concentrated in a certain region and we wish to determine the overall effect of their pumping on the water table. Often we do not know the exact pumping rate of the individual wells.

As an example, consider a certain portion of an aquifer of area A which contains a large number of pumping wells such that we may assume that the rate of pumping is more or less uniformly distributed over A. If the total pumping rate is $\sum Q$, the problem may be simplified by considering the influence of a withdrawal per unit area of $(1/A)\sum Q$, rather than the effect of each individual well in A separately. For flow regimes where superposition is permissible, the drawdown $s(P, t)$ at an observation point P at time t is given by

$$s(P, t) = \frac{1}{4\pi T} \int_{(A)} \frac{\sum Q}{A} Z(P, P', t)\, dA(P') \qquad (8\text{-}150)$$

where point P' denotes the location of an elementary "well" discharging at a rate $(\sum Q/A)\, dA$, and Z is the well function corresponding to the type of aquifer under consideration.

When A has the form a circle of radius R centered at the origin, P has the polar coordinates (r, θ) and P' has the polar coordinates (r', θ') we obtain:

$$s(r, t) = \frac{1}{4\pi T} \int_{(A)} \frac{\sum Q}{\pi R^2} Z(r, \theta; r', \theta'; t)\, r'\, d\theta'\, dr' \qquad (8\text{-}151)$$

For a confined aquifer, Z becomes the well function $W(u_\rho)$, where $u_\rho = S\rho^2/4Tt$, $\rho^2 = r^2 + r'^2 - 2rr'\cos(\theta - \theta')$. For this case, Hantush (1964) gives for $t > 0.4r^2 S/T$ and $r < R$

$$s(r, t) = \frac{\sum Q}{4\pi T} \left\{ W(u_R) + (1/u_R)[1 - \exp(-u_R)] - \left(\frac{r}{R}\right)^2 \exp(-u_R) \right\} \qquad (8\text{-}152)$$

while for $t > 0.4R^2S/T$ and $r > R$ he gives

$$s(r, t) = \frac{\sum Q}{4\pi T}\{W(u) + (0.5 \, u_R)\exp(-u)\} \tag{8-153}$$

where $u = Sr^2/4Tt$, $u_R = SR^2/4Tt$.

If the aquifer is a phreatic one, but drawdowns are relatively small so that the Dupuit assumptions are applicable, we have to replace $4\pi Ts$ in (8-152) and (8-153) by $2\pi K(H^2 - h^2)$.

8-10 WELLS NEAR BOUNDARIES TREATED BY THE METHOD OF IMAGES

So far we have considered, and derived solutions, in this section only for a single well in an aquifer of infinite areal extent, or at the center of a circular aquifer, on the external boundary of which homogeneous boundary conditions are specified. However, very extensive aquifers are rarely encountered in nature. In general, discontinuities in geological formations (e.g., caused by faults, streams, lakes, or seas) bound every aquifer. Such boundaries are of interest whenever they are located within the zone of influence of considered wells. Hence, we need solutions for wells in bounded domains.

Figure 8-22 shows examples of aquifer boundaries resulting from geological discontinuities or from the presence of streams (influent or effluent) intersecting aquifers. The more common types of boundaries are the impervious boundary and the equipotential boundary. We may also have as boundary an interface separating regions of different aquifer properties (e.g., aquifer transmissivity).

Often the boundaries mentioned above may be regarded as abrupt discontinuities. This greatly simplifies the mathematical treatment of determining the potential distribution in bounded regions. This approximation is especially justified when wells are located at a sufficient distance from a considered boundary. Under certain conditions curved boundaries may be approximated by planes or straight lines.

The *method of images* (e.g., Bear, 1972, p. 304) is most useful in the treatment of wells near *straight line boundaries* of the two types mentioned above. According to this method, the real bounded field of flow is replaced by a *fictitious* field with simpler boundary conditions, but, in general, of a larger area. Image wells are located in the image domain such that the sought flow pattern produced in the (bounded) real flow domain by the real wells is the same as that produced in the same domain, now as part of the expanded real + image domains, by the ensemble of real and *image* wells. We refer to the expanded domain as the *fictitious* one. The identity of flow patterns means that although a boundary is removed and does not appear in the fictitious domain, the flow pattern has still to satisfy the condition imposed along that line, or line segment. In addition to satisfying boundary conditions, the ensemble of real and imaginary wells should produce a potential distribution such that the appropriate continuity equation will be satisfied in the real domain.

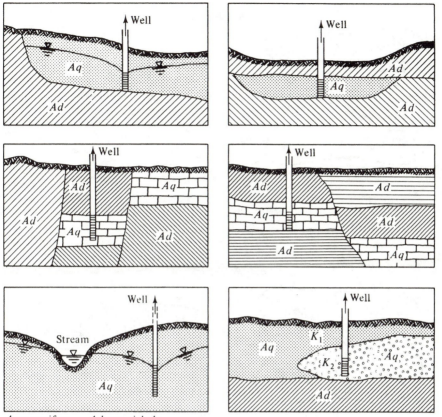

Aq = aquifer; Ad = aquiclude

Figure 8-22 Boundaries resulting from geological discontinuities or from streams.

The principles of the method of images are summarized in Fig. 8-23. The location of a well which acts as an image well to a real one is a mirror image, with respect to the straight line boundary, of the real well. In Fig. 8-23a, for a constant potential boundary, we see that the image of a pumping well, $+Q_w$ is a recharge well, $-Q_w$ having the same strength. For an impervious boundary (Fig. 8-23b), the image well has the same sign. Several examples are given below. In examples 1 and 2, a physical interpretation is given of the effect of the image wells rather than a mathematical justification (Bear, 1972, p. 304). In all cases, unless otherwise specified, pumping wells are (mathematical) point sinks and recharge wells are (mathematical) point sources. In some cases, the radius of the well, r_w, is taken into account. We shall consider only flow to a fully penetrating well in a homogeneous isotropic aquifer of constant thickness. As we are not interested in the piezometric head $\phi\,(= -\infty$ at a point sink), but in the drawdown $s\,(= \phi_0 - \phi = +\infty$ at a point sink), the constant discharge rate of a pumping well will be taken as $+Q_w$, whereas the constant recharge rate of a recharge well will be $-Q_w$.

Example 1 Steady flow to a single well located at point $(x_0, 0)$ in a semi-infinite aquifer, $x > 0$ (Fig. 8-24). The line $x = 0$ is a boundary of zero draw-

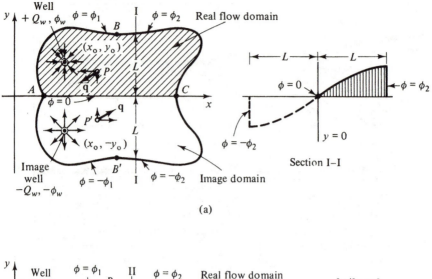

(a)

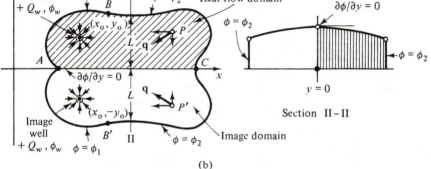

(b)

Figure 8-23 The method of images. (a) Image domain for $\phi = 0$ along $y = 0$. (b) Image domain for $\partial\phi/\partial y = 0$ along $y = 0$.

down ($s = 0$; equipotential boundary). It corresponds to an unclogged stream, with a practically horizontal water table, in contact with the aquifer.

The fictitious field here is the infinite field, $-\infty < x < +\infty$, with two wells: the real pumping well $+Q_w$ at $(x_0, 0)$ and a recharge (negative) well $-Q_w$ at $(-x_0, 0)$. In other words, the real semi-infinite field with a single well is replaced by a fictitious infinite field with two wells. Because of symmetry, at every point along $x = 0$ ($r = r'$) we have $s = 0$. Thus the boundary condition along the stream is satisfied by the combined drawdown and build-up of the two wells in the infinite field.

By superposition (Sec. 5-7), the drawdown s at any observation point (x, y) in the real region, $x > 0$, is given by the sum of drawdowns (actually drawdown plus build-up) of the two wells, each operating in the fictitious infinite field. The single well equation for this case is (8-6). The combined

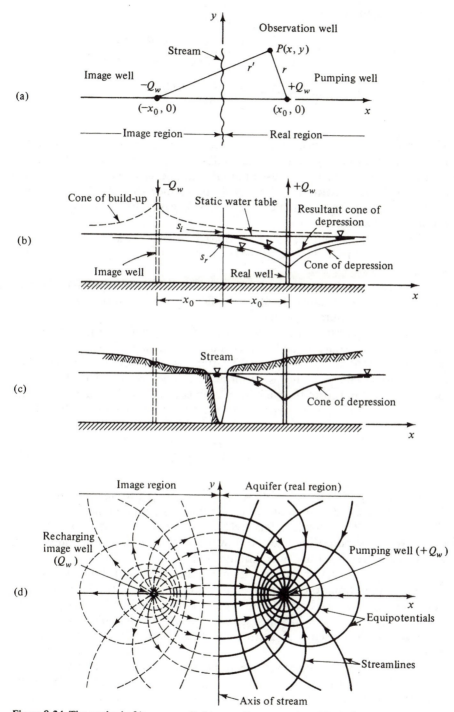

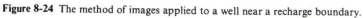

Figure 8-24 The method of images applied to a well near a recharge boundary.

drawdown is, therefore

$$s(x, y) = \phi_0 - \phi(r) = \frac{Q_w}{2\pi T} \ln \frac{R}{r} + \frac{(-Q_w)}{2\pi T} \ln \frac{R}{r'} = \frac{Q_w}{2\pi T} \ln \frac{r'}{r}$$

$$= \frac{Q_w}{4\pi T} \ln \frac{(x + x_0)^2 + y^2}{(x - x_0)^2 + y^2} \tag{8-154}$$

where r and r' are distances from $P(x, y)$ to the real pumping well and to the image well, respectively, and R is the radius of influence of each well when it operates in an infinite aquifer (assuming $R > x_0$; otherwise the boundary will not affect the flow pattern). It can easily be verified that (8-154) (i) satisfies the Laplace equation (8-1) in the domain $x > 0$, (ii) satisfies the boundary conditions $s = 0$ along $x = 0$.

Lines of equal drawdown ($s = $ const.) are the circles $(r'/r)^2 = C = \exp(4\pi Ts/Q_w)$ of radii $2x_0 \sqrt{C}/(C - 1)$ and centers at $x = x_0(C + 1)/(C - 1)$, $y = 0$.

It is of interest to note that from (8-154) it follows that as $r, r' \to \infty$, $\phi(r) \to 0$. In this case, unlike (8-6), steady flow is possible as the stream along $x = 0$ serves as a source of water.

The above considerations are exact for wells which are mathematical point sinks and sources, and approximate for actual wells with a finite radius r_w, as long as $r_w \ll x_0$. In the latter case, theoretically, the circular pumping well's circumference is no more an equipotential surface. However, for all practical purposes the well's face is an equipotential surface. Consider a point $y = 0$, $x = r_w + x_0$, with $r_w \ll x_0$. Then from (8-154) it follows that

$$s(r_w) \equiv s_w \approx \frac{Q_w}{2\pi T} \ln \frac{2x_0}{r_w} \tag{8-155}$$

Comparing (8-155) with (8-6) with $s_w = \phi_0 - \phi_w$, we may interpret the distance $2x_0$ as a radius of influence of a well near an equipotential boundary. The flownet in the region $x > 0$ is given by the right-hand side of Fig. 8-24d, obtained as a superposition of the flownets corresponding to a source and a sink each operating alone in an infinite field.

Although this example deals with a confined aquifer, (8-154) and (8-155), are good approximations also for a phreatic aquifer with relatively small drawdowns ($s_w \ll H$). It is with this approximation in mind that we demonstrate this case by Fig. 8-24 which actually shows flow in a phreatic aquifer; the effect of the presence of a stream is better understood in this type of aquifer.

Finally, as the stream along $x = 0$ is the only source of water in the field $x > 0$, it is interesting to determine the portion of the pumped water taken from the stream between two points, say at $y = -d$ and $y = +d$. Using (8-154) for $\phi(r)$, we obtain

$$\frac{Q}{Q_w} = \frac{1}{Q_w} \int_{-d}^{+d} \left(-T \frac{\partial \phi}{\partial x} \bigg|_{x=0} \right) dy = \frac{T}{Q_w} \int_{-d}^{+d} \left(\frac{Q_w}{4\pi T} \cdot \frac{4x_0}{x_0^2 + y^2} \right) dy$$

$$= \frac{2}{\pi} \arctan \frac{d}{x_0} \qquad (8\text{-}156)$$

Example 2 Let a well be located at point $(x_0, 0)$ in a semi-infinite aquifer $x > 0$ (Fig. 8-25). The line $x = 0$ is an impervious boundary, and the condition along it is $\partial \phi / \partial n \equiv \partial \phi / \partial x = 0$, where n is in the direction of the normal to the boundary. In this case the image well has to be located at $(-x_0, 0)$; it is a pumping well discharging at a constant rate $+Q_w$. By symmetry, along the boundary $x = 0$ we have $\partial \phi / \partial n \equiv \partial \phi / \partial x = 0$. When the specific discharge vectors at points along this boundary are summed vectorially, they give a resultant which has a component tangent to the boundary, while the component normal to the boundary vanishes. Thus a single well in a semi-infinite aquifer is replaced by two wells in an infinite one. For steady flow, using (8-3), we have

$$\phi(r) - \phi_w = \frac{Q_w}{2\pi T} \ln \frac{r}{r_w} + \frac{Q_w}{2\pi T} \ln \frac{r'}{r_w} = \frac{Q_w}{2\pi T} \ln \frac{rr'}{r_w^2} = \frac{Q_w}{\pi T} \ln \frac{(rr')^{1/2}}{r_w}$$

$$(8\text{-}157)$$

Figure 8-25 (with the comments on a phreatic aquifer and a confined one of Ex. 1 shows this situation. We could also use (8-6).

At a large distance from the pumping well, $r' \approx r$ and (8-157) becomes

$$\phi(r) - \phi_w = \frac{Q_w}{\pi T} \ln \frac{r}{r_w} ; \qquad \phi(r) = \phi_w + \frac{Q_w}{\pi T} \ln \frac{r}{r_w} \qquad (8\text{-}158)$$

that is, twice the drawdown produced by a single well pumping from an infinite aquifer. As in the case of a single well in an infinite aquifer, steady flow is not possible.

The method of images is also applicable to fields bounded by two straight lines at an angle less than 180°, to rectangular closed domains, to infinite and semi-infinite strips, etc. Sometimes additional image wells are required in order to balance the drawdowns or the normal fluxes along the continuation of the boundaries and thus satisfy the appropriate boundary conditions on them. Sometimes, an infinite number of image wells is needed to achieve this purpose. Several additional examples of wells near various boundaries are given in Figs 8-26 and 8-27 (Bear, 1972, p. 309).

In each case a system of image wells is introduced such that the specified conditions are satisfied along the given boundaries. The drawdown in the region of interest is obtained by summing the effect of *all* wells (both image and real), as if each of them was operating alone in an infinite field.

For example, for a well in an infinite quadrant bounded by a stream

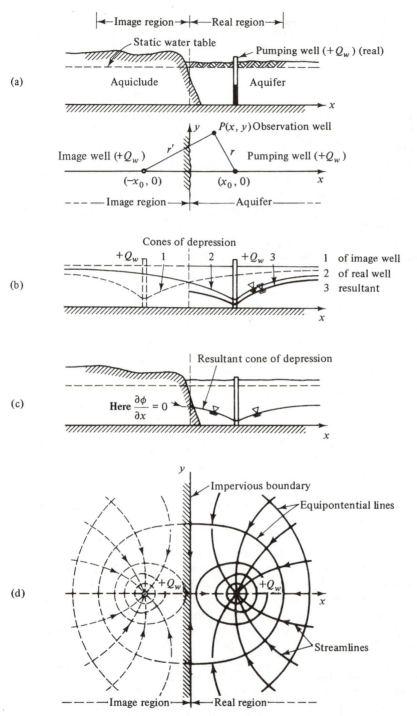

Figure 8-25 The method of images applied to a well near an impervious boundary.

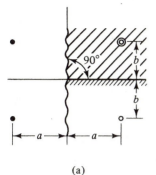
(a)

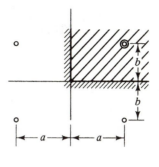
(b)

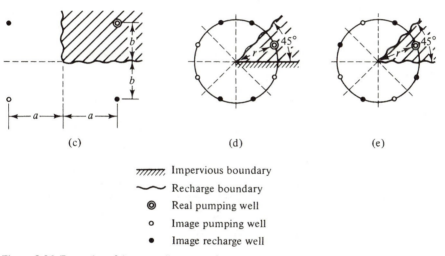

(c)

(d)

(e)

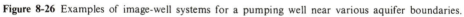

/////	Impervious boundary
⌒	Recharge boundary
◎	Real pumping well
○	Image pumping well
●	Image recharge well

Figure 8-26 Examples of image-well systems for a pumping well near various aquifer boundaries.

(Fig. 8-26c), we obtain by using (8-6)

$$s = \frac{Q_w}{4\pi T} \ln \frac{[(x-a)^2 + (y+b)^2][(x+a)^2 + (y-b)^2]}{[(x-a)^2 + (y-b)^2][(x+a)^2 + (y+b)^2]} \quad (8\text{-}159)$$

For a well in an infinite strip between two streams (Fig. 8-27b) we obtain

$$s = \frac{Q_w}{4\pi T} \sum_{n=-\infty}^{\infty} \ln \frac{(x-x_0-2nd)^2 + y^2}{(x+x_0-2nd)^2 + y^2} \quad (8\text{-}160)$$

which can be shown to be equal to

$$\frac{Q_w}{2\pi T} \ln \frac{1 - 2\exp\{-\pi|y|/b\} \cdot \cos\{\pi(x-x_0)/b\} + \exp\{-2\pi|y|/b\}}{1 - 2\exp\{-\pi|y|/b\} \cdot \cos\{\pi(x+x_0)/b\} + \exp\{-2\pi|y|/b\}} \quad (8\text{-}161)$$

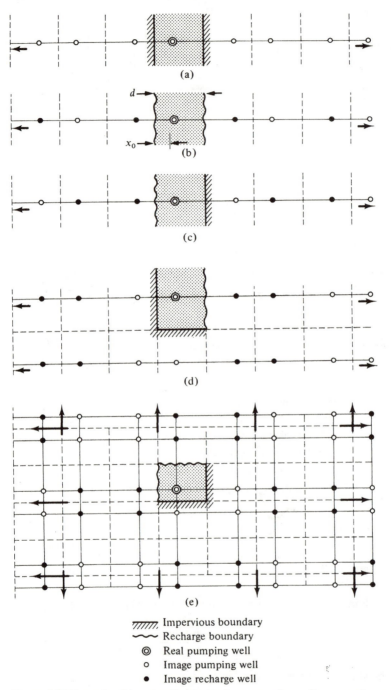

Figure 8-27 Example of image-well systems for a pumping well near various aquifer boundaries (arrows indicate that image-well system continues to infinity). (a) Infinite strip. (b) Infinite strip. (c) Infinite strip. (d) Semi-infinite strip. (e) Rectangle.

On the face of the pumping well, we have approximately

$$s_w = \frac{Q_w}{2\pi T} \ln \frac{\pi r_w}{2b \sin(\pi x_0/b)} \tag{8-162}$$

Details of other examples are worked out in Hydrologisch Colloquium (1964).

The method of images is also applicable to unsteady flow. In each case, the appropriate equation for the drawdown caused by a single well in an infinite aquifer is used. For example, for a well pumping at a constant rate Q_w near a stream ($=$ boundary of zero drawdown) in a confined aquifer (Fig. 8-24), Hantush (1959a) gives

$$s = \frac{Q_w}{4\pi T} \left[W\left(\frac{Sr^2}{4Tt}\right) - W\left(\frac{Sr'^2}{4Tt}\right) \right] = \frac{Q_w}{4\pi T} [W(u) - W(u')]$$

$$= \frac{Q_w}{4\pi T} M(u, \beta) \tag{8-163}$$

where $M(u, \beta)$ is a tabulated function, $\beta = r'/r$, r' is the distance of observation point to image well, $u = Sr^2/4Tt$, $u' = Sr'^2/4Tt$. For $\beta^2 u < 0.05$, we obtain

$$s \approx \frac{Q_w}{4\pi T} [\ln \beta^2 - u(\beta^2 - 1)] \tag{8-164}$$

Also, for $u, u' < 0.01$

$$s \approx \frac{Q_w}{4\pi T} \left[\ln \frac{2.25\, Tt}{Sr^2} - \ln \frac{2.25\, Tt}{Sr'^2} \right] = \frac{Q_w}{2\pi T} \ln \frac{r'}{r} \tag{8-165}$$

which describe the ultimate steady state.

Hantush (1965) also solves the case of unsteady flow to a well near a river with a clogged bed.

The method is applicable also to any type of aquifer as long as we employ the drawdown formula corresponding to a single well operating in an infinite aquifer of that type.

A boundary of special interest is that separating two domains of different aquifer transmissivity.

Consider an infinite confined aquifer in the xy plane, with a discontinuity in transmissivity: for $x > 0$, $T = T'$ and for $x < 0$, $T = T''$ (Fig. 8-28; Bear, 1972, p. 311). A well at $(x_0, 0)$ is pumping at a constant rate Q_w.

In order to determine the drawdown, s, everywhere in the xy plane, we have to determine the two head distributions ϕ' for $x > 0$ and ϕ'' for $x < 0$, satisfying the partial differential equations

$$\nabla^2 \phi' = 0 \quad \text{for } x > 0 \quad \text{and} \quad \nabla^2 \phi'' = 0 \quad \text{for } x < 0$$

and the boundary conditions on $x = 0$

$$\phi' = \phi'' \quad \text{and} \quad T'\partial\phi'/\partial x = T''\partial\phi''/\partial x$$

One way to solve this problem is to use the method of images and

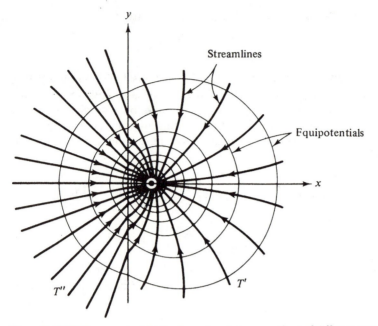

Figure 8-28 Well near a straight line boundary between regions of different transmissivities.

consider an image well at $(-x_0, 0)$ of a constant rate CQ_w, where C is yet an unknown constant. Using superposition, the drawdown produced at every point is

$$s' = \phi_0 - \phi' = \frac{Q_w}{4\pi T'} \ln \frac{R^2}{(x - x_0)^2 + y^2} + \frac{CQ_w}{4\pi T'} \ln \frac{R^2}{(x + x_0)^2 + y^2};$$
$$x > 0 \qquad (8\text{-}166)$$

$$s'' = \phi_0 - \phi'' = \frac{BQ_w}{4\pi T''} \ln \frac{R^2}{(x - x_0)^2 + y^2}; \qquad x < 0 \qquad (8\text{-}167)$$

where B is another, yet unknown, constant, R is the radius of influence of the wells, and ϕ_0 is the undisturbed potential. Both B and C are determined by inserting (8-166) and (8-167) into the boundary conditions. This yields

$$B = 2N(N + 1); \qquad C = (1 - N)/(1 + N)$$

where $N = T''/T'$. Hence, the drawdown in the two regions is given by

$$s' = \phi_0 - \phi' = \frac{Q_w}{4\pi T'} \ln \frac{R^2}{(x - x_0)^2 + y^2}$$

$$+ \frac{(1 - N)}{(1 + N)} \cdot \frac{Q_w}{4\pi T'} \ln \frac{R^2}{(x + x_0)^2 + y^2}; \qquad x > 0$$

$$s'' = \phi_0 - \phi'' = \frac{2N}{(N + 1)} \frac{Q_w}{4\pi T''} \ln \frac{R^2}{(x - x_0)^2 + y^2}; \qquad x < 0 \qquad (8\text{-}168)$$

In the region $x < 0$, the effect of the discontinuity is equivalent to replacing the transmissivity T'' by the arithmetic mean $(T'' + T')/2$: the flownet itself is unchanged. In the region $x > 0$ the flownet itself is changed (Fig. 8-28): streamlines are now curved due to the presence of the image well.

The special case of flow in the $x > 0$ region near an impervious boundary at $x = 0$ may be obtained from (8-168) by setting in it $T'' = 0$ or $N = 0$. Flow to a well near a boundary of constant head is given by inserting $T'' \to \infty$ or $N \to \infty$ in (8-168).

8-11 RECHARGING AND PUMPING WELLS IN UNIFORM FLOW

A natural, approximately uniform, flow (at least locally) exists in most undisturbed aquifers. When pumping or recharging wells are introduced into aquifers with such flows, the method of superposition (Sec. 5-7) may be employed to determine the resulting flownets. Obviously, in a phreatic aquifer, superposition is justified only as an approximation.

Two examples are given below.

Example 1 A single pumping well in uniform flow
Consider a single, infinitesimally narrow well of constant discharge Q_w located at $x = 0$, $y = 0$ in a homogeneous ($T = $ const.) isotropic confined aquifer of constant thickness B. Uniform flow at a constant specific discharge rate q_0 in the $-x$ direction takes place in this aquifer. The piezometric head ϕ and stream function ψ ($\equiv \Psi/K$; Sec. 5-8) for this case are given by (Bear, 1972, p. 324)

$$\phi = \frac{q_0 B}{T} x + \frac{Q_w}{4\pi T} \ln(x^2 + y^2); \qquad \psi = \frac{q_0 B}{T} y + \frac{Q_w}{2\pi T} \tan^{-1}\left(\frac{y}{x}\right)$$

$$(8\text{-}169)$$

(see (7-178) for a flow in the $+x$ direction and a recharging well). For a well of finite radius r_w

$$\phi(r) - \phi(r_w) = \frac{q_0 B}{T} x + \frac{Q_w}{2\pi T} \ln \frac{r}{r_w}, \qquad x^2 + y^2 = r^2 \qquad (8\text{-}170)$$

where the second term on the right-hand side is nothing but (8-3) for steady flow. However, a certain, rather very small, error is introduced by doing so as the circumference of the well is actually no more an equipotential surface.

The flownet for this case is shown in Fig. 8-29. The two main features of this flownet are a *groundwater divide* and a *stagnation point*. The groundwater divide bounds the region of the aquifer supplying the entire well's discharge. It is a streamline whose shape is defined by

$$y/x = \pm \tan(2\pi q_0 By/Q_w), \qquad \begin{matrix} + \text{ for } y > 0 \\ - \text{ for } y < 0 \end{matrix} \qquad (8\text{-}171)$$

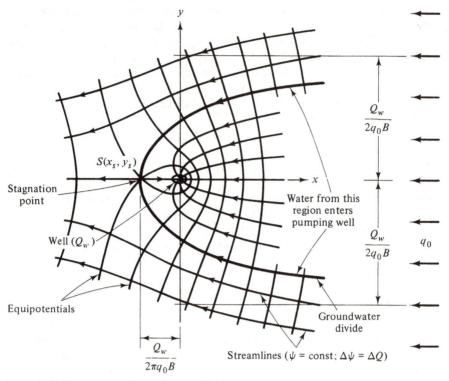

Figure 8-29 A single pumping well in uniform flow.

The water divide approaches asymptotically the lines $y = \pm Q_w/2q_0B$. *A stagnation point*, *S*, which is the point where the resultant velocity, produced by both the pumping well and the natural flow in the aquifer, vanishes, occurs at a point whose coordinates are

$$x_s = -Q_w/2\pi Bq_0, \qquad y_s = 0 \qquad (8\text{-}172)$$

Figure 8-30 shows the case of a recharge well in uniform flow. Again we have a water divide which delineates the aquifer region in which indigenous water will eventually be replaced by the recharged water. We also have a stagnation point $S(x_s, y_s)$, this time upstream of the recharging well. Note that in order to show the similarity of the two cases, we have reversed the direction of the uniform flow in the aquifer. The spreading of the injected water body is discussed in Sec. 7-10.

Figure 8-29 also shows the potential distribution described by (8-170). For the potential distribution shown in Fig. 8-30, we have to replace q_0 by $-q_0$ and Q_w by $-Q_w$ in (8-170).

When pumping produces an unsteady flow regime in an aquifer (i.e., aquifer storativity is taken into account), superimposed on a steady uniform flow, the appropriate drawdown equation should be used. For example, for a confined aquifer, the potential distribution for a single pumping well is

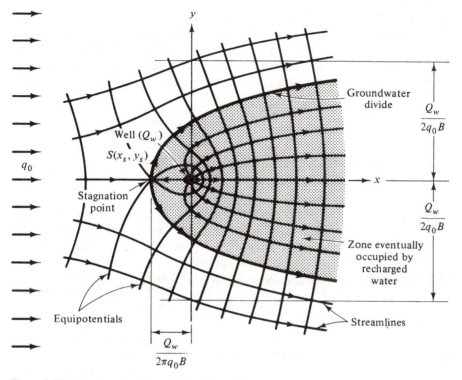

Figure 8-30 A single recharging well in uniform flow.

given by

$$\phi = \frac{q_0 B}{T} x + \frac{Q_w}{4\pi T} W(u) \tag{8-173}$$

where $\phi = 0$ at $x = 0$, $t = 0$. In this case, the locations of the water divide and of the stagnation point are time dependent. For example, the stagnation point $(x_s, 0)$ may be obtained from the specific discharge component q_x along the x axis

$$q_x = -K \frac{\partial \phi}{\partial x} = -q_0 - \frac{Q_w x}{2\pi (x^2 + y^2)^{1/2} B} \exp \left\{ -\frac{S(x^2 + y^2)}{4Tt} \right\} \tag{8-174}$$

From $q_x = 0$ at $x = x_s$, $y_s = 0$ (obtained from the requirement $q_y = 0$), we obtain

$$q_0 + \frac{Q_w}{2\pi x_s B} \exp \left\{ -\frac{S x_s^2}{4Tt} \right\} = 0 \tag{8-175}$$

the solution of which gives the value of $x_s = x_s(t)$, that is the translation of the stagnation point with time. Another form of (8-175) is

$$t = \frac{S x_s^2}{4T \ln(-Q_w/2\pi q_0 B x_s)} \tag{8-176}$$

Example 2 A pumping and recharging pair of wells in uniform flow

Consider a recharging well at $(+d, 0)$ and a pumping well at $(-d, 0)$ in a homogeneous isotropic aquifer in which flow takes place at a constant specific discharge q_0 in a direction making an angle α with the $+x$ axis. Both wells are of equal strength $Q_w = $ const. For this case

$$\left.\begin{array}{l} \phi = -\dfrac{q_0 B}{T}(x \cos \alpha + y \sin \alpha) + \dfrac{Q_w}{4\pi T} \ln \dfrac{(x+d)^2 + y^2}{(x-d)^2 + y^2} \\[4mm] \psi = -\dfrac{q_0 B}{T}(y \cos \alpha - x \sin \alpha) + \dfrac{Q_w}{2\pi T}\left\{ \tan^{-1} \dfrac{y}{x+d} - \tan^{-1} \dfrac{y}{x-d}\right\} \end{array}\right\}$$

(8-177)

Several examples of detailed flownets are described in Fig. 8-31 for different values of α. One may observe that under certain conditions (determined by the relationships between q_0, α, and Q_w) no streamline emerging from the recharging well terminates in the pumping well. This means that no injected fluid will ever reach the pumping well. Dacosta and Bennett (1960) study this problem in detail in connection with artificial recharge operations. They also determine the location of stagnation points and the amounts of interflow between the wells by taking twice the difference between the value of ψ passing through the origin of coordinates, and the value of ψ passing through one of the stagnation points (multiplied by K).

The shaded areas in Fig. 8-31 (pages 371–372) indicate regions of interflow. Groundwater divides and stagnation points can easily be determined for each case from (8-177).

The situations shown in Figs. 8-31a through d are not the only possible ones for the respective cases. As already indicated above, the resulting flownet depends in each case on the relationships between q_0, α and Q_w, with a possibility of different values of Q_w for the two wells. To illustrate this point, let us consider the case shown in Fig. 8-31a in which the shaded diamond-shaped area shows where recirculation takes place between the wells (with the pumping well located upstream of the recharging one). If however, the distance between the wells is made sufficiently large for a given well discharge, Q_w (equal to the rate of recharge) and a uniform specific discharge q_0, recirculation can be prevented entirely. This case is shown in Fig. 8-32a. As pumping and recharging rates increase, for the same distance, $2d$, and uniform specific discharge, q_0, a value of Q_w is reached such that the uniform groundwater flow is just balanced by the opposing flows produced by the two wells at a point midway between them (again for equal values of pumping and recharge) as shown in Fig. 8-31b. A further increase in Q_w will then produce the situation shown in Fig. 8-32a. In order to obtain the critical value of Q_w, we have to equate q_0 to the sum of the specific discharges induced by the two wells at that point

$$q_0 = \frac{Q_w}{2\pi d B} + \frac{Q_w}{2\pi d B} = \frac{Q_w}{\pi d B}$$

(8-178)

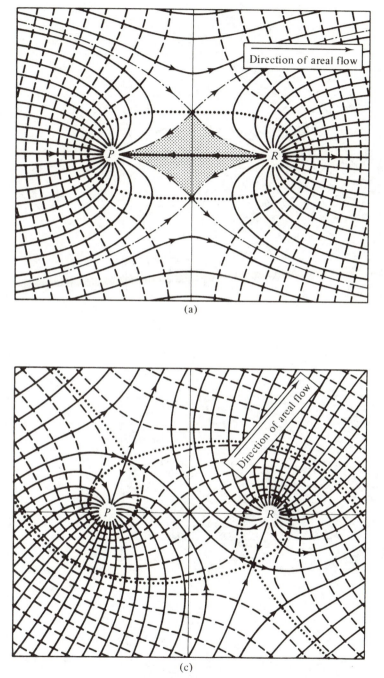

(a)

(c)

P = pumping well; R = recharge well

Figure 8-31 Examples of flownets and regions of interflow for a pair of pumping and recharging wells in uniform flow *(Dacosta and Bennett, 1960)*.

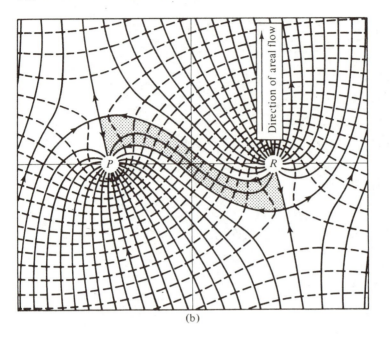

(b)

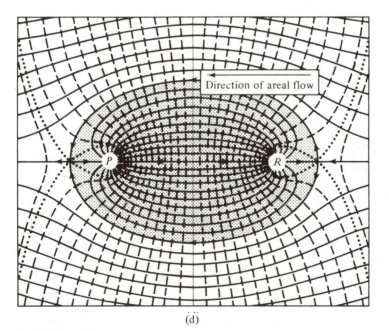

(d)

Fig. 8-31 (contd.)

Thus there is no recirculation for $Q_w \le \pi dB q_0$. We can determine the two stagnation points along $y = 0$ for this case (Fig. 8-32a)

$$- q_0 + \frac{Q_w}{2\pi(d - x_s) B} + \frac{Q_w}{2\pi(d + x_s) B} = 0; \qquad x_s = \pm d \sqrt{1 - \frac{Q_w}{\pi q_0 dB}}$$

(8-179)

At the critical situation, these two stagnation points merge at the origin as shown in Fig. 8-32b. For $Q_w > \pi dB q_0$ (Fig. 8-31a), we have again two stagnation points, this time along $y = 0$, with

$$y_s = \pm d \sqrt{\frac{Q_w}{\pi d q_0 B} - 1}$$

(8-180)

By taking the difference between the values of the stream function, corresponding to streamlines, passing through these two stagnation points, one can determine the proportion of recirculation, Q_{wr}, between the two wells

$$\frac{Q_{wr}}{Q_w} = \frac{2}{\pi}\left\{ \tan^{-1} \sqrt{\frac{Q_w}{\pi d q_0 B} - 1} - \frac{\pi d q_0 B}{Q_w} \sqrt{\frac{Q_w}{\pi d q_0 B} - 1} \right\}$$

(8-181)

A similar analysis may be carried out for other cases of practical interest. For example

(a) The two wells are of different strengths,
(b) more than two wells operate in the field, some of them are pumping wells, the others recharge ones, and
(c) the aquifer is phreatic (but it is assumed that superposition is permitted), or leaky.

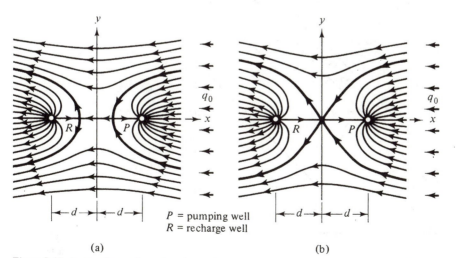

P = pumping well
R = recharge well

(a)   (b)

Figure 8-32 A pumping and a recharging well in uniform flow. a() No recirculation. (b) Limiting case.

In all these cases, the piezometric head distribution $\phi = \phi(x, y)$, or $\phi = \phi(x, y, t)$ for unsteady flow, is obtained by superposition.

In Sec. 7-10, use is made of the material presented in this section in cases where the injected water is of a quality which is different from the indigeneous one in the aquifer and it is assumed that a sharp front separates the two kinds of water.

8-12 WELL LOSSES AND SPECIFIC WELL DISCHARGE

In the discussion in the previous sections it was implicitly assumed that the drawdown in a pumping well (as in an observation well) is due only to head losses in the aquifer through which the flow towards the pumping well takes place (= *formation loss*). In the practice, however, in a pumping well an additional drawdown results from a head loss caused by the flow through the well's screen or perforated casing, whenever they exist, and by the (often turbulent) flow of the water inside the well to the pump's intake (= *well loss*). A brief discussion on this point has already been presented in Sec. 8-1 (see Fig. 8-3).

Because water velocities increase as the well is approached (as $V_r = Q_w/2\pi r B n$), it is possible that close to a well, the flow in an aquifer will become turbulent. Jacob (1947) takes care of this possibility by introducing the concept of an *effective radius* of a well.

For a well pumping at a rate Q_w, Jacob (1950) expresses the well loss as CQ_w^2, where C is a constant, referred to as the *well loss constant*. It characterizes the quality of the well's completion, its screen, its gravel pack (if present), etc. Thus,

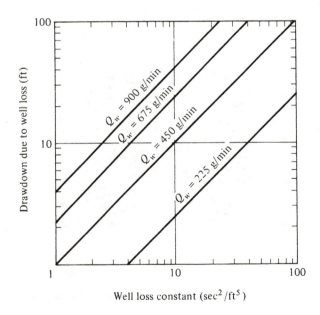

Figure 8-33 Typical relationship between well loss constant and drawdown (*Walton, 1970, p. 313*).

the total drawdown, s_{tw} at a pumping well is given by

$$s_{tw} = s_w(r_w, t) + CQ_w^2 = B(r_w, t)Q_w + CQ_w^2 \qquad (8\text{-}182)$$

where $s_w(r_w, t) = B(r_w, t)Q_w$ is the *formation loss* (through the aquifer) expressed by one of the drawdown equations (depending on the type of aquifer) developed in the previous sections. Figure 8-33 shows C as a function of drawdown due to well loss for selected pumping rates. The term s_w already takes into account the type of aquifer, effect of boundaries in the vicinity, partial penetration, etc. For example, for steady flow to a well in a confined aquifer, we have from (8-4)

$$B = [\ln(R/r_w)]/2\pi T \qquad (8\text{-}183)$$

For unsteady flow to a well in a confined aquifer, we have from (8-61)

$$B = [W(Sr_w^2/4Tt)]/4\pi T \qquad (8\text{-}184)$$

etc. Rorabaugh (1953) suggested that the head loss associated with turbulent flow in the formation, as well as with flow into and inside a well, may be expressed as CQ_w^n, where n may deviate significantly from 2. Figure 8-34 shows variations of the well loss, CQ_w^n for $B = $ const. (i.e., steady flow).

Hitherto we have assumed that, in general, variations in the well's radius r_w do not appreciably affect the drawdown at a pumping well, i.e., the value of B. However, they may appreciably affect the contribution of the term CQ_w^n to the well's drawdown as the entrance velocity depends on the entrance area, which, in turn, depends on the square of the well's radius. In general, for relatively low pumping rates, the well loss may be neglected. For a high pumping rate, or for a low quality of well completion, it may represent a significant portion of the total loss.

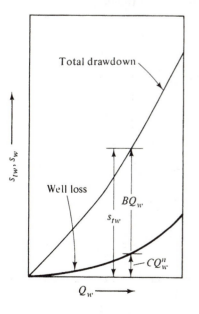

Figure 8-34 Variation of total drawdown with well discharge for $B = $ const.

In order to determine the values of C and n, and by doing so evaluate the quality of the well's completion, a step-drawdown test (Sec. 11-1) has to be performed.

The *specific discharge of a well* (also called *specific capacity of a well*) is defined as the ratio of its discharge to its total drawdown (= discharge per unit drawdown), taking into account both formation and well losses. From (8-182) it follows that

$$\frac{Q_w}{s_{wt}} = \frac{1}{B(r_w, t) + CQ_w} \tag{8-185}$$

or, with CQ_w^n expressing the well's loss

$$\frac{Q_w}{s_{wt}} = \frac{1}{B(r_w, t) + CQ_w^{n-1}} \tag{8-186}$$

Both equations show that the specific discharge decreases with time and with rate of discharge. Although B depends on r_w, the effect of r_w on B is very small since the dependence is, in general, on $\ln r_w$. Figure 8-35 shows the variation of specific discharge of a well with time.

Another constant of interest is the *specific drawdown* which gives the drawdown per unit discharge

$$\frac{s_{wt}}{Q_w} = CQ_w + B(r_w, t) \tag{8-187}$$

It is described by a straight line when plotted against Q_w.

In many regions, studies are carried out on the correlation between the last two characteristics and various aquifer and well parameters (e.g., depth of screen) in order to learn how to improve the yield from wells in the aquifer.

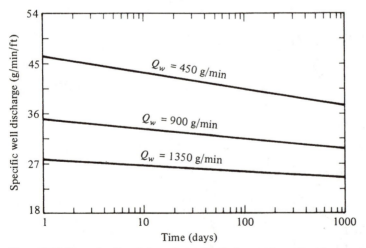

Figure 8-35 Example of variation of specific discharge of a well with pumping rate and time *(after Jacob, 1950)*.

8-13 HYDRAULICS OF RECHARGING WELLS

This brief section is introduced here in order to emphasize that, in general, the entire discussion presented in this chapter is applicable also to recharging wells through which water is introduced into an aquifer as part of artificial recharge operations (Sec. 3-4). As always, for a phreatic aquifer, this is an approximation. Marino and Yeh (1972) treat unsteady flow in a recharge well–unconfined aquifer system.

Whereas pumping produces a pattern of a radially converging flow to the well, with a drawdown of the water table, or the piezometric surface, a recharging well produces a pattern of radially diverging flow from the well and a build-up of the water table, or the piezometric surface. The modifications which have to be introduced in all the equations and formulas presented in this chapter are obvious and require no further discussion.

Yet, we should emphasize an important difference which sometimes cannot be overlooked. When pumping takes place, silt and fine material, if present in the formation, is removed from it in the vicinity of the well, where the average flow velocity ($V_r = Q_w/2\pi r Bn$) is sufficiently high. These fines are removed with the pumped water. A zone of increased permeability is thus created around the well. Figure 8-3c shows this case. Actually this is a standard procedure (called *well development*) in a newly drilled well. Pumping operations at a rate which is higher than that normally planned for the well, are carried out for some time, until the pumped water is clear, containing no fine material.

In a recharging well, however, we *always* bring impurities into the formation. These impurities may include fine material, organic matter, and air. At some distance from a recharging well, the velocity of the water reduces to the point which results in the settlement of silt and fine particles carried with the water. The permeability of the formation is reduced. Dissolved contaminants carried with the water may interact with the solid skeleton (e.g., base exchange in soils containing clay and silt), and/or with the indigenous water in the aquifer and produce clogging of the formation. Air bubbles are carried with the water and lodged in small pores. Air, dissolved in the water is released from solution as the pressure drops in the casing or in the formation itself, forming bubbles which are also lodged in the small pores. Organic matter and bacteria may produce a growth on the well's screen. Altogether, we have here a phenomenon of reduction of effective flow area and clogging. Because of the higher velocity and pressure gradients near the well, part of the reduction of permeability occurs at some distance from the well.

We thus have a zone of reduced permeability near the well. The permeability continues to drop with the increase in volume of recharged water. This produces an additional build-up which has to be taken into account when calculating build-up recharge relationships.

For the sake of completeness, we should also mention the possibility that as a result of the method of drilling, the permeability in the immediate vicinity of a well is *reduced*. For example, when the drilling fluid (*mud*) used in conventional rotary techniques penetrates the formation and is not completely removed upon completion of drilling. In (oil) reservoir engineering, this damage to the permeability in the proximity of a well is called "*skin effect.*"

As an example, consider the case where the reduced permeability is expressed by

$$
\left.
\begin{aligned}
K(r) &= K_0 - (K_0 - K_r)\exp\left\{- \alpha \left(\frac{r - r_w}{r_e - r}\right)\right\} ; \quad r_w < r \leq r_e, \\
\\
K(r) &= K_0 = \text{const.}; \quad r > r_e
\end{aligned}
\right\}
\tag{8-188}
$$

where K_0 and K_r are the aquifer's undamaged permeability and the reduced one at the well's surface by clogging, respectively, and α is a coefficient. Other expressions for $K(r)$ are also possible (e.g., Karanjac, 1972). It is also possible to introduce the effect of time by letting K_r be a function of the recharged volume (e.g. $K_r(t) = \beta \int_0^t Q_w(t)\,dt$; β = a coefficient).

Assuming steady flow (and build-up), we obtain from continuity considerations

$$
Q_w = - 2\pi r K(r) B \frac{d\phi}{dr} = \text{const.}
\tag{8-189}
$$

where Q_w is the constant rate of recharge. Hence, by integration, we obtain

$$
\phi_w - \phi(r) = \frac{Q_w}{2\pi B} \int_{r_w}^{r} \frac{dr}{rK(r)}
\tag{8-190}
$$

$$
\phi_w - \phi(r_e) = \frac{Q_w}{2\pi B} \int_{r_w}^{r_e} \frac{dr}{rK(r)}
\tag{8-191}
$$

$$
\phi(r_e) - \phi(R) = \frac{Q_w}{2\pi B} \int_{r_e}^{R} \frac{dr}{K_0 r} = \frac{Q_w}{2\pi K_0 B} \ln \frac{R}{r_e}
\tag{8-192}
$$

$$
\phi_w - \phi(R) = \frac{Q_w}{2\pi B} \left\{ \int_{r_w}^{r_e} \frac{dr}{rK(r)} + \ln \frac{R}{r_e} \right\}
\tag{8-193}
$$

into which we insert the expression (8-188) for $K(r)$. The build-up as obtained by integrating (8-193) is to be compared with the build-up

$$
\phi_w - \phi(R) = \frac{Q_w}{2\pi B K_0} \ln \frac{R}{r_w}
\tag{8-194}
$$

which corresponds to a constant $K = K_0$ for the entire region $r_w < r < R$; R is the radius of influence where practically no build-up is observed. The additional build-up thus obtained is due to clogging. When the permissible build-up is limited, this means that the recharge rate Q_w has to be reduced. When the reduced recharge rates become uneconomic, cleaning operations have to be undertaken in order to restore the recharge capacity of the well.

NINE

FRESH WATER–SALT WATER INTERFACE IN COASTAL AQUIFERS

Coastal aquifers constitute an important source for water, especially in arid and semi-arid zones which border the sea. Many coastal areas are also heavily urbanized, a fact which makes the need for fresh water even more acute. However, the proximity of the sea requires special attention and special management techniques.

The objective of this chapter is to present the hydrologic information required for managing coastal aquifers, in view of the danger of sea water intrusion and the relationship that exists between the outflow of fresh water from the aquifer to the sea and the extent of sea water intrusion.

9-1 OCCURRENCE

In general, in coastal aquifers a hydraulic gradient exists toward the sea that serves as a recipient for the excess of their fresh water (replenishment minus pumpage). Owing to the presence of sea water in the aquifer formation under the sea bottom, a zone of contact is formed between the lighter fresh water (specific weight γ_f) flowing to the sea and the heavier, underlying, sea water (specific weight $\gamma_s > \gamma_f$). Typical cross sections with interfaces under natural conditions are shown in Fig. 9-1. Figure 9-2 gives a typical cross section of a coastal phreatic aquifer with exploitation. In all cases, there exists a body of sea water, often in the form of a wedge, underneath the fresh water. One should note that, like most figures describing flow in aquifers, these are also *highly distorted figures*, not drawn to scale.

Fresh water and sea water are actually miscible fluids and therefore the zone of contact between them takes the form of a transition zone caused by hydrodynamic dispersion (Chap. 7). Across this zone, the density of the mixed water varies from that of fresh water to that of sea water. However, under certain

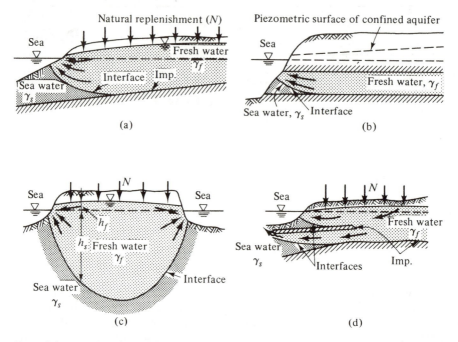

Figure 9-1 Examples of interfaces in coastal aquifers.

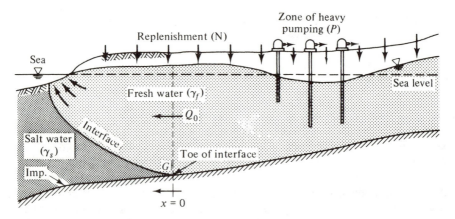

Figure 9-2 A typical cross section of a coastal aquifer with pumping.

conditions, the width of this zone is relatively small (e.g., when compared with the thickness of the aquifer), so that an *abrupt interface approximation* can be introduced. In a way, this is similar to the introduction of a phreatic surface as an approximation for the moisture distribution in the soil (Sec. 2-2). For example, observations (Jacobs and Schmorak, 1960; Schmorak, 1967) along the coast of Israel indicate that indeed this assumption of an abrupt interface is justified.

On the other hand, Cooper (1959) describes a case where the transition zone is very wide so that the interface approximation is no longer valid.

In what follows, we shall assume that an abrupt interface separates the regions occupied by the two fluids.

Under natural undisturbed conditions in a coastal aquifer, a state of equilibrium is maintained, with a stationary interface and a fresh water flow to the sea above it. At every point on this interface, the elevation and the slope are determined by the fresh water potential and gradient (or flow velocity). The continuous change of slope results from the fact that as the sea is approached, the specific discharge of fresh water tangent to the interface increases. By pumping from a coastal aquifer in excess of replenishment, the water table (or the piezometric surface) in the vicinity of the coast is lowered to the extent that the piezometric head in the fresh water body becomes less than in the adjacent sea water wedge, and the interface starts to advance inland until a new equilibrium is reached. This phenomenon is called *sea water intrusion* (or *encroachment*). As the interface advances, the transition zone widens; nevertheless, we shall assume that the abrupt interface approximation remains valid. When the advancing interface reaches inland pumping wells, the latter become contaminated. When pumping takes place in a well located above the interface, the latter upcones towards the pumping well. Unless the rate of pumping is carefully controlled, sea water will eventually enter the pumped well. Actually the real situation is even more dangerous in view of the presence of a wide transition zone rather than an abrupt interface.

Rather than treat the problem as one with an abrupt interface, we can always consider it as one with a continuous variation of salt concentration and density, using the material presented in Chap. 7 on hydrodynamic dispersion. However, the assumption of an abrupt interface—especially when certain assumptions related to horizontal flow are also introduced—greatly simplifies the problem in many cases of practical interest.

As we shall see below, there exists a relationship between the rate of fresh water discharge to the sea and the extent of sea water intrusion. This makes sea water encroachment a *management problem* as the fresh water discharge to the sea is the difference between the rate of natural and artificial recharge and that of pumping. Extensive research is being carried out in many parts of the world with the objectives of understanding the mechanism of sea water intrusion and learning to control it in order to improve the yield of coastal aquifers. Some elements of these investigations are reviewed here.

In Sec. 7-10 we have discussed an abrupt interface between two liquids in a horizontal plane. Here we consider the effect of the density difference between the two liquids.

9-2 EXACT MATHEMATICAL STATEMENT OF THE PROBLEM

We are dealing here with the mathematical statement of a flow problem involving two liquids, assuming that an abrupt interface (always) separates them, such that each liquid occupies a separate part of the entire flow domain (here, the aquifer). In general, the interface is not stationary. Sources and sinks of liquid (i.e., pumping and artificial recharge) may exist in both regions.

In each of the two regions we may define a piezometric head. Assuming that water is incompressible, we have in the fresh water region: $\phi_f = z + p/\gamma_f$ (or $\phi_f^* = z + \int_{p_0}^p dp/\gamma_f(p)$ for a compressible fluid), and in the salt water region: $\phi_s = z + p/\gamma_s$ (or $\phi_s^* = z + \int_{p_0}^p dp/\gamma_s(p)$ for a compressible fluid).

Let Fig. 9-3 represent the two subdomains, R_1 and R_2, occupied by fresh water and salt water, respectively. Then the problem can be stated in the following way.

Determine ϕ_f in R_1 and ϕ_s in R_2 such that

$$\nabla \cdot (K_f \cdot \nabla \phi_f) = S_0 \frac{\partial \phi_f}{\partial t} \quad \text{in } R_1; \qquad \nabla \cdot (K_s \cdot \nabla \phi_s) = S_0 \frac{\partial \phi_s}{\partial t} \quad \text{in } R_2 \qquad (9\text{-}1)$$

where $K_f = k\gamma_f/\mu_f$; $K_s = k\gamma_s/\mu_s$ and the specific storativity, S_0, is assumed the same for R_1 and R_2. Obviously, other equations can be used when necessary, depending on the assumptions we make with respect to the medium and the fluids.

In addition, we have to specify initial conditions for ϕ_f in R_1 and for ϕ_s in R_2. Boundary conditions for ϕ_f on B_1 and ϕ_s on B_2 are the usual ones encountered in the flow of a single fluid (Sec. 5-3). The boundary condition on the interface requires special attention. Moreover, as in the case of a phreatic surface, the location of an interface is unknown until the problem is solved. In fact, the location and shape of an interface, say, expressed in the form of

$$F(x, y, z, t) = 0 \qquad (9\text{-}2)$$

is what we are looking for.

Denoting the elevation of points on the interface by $\zeta = \zeta(x, y, t)$, the relationship for F becomes

$$z = \zeta(x, y, t), \quad \text{or}: F \equiv z - \zeta(x, y, t) = 0 \qquad (9\text{-}3)$$

The pressure at a point $P(x, y, \zeta)$ on the interface is the same when approached

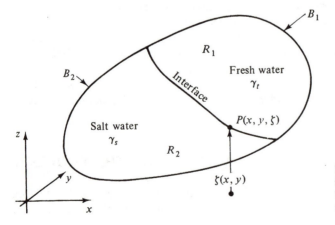

Figure 9-3 An abrupt interface between regions occupied by fresh water and by salt water.

from both sides. Hence, from the definitions of ϕ_f and ϕ_s, we have

$$\gamma_f(\phi_f - \zeta) = \gamma_s(\phi_s - \zeta)$$

or

$$\zeta(x, y, t) = \phi_s \frac{\gamma_s}{\gamma_s - \gamma_f} - \phi_f \frac{\gamma_f}{\gamma_s - \gamma_f}$$
$$= \phi_s(1 + \delta) - \phi_f \delta; \qquad \delta = \gamma_f/(\gamma_s - \gamma_f) \qquad (9\text{-}4)$$

Once we know the distributions $\phi_f = \phi_f(x, y, z, t)$ and $\phi_s = \phi_s(x, y, z, t)$, (9-3) becomes the sought equation for $F(x, y, z, t)$

$$F \equiv z - \phi_s(1 + \delta) + \phi_f \delta = 0 \qquad (9\text{-}5)$$

The boundary conditions on the interface are as follows.

(*a*) Same specific discharge on both sides: $(q_n)_f = (q_n)_s$ on F.
(*b*) Same pressures on both sides: $\gamma_f(\phi_f - \zeta) = \gamma_s(\phi_s - \zeta)$ on F.

Since the interface is a material surface, with fluid particles remaining always on it, we have for F

$$dF/dt \equiv \partial F/\partial t + \mathbf{V}_f \cdot \nabla F = 0; \qquad \partial F/\partial t + \mathbf{V}_s \cdot \nabla F = 0 \qquad (9\text{-}6)$$

where $n\mathbf{V}_f = -\mathbf{K}_f \cdot \nabla \phi_f$; $n\mathbf{V}_s = -\mathbf{K}_s \cdot \nabla \phi_s$. By combining (9-5) with (9-6), and neglecting solid velocity we obtain:

$$n\delta \frac{\partial \phi_f}{\partial t} - n(1 + \delta) \frac{\partial \phi_s}{\partial t} - \mathbf{K}_f \cdot [\nabla z - (1 + \delta) \nabla \phi_s + \delta \nabla \phi_f] \cdot \nabla \phi_f = 0 \qquad (9\text{-}7)$$

$$n\delta \frac{\partial \phi_f}{\partial t} - n(1 + \delta) \frac{\partial \phi_s}{\partial t} - \mathbf{K}_s \cdot [\nabla z - (1 + \delta) \nabla \phi_s + \delta \nabla \phi_f] \cdot \nabla \phi_s = 0 \qquad (9\text{-}8)$$

Thus, the boundary conditions on an interface take the form of two *nonlinear partial differential equations* in ϕ_f and ϕ_s. This is the reason why the derivation of the shape and position of an interface by solving the partial differential equations (9-1) subject to the boundary conditions (9-7) and (9-8) on the surface defined by (9-3), is practically impossible. Even numerical methods fail here. In Sec. 9-4, the *hydraulic approach* is employed to reduce the interface problem essentially to one of flow in a plane, thus eliminating the boundary conditions on the interface. Other approximations are discussed in Secs. 9-3, 9-6, and 9-7.

It is of interest to determine the slope at a point on a stationary interface. Figure 9-4 shows an element AB along an interface in two-dimensional flow in the vertical xz plane. The components of the specific discharge tangential to the interface in the two regions are given by

$$(q_f)_s = -\frac{k\gamma_f}{\mu_f} \frac{\partial \phi_f}{\partial s} = -\frac{k}{\mu_f} \left(\frac{\partial p}{\partial s} + \gamma_f \frac{\partial z}{\partial s} \right) \qquad (9\text{-}9)$$

$$(q_s)_s = -\frac{k\gamma_s}{\mu_s} \frac{\partial \phi_s}{\partial s} = -\frac{k}{\mu_s} \left(\frac{\partial p}{\partial s} + \gamma_s \frac{\partial z}{\partial s} \right) \qquad (9\text{-}10)$$

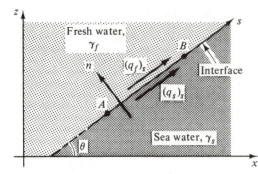

Figure 9-4 Dynamic equilibrium at an interface.

where $k = $ const. By eliminating $\partial p/\partial s$ from both equations, we obtain

$$\sin \theta = \frac{\partial z}{\partial s} = \frac{(q_f)_s \, \mu_f - (q_s)_s \, \mu_s}{k(\gamma_s - \gamma_f)} \tag{9-11}$$

where θ is the angle that the interface makes with the $+x$ direction. For $q_s = 0$ (i.e., stationary sea water), as $(q_f)_s$ increases, θ also increases. The shape of a stationary interface as a coast is approached follows from this conclusion: q_f increases as the coast is approached and hence θ also increases.

9-3 THE GHYBEN–HERZBERG APPROXIMATION

Beginning with Badon–Ghyben (1888) and Herzberg (1901), investigations of the interface in a coastal aquifer have aimed at determining the relationship between its shape and position, and the various hydrologic components of the groundwater balance in the region near the coast.

Figure 9-5 shows the idealized Ghyben–Herzberg model of an interface in a coastal phreatic aquifer. Essentially, Ghyben and Herzberg assume static

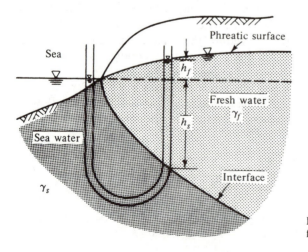

Figure 9-5 Ghyben–Herzberg interface model.

equilibrium and a hydrostatic pressure distribution in the fresh water region, with *stationary sea water*. Instead, we may assume dynamic equilibrium, i.e., steady flow, but with horizontal flow in the fresh water region. This means that equipotentials are vertical lines or surfaces (identical to the Dupuit assumption). With the nomenclature of Fig. 9-5, we have under these conditions

$$h_s = [\gamma_f/(\gamma_s - \gamma_f)] h_f \equiv \delta h_f; \qquad \delta = \gamma_f/(\gamma_s - \gamma_f) \qquad (9\text{-}12)$$

For example, for $\gamma_s = 1.025$ gr/cm³, $\gamma_f = 1.000$ gr/cm³, $\delta = 40$, and $h_s = 40 h_f$, that is, at any distance from the sea, the depth of a *stationary interface* below sea level is 40 times the height of the fresh water table above it. Obviously, as the sea is approached, the assumption of horizontal flow is no longer valid, because vertical flow components can no longer be neglected. Moreover, in Fig. 9-5, no outlet is left for the fresh water flow to the sea. Figure 9-6 shows the actual flow conditions near the coast. Point A on the interface indicates the actual depth of the interface at that distance from the coast. Point B is located at the intersection of the interface and the fresh water equipotential $\phi_f = h_f$. Accordingly, point B is at a depth equal to δh_f which is the depth predicted by the Ghyben–Herzberg relationship for the interface corresponding to h_f. The actual depth (point A) is thus greater than that predicted by the Ghyben–Herzberg relationship.

If we set $\phi_s = \text{const.} = 0$ (i.e., immobile sea water) in (9-4), we see that the difference between the Ghyben–Herzberg approximation (9-12) and the exact expression (9-4) stems from the difference between h_f and ϕ_f (that is, between the assumed vertical equipotential with $\phi_f = h_f$ and the actual curved one). In a confined aquifer, h_s in (9-12) is the depth of a point on the interface below sea level, whereas h_f is the (fresh water) piezometric head.

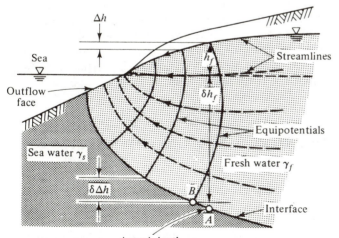

Figure 9-6 Actual flow pattern near the coast.

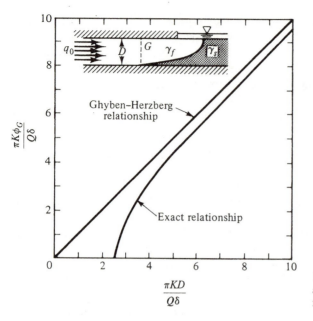

From (9-4) it follows that when the interface is in motion, ϕ_s ($\neq$ const.) also affects the shape of the interface.

Bear and Dagan (1964) investigate the validity of the Ghyben–Herzberg relationship. They find that in a confined horizontal aquifer of constant thickness D, the approximation is good, within an error of 5 percent, for determining the depth of the interface toe (point G in Fig. 9-7) in steady flow, and hence also the length of the intruding sea water wedge, provided $\pi KD/Q\delta > 8$, where Q is the fresh water discharge to the sea. Figure 9-7 shows this relationship; the exact solution is obtained by the hodograph method.

As the coast is approached, the depth of the interface is greater than that given by the Ghyben–Herzberg relationship (Fig. 9-6). In the case of a phreatic aquifer, a seepage face is also present above sea level. One should also note that for a downward sloping flat sea bottom it follows from the exact solution that the interface terminates at the sea bottom with a tangent in the vertical direction. The interface always terminates on the sea bottom at some distance from the coast (Fig. 9-6).

9-4 CONTINUITY EQUATION BASED ON THE DUPUIT ASSUMPTION

At the end of Sec. 9-2 we reached the conclusion that attempting to derive a solution for the shape and position of an interface in a coastal aquifer by solving the balance equations in the three-dimensional space with the nonlinear interface boundary conditions is impractical. Instead, similar to the derivation of the equations which govern the flow in a phreatic aquifer (Sec. 5-4), we shall employ the hydraulic approach and average the three-dimensional balance equation $\nabla \cdot (\mathbf{K} \cdot \nabla \phi) = S_0 \partial \phi / \partial t$, separately for each region, over the vertical (see App. B). We shall use the nomenclature of Fig. 9-8, which shows an interface in a phreatic aquifer.

For the fresh water region, bounded from below by an interface at $\zeta_1(x, y, t)$ and from above by a phreatic surface with accretion at $\zeta_2(x, y, t)$ and the sea water region bounded from above by the interface at $\zeta_1(x, y, t)$ and from below by an impervious bottom at $\zeta_0(x, y)$, we have from (5-26)

$$\nabla \cdot \mathbf{q}_f + S_{0f} \frac{\partial \phi_f}{\partial t} = 0; \qquad \nabla \cdot \mathbf{q}_s + S_{0s} \frac{\partial \phi_s}{\partial t} = 0 \qquad (9\text{-}13)$$

where

$$\mathbf{q}_f = -\mathbf{K}_f \cdot \nabla \phi_f; \qquad \mathbf{q}_s = -\mathbf{K}_s \cdot \nabla \phi_s \qquad (9\text{-}14)$$

By integrating along the vertical, making use of Leibnitz rule (see App. B), we obtain for the sea water region

$$\int_{\zeta_0}^{\zeta_1} \left(\nabla \cdot \mathbf{q}_s + S_{0s} \frac{\partial \phi_s}{\partial t} \right) dz = \nabla' \cdot \int_{\zeta_0}^{\zeta_1} \mathbf{q}'_s \, dz - \mathbf{q}'_s \Big|_{\zeta_1} \cdot \nabla' \zeta_1 + \mathbf{q}'_s \Big|_{\zeta_0} \cdot \nabla' \zeta_0$$

$$+ \int_{\zeta_0}^{\zeta_1} \frac{\partial q_{sz}}{\partial z} \, dz + S_{0s} \left(\frac{\partial}{\partial t} \int_{\zeta_0}^{\zeta_1} \phi_s \, dz - \phi_s \Big|_{\zeta_1} \frac{\partial \zeta_1}{\partial t} \right) = 0 \quad (9\text{-}15)$$

where $\nabla'(\ ) \equiv [\partial(\ )/\partial x] \, \mathbf{1x} + [\partial(\ )/\partial y] \, \mathbf{1y}$; $\mathbf{q}' = q_x \mathbf{1x} + q_y \mathbf{1y}$. Now

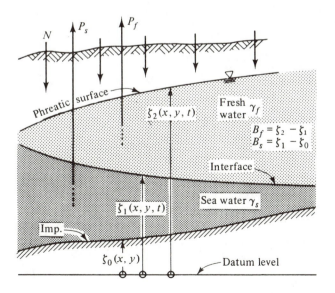

Figure 9-8 Nomenclature for an interface in a phreatic coastal aquifer.

$$\frac{\partial}{\partial t} \int_{\zeta_0}^{\zeta_1} \phi_s \, dz = \frac{\partial}{\partial t} [\tilde{\phi}_s B_s] = \tilde{\phi}_s \frac{\partial B_s}{\partial t} + B_s \frac{\partial \tilde{\phi}_s}{\partial t} = \tilde{\phi}_s \frac{\partial \zeta_1}{\partial t} + B_s \frac{\partial \tilde{\phi}_s}{\partial t},$$

where

$$\tilde{\phi}_s = \frac{1}{B_s} \int_{\zeta_0}^{\zeta_1} \phi_s \, dz; \qquad B_s = \zeta_1 - \zeta_0$$

Assuming $\tilde{\phi}_s \approx \phi_s|_{\zeta_1}$, we obtain from (9-15)

$$\nabla' \cdot (B_s \tilde{\mathbf{q}}'_s) - \mathbf{q}'_s\Big|_{\zeta_1} \cdot \nabla'\zeta_1 + \mathbf{q}'_s\Big|_{\zeta_0} \cdot \nabla'\zeta_0 + q_{sz}\Big|_{\zeta_1} - q_{sz}\Big|_{\zeta_0} + S_{0s}B_s \frac{\partial \tilde{\phi}_s}{\partial t} = 0 \quad (9\text{-}16)$$

where

$$B_s \tilde{\mathbf{q}}'_s = \int_{\zeta_0}^{\zeta_1} \mathbf{q}'_s \, dz \equiv \mathbf{Q}'_s$$

is the flow of sea water per unit width of aquifer.

For the fresh water zone, we obtain in a similar way

$$\int_{\zeta_1}^{\zeta_2} \left(\nabla \cdot \mathbf{q}_f + S_{0f} \frac{\partial \phi_f}{\partial t} \right) dz = \nabla' \cdot (B_f \tilde{\mathbf{q}}'_f) - \mathbf{q}'_f\Big|_{\zeta_2} \cdot \nabla'\zeta_2 + \mathbf{q}'_f\Big|_{\zeta_1} \cdot \nabla'\zeta_1 + q_{fz}\Big|_{\zeta_2}$$

$$- q_{fz}\Big|_{\zeta_1} + S_{0f}B_f \frac{\partial \tilde{\phi}_f}{\partial t} = 0 \qquad (9\text{-}17)$$

where $B_f = \zeta_2 - \zeta_1$, and we have assumed $\tilde{\phi}_f = \phi_f|_{\zeta_2} = \phi_f|_{\zeta_1}$, that is, vertical equipotentials, which is equivalent to the Dupuit assumption. With this approximation, we also have from (9-4)

$$\zeta_1 = (1 + \delta) \tilde{\phi}_s - \delta \tilde{\phi}_f \qquad (9\text{-}18)$$

and

$$F \equiv z - \zeta_1 = z - (1 + \delta) \tilde{\phi}_s + \delta \tilde{\phi}_f = 0 \qquad (9\text{-}19)$$

as the equation describing the interface and satisfying

$$n \, \partial F/\partial t + \mathbf{q}_s \cdot \nabla F = 0; \qquad n \, \partial F/\partial t + \mathbf{q}_f \cdot \nabla F = 0 \qquad (9\text{-}20)$$

or

$$n(1 + \delta) \, \partial \tilde{\phi}_s/\partial t - n\delta \, \partial \tilde{\phi}_f/\partial t = \mathbf{q}_s\Big|_{\zeta_1} \cdot \nabla(z - \zeta_1) = q_{sz}\Big|_{\zeta_1} - \mathbf{q}'_s\Big|_{\zeta_1} \cdot \nabla'\zeta_1 \quad (9\text{-}21)$$

$$n(1 + \delta) \, \partial \tilde{\phi}_s/\partial t - n\delta \, \partial \tilde{\phi}_f/\partial t = \mathbf{q}_f\Big|_{\zeta_1} \cdot \nabla(z - \zeta_1) = q_{fz}\Big|_{\zeta_1} - \mathbf{q}'_f\Big|_{\zeta_1} \cdot \nabla'\zeta_1 \quad (9\text{-}22)$$

By combining (9-16) and (9-21), we obtain for the sea water zone

$$\nabla' \cdot (B_s \tilde{\mathbf{q}}'_s) + [S_{0s}B_s + n(1 + \delta)] \, \partial \tilde{\phi}_s/\partial t - n\delta \, \partial \tilde{\phi}_f/\partial t + \mathbf{q}'_s\Big|_{\zeta_0} \cdot \nabla'\zeta_0 - q_{sz}\Big|_{\zeta_0} = 0$$

$$(9\text{-}23)$$

Then, we derive $\tilde{\mathbf{q}}_s$

$$\tilde{\mathbf{q}}_s = \frac{1}{B_s} \int_{\zeta_0}^{\zeta_1} \mathbf{q}_s\, dz = -\frac{1}{B_s} \int_{\zeta_0}^{\zeta_1} K_s \cdot \nabla \phi_s\, dz$$

$$= -\frac{K_s}{B_s} \cdot \left[\nabla' \int_{\zeta_0}^{\zeta_1} \phi_s dz - \phi_s \Big|_{\zeta_1} \nabla'\zeta_1 + \phi_s \Big|_{\zeta_0} \nabla'\zeta_0 + \left(\phi_s \Big|_{\zeta_1} - \phi_s \Big|_{\zeta_0} \right) \mathbf{1z} \right]$$

$$\approx -K_s \cdot \nabla' \phi_s \tag{9-24}$$

where K_s' includes only components in the x and y directions, and we have assumed that $K_s' = K_s'(x, y)$, independent of z, and $\phi_s|_{\zeta_1} \approx \phi_s|_{\zeta_0} \approx \tilde{\phi}_s$. From (9-23) and (9-24) we obtain for the sea water zone

$$\nabla' \cdot (B_s K_s' \cdot \nabla' \tilde{\phi}_s) - \mathbf{q}_s' \Big|_{\zeta_0} \cdot \nabla'\zeta_0 + q_{sz} \Big|_{\zeta_0} - [S_{0s}B_s + n(1 + \delta)]\, \partial\tilde{\phi}_s/\partial t + n\delta\, \partial\tilde{\phi}_f/\partial t = 0$$

$$\tag{9-25}$$

For an impervious bottom, by (5-66), the sum of the second and third terms on the left-hand side of (9-25) vanish, and we obtain for the sea water zone

$$\nabla' \cdot (B_s K_s' \cdot \nabla' \tilde{\phi}_s) - [S_{0s}B_s + n(1 + \delta)]\, \partial\tilde{\phi}_s/\partial t + n\delta\, \partial\tilde{\phi}_f/\partial t = 0 \tag{9-26}$$

If sinks (e.g., wells) are located in the sea water zone, we add the term $-P_s(x, y, t)$, expressing distributed pumping, or $-P_s(x_i, y_i, t)\, \delta(x - x_i, y - y_i)$, expressing localized pumping at (x_i, y_i) on the left-hand side of (9-25) or (9-26); $\delta(x, y)$ here is the *Dirac delta function* (see comment following (5-57)).

For the phreatic surface, it follows from (5-46) and (9-19) that

$$n\, \partial F/\partial t + (\mathbf{q} - \mathbf{N}) \cdot \nabla F = 0; \qquad \mathbf{N} = -N\mathbf{1z}$$

$$F = z - \zeta_2 = z - \phi_f \Big|_{\zeta_2} \approx z - \tilde{\phi}_f$$

$$n\, \partial\tilde{\phi}_f/\partial t = (\mathbf{q}_f - \mathbf{N}) \cdot \nabla(z - \zeta_2)$$

$$= q_{fz} \Big|_{\zeta_2} + N - \mathbf{q}_f' \Big|_{\zeta_2} \cdot \nabla'\zeta_2 \tag{9-27}$$

For $\tilde{\mathbf{q}}_f$ we obtain

$$\tilde{\mathbf{q}}_f = \frac{1}{B_f} \int_{\zeta_1}^{\zeta_2} \mathbf{q}_f dz = -\frac{1}{B_f} \int_{\zeta_1}^{\zeta_2} K_f \cdot \nabla \phi_f dz =$$

$$-\frac{K_f'}{B_f} \cdot \left[\nabla' \int_{\zeta_e}^{\zeta_2} \phi_f dz - \phi_f \Big|_{\zeta_2} \nabla'\zeta_2 + \phi_f \Big|_{\zeta_1} \nabla'\zeta_1 + \phi_f \Big|_{\zeta_2} - \phi_f \Big|_{\zeta_e} \right] \approx -K_f' \cdot \nabla'\tilde{\phi}_f$$

$$\tag{9-28}$$

By combining (9-17), (9-22), and (9-27) we obtain for the fresh water region

$$\nabla' \cdot (B_f K_f' \cdot \nabla' \tilde{\phi}_f) + n(1 + \delta)\, \partial\tilde{\phi}_s/\partial t - [n(1 + \delta) + S_{0f}B_f]\, \partial\tilde{\phi}_f/\partial t + N = 0 \tag{9-29}$$

If sinks (e.g., wells) are located in the fresh water region, we add the term $-P_f(x, y, t)$ or $-P_f(x_i, y_i, t)\,\delta(x - x_i, y - y_i)$ on the left-hand side of (9-29).

Another possibility for deriving expressions for $\mathbf{q}_s$ and $\mathbf{q}_f$ is to assume that $\phi_f(x, y, z, t) \cong \tilde{\phi}_f(x, y, t)$ and $\tilde{\phi}_s(x, y, z, t) \cong \tilde{\phi}_s(x, y, t)$, while $\mathbf{K}_f = \mathbf{K}_f(x, y, z)$ and $\mathbf{K}_s = \mathbf{K}_s(x, y, z)$. Then (9-24) and (9-28) are replaced by

$$B_f \tilde{\mathbf{q}}_f = \int_{\zeta_1}^{\zeta_2} \mathbf{q}_f\, dz = -\nabla' \tilde{\phi}_f \cdot \int_{\zeta_1}^{\zeta_2} \mathbf{K}(x, y, z)\, dz = -B_f \tilde{\mathbf{K}}'_f(x, y) \cdot \nabla' \tilde{\phi}_f \quad (9\text{-}30)$$

and a similar expression for $B_s \tilde{\mathbf{q}}_s$.

In general, $S_{0f} B_f \ll n$; $S_{0s} B_s \ll n$, so that the elastic storativity expressed by $S_{0f} B_f$ and $S_{0s} B_s$ may be neglected in (9-25) and (9-29).

For a confined aquifer, $\zeta_2 = \zeta_2(x, y)$ and the fresh water equation (9-29) reduces to

$$\nabla' \cdot (B_f \mathbf{K}'_f \cdot \nabla' \tilde{\phi}_f) + \mathbf{q}'_f \Big|_{\zeta_2} \cdot \nabla' \zeta_2 - q_{fz} \Big|_{\zeta_2} + n(1 + \delta)\, \partial \tilde{\phi}_s / \partial t$$

$$- (n\delta + S_{0f} B_f)\, \partial \tilde{\phi}_f / \partial t = 0 \quad (9\text{-}31)$$

Since at an impervious boundary

$$-\mathbf{q}' \Big|_{\zeta_2} \cdot \nabla' \zeta_2 + q_z \Big|_{\zeta_2} \equiv \mathbf{q} \Big|_{\zeta_2} \cdot \nabla(z - \zeta_2) = 0$$

(9-31) reduces to

$$\nabla' (B_f \mathbf{K}'_f \cdot \nabla \tilde{\phi}_f) + n(1 + \delta)\, \partial \tilde{\phi}_s / \partial t - (n\delta + S_{0f} B_f)\, \partial \tilde{\phi}_f / \partial t = 0 \quad (9\text{-}32)$$

Pinder and Page (1976) solve (9-25) and (9-29), numerically, for the interface under an island.

Essentially, we have assumed here horizontal flow in both the fresh water and the sea water regions. With this assumption, and with $S_{0f} B_f \ll n$, $S_{0s} B_s \ll n$, we could have written directly the two water balances for a phreatic aquifer.

$$-\nabla \cdot \mathbf{Q}'_f + N - n\frac{\partial(\zeta_2 - \zeta_1)}{\partial t} = 0; \qquad -\nabla \cdot \mathbf{Q}'_s - n\frac{\partial \zeta_1}{\partial t} = 0 \quad (9\text{-}33)$$

(compare with (9-25) and (9-29)), where we may introduce

$$\mathbf{Q}'_f = -B_f \mathbf{K}'_f \cdot \nabla' \tilde{\phi}_f; \qquad \mathbf{Q}'_s = -B_s \mathbf{K}'_s \cdot \nabla' \tilde{\phi}_s, \qquad \zeta_2 = \tilde{\phi}_f, \qquad \zeta_1 = (1 + \delta)\tilde{\phi}_s - \delta \tilde{\phi}_f$$

Initial and Boundary Conditions

We have to specify initial and boundary conditions for both the fresh water region and the sea water one. They are not independent in view of (9-18). One should recall that the flow is assumed essentially horizontal in both regions.

If we specify initial fresh water levels $\tilde{\phi}_f = \tilde{\phi}_f(x, y, 0) \equiv \zeta_2(x, y, 0)$ and sea water heads $\tilde{\phi}_s = \tilde{\phi}_s(x, y, 0)$, then the interface elevation $\zeta_1(x, y, 0)$ is dictated by (9-18). One way of obtaining initial conditions is from field observations.

With respect to the boundary conditions along the coast, we have first to

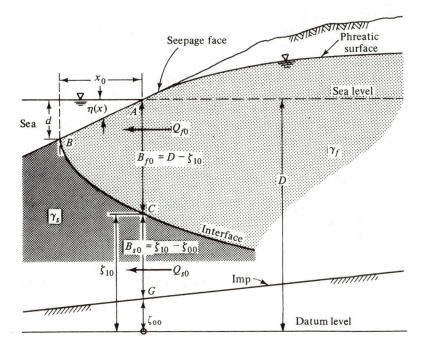

Figure 9-9 Nomenclature for boundary conditions along the coast.

decide where to locate this boundary. As mentioned above (e.g., end of Sec. 9-3), we have always an *outflow face* on the bottom of the sea, through which the fresh water leaves the aquifer (*AB* in Fig. 9-9). From the definition of ϕ_f, it follows that along the bottom of the sea (measured with respect to sea level):

$$\phi_f = z + \frac{p}{\gamma_f} = -\eta + \frac{\eta \gamma_s}{\gamma_f} = \eta \frac{\gamma_s - \gamma_f}{\gamma_f} = \eta/\delta > 0 \qquad (9\text{-}34)$$

where $\eta(x)$ is the depth of the sea bottom. Hence, $\phi_f|_A < \phi_f|_B$.

Usually, the vertical surface through point A (that is, AG) is taken as the boundary of both the fresh water and the salt water regions. The length x_0 of the outflow face can be estimated from the parabolic shape of a steady interface in an infinitely thick aquifer (Fig. 9-10; Glover, 1959)

$$y^2 = (2Q\delta/K)x + (Q\delta/K)^2; \qquad \delta = \gamma_f/(\gamma_s - \gamma_f)$$
$$x = 0, \ y = \delta Q/K; \qquad y = 0, \ x = -x_0 = -\delta Q/2K \qquad (9\text{-}35)$$

For a steady flow and a vertical outflow face (i.e., assuming that AG is the sea boundary), Henry (1959) obtained $y|_{x=0} = 0.741\delta Q/K$.

Because of the approximation involved, there is no unique way of expressing the boundary conditions for the fresh water and the sea water flow domains along AG. The following considerations may be useful.

1. For the fresh water we may assume $\phi_f|_{AC} = 0$ (measured with respect to sea level). However, as we know that $\phi_f|_C > \phi_f|_B > \phi_f|_A = 0$, that is, equipotentials are not vertical in the domain ABC, a certain error is introduced. We may reduce it by taking some higher value for $\phi_f|_{AC}$.
2. If we know the fresh water discharge rate, Q_{f0}, to the sea through AC (e.g., in a two-dimensional case), we can estimate $AC = B_{f0} = \delta Q_{f0}/K$. Then, knowing $\phi_f|_{AC}$ and B_{f0}, $\phi_s|_{CG}$ is determined from (9-18).
3. In general, Q_{f0} and Q_{s0} (per unit length of coast) vary along the boundary and are unknown. If we assume $\phi_f|_{AB} \cong d/2\delta \approx 0$, then

$$\frac{\phi_f|_{AC}}{R} = Q_{f0}; \qquad Q_{f0} = +K_f B_{f0} \left.\frac{\partial \phi_f}{\partial x}\right|_{AC} \qquad (9\text{-}36)$$

where R is the resistance to the flow to the sea through ACB. One may estimate $R \approx \bar{L}/K\bar{A} = \alpha/K$ where $\bar{L}$ is an average length of flow, $\bar{A}$ is an average flow cross section and α ($=\bar{L}/\bar{A}$) is a dimensionless coefficient which has to be estimated. With these considerations, (9-36) becomes

$$\frac{\phi_f}{\alpha} - B_{f0}\frac{\partial \phi_f}{\partial x} = 0 \quad \text{on } AC \qquad (9\text{-}37)$$

which is a third type boundary condition.

A similar expression, with a coefficient β, can be written for the sea water boundary CG

$$\frac{\phi_s}{\beta} - B_{s0}\frac{\partial \phi_s}{\partial x} = 0 \quad \text{on } CG. \qquad (9\text{-}38)$$

In addition, we have from (9-18) the condition

$$z\Big|_c = -B_{f0} = \phi_s(1 + \delta) - \phi_f\delta \qquad (9\text{-}39)$$

and

$$B_{s0} + B_{f0} = D - \zeta_{00} \quad \text{is a known value.}$$

With these, (9-37) and (9-39) become

$$\frac{\phi_f}{\alpha} + [\phi_s(1 + \delta) - \phi_f\delta]\frac{\partial \phi_f}{\partial x} = 0 \qquad (9\text{-}40)$$

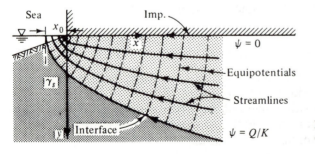

Figure 9-10 The flow net near the coast *(Glover, 1959)*.

$$\frac{\phi_s}{\beta} + [D - \zeta_{00} + \phi_s(1 + \delta) - \phi_f\delta]\frac{\partial\phi_s}{\partial x} = 0 \qquad (9\text{-}41)$$

which are the boundary conditions in terms of ϕ_f and ϕ_s. The coefficients α and β have to be estimated (possibly as part of the procedure of parameter identification) and the depth of the interface (point C) will adjust itself such that (9-18) is satisfied.

At no point in the development of the equations and boundary conditions presented above, was use made of the Ghyben–Herzberg relationship (9-12), which corresponds to steady flow with a stationary interface and $\phi_s = $ const., $\phi_f = h_f$. Instead, we have used the relationship (9-4) based only on the identity of pressure as the interface is approached from both sides.

Equations (9-26) and (9-29) or (9-32) are two equations in $\tilde{\phi}_f(x, y, t)$ and $\tilde{\phi}_s(x, y, t)$ which have to be solved simultaneously, subject to appropriate boundary conditions on $\tilde{\phi}_s$ and $\tilde{\phi}_f$ in the xy plane. For most problems of practical interest, the solution will have to be derived numerically (e.g., Pinder and Page, 1976; Kapuler and Bear, 1975). Once $\tilde{\phi}_s$ and $\tilde{\phi}_f$ are known, the position and shape of the interface are obtained from (9-18). However, one should note that in (9-25) and (9-29) we have $B_f = \zeta_2 - \zeta_1$ and $B_s = \zeta_1 - \zeta_0$, where ζ_0 is the known bottom elevation, $\zeta_2 \equiv \tilde{\phi}_f$, while ζ_1 is related to $\tilde{\phi}_f$ and $\tilde{\phi}_s$ by (9-18). In an iterative numerical scheme of solution, this poses no special difficulty. Also, special attention should be given in the solution scheme to the possibility that the sea water region disappears at some distance from the coast (Fig. 9-2), where the interface intersects the impervious bottom of the aquifer.

9-5 STATIONARY INTERFACE

To demonstrate the shape of the interface and the relationship which exists between the extent of sea water intrusion and the flow of fresh water to the sea, consider a stationary interface, i.e., steady flow, in a case where the flow is everywhere perpendicular to the coast line. Figure 9-2 shows such a cross section in a phreatic aquifer. For the sake of simplicity we shall assume a horizontal bottom and a constant thickness B for a confined aquifer (Fig. 9-11a). Let the origin $x = 0$ be located at the interface toe (point G). The seaward fresh water flow at this point is Q_0. It is the difference between the total replenishment of the aquifer and the withdrawal, in the coastal strip to the right of point G.

Consider first the confined aquifer shown in Fig. 9-11a. Using Dupuit's assumption of horizontal flow of fresh water (and vertical equipotentials), continuity leads to

$$Q_0 = \text{const.} = -K_f h(x)\,\partial\phi(x)/\partial x; \quad K \equiv K_f = k\gamma_f/\mu_f, \quad \phi \equiv \phi_f \qquad (9\text{-}42)$$

Alternatively, we could obtain (9-42) as a reduction of the equations developed in the previous section, with $\tilde{\phi}_s = $ const. and $\tilde{\phi}_f = \phi_f$.

Since $h_s = d + h(x) = \delta\phi$, and $\delta\phi_0 = d + B$, (9-42) becomes

$$Q_0 = -\frac{Kh}{\delta}\frac{dh(x)}{dx} \quad \text{or} \quad Q_0 = -K[\delta\phi(x) - d]\frac{d\phi(x)}{dx} \qquad (9\text{-}43)$$

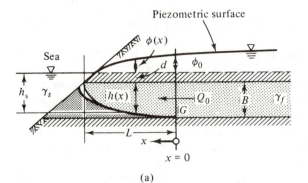

(a)

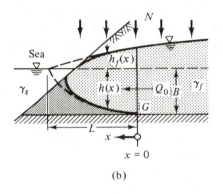

(b)

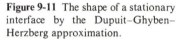

Figure 9-11 The shape of a stationary interface by the Dupuit–Ghyben–Herzberg approximation.

Because the flow is steady and assumed horizontal, we have used the Ghyben–Herzberg relationship (9-12).

It may be of interest to compare the expression for Q_0 in (9-43), based on the Dupuit assumptions, with the one derived by integration

$$
\begin{aligned}
Q &= -K \int_{-(h+d)}^{-d} \frac{\partial \phi}{\partial x} \, dz \\
&= -K \left[\frac{\partial}{\partial x} \int_{-(h+d)}^{-d} \phi \, dz - \phi \Big|_{-d} \frac{\partial(-d)}{\partial x} + \phi \Big|_{-(d+h)} \frac{\partial\{-(d+h)\}}{\partial x} \right] \\
&= -K \left[\frac{\partial}{\partial x} (h\tilde{\phi}) - \phi \Big|_{-(d+h)} \frac{\partial h}{\partial x} \right] = -K \left[\tilde{\phi} \frac{\partial h}{\partial x} + h \frac{\partial \tilde{\phi}}{\partial x} - \phi \Big|_{-(h+d)} \frac{\partial h}{\partial x} \right] \\
&= -Kh \frac{\partial \tilde{\phi}}{\partial x} + \left[\tilde{\phi} - \phi \Big|_{-(h+d)} \right] \frac{\partial h}{\partial x} \approx -Kh \frac{\partial \tilde{\phi}}{\partial x}
\end{aligned}
\tag{9-44}
$$

where the approximation is based on the hydrostatic pressure distribution in the stationary sea water, that is, $\tilde{\phi}|_{-(h+d)} = -(h+d) + (h+d)\gamma_s/\gamma_f = (h+d)/\delta$, and $(h+d)/\delta = \phi|_{-(h+d)} \approx \tilde{\phi}(x) = \phi(x)$.

By integrating (9-43), with $x = 0$, $\phi = \phi_0$ (or $h = B$), we obtain

$$Q_0 x = K[B^2 - h^2(x)]/2\delta; \qquad Q_0 x = K\delta(\phi_0^2 - \phi^2)/2 - Kd(\phi_0 - \phi) \quad (9\text{-}45)$$

which shows that the interface has the form of a parabola.

At $x = L$ we set $h = 0$, $\phi = d/\delta$. Then, with $\phi_0 = (B + d)/\delta$ we obtain

$$Q_0 L = \frac{K\phi_0}{2}(\delta\phi_0 - 2d) + Kd^2/2\delta = \frac{K}{2\delta}B^2 \qquad (9\text{-}46)$$

The relationship among the length L of sea water intrusion, the discharge Q_0 to the sea and piezometric head ϕ_0 above the toe, is clearly expressed by this equation.

As Q_0 increases, L decreases. This means that *the extent of sea water intrusion, expressed by* L, *is a decision variable in the management of the coastal aquifer*; it is controlled by controlling Q_0, or alternatively by controlling the recharge and/or pumping in the coastal aquifer strip.

Figure 9-11b shows a phreatic aquifer with uniform accretion N (say, natural replenishment from precipitation). Again, assuming that the steady flow in the aquifer is essentially horizontal and $h(x) = \delta h_f(x)$, continuity leads to

$$Q_0 + Nx = -K(h + h_f)\,\partial h_f/\partial x = -K(1 + \delta)\,h_f\partial h_f/\partial x \qquad (9\text{-}47)$$

Integrating (9-47) from $x = 0$, $h_f = \phi_0$, $h = B$, leads to

$$\phi_0^2 - h_f^2 = (2Q_0 x + Nx^2)/K(1 + \delta) \qquad (9\text{-}48)$$

If $N = N(x) \neq$ const., we have to start from a continuity equation in the form

$$-\frac{\partial Q}{\partial x} + N(x) = 0; \qquad Q = K(h + h_f)\,\partial h_f/\partial x \qquad (9\text{-}49)$$

It may, again, be interesting to derive the expression for Q in (9-49) by integration

$$Q = -K\int_{-h}^{h_f} \frac{\partial \phi}{\partial x}\,dz = -K\left[\int_{-h}^{h_f} \phi\,dz - \phi\bigg|_{h_f}\frac{\partial h_f}{\partial x} + \phi\bigg|_{-h}\frac{\partial(-h)}{\partial x}\right]$$

$$= -K\frac{\partial}{\partial x}\left[(h_f + h)\,\tilde{\phi}\right] - K\left[-h_f\frac{\partial h_f}{\partial x} - \phi\bigg|_{-h}\frac{\partial h}{\partial x}\right]$$

Since $\phi|_{-h} = -h + \gamma_s h/\gamma_f = h/\delta$, we obtain

$$Q = -K(h_f + h)\frac{\partial\tilde{\phi}}{\partial x} - K\left[(\tilde{\phi} - h_f)\frac{\partial h_f}{\partial x} + \left(\tilde{\phi} - \frac{h}{\delta}\right)\frac{\partial h}{\partial x}\right]$$

Assuming $h_f \approx \tilde{\phi} \approx \phi|_{-h} = h/\delta$, we obtain

$$Q = -K(h + h_f)\frac{\partial h_f}{\partial x}$$

as in (9-49).

The rate of fresh water flowing into the sea in a phreatic aquifer with constant accretion is $Q_0 + NL$. Bear and Dagan (1962, 1963, 1964a, and 1966) and Strack

(1973) studied the possibility of intercepting part of the fresh water flowing to the sea in a coastal aquifer. The field technique, developed by Water Planning for Israel, Tel-Aviv, in the early 60's is called the *coastal collector*. According to this technique, an array of shallow wells is placed along a line parallel to the coast, and not far from it, in order to intercept part of the fresh water flow to the sea, without causing the toe of the interface to advance farther inland. The method is implemented in the coastal aquifer in Israel.

At $x = L$, $h_f = 0$, hence we obtain from (9-48)

$$\phi_0^2 = \frac{2Q_0L + NL^2}{K(1 + \delta)} \quad \text{or} \quad Q_0 = \frac{KB^2}{2L} \frac{(1 + \delta)}{\delta^2} - \frac{NL}{2}, \qquad \phi_0 = \frac{B}{\delta} \qquad (9\text{-}50)$$

For $N = 0$

$$\phi_0^2 = 2Q_0L/K(1 + \delta) = B^2/\delta^2 \qquad (9\text{-}51)$$

Again, the interface has a parabolic shape and we have here a relationship between L and Q_0. Equation (9-51) relates ϕ_0 (piezometric head above the toe) to L. By controlling ϕ_0 (say, by means of artificial recharge), the water table may be lowered both landward and seaward of the toe, without causing any additional sea water intrusion. Landward of the toe, water levels may fluctuate as a result of some optimal management scheme. When pumpage takes place seaward of the toe, the interface there will rise and may contaminate wells if their screened portion is not at a sufficient distance above it (Sec. 9-7).

It may be of interest to note that when $Q_0 = 0$, that is, no seaward fresh water flow takes place above the toe, $L = (B/\delta)[K(1 + \delta)/N]^{1/2}$. This case corresponds to the lowest value of fresh water discharge to the sea, $Q_{fL} (= NL)$, with seaward fresh water flow everywhere above the interface.

Instead of $\phi = 0$ at $x = L$, we could use another boundary condition for the confined aquifer

$$x = L, \qquad h = \beta \delta Q/K \qquad (9\text{-}52)$$

For $\beta = 1$, this is the exact solution for the depth of the interface as derived by Glover (1959), assuming that we have for the interface the parabola (9-35)

$$y^2 = (2\delta Q/K) x + (\delta Q/K)^2 \qquad (9\text{-}53)$$

shown in Fig. 9-10 (Bear, 1972, p. 552). Henry (1959) obtained (9-52) with $\beta = 0.741$. Columbus (1965), Vappicha and Nagaraja (1976), among others, used this boundary condition.

The same boundary condition, with $Q = Q_{fL} = $ flow of fresh water to the sea, can be used as an approximation for the phreatic aquifer; Q_{fL} is the difference between total recharge and total pumping in the coastal aquifer strip.

A Two Layered Coastal Aquifer

Often a coastal aquifer is divided by impervious (or semipervious), relatively thin, layers into a number of subaquifers. Figure 9-1d shows an example of an upper phreatic aquifer and a lower confined one.

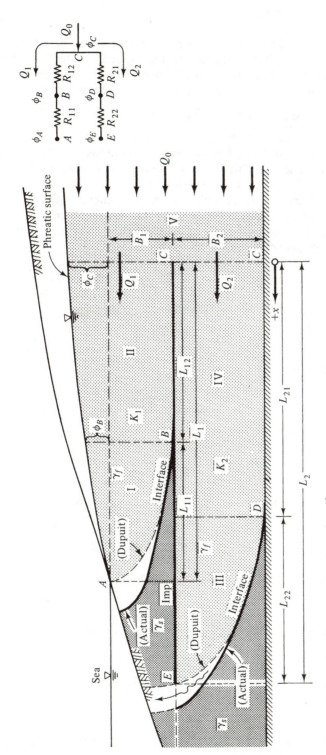

Figure 9-12 Sea water intrusion in a two layered aquifer.

As an example, consider the case of steady flow shown in the coastal aquifer shown in Fig. 9-12. We shall make use of the Dupuit assumption of vertical equipotentials. However, although this is a fairly good approximation with respect to potentials, deducing from it that the discharge rates are distributed according to

$$Q_1/Q_2 = (\phi_C + B_1)/B_2, \qquad Q_1 + Q_2 = Q_0 \qquad (9\text{-}54)$$

would be very erroneous. It is obvious that the rate of flow in each subaquifer depends on the resistance that the water encounters along its flow path (which, in turn, depends on the length of the flow, cross section and hydraulic conductivity) and on the total drop of head in that subaquifer.

Let us divide the fresh water domain in each subaquifer into two subdomains: one above the stationary interface (i.e., no flow in the salt water region) and the other from the interface toe to the end of the separating layer. It is convenient to visualize these subdomains as resistances to the flow

$$R_{12} = \frac{\phi_C - \phi_B}{Q_1}, \quad R_{11} = \frac{\phi_B - \phi_A}{Q_1}, \quad R_{21} = \frac{\phi_C - \phi_D}{Q_2}; \quad R_{22} = \frac{\phi_D - \phi_E}{Q_2} \qquad (9\text{-}55)$$

Hence

$$Q_1 = \frac{\phi_C - \phi_A}{R_{11} + R_{12}}; \quad Q_2 = \frac{\phi_C - \phi_E}{R_{21} + R_{22}}; \quad \frac{Q_1}{Q_2} = \frac{\phi_C - \phi_A}{\phi_C - \phi_E} \frac{R_{21} + R_{22}}{R_{11} + R_{12}} \qquad (9\text{-}56)$$

Now, for $\phi_C \ll B_1$

$$R_{12} = \frac{\phi_C - \phi_B}{Q_1} \approx \frac{L_{12}}{K_1 B_1}$$

or, more generally

$$R_{12} = \int_0^{L_{12}} \frac{dx}{K_1(x) B_1(x)} \qquad (9\text{-}57)$$

From (9-50) written for $N = 0$, $Q_0 = Q_1$, $\phi_B - \phi_A = \phi_0 = B_1/\delta$, we obtain

$$R_{11} = \frac{\phi_B - \phi_A}{Q_1} \approx \frac{2L_{11}}{KB_1} \cdot \frac{\delta}{1 + \delta} \approx \frac{L_{11}}{K_1 B_1/2} \qquad (9\text{-}58)$$

Similarly

$$R_{21} = \frac{\phi_C - \phi_D}{Q_2} = \frac{L_{21}}{K_2 B_2} \qquad (9\text{-}59)$$

and from (9-46), with $\phi_D - \phi_E = \phi_0 = B_2/\delta$, it follows that

$$R_{22} = \frac{\phi_D - \phi_E}{Q_2} = \frac{L_{22}}{K_2 B_2/2} \qquad (9\text{-}60)$$

Hence

$$\frac{Q_1}{Q_2} = \frac{\phi_C}{\phi_C - B_1/\delta} \cdot \frac{(2L_{22} + L_{21})/K_2 B_2}{(2L_{11} + L_{12})/K_1 B_1} = \frac{\phi_C}{\phi_C - B_1/\delta} \cdot \frac{K_1 B_1}{K_2 B_2} \cdot \frac{L_2 + L_{22}}{L_1 + L_{11}} \qquad (9\text{-}61)$$

Altogether, we have to solve for: ϕ_C, ϕ_B, ϕ_D, L_{11}, L_{12}, L_{21}, L_{22}, Q_1 and Q_2. We have Eqs. (9-57) through (9-60) and

$$L_{11} + L_{12} = L_1; \qquad L_{21} + L_{22} = L_2; \qquad Q_1 + Q_2 = Q_0$$

$$\phi_B = B_1/\delta; \qquad \phi_D = (B_1 + B_2)/\delta \qquad (9\text{-}62)$$

so that a solution in terms of Q_0 can be obtained. By adding (9-57) and (9-58), we obtain

$$Q_1 = \frac{T_1}{L_1}\left(\phi_C - \frac{B_1}{2\delta}\right) \qquad (9\text{-}63)$$

Similarly, from (9-59) and (9-60)

$$Q_2 = \frac{T_2}{L_2}\left(\phi_C - \frac{B_2}{2\delta} - \frac{B_1}{\delta}\right) \qquad (9\text{-}64)$$

$$Q_1 + Q_2 = Q_0; \qquad \phi_C = \frac{T_2 B_2/2L_2\delta + B_1(T_1/2L_1 + T_2/L_2)/\delta + Q_0}{T_1/L_1 + T_2/L_2} \qquad (9\text{-}65)$$

$$\frac{Q_1}{Q_2} = \frac{T_1 L_2}{T_2 L_1} \cdot \frac{Q_0 + (T_2/L_2)(B_1 + B_2)/2\delta}{Q_0 - (T_1/L_1)(B_1 + B_2)/2\delta}, \quad \text{etc.} \qquad (9\text{-}66)$$

The same method can be applied to the case where replenishment of the phreatic aquifer takes place and to cases with a larger number of layers.

Table 9-1 shows some results of an example computed by the above formulas. In all cases: $Q_0 = 0.2 \times 10^6$ m³/year/km; no natural replenishment, and $L_1 = L_2 = L$; the upper and lower numbers indicate upper and lower subaquifers. Figure 9-13 shows some of these results graphically.

The following conclusions may be drawn from these results.

(a) As $(B_1 + B_2)$ increases, so does the length of sea water intrusion, with the intrusion in the lower aquifer being larger than in the upper one.
(b) A similar conclusion may be drawn with respect to K (in a homogeneous aquifer).
(c) As Q_0 increases, the length of sea water intrusion decreases.
(d) As the length of the separating layer (measured from the coast) increases, we will have more sea water intrusion in the upper aquifer and less in the lower one.
(e) As the ratio B_1/B_2 becomes larger, so does the ratio Q_1/Q_2, but the rate of growth of the latter is faster.

Mualem and Bear (1974) discuss the case where the separating layer is semipervious, using the analogy that exists between a phreatic surface and an interface (Bear, 1972, p. 522). Figure 9-14 shows some of the cases for which they present solutions for $h(x)$ and for Q_{u0}/Q_{10}.

Effect of Wells on Sea Water Intrusion

Strack (1976) develops a technique for determining the shape and position of an interface in steady, essentially horizontal fresh water flow in a coastal aquifer,

Table 9-1 Examples of sea water intrusion in two subaquifers

Case	Hydraulic conductivity K (m/year)	Thickness of aquifer B (m)	Distance of separation L_i (m)	Inflow Q_i (m/year/km)	Interface intrusion L_{ii} (m)
1	2500	40	3000	120	500
	2500	40		80	750
2	5000	40	3000	140	857
	5000	40		60	1998
3	3600	40	3000	128	671
	3600	40		72	1213
4	3600	20	3000	107	202
	3600	20		93	233
5	3600	40	2000	143	604
	3600	40		57	1520
6	3600	40	4000	122	711
	3600	40		78	1102
7	3600	60	3000	172	1133
	3600	20		28	760
8	3600	20	3000	72	302
	3600	60		128	1514

assuming no flow in the salt water zone. The aquifer is isotropic, homogeneous, and with a horizontal impervious base. His objective is to introduce a *single harmonic potential* (i.e., satisfying the Laplace equation) which is defined for the fresh water zone, *both* for the zone above the interface and that beyond it (zones I and II, respectively, in Fig. 9-15). The single valued potential is continuous throughout the multiple zone aquifer. Figure 9-15 illustrates a phreatic aquifer case. Strack (1976) also considers flow in a confined quifer.

For the flow in Zone I (phreatic flow above an interface, Fig. 9-15), Strack (1976) introduces a potential ϕ^s defined by

$$\phi^s = \tfrac{1}{2}(h_f + h_s)h_f = \tfrac{1}{2}(1 + \delta)h_f^2 \tag{9-67}$$

$$\mathbf{Q'} = -K\nabla'\phi^s = -K(1 + \delta)h_f\nabla'h_f \tag{9-68}$$

where symbols are explained in Fig. 9-15 and $\mathbf{Q'}$ is fresh water discharge per unit width through the entire thickness of the aquifer.

For the flow in zone II (phreatic flow without an interface), he introduces the potential

$$\phi^s = \tfrac{1}{2}(\zeta_2^2 - B^2\gamma_s/\gamma_f) = \tfrac{1}{2}\{h_f^2 + B(2\delta h_f - B)/\delta\} \tag{9-69}$$

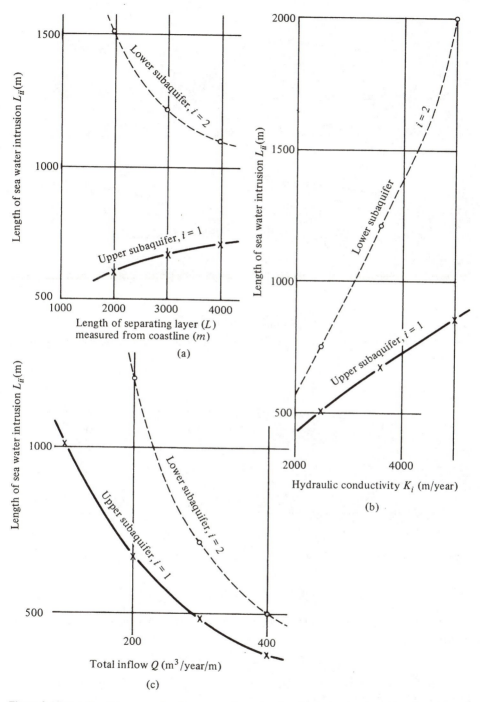

Figure 9-13 Results of an example of sea water intrusion in a layered coastal aquifer *(Kapuler and Bear, 1975)*.

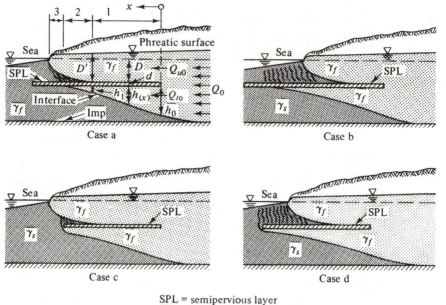

SPL = semipervious layer
🟦🟦🟦 = Zone of mixed water

Figure 9-14 The interface in a coastal aquifer in the presence of a semipervious layer *(Mualem and Bear, 1974).*

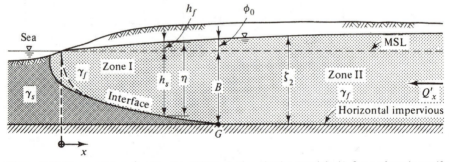

Figure 9-15 Nomenclature for Strack's (1976) single valued potential, ϕ^s, for a phreatic aquifer.

$$\mathbf{Q}' = -K\,\nabla'\phi^s = -K\zeta_2\nabla'\zeta_2 = -K\zeta_2\nabla'h_f \qquad (9\text{-}70)$$

At point G (= toe of interface), $B = \delta h_f$ and the two values of ϕ^s become identical

$$\phi^s = \tfrac{1}{2}(1 + \delta)\,\phi_0^2 = \tfrac{1}{2}(1 + \delta)\,(B/\delta)^2 \qquad (9\text{-}71)$$

Also, the flows at that point, when approached from both sides, become identical. Since $\nabla'\cdot\mathbf{Q}' = 0$ in each zone, we have

$$\partial^2\phi^s/\partial x^2 + \partial^2\phi^s/\partial y^2 = 0 \qquad (9\text{-}72)$$

which means that Strack's potentials ϕ^s are harmonic functions in the respective zones.

Along the interface toe (point G in Fig. 9-16a and the dashed line in Fig. 9-16b), $h_s = B = \delta h_f$. Hence

$$\text{for zone I: } \phi^s = \frac{1}{2}(1 + \delta)\frac{B^2}{\delta^2}; \qquad \text{for zone II: } \phi^s = \frac{1}{2}(1 + \delta)\frac{B^2}{\delta^2} \qquad (9\text{-}73)$$

i.e., the interface toe corresponds to an equipotential in both cases.

Strack (1976) demonstrates the application of ϕ^s by considering the case of a single well located at a distance x_w from the coast (Fig. 9-16). The well is pumping at a rate $Q_w = $ const., superimposed on uniform flow from infinity to the coast at a specific discharge Q'_{0x} ($=$ discharge per unit width through the entire thickness of the aquifer).

For this case, the potential ϕ^s (defined such that $\mathbf{Q}' = -K\nabla'\phi^s$) for zone II is (Sec. 8-10)

$$\phi^s = \frac{Q'_{0x}}{K}x + \frac{Q_w}{4\pi K}\ln\frac{(x - x_w)^2 + y^2}{(x + x_w)^2 + y^2} \qquad (9\text{-}74)$$

It may be verified that this function is indeed harmonic, that it incorporates a well of discharge Q_w at $(x_w, 0)$ and that the boundary condition of $h_f = 0$, that is, $\phi^s = 0$, is satisfied along the coast ($x = 0$).

The equation for the toe of the interface in the xy plane can now be found by setting $\phi^s = \frac{1}{2}(1 + \delta)B^2/\delta^2$ in (9-74)

$$\frac{1}{2}(1 + \delta)\frac{B^2}{\delta^2} = \frac{Q'_{0x}}{K}x + \frac{Q_w}{4\pi K}\ln\frac{(x - x_w)^2 + y^2}{(x + x_w)^2 + y^2} \qquad (9\text{-}75)$$

Figure 9-16b (dashed curve) shows the location of the toe as defined by (9-75).

Actually when we have at a distance x_w from the coast a well pumping at a rate Q_w, we do not know a priori whether we have the situation shown in Fig. 9-16, that is, with the well in the fresh water zone landward of the interface toe, or that shown in Fig. 9-17, where the interface has advanced beyond the well. Both figures show a vertical cross section through the well along a line which is perpendicular to the coast.

Recalling that the elevation of the water table above the toe (point G) is $\phi_0 = B/\delta$ (in the phreatic aquifer considered here), the pumping rate Q_w may be such that along x_w we have *everywhere* $h_f < B/\delta$. Then there is no *barrier* to prevent the interface from advancing landward until a head $\phi_0 = B/\delta$, corresponding to an equilibrium with a stationary interface, is reached. The well will then operate above the interface, causing upconing. The problem of interface upconing is considered in detail in Sec. 9-7.

In order to ensure that the interface toe will be arrested at some distance seaward of the well, we must maintain a *barrier* at an elevation $\phi_0 = B/\delta$. Beyond such a barrier, water table elevations may drop again as in Fig. 9-16a (even

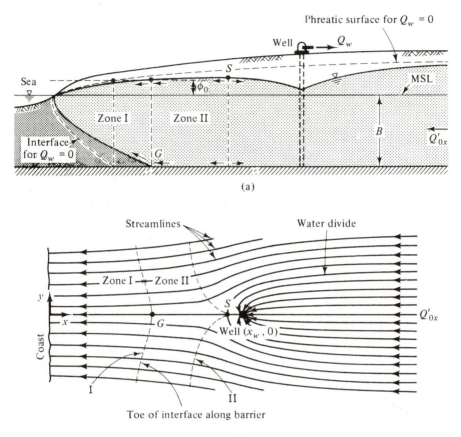

(a)

(b)

Figure 9-16 Location of the interface for a single well near the coast *(Strack, 1976).*

below sea level!). Sometimes, artificial recharge (Sec. 3-4) is practised in order to maintain such a barrier. This statement is true for a regional situation (in the xy plane) and not only for the cross section represented in Figs 9-16a and 9-17.

Figure 9-16b shows two significant points along x_w: point G where the water table elevation is $\phi_0 = B/\delta$, and point S, which is a *stagnation point* (see Sec. 8-11) on the *water divide* defining the region supplying water to the well. Note that the distance x_w from the sea is different in Fig. 9-16a and in Fig. 9-16b. The water velocity at S is zero. Landward of S the flow is towards the well; seaward of S it is towards the sea. From the discussion above, it follows that for low values of Q_w, points S and G in Fig. 9-16b are distinct and at some distance apart. As Q_w increases, they approach one another until they merge to a single point which is a stagnation point with head $\phi_0 = B/\delta$. The interface will advance (along the cross section and elsewhere) until it reaches this point (Fig. 9-18 and dashed line II in Fig. 9-16a). This is an unstable critical situation. The curve representing the toe of the interface corresponds to the equipotential $\phi_0 = B/\delta$. Both seaward

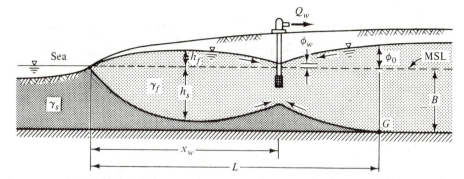

Figure 9-17 A well pumping above the interface in a coastal aquifer.

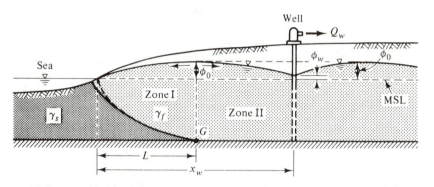

Figure 9-18 A single well near the coast—critical situation *(Strack, 1976).*

and landward of this curve, potentials are lower. Any further increase in pumping rate, Q_w, will produce a further drawdown of the water table, resulting in a rapid advance of the interface until a new equilibrium is reached, with the interface toe landward of the well, as shown in Fig. 9-17. A safe and stable situation is when a band of some width, where water table elevations are higher than B/δ, exists between the pumping well and the coast.

Figure 9-19 summarizes the different situations as Q_w increases. Strack (1976) calculates the critical Q_w by requiring that the toe of the interface coincides with the stagnation point (points G and S, respectively, in Fig. 9-16b). For the stagnation point he obtains (see Sec. 8-11)

$$x_s = x_w \left\{ 1 - \frac{Q_w}{\pi Q'_{0x} x_w} \right\}^{1/2} ; \qquad y_s = 0 \qquad (9\text{-}76)$$

He then inserts these coordinates in (9-75), thus requiring that the interface passes through the stagnation point, and obtains an expression for the critical Q_w

$$\lambda = 2 \left\{ 1 - \frac{\mu}{\pi} \right\}^{1/2} + \frac{\mu}{\pi} \ln \frac{1 - (1 - \mu/\pi)^{1/2}}{1 + (1 - \mu/\pi)^{1/2}} \qquad (9\text{-}77)$$

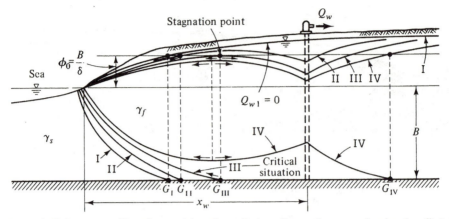

Figure 9-19 Summary of interface positions for a single well near the coast, showing the effect of a barrier.

where

$$\lambda = \frac{KB^2}{Q_{0x}'x_w} \cdot \frac{1+\delta}{\delta^2}; \qquad \mu = \frac{Q_w}{Q_{0x}'x_w} \qquad (9\text{-}78)$$

Strack (1976) also develops expressions for the location of the interface toe (in the xy plane) and the critical Q_w for a confined aquifer.

9-6 APPROXIMATE SOLUTIONS FOR A MOVING INTERFACE

The exact mathematical statement for a moving interface problem is actually presented in Sec. 9-2, with R_1 and R_2 denoting the fresh water and the salt water zones, above and below the interface, respectively. It is shown there that this statement consists of solving the appropriate partial differential equation of continuity, e.g., (9-1), in the two respective zones, such that ϕ_f and ϕ_s satisfy the boundary conditions (9-7) and (9-8) on the interface (in addition to other conditions along the external boundaries of the flow domains). However, due to the mathematical complexity of the problem as stated in this way, no analytical solution is known (except for a very small number of special cases) and approximations of different kinds have been employed by various investigators. In Sec. 9-4, averaged continuity equations—one for each water zone—are introduced, thus eliminating the interface boundary conditions by incorporating them in the equations themselves. By solving the two equations for $\tilde{\phi}_f$ and $\tilde{\phi}_s$, the elevation of the time-dependent interface can be obtained from (9-18).

In spite of the simplification achieved by introducing these approximations, the problem still requires a numerical solution for problems of practical interest.

When we consider sea water intrusion into a coastal aquifer, where the zone underlain by sea water wedge is limited, the problem is even more complicated, as we have on hand a problem with a moving boundary. We may illustrate this point by referring to Fig. 9-20 which shows a cross section perpendicular to the coast. Using the nomenclature of this figure, with all discharge rates being per unit length parallel to the coast, we obtain from (9-33) for $x < L$

$$\frac{\partial Q_f}{\partial x} - N + n_e \frac{\partial \phi_f}{\partial t} + n \frac{\partial \eta}{\partial t} = 0; \qquad \zeta_2 - \zeta_1 \equiv \eta + \phi_f \qquad (9\text{-}79)$$

where we have distinguished between (total) porosity, n, and specific yield, n_e. For the salt water in the zone $x < L$, we obtain

$$\frac{\partial Q_s}{\partial x} - n \frac{\partial \eta}{\partial t} = 0 \qquad (9\text{-}80)$$

Inserting

$$Q_f = -K_f(\eta + \phi_f) \frac{\partial \phi_f}{\partial x}; \qquad Q_s = -K_s(D - \eta) \frac{\partial \phi_s}{\partial x}; \qquad \eta = \delta \phi_f - (1 + \delta) \phi_s \qquad (9\text{-}81)$$

into (9-79) and (9-80) we obtain

$$\frac{\partial}{\partial x} \left[K_f(\eta + \phi_f) \frac{\partial \phi_f}{\partial x} \right] + N - n_e \frac{\partial \phi_f}{\partial t} - n \frac{\partial \eta}{\partial t} = 0 \qquad (9\text{-}82)$$

$$\frac{\partial}{\partial x} \left[K_s(D - \eta) \frac{\partial}{\partial x} \left(\frac{\delta}{1 + \delta} \phi_f + \frac{\eta}{1 + \delta} \right) \right] + n \frac{\partial \eta}{\partial t} = 0 \qquad (9\text{-}83)$$

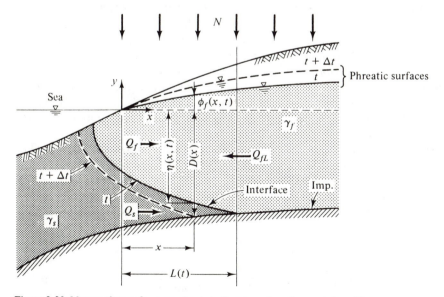

Figure 9-20 Nomenclature for a moving interface in a phreatic coastal aquifer.

We thus obtain two equations in $\phi_f(x, t)$ and $\eta(x, t)$ in the region $0 \le x \le L$, with $x = 0$, $\phi_f = 0$, $\eta = 0$; $x = L$, $\eta = D$, $Q_{fL} = -K_f(D + \phi_f)(\partial\phi_f/\partial x)$; Q_{fL} is a known function derived from the flow in the region $x > L$.

The difficulties involved in solving (9-82) and (9-83) stem from the fact that these are nonlinear equations and that the boundary at $x = L$ is moving. Thus, although the Dupuit assumptions have helped to simplify the problem, the resulting equations cannot be solved analytically and still require a numerical solution.

For the two-dimensional case, the Hele–Shaw analog (Sec. 5-6; Bear, 1972; pp. 687–697) is a relatively simple tool which gives direct visual solutions for a moving interface in rather complicated aquifer geometries.

Equations (9-82) and (9-83) can be extended to describe a regional case, with $\phi_f = \phi_f(x, y, t)$ and $\eta = \eta(x, y, t)$.

Solution by the Method of Successive Approximations

Bear and Dagan (1964b) use the method of successive steady states (or quasi-steady states) as proposed by Polubarinova-Kochina (1952, 1962; Bear, 1972, par. 8.4.4). According to this approach, we assume that at every instant, the moving interface maintains its steady state parabolic shape, e.g., (9-45), and that the unsteady process may be regarded as a sequence of steady states. A partial justification for this assumption stems from the observation that temporal variations of flow characteristics are much smaller than spatial ones.

Figure 9-21 summarizes the nomenclature that Bear and Dagan (1964b) use for solving the problem of a moving interface in a confined coastal aquifer. Initially, the seaward fresh water discharge, Q_{fL}^0, above the toe is equal to that flowing into the sea. At $t = 0$ the discharge Q_{fL}^0 above the toe is suddenly reduced to a new constant value $0 < Q_{fL}^* < Q_{fL}^0$, causing the interface to start advancing landward until a new equilibrium is reached. During the transition period, the fresh water outflow to the sea at $x = 0$ is gradually reduced from the initial Q_{fL}^0 to Q_{fL}^*. If the reduction of Q_{fL}^0 above the toe is such that $Q_{fL}^* < 0$, that is, the flow above the toe is in a landward direction, no equilibrium can be reached and the interface will continue to advance landward. In this case, the fresh water discharge at $x = 0$ gradually decreased from $Q_{f0}^0 > 0$, reaches zero and becomes negative, causing sea water to enter the aquifer.

In order to solve the problem, Bear and Dagan (1964b) assume specific modes of variation of $Q_f(x, t)$ in the region $0 \le x \le L$ from Q_{fL}^* at $x = L$ to $Q_{f0}(t)$ at $x = 0$. They thus regard $Q_{f0}(t) \equiv Q_f(0, t)$ as an unknown variable of the problem. Assuming that elastic storativity may be neglected, continuity considerations applied to the volume of sea water, U_{sw}, in the intruding wedge lead to

$$U_{sw}(t) = \int_{\eta=0}^{B} x(\eta, t)\, d\eta \qquad (9\text{-}84)$$

$$n\, dU_{sw}(t)/dt = -Q_{s0}(t) = Q_{f0} - Q_{fL} \qquad (9\text{-}85)$$

For the two variables of the problem: $\eta = \eta(x, t)$ and $Q_{f0}(t)$, we now have two equations modified for the confined case studied here. From (9-82), modified for the changed direction of the flows in Fig. 9-21, we obtain

$$\frac{\partial \eta}{\partial x} = \delta \frac{\partial \phi_f}{\partial x} - (1 + \delta) \frac{\partial \phi_s}{\partial x} = \delta \frac{Q_f}{K_f \eta} - (1 + \delta) \frac{Q_s}{K_s(B - \eta)} \tag{9-86}$$

Since at every instant $Q_f + Q_s = Q_{fL}^*$ and assuming $K_f/\delta \approx K_s/(1 + \delta) = K/\delta$, (9-86) reduces to

$$\frac{\partial \eta}{\partial x} = \frac{\delta}{K} \left[\frac{Q_f B}{\eta(B - \eta)} - \frac{Q_{fL}^*}{B - \eta} \right] \tag{9-87}$$

The second equation is obtained from (9-85)

$$n \frac{d}{dt} \int_0^B x(\eta, t) \, d\eta = Q_{f0} - Q_{fL}^* \tag{9-88}$$

Note that Q_f is also unknown.

Bear and Dagan (1964b) propose the following procedure in order to solve the two equations (9-87) and (9-88) for Q_{f0} and η.

(a) An assumption is made regarding the variations in $Q_f(x, t)$ as a function of Q_{fL}^*, $Q_{f0}(t)$ and $\eta(x, t)$. The assumed distribution should satisfy the conditions: $x = 0, \eta = 0, Q_f = Q_{f0}(t)$, and $x = L, \eta = B, Q_f = Q_{fL}^*$.
(b) With Q_f of (a), (9-87) is integrated to yield $\eta = \eta [x, Q_{f0}(t)]$.
(c) η of (b) is inserted in (9-88), which is then solved for $Q_{f0}(t)$.

The accuracy of the solution depends on that of the assumed variations of Q_f. Bear and Dagan (1964b) consider two approximations.

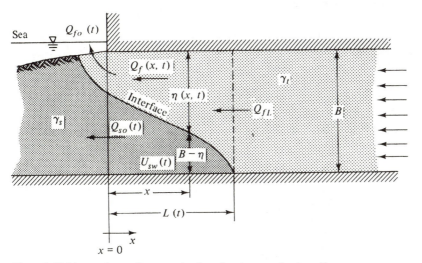

Figure 9-21 Nomenclature for a moving interface in a confined aquifer.

Case 1 A linear variation of Q with η (Fig. 9-22a), i.e.,

$$Q_f(\eta, t) = Q_{fL}^* + [Q_{fo}(t) - Q_{fL}^*](B - \eta)/B = Q_{fo}(t) - [Q_{fo}(t) - Q_{fL}^*]\eta/B \tag{9-89}$$

By inserting (9-89) in (9-87), we obtain

$$\frac{\partial \eta}{\partial x} = \frac{\delta}{K}\frac{Q_{fo}(t)}{\eta} \tag{9-90}$$

This equation may be interpreted as describing an instantaneous steady state, with stationary sea water and $Q_f = Q_{fo} = \text{const}$. Its integration yields the parabolic interface

$$\eta^2(x, t) = \frac{2\delta Q_{fo}(t)}{K}x \tag{9-91}$$

By inserting (9-91) into (9-88), we obtain

$$\frac{6\delta\, dt}{nKB^3} = \frac{dQ_{fo}}{(Q_{fo})^2(Q_{fL}^* - Q_{fo})} \tag{9-92}$$

Integrating (9-92), with $t = 0$, $Q_{fo} = Q_{fo}^0$, $Q_{fL}^* \neq 0$, gives

$$\frac{6\delta}{nKB^3}t = \frac{1}{(Q_{fL}^*)^2}\ln\frac{Q_{fo}}{Q_{fo}^0} - \frac{1}{Q_{fL}^*}\left(\frac{1}{Q_{fo}} - \frac{1}{Q_{fo}^0}\right) - \frac{1}{(Q_{fL}^*)^2}\ln\frac{Q_{fL}^* - Q_{fo}}{Q_{fL}^* - Q_{fo}^0} \tag{9-93}$$

For the special case of $Q_{fL}^* = 0$, (9-93) reduces to

$$Q_{fo} = [12\delta t/nKB^3 + 1/(Q_{fo}^0)^2]^{-1/2} \tag{9-94}$$

The last two equations describe the temporal variations of the fresh water discharge to the sea, when at $t = 0$ the seaward fresh water flow above the toe is changed from Q_{fL}^0 to Q_{fL}^*. Experiments (Bear and Dagan, 1964b) seem to indicate that these equations are valid only for $Q_{fL}^* > Q_{fL}^0$, that is, the case of a seaward movement of the toe (Fig. 9-22).

Case 2 When the discharge rate at $x = L$ is suddenly changed from Q_{fL}^0 to Q_{fL}^*, the variations of Q_f are better approximated by the dashed line of Fig. 9-22a. Hence, Bear and Dagan investigate a second approximation shown in Fig. 9-22c

$$Q_f(\eta, t) = Q_{fo} - (1 - c)(Q_{fo} - Q_{fL}^*)\eta/B \tag{9-95}$$

where c is a dimensionless constant to be determined below. For $c = 0$ we obtain the first approximation (9-89).

By inserting (9-95) in (9-87), we obtain

$$\frac{\partial \eta}{\partial x} = \frac{\delta}{K}\left[\frac{Q_{fo}}{\eta} + \frac{c(Q_{fo} - Q_{fL}^*)}{B - \eta}\right] \tag{9-96}$$

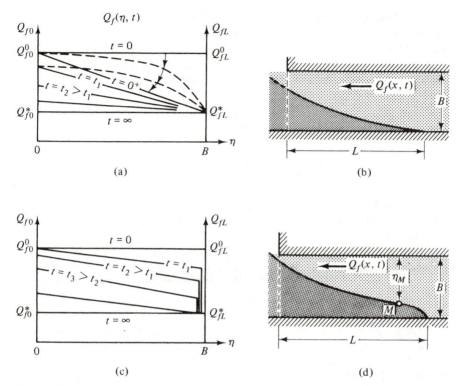

Figure 9-22 First and second approximations for $Q_{f0} = Q_{f0}(\eta, t)$ and resulting interfaces *(Bear and Dagan, 1964b)*.

which can also be interpreted as describing instantaneously a steady state with a constant fresh water flow Q_{f0} and sea water flow of $c(Q_{f0} - Q^*_{fL})$. The corresponding shape of the interface (Bear and Dagan, 1964b) is given in Fig. 9-22d. It has an inflection point M (obtained for $\partial^2\eta/\partial x^2 = 0$) at

$$\eta_M = B/\{1 + [c(1 - Q^*_{fL}/Q_{f0})]^{1/2}\} \qquad (9\text{-}97)$$

Experiments in a Hele–Shaw analog of cases where the interface moves landward as a result of changing Q^0_{fL} show that an inflection point indeed occurs at $\eta_M > 0.75B$, when $Q^0_{fL} = 0$. This corresponds approximately to $c \cong 0.1$. As small variations in the value of c do not appreciably affect the result, Bear and Dagan (1964b) integrate (9-96) with $c = 0.1$ and give the shape of the moving interface in the form of $x = x(\eta, Q_{f0}, Q^*_{fL})$. They also derive $Q_{f0}(t)$ from (9-88). Their results are given in a graphic form (since numerical integration is used) both for $Q^*_{fL} > 0$ and $Q^*_{fL} < 0$; in both cases $Q^*_{fL} < Q^0_{fL}$. For the special case $Q^*_{fL} = 0$, they obtain

$$Q_{f0}/Q^0_{fL} = \{1/[1 + 14.3\delta t Q^0_{fL}/nKB^3]\}^{1/2} \qquad (9\text{-}98)$$

According to Bear and Dagan (1964b), experiments indicate that the second approximation is applicable only to cases of landward interface movement caused by a reduction of seaward fresh water flow, while the first approximation is sufficiently accurate for practical purposes, only for a seaward moving interface. Both approximations seem to be valid only as long as Q_{f0} is sufficiently large.

For $Q_{fL}^* = 0$, that is, stopping all seaward fresh water flow above the toe, it is possible to derive the distance $x|_{\eta=B}$ of the moving toe for both approximations

$$x\bigg|_{\eta=B} = (mKBt/\delta n + x_0^2)^{1/2} \tag{9-99}$$

where x_0 is the value of L at $t = 0$. For the first approximation $m = 3$; for the second one, $m = 1.75$. Rumer and Harleman (1963) obtain the same result with $m = 1$.

Shima (1969) derives a solution for a moving interface employing an approach similar to that of Bear and Dagan (1964b).

Vappicha and Nagaraja (1976) also use the method of successive steady states in order to determine the transient profile of an interface in a coastal aquifer for different boundary conditions at $x = L$ and with sloping or horizontal impervious bottoms. They compare their computations with results of experiments conducted in a Hele–Shaw analog (Bear, 1972, pp. 687–697). They observed errors of up to 15% in the length of sea water intrusion and 3–4% in the interface profile, but conclude that the solution is of an accuracy sufficient to serve as a rough estimate of the interface movement.

Let us consider the case shown in Fig. 9-23. The steady state interface is described by

$$Q_{fL} = -Q_f = \text{const.}$$

$$-Q_f = K(h + h_f)\frac{\partial h_f}{\partial x} = K\frac{1+\delta}{2\delta^2}\frac{\partial h^2}{\partial x} = Q_{fL} \tag{9-100}$$

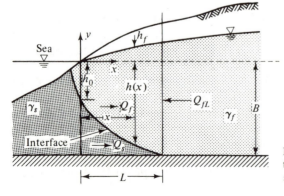

Figure 9-23 Determining the moving interface by the method of successive steady states.

Integrating this equation from $x = 0$, $h = h_0$; $x = L$, $h = B$, we obtain an expression for the parabolic shape of the steady interface

$$Q_{fL}x = K\frac{1+\delta}{2\delta^2}(h^2 - h_0^2); \qquad Q_{fL}L = K\frac{1+\delta}{2\delta^2}(B^2 - h_0^2) \quad (9\text{-}101)$$

(compare with (9-50)), or

$$x = \frac{\acute{K}(1+\delta)}{2Q_{fL}\delta^2}(h^2 - h_0^2) \quad (9\text{-}102)$$

From (9-35) it follows that $h_0 = \delta Q_f|_{x=0}/K$, where $-Q_f|_{x=0}$ means seaward flow. Henry (1959) obtains $h_0 = -0.741\delta Q_f|_{x=0}/K$ for steady flow with $-Q_f|_{x=0} = Q_{fL} = $ const. Let us express h_0 by $h_0 = -\beta\delta Q_f|_{x=0}/K$, where $0.5 < \beta < 1$ is a coefficient. One should note that the above expression for h_0 was derived for steady state with *seaward* flow. Here we shall assume that it is also valid for landward flow at $x = 0$.

The volume of the sea water wedge, U_{sw}, is given by

$$U_{sw} = n\int_{h_0}^{B} x\, dh = n\int_{h_0}^{B} \frac{K(1+\delta)}{2Q_{fL}\delta^2}(h^2 - h_0^2)\, dh = \frac{nK(1+\delta)}{6Q_{fL}\delta^2}(B^3 - 3h_0^2 B + 2h_0^3)$$

$$= \frac{n(1+\delta)\beta}{6\delta}\left(\frac{B^3}{h_0} - 3h_0 B + 2h_0^2\right) \quad (9\text{-}103)$$

As the interface moves, this volume changes such that

$$\frac{dU_{sw}}{dt} = Q_s\bigg|_{x=0} \quad (9\text{-}104)$$

Continuity requires that if both elastic and phreatic storages are neglected, then always

$$Q_s\bigg|_{x=0} + Q_f\bigg|_{x=0} = -Q_{fL} \quad (9\text{-}105)$$

From (9-103) through (9-105), we obtain for $h_0 = h_0(t)$

$$\frac{n(1+\delta)\beta}{6\delta}\left(-\frac{B^3}{h_0^2} - 3B + 4h_0\right)\frac{dh_0}{dt} = Q_s\bigg|_{x=0} = -Q_f\bigg|_{x=0} - Q_{fL}$$

$$= \frac{Kh_0}{\beta\delta} - Q_{fL} \quad (9\text{-}106)$$

$$\frac{dh_0}{dt} = \frac{Q_{fL} - Kh_0/\beta\delta}{n(1+\delta)\beta[B^3/h_0^2 + 3B - 4h_0]/6\delta} \quad (9\text{-}107)$$

This is an ordinary differential equation which can be solved for $h_0 = h_0(t)$, given $Q_{fL}(t)$ determined by analyzing the water balance landward of the toe. Once $h_0(t)$ is derived, (9-102) can be used for determining the entire (parabolic) profile of the moving interface. A numerical technique is often employed in solving (9-107). Vappicha and Nagaraja (1976) present solutions of (9-107) for three different boundary conditions at $x = L$.

Bear and Dagan (1964a; Bear, 1972, p. 539; Dagan and Bear, 1968) employ the method of *small perturbations* in an attempt to linearize the moving interface problem. In Sec. 9-7, these techniques are presented for solving the problem of interface upconing.

9-7 UPCONING BELOW A WELL PUMPING ABOVE AN INTERFACE

Sometimes, water supply wells pump water from a fresh water zone that is underlain by sea water, with the two zones separated by an (assumed) sharp interface. A situation like this is common in oceanic islands (Fig. 9-1c) and in elongated peninsulas. It occurs also in a coastal aquifer whenever wells are placed seaward of an interface toe, or when an advancing interface bypasses existing wells. As a result of the pumping, and the potential distributions established in the two (fresh water and salt water) zones, the interface rises towards the pumping well. This phenomenon is called *interface upconing*.

The phenomenon of upconing is a rate sensitive one. Up to a certain critical pumping rate, an equilibrium with an upconed interface is possible. When the pumping rate in a well is raised from one (steady) level to a higher (steady) one, yet below the critical pumping rate, a new equilibrium, with a higher upconed interface, is established following a transition period. At the critical pumping rate, the interface is very unstable and any increase in pumping rate will immediately bring the interface, and with it sea water, into the pumping well. The fast rising upconed interface will reach the pumping well in a cusp-like form. When pumping stops, the upconed interface (in the form of a local mound superimposed on the regional interface) undergoes decay towards the initial steady state interface without pumping. When, at the same time, seaward flow of fresh water takes place above the interface, the upconed interface is displaced seaward as it rises and decays.

One should keep in mind that the entire discussion in this chapter (except for Sec. 9-8) is based on the simplified assumption of a sharp interface, while in reality we have a *transition zone* from completely fresh water to completely sea water, due to the phenomena of molecular diffusion and, especially, hydrodynamic dispersion (see Chap. 7 and Sec. 9-8). The transition zone becomes even wider as the interface rises.

The presence of a transition zone means that relatively saline water may enter and contaminate a pumping well long before the calculated sharp interface reaches it.

Figure 9-24 shows, qualitatively, several typical situations of a well pumping above the interface in a coastal aquifer. At a low pumping rate, part of the fresh water, originally flowing to the sea (Fig. 9-24a) is intercepted by the well (Fig. 9-24b). The zone contributing fresh water to the well is bounded by a water divide. A stagnation point (A) exists seaward of the well, where the specific discharge produced by the pumping is equal in magnitude and opposite in direction to

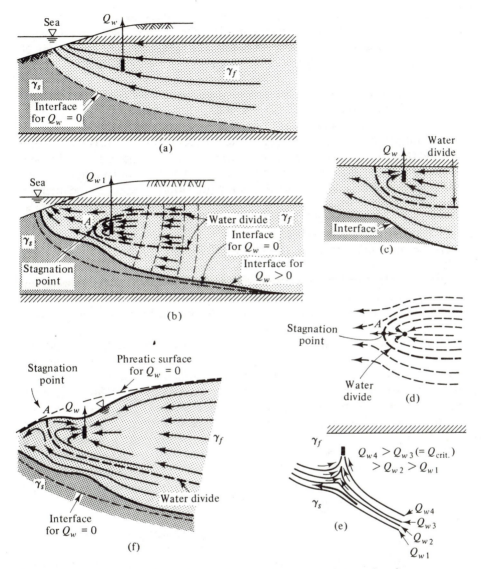

Figure 9-24 The fresh water flow in the vicinity of a well pumping above an interface.

that of the undisturbed seaward flow (Fig. 9-24b and d). As a result of the pumping the interface cones up a certain distance towards the pumping well. At the same time, the entire interface rises and its toe advances a certain distance landward.

The water divide and the corresponding stagnation point may also take the form shown in Fig. 9-24c. For this case, Fig. 9-24d shows the trace of the water divide on the upper confining layer. Similar situations arise in a phreatic aquifer. For example, Fig. 9-24f shows a situation which corresponds to that shown in Fig. 9-24c.

The shape of the water divide and the location of the stagnation point depend on the location and length of the well's opening. One should recall that in reality, the flow is in a three-dimensional space (unless the well is replaced by an infinitely long drain, in which case the situation is one of two-dimensional flow in the vertical plane).

For a single well, the phenomenon of upconing disappears at some distance from the well, measured in a direction parallel to the coast.

From (9-11) it follows that (at least for steady flow and stationary sea water) the interface always rises in the direction of the fresh water flow. Hence, as long as the interface, in spite of its upconed shape, continuously rises in the direction of the sea, at least some fresh water will continue to flow everywhere above it towards the sea. Recalling the presence of the transition zone above the interface, this situation reduces the chance of any sea water (even in the diluted form present in the transition zone) reaching the pumping well. Obviously the situation improves as the well is more shallow, i.e., closer to the upper confining layer.

As the pumping rate is increased, a landward flow along the interface may appear (Fig. 9-24e). Under such conditions we *always* have contamination of the pumped water from the transition zone. Finally, beyond the critical pumping rate ($Q_{\text{crit.}}$), the interface and sea water reach the well.

The determination of the exact shape of the upconed interface and the value of the critical pumping rate, beyond which sea water will enter the pumping well, as affected by various hydrological (e.g., rate of accretion, aquifer permeability, rate of seaward flow of fresh water above the toe) and geometrical (e.g., distance of well from coast, depth of well, thickness of aquifer) parameters, requires, in principle, the solution of the general three-dimensional flow problem as presented in Sec. 9-2. When we are interested in the regional shape of the interface, and are ready to accept only an approximate shape of the upconed interface in the close proximity of pumping wells, we can use the averaged equations as presented in Sec. 9-4, written for the dependent variables $\phi_f(x, y, t)$ and $\phi_s(x, y, t)$. Once ϕ_f and ϕ_s are known, the elevation of the interface is determined by (9-18). Pinder and Page (1976) obtain the local upconings under pumping wells, while solving for the regional interface under an island. They obtain their solution by solving (9-26) and (9-29) numerically.

It can be safely stated that most upconing problems of practical interest, like most regional interface ones, have to be solved numerically by one of the techniques described in Sec. 5-6. Several exact and approximate analytical solutions for local upconing appear in the literature for the case of steady flow in a vertical two-dimensional field in which fresh water is intercepted by a drain (rather than by a well). For example, Bear and Dagan (1964) and Yih (1964) present solutions obtained by the hodograph method (Bear, 1972). Their results for a single drain in infinite and semi-infinite domains are summarized in Fig. 9-25. In order to use these results we have to know one point on the interface in the xy plane. Strack (1972, 1973) extends the theory of the hodograph in order to deal with steady flow to a single drain and to a multiple drain system, operating above an interface in a coastal aquifer. He considered both pumping and recharging drains in the vertical two-dimensional field of flow.

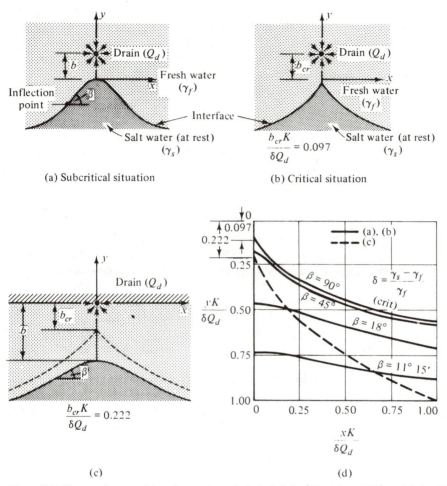

Figure 9-25 Shapes of upconed interface under a drain in infinite (Figs (a) and (b)), and finite (Fig. c) flow domains *(Bear and Dagan, 1964)*.

Upconing by Muskat's Approximation

Muskat (1937) presents a simple approximation for handling steady flow to a sink operating above a horizontal interface (or approximately so), provided the interface rises only a relatively small portion of the distance between its initial position and the sink. Under these conditions, he assumes that the sea water is stationary so that the interface behaves as an impervious boundary. In addition, he assumes that the (a priori unknown) shape of this local upconed interface beneath the sink (well or drain) does not result in a potential distribution (in the region occupied by fresh water) which is much different from that derived for a real fixed impervious boundary. With these assumptions, he derives the shape of the upconed interface from the potential distribution on some hypothetical horizontal plane introduced as an impervious boundary to the fresh water flow domain. Usually the initial horizontal interface is taken as that boundary.

Figure 9-26 shows a point sink that withdraws water at a constant rate Q_w from a confined homogeneous and isotropic aquifer of constant thickness B. From Muskat's assumptions it follows that

$$\Delta\zeta \equiv \zeta_M - \zeta_N = \frac{p_N - p_M}{\gamma_s} = \frac{\gamma_f}{\gamma_s}[(\phi_f)_N - (\phi_f)_M + \Delta\zeta]$$

$$\Delta\zeta = \frac{\phi_N - \phi_M}{(\gamma_s - \gamma_f)/\gamma_f} \cong \frac{\phi_{N'} - \phi_{M'}}{(\gamma_s - \gamma_f)/\gamma_f} \qquad (9\text{-}108)$$

where $\phi_{M'}$ and $\phi_{N'}$ are the fresh water piezometric heads at point M' and N' on the xy plane, respectively.

If point $N' \equiv N$ at some distance x_0 from the well is assumed unaffected by the pumping, we insert $\zeta_N = 0$ in (9-108).

It can be shown (Bear and Dagan, 1964; Bear, 1972, par. 9.5.7) that Muskat's model can be derived as a linearized approximation based on small perturbations.

The potential distribution produced by the sink, assuming that the initial (undisturbed) interface is an impervious boundary, can be obtained analytically (for simple cases), numerically, or by means of an electric analog. Bear and Dagan (1966) used a three-dimensional electrolytic tank to determine $\phi_f(x, y)$ for a point sink above a sloping planar interface and from it the shape of the upconed interface.

As a first example, consider a drain above a horizontal interface, in an infinite two-dimensional flow domain in the vertical xy plane (Fig. 9-27a). According to Muskat's method, we assume that the undisturbed interface acts as an impervious boundary to the semi-infinite flow domain $y > 0$. Hence, using the method of images, with point C a known point on the interface, we obtain the potential ϕ for a point in the half space $y > 0$

$$\phi = \frac{Q_d}{2\pi K}\ln rr' + \text{const.} \qquad (9\text{-}109)$$

where r and r' are distances of that point from the real and image drains, respectively. Assuming now that point C on the interface is not displaced, we

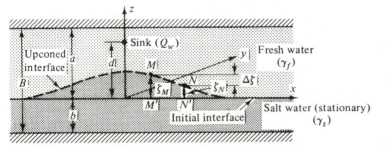

Figure 9-26 Upconing by Muskat's approach.

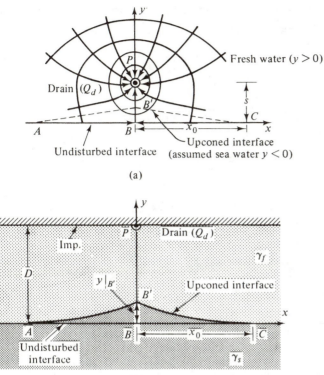

Figure 9-27 Upconing under a drain by Muskat's approximate method. (a) Infinite aquifer. (b) Semi-infinite aquifer.

obtain for point B' on the upconed interface $AB'C$

$$\phi_C - \phi_{B'} = \frac{Q_d}{2\pi K} \ln \frac{s^2 + x_0^2}{s^2 - (y|_{B'})^2} \qquad (9\text{-}110)$$

At the same time it follows from (9-108) that $\phi_C - \phi_{B'} = y|_{B'}(\gamma_s - \gamma_f/\gamma_f)$. Hence

$$\frac{Q_d}{2\pi K} \ln \frac{s^2 + x_0^2}{s^2 - (y|_{B'})^2} = y|_{B'} \frac{\gamma_s - \gamma_f}{\gamma_f} \qquad (9\text{-}111)$$

$$\frac{2\pi K}{\delta Q_d} s \frac{y|_{B'}}{s} = \ln \frac{1 + (x_0/s)^2}{1 - (y|_{B'}/s)^2} \qquad \delta = \gamma_f/(\gamma_s - \gamma_f) \qquad (9\text{-}112)$$

gives the relationship between $y|_{B'}$ and Q_d. given x_0 and s. One should recall that the whole approach is valid only for $y|_{B'}/s \ll 1$, say $y|_{B'} < s/3$. By differentiating (9-112), we obtain the critical values of $y|_{B'}$ for a given x_0/s .

$$\frac{y|_{B'}/s}{1 - (y|_{B'}/s)^2} = \frac{\pi K s}{\delta Q_d}; \qquad Q_d \equiv (Q_d)_{\text{crit.}} \qquad (9\text{-}113)$$

When the aquifer is a confined one (Fig. 9-27b), we obtain for the potential difference $\phi_C - \phi_B$, using images

$$\phi_C - \phi_{B'} = \frac{Q_d}{2\pi K} \ln \frac{\cosh(\pi x_0/D) + 1}{\cos(\pi y|_{B'}/D) + 1} \tag{9-114}$$

Hence

$$\frac{Q_d}{2\pi K} \ln \frac{\cosh(\pi x_0/D) + 1}{\cos(\pi y|_{B'}/D) + 1} = y|_{B'} \frac{\gamma_s - \gamma_f}{\gamma_f} \tag{9-115}$$

For the critical values of $y|_{B'}$ and Q_d, we obtain

$$\frac{\sin(\pi y|_{B'}/D)}{\cosh(\pi y|_{B'}/D) + 1} = \frac{2KD}{\delta Q_d} \tag{9-116}$$

Bear and Dagan (1962) show that in this case, Muskat's approximation introduces rather large errors with respect to an exact solution.

For steady flow to a point sink in an infinite aquifer (Fig. 9-28a), the potential at some point (x, y, z) in the flow domain $z > 0$ (using the method of images with respect to the horizontal plane) is given by

$$\phi = -\frac{Q}{4\pi K} \left\{ \frac{1}{[x^2 + y^2 + (z - d)^2]^{1/2}} + \frac{1}{[x^2 + y^2 + (z + d)^2]^{1/2}} \right\} + \text{const.} \tag{9-117}$$

For a point on the plane $z = 0$, we obtain

$$\phi\Big|_{z=0} = -\frac{Q}{2\pi K [x^2 + y^2 + d^2]^{1/2}} + \text{const.} \tag{9-118}$$

Equation (9-108) can now be used to determine $z|_{B'}$ assuming, for example, that on the circle $x^2 + y^2 = x_0^2$ the interface remains undisturbed.

Another approximate assumption is that the velocity of the fresh water just above the interface $(z = 0)$ can be determined from the potential distribution as

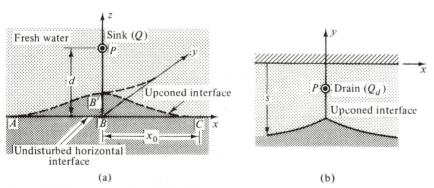

(a)   (b)

Figure 9-28 Upconing under a sink by Muskat's approximate method.

expressed by (9-118). For example, the velocity in the $+x$ direction is given by

$$V_x = -\frac{K}{n}\frac{\partial \phi}{\partial x}\bigg|_{z=0} = -\frac{Q}{2\pi n}\frac{x}{[x^2 + y^2 + d^2]^{3/2}} \tag{9-119}$$

If the fresh flow to the sea (in the absence of upconing) is $-V_{\text{sea}}$ (in the $-x$ direction), we may assume that in the linearized model employed here, super-position is permitted so that the resultant velocity V_x in the $+x$ direction, with upconing, is given by

$$V_x = -\frac{Q}{2\pi n}\frac{x}{[x^2 + y^2 + d^2]^{3/2}} - V_{\text{sea}} \tag{9-120}$$

From (9-120) it is possible to determine the maximum value of V_x, stagnation points ($V_x = 0$), etc. When Q is such that at all points of the plane $z = 0$, the resultant velocity is everywhere towards the sea (that is, $V_x < 0$), no streamline from this plane reaches the sink. We say that flushing of the transition zone always occurs. Otherwise, there exists a region in the $z = 0$ plane from which streamlines reach the sink, carrying with them saline water. Bear and Dagan (1964) study also up-coning under an array of wells along a line parallel to the coast and the possibility of using a combination of pumping and recharging wells to suppress upconing (Bear and Dagan, 1966).

Upconing by the Dupuit Assumption

As another approximate approach, we can also make use of the Dupuit assump-tion, i.e. assuming that in spite of the upconing, the flow in the vicinity of a well pumping above the interface is essentially horizontal. Again, we should emphasize that we deal with interface rises which are relatively small (or we are ready to accept rather large errors). For example, for the case shown in Fig. 9-29a, we have

$$s^2 - h^2 = \frac{Q_d \delta}{K}(x_0 - x) \tag{9-121}$$

where $\delta = \gamma_f/(\gamma_s - \gamma_f)$. Obviously the largest error is directly under the drain. For the two-dimensional flow to a drain in a phreatic aquifer shown in Fig. 9-29b, we obtain

$$\frac{Q_d}{2K} = (\eta_1 + \eta_2)\frac{d\phi}{dx} = (\eta_1 + \eta_2)\frac{\partial \eta_1}{\partial x}; \qquad \delta(D - s - \eta_1) = s - \eta_2 \tag{9-122}$$

$$\frac{Q_d}{2K} = \frac{1}{\delta^2}[\delta D - s(1 + \delta) + \eta_2(1 + \delta)]\frac{d\eta_2}{dx} \tag{9-123}$$

$$\frac{Q_d}{2K}(x_0 - x) = \frac{1}{\delta^2}\{[\delta D - s(1 + \delta)](s - \eta_2) + \tfrac{1}{2}(1 + \delta)(s^2 - \eta_2^2)\} \tag{9-124}$$

Bear and Dagan (1962) indicate that experimental results justify the use of the Dupuit assumption.

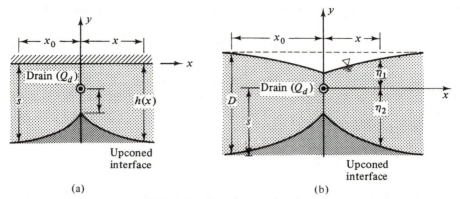

(a)	(b)

Figure 9-29 Upconing under a drain, using Dupuit approximation.

Dupuit–Ghyben–Herzberg assumptions may also be employed for a drain operating above an interface in a coastal aquifer. In the case of the one-dimensional steady flow (towards the sea) shown in Fig. 9-30, the fresh water flow domain is divided into three zones, in each of which the flow is assumed to be essentially horizontal. A continuity equation is written for each zone, taking into account the drain operating between zones III and II.

The same approach may also be applied to the case of steady flow to an array of wells parallel to the coast, which intercept part of the seaward flow (Bear and Dagan, 1963). Figure 9-31 shows this case. Fig. 9-30 may be considered as representing the vertical cross section perpendicular to the coast through the well in Fig. 9-31. From the Dupuit–Ghyben–Herzberg assumptions, justified when $B \ll L_1$ and $B \ll L_2$, we obtain for $h = h(x, y)$

$$\frac{\partial^2 h^2(x, y)}{\partial x^2} + \frac{\partial^2 h^2(x, y)}{\partial y^2} = 0 \qquad (9\text{-}125)$$

which is linear in h^2 and hence superposition with respect to h^2 is permitted.

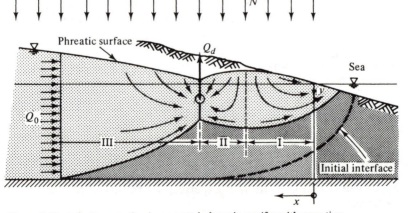

Figure 9-30 A drain operating in a coastal phreatic aquifer with accretion.

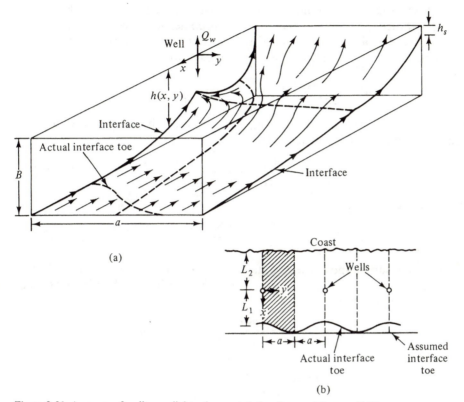

(a)

(b)

Figure 9-31 An array of wells parallel to the coast *(after Bear and Dagan, 1963)*.

For an infinite array of wells, Muskat's (1937, p. 525; see also (8-137) and Fig. 8-20)) development leads to

$$h^2(x, y) = \frac{\delta Q_w}{2\pi K} \ln\left(\cosh \frac{\pi x}{a} - \cos \frac{\pi y}{a} \right) + \text{const.} \qquad (9\text{-}126)$$

With $h = h_1$ at $x = L_1$ (i.e., along the toe of the interface), we obtain from (9-109)

$$h_1^2 = \frac{\delta Q_w}{2\pi K} \ln\left(\cosh \frac{\pi L_1}{a} - \cos \frac{\pi y}{a} \right) + \text{const.} \qquad (9\text{-}127)$$

Since

$$L_1 \gg 2a, \qquad \cosh \pi L_1/a > 270, \qquad \cosh \pi L_1/a \gg \cos \pi y/a \qquad (9\text{-}128)$$

we obtain

$$h_1^2 \cong \frac{\delta Q_w}{2\pi K}\left(\frac{\pi L_1}{a} - \ln 2 \right) + \text{const.} \qquad (9\text{-}129)$$

Similarly, with $x = -L_2$, $h = h_2$

$$h_2^2 \cong \frac{\delta Q_w}{2\pi K}\left(\frac{\pi L_2}{a} - \ln 2 \right) + \text{const.} \qquad (9\text{-}130)$$

From these equations it follows that for distances $x \gg 2a$, lines $h = $ const. become parallel to the coast and the flow becomes everywhere perpendicular to the coast. To obtain the value of the constant of integration in the above equations, we set $h = h_w$ at $x = y = r_w = $ radius of well

$$h_w^2 = \frac{\delta Q_w}{2\pi K} \ln\left(\cosh \frac{\pi r_w}{a} - \cos \frac{\pi r_w}{a}\right) + \text{const.} \qquad (9\text{-}131)$$

Since $r_w \ll a$, we may replace $\cosh \pi r_w/a$ and $\cos \pi r_w/a$ by the first three terms of their respective series expansions. This leads to

$$h_w^2 = \frac{\delta Q_w}{\pi K} \ln \frac{\pi r_w}{a} + \text{const.} \qquad (9\text{-}132)$$

and (9-126) reduces to

$$h^2(x, y) - h_w^2 = \frac{\delta Q_w}{2\pi K} \ln \frac{\cosh(\pi x/a) - \cos(\pi y/a)}{(\pi r_w/a)^2} \qquad (9\text{-}133)$$

Now we superimpose the results derived above on steady seaward fresh water flow without wells. Again using the Dupuit–Ghyben–Herzberg assumptions (Sec. 9-5), we obtain

$$x > 0: \qquad h^2 = -\frac{Q - Q_w/2}{aK/\delta}(L_1 - x) + \text{const.} \qquad (9\text{-}134)$$

$$x < 0: \qquad h^2 = \frac{Q - Q_w/2}{aK/\delta}(L_2 + x) + \text{const.,} \qquad (9\text{-}135)$$

where Q is the total seaward flow in a strip of width $2a$. The outflow to the sea is $Q_s = Q - Q_w$.

With $h = B$ at $x = L$, and $H = h_s$ at $x = -L_2$, we obtain

$$B^2 - h^2(x, y) = \frac{Q - Q_w/2}{aK/\delta}(L_1 - x) + \frac{\delta Q_w}{2\pi K}\left[\frac{\pi L_1}{a} - \ln 2\left(\cosh \frac{\pi x}{a} - \cos \frac{\pi y}{a}\right)\right] \qquad (9\text{-}136)$$

$$h^2(x, y) - h_s^2 = \frac{Q - Q_w/2}{aK/\delta}(L_2 + x) - \frac{\delta Q_w}{2\pi K}\left[\frac{\pi L_2}{a} - \ln 2\left(\cosh \frac{\pi x}{a} - \cos \frac{\pi y}{a}\right)\right] \qquad (9\text{-}137)$$

At the well, $x = y = r_w$, $h = h_w$. With $r_w \ll L_1$, $r_w \ll L_2$, $r \ll a$, we obtain from (9-136) and (9-137)

$$B^2 - h_w^2 = \frac{Q - Q_w/2}{aK/\delta}L_1 + \frac{\delta Q_w}{2\pi K}\left[\frac{\pi L_1}{a} - \ln 2\left(\frac{\pi r_w}{a}\right)^2\right] \qquad (9\text{-}138)$$

$$h_w^2 - h_s^2 = \frac{Q - Q_w/2}{aK/\delta}L_2 - \frac{\delta Q_w}{2\pi K}\left[\frac{\pi L_1}{a} - \ln 2\left(\frac{\pi r_w}{a}\right)^2\right] \qquad (9\text{-}139)$$

These equations may be used to estimate Q_w and L_1, knowing (or estimating) all other variables. Bear and Dagan (1963) use (9-138) and (9-139) to determine the efficiency of fresh water interception by a coastal collector as a means for reducing fresh water flow to the sea. They also modify the above relationships for a phreatic aquifer with accretion.

Upconing by the Method of Small Perturbations

Bear and Dagan (1964a; Dagan and Bear, 1968; Bear, 1972, p. 539) also present solutions for time dependent interface upconing, using the *method of small perturbations*. Again, this means that only relatively small interface rises (e.g., less than 1/3 the distance from the initial interface to the sink) may be considered. Essentially, they attempt to solve the interface problem as stated mathematically in Sec. 9-2.

For a homogeneous aquifer, $K_f \cong K_s = K = \text{const.}$, with no elastic storage ($S_0 = 0$), this problem reduces to:

Determine ϕ_f in the fresh water zone R_f, and ϕ_s in the salt water zone R_s, such that

$$\nabla^2 \phi_f = 0 \quad \text{in } R_f, \quad \text{and} \quad \nabla^2 \phi_s = 0 \quad \text{in } R_s \tag{9-140}$$

and, from (9-7) and (9-8), the following conditions are satisfied on the interface

$$n\delta \frac{\partial \phi_f}{\partial t} - n(1 + \delta) \frac{\partial \phi_s}{\partial t} - K\{\nabla z - (1 + \delta)\nabla\phi_s + \delta\nabla\phi_f\}\nabla\phi_f = 0 \tag{9-141}$$

$$n\delta \frac{\partial \phi_f}{\partial t} - n(1 + \delta) \frac{\partial \phi_s}{\partial t} - K\{\nabla z - (1 + \delta)\nabla\phi_s + \delta\nabla\phi_f\}\nabla\phi_s = 0 \tag{9-142}$$

The elevation of the interface is defined by

$$\zeta = (1 + \delta)\phi_s - \delta\phi_f \tag{9-143}$$

Referring to Fig. 9-32, we assume that both ϕ_f and ϕ_s can be expressed as a sum of a power series of a small parameter ε

$$\phi_i(x, y, z, t) = \phi_i^0(x, y, z) + \varepsilon\phi_i'(x, y, z, t) + \varepsilon^2 \phi_i''(x, y, z, t) + \cdots, \qquad i = f, s \tag{9-144}$$

where ϕ_f^0 and ϕ_s^0 are steady state potential distributions describing an average flow pattern, or an initial steady one. The remaining terms of the series in (9-144) describe deviations from ϕ_i^0, with ε having the character of a perturbation. An assumption commonly used in the solution of similar problems in physics and mathematics is that $\varepsilon \ll 1$. This assumption is obviously valid when the deviations of the unsteady flow from the steady one are indeed small.

Although the method is applicable also to second (Dagan, 1964) and higher order linearizations, it is described and employed here only to derive a first order linearized solution. This means that only the first two terms on the right-hand side of (9-144) are used.

By inserting (9-144) into (9-140) and recalling that by definition

$$\nabla^2 \phi_f^0 = 0 \quad \text{in } R_f; \qquad \nabla^2 \phi_s^0 = 0 \quad \text{in } R_s \tag{9-145}$$

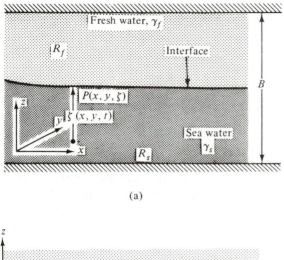

(a)

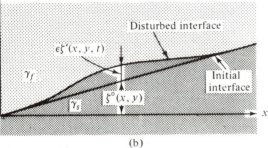

(b)

Figure 9-32 Nomenclature for the method of small perturbations (*Dagan and Bear, 1968*).

we obtain

$$\nabla^2 \phi'_f = 0 \quad \text{in } R_f; \qquad \nabla^2 \phi'_s = 0 \quad \text{in } R_s \qquad (9\text{-}146)$$

In a similar manner we assume that the elevation ζ of points along the interface may be expanded in a power series, the first two terms of which are

$$\zeta(x, y, t) = \zeta^0(x, y) + \varepsilon \zeta'(x, y, t) \qquad (9\text{-}147)$$

In (9-147), $\varepsilon \zeta'$ is the (small) displacement of the interface, caused by the unsteady flow (Fig. 9-32b), whereas ζ^0 describes some average interface position, or an initial steady-state one. Rewriting (9-144) for points on the interface ($z = \zeta$) and expanding in a Taylor series, we obtain

$$\phi_i^0 \bigg|_{z = \zeta = \zeta^0 + \varepsilon \zeta'} = \phi_i^0 \bigg|_{z = \zeta^0} + \varepsilon \zeta' \frac{\partial \phi_i^0}{\partial z} + 0(\varepsilon^2), \qquad i = f, s$$

Similar expressions may be derived for

$$\phi_f \bigg|_{z = \zeta}, \qquad \phi_s \bigg|_{z = \zeta}, \qquad \frac{\partial \phi_f}{\partial x} \bigg|_{z = \zeta}, \qquad \frac{\partial \phi_s}{\partial x} \bigg|_{z = \zeta}$$

and for all other expressions appearing in (9-141) and (9-142). For example

$$\left.\frac{\partial \phi_f}{\partial x}\right|_{z=\zeta} = \left.\frac{\partial \phi_f^0}{\partial x}\right|_{z=\zeta^0} + \varepsilon\left(\frac{\partial \phi_f'}{\partial x} + \zeta'\frac{\partial^2 \phi_f^0}{\partial x \partial z}\right)\Bigg|_{z=\zeta^0} + 0(\varepsilon^2) \qquad (9\text{-}148)$$

In this way, all terms are expressed *for points on the average* (or initial) *interface whose elevations are* $\zeta^0(x, y)$.

For the sake of simplicity, let us assume that the aquifer is homogeneous and isotropic so that a specific discharge potential $\Phi_i = K_i\phi_i$, $i = f, s$, may be introduced in the two regions, R_f and R_s, respectively. We shall also use the notation $\alpha_f = \delta/K_f$ and $\alpha_s = (1 + \delta)/K_s$.

By introducing the above expansions into the interface boundary conditions (9-141) through (9-143) and collecting terms of the same power of ε, we obtain the following relationships for ζ', Φ_f', and Φ_s'

$$\zeta' = (\alpha_s\Phi_s' - \alpha_f\Phi_f')/(1 - \alpha_s\partial\Phi_s^0/\partial z + \alpha_f\partial\Phi_f^0/\partial z) \quad \text{on } z = \zeta^0 \qquad (9\text{-}149)$$

$$n(\alpha_f\partial\Phi_f'/\partial t - \alpha_s\partial\Phi_s'/\partial t) - 2\alpha_f[\nabla\Phi_f^0 \cdot \nabla\Phi_f' + \zeta'\nabla\Phi_f^0 \cdot \nabla(\partial\Phi_f^0/\partial z)]$$
$$+ \alpha_s[\nabla\Phi_s^0 \cdot \nabla\Phi_s' + \zeta'\nabla\Phi_s^0 \cdot \nabla(\partial\Phi_s^0/\partial z) + \nabla\Phi_s^0 \cdot \nabla\Phi_f'$$
$$+ \zeta'\nabla\Phi_s^0 \cdot \nabla(\partial\Phi_f^0/\partial z)] - (\partial\Phi_f'/\partial z + \zeta'\,\partial^2\Phi_f^0/\partial z^2) = 0 \qquad \text{on } z = \zeta^0$$
$$n(\alpha_s\partial\Phi_s'/\partial t - \alpha_f\partial\Phi_f'/\partial t) - 2\alpha_s[\nabla\Phi_s^0 \cdot \nabla\Phi_s' + \zeta'\nabla\Phi_s^0 \cdot \nabla(\partial\Phi_s^0/\partial z)]$$
$$+ \alpha_f[\nabla\Phi_s^0 \cdot \nabla\Phi_f' + \zeta'\nabla\Phi_s^0 \cdot \nabla(\partial\Phi_f^0/\partial z) + \nabla\Phi_f^0 \cdot \partial\Phi_s' + \zeta'\nabla\Phi_f^0 \cdot \nabla(\partial\Phi_s^0/\partial z)]$$
$$+ (\partial\Phi_s'/\partial z + \zeta'\,\partial^2\Phi_s^0/\partial z^2) = 0 \qquad \text{on } z = \zeta^0 \qquad (9\text{-}150)$$

The advantages of replacing (9-141) through (9-143) by (9-149) and (9-150) are that the latter are written for the fixed (in time) interface $z = \zeta^0$ and that the boundary conditions (9-150) are *linear* in Φ_f' and Φ_s'.

For the steady state solution we have to solve (9-145) with the boundary conditions

$$\left.\begin{aligned}\zeta^0 &= \alpha_s\Phi_s^0 - \alpha_f\Phi_f^0 \\ \alpha_f(\nabla\Phi_f^0)^2 - \alpha_s\nabla\Phi_f^0 \cdot \nabla\Phi_s^0 + \partial\Phi_f^0/\partial z &= 0 \\ \alpha_s(\nabla\Phi_s^0)^2 - \alpha_f\nabla\Phi_f^0 \cdot \nabla\Phi_s^0 - \partial\Phi_s^0/\partial z &= 0\end{aligned}\right\} \quad \text{on } z = \zeta^0 \qquad (9\text{-}151)$$

Thus the original problem has been transformed into one of first solving a steady flow problem for $\Phi_f^0(x, y, z)$ and $\Phi_s^0(x, y, z)$, satisfying (9-145) with a fixed interface at elevations $\zeta^0(x, y)$, and then solving (9-146), (9-149), and (9-150) for ζ', Φ_f', and Φ_s'. No advantage has been gained with respect to the steady flow problem; the boundary conditions (9-151) in this case are still nonlinear. However, in certain cases the initial steady state solution is known.

Bear and Dagan (1964a; Dagan and Bear, 1968) present solutions for several cases.

Returning now to the problem of unsteady interface upconing, consider the case of a drain (in two-dimensional flow in the vertical xz plane) above an initially horizontal interface (Fig. 9-33).

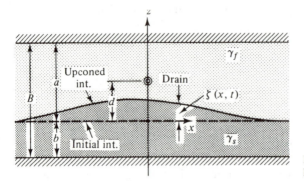

Figure 9-33 A drain above an initially horizontal interface.

Mathematically stated, the problem is to solve (9-140) for ϕ_f and ϕ_s (or, since the medium is assumed homogeneous and isotropic, for Φ_f and Φ_s) in the fresh and sea water regions, respectively. Since in this case ϕ_f^0, ϕ_s^0, and ζ^0 are identically zero, we have only to solve for the unsteady parts ϕ_f', ϕ_s', and ζ'. For the sake of simplicity, we shall denote henceforth $\varepsilon\phi_f'$, $\varepsilon\phi_s'$, and $\varepsilon\zeta'$ by ϕ_f, ϕ_s and ζ. The heads ϕ_f and ϕ_s have to satisfy the following boundary conditions

$$z = a, \qquad \partial\Phi_f/\partial z = 0, \Bigg\} \, t \geq 0, \quad -\infty < x < \infty \qquad (9\text{-}152)$$
$$z = -b, \qquad \partial\Phi_s/\partial z = 0, \Bigg\} \qquad\qquad\qquad\qquad\qquad\quad (9\text{-}153)$$

$$t \geq 0, \qquad \lim_{r \to 0} r(\partial\Phi_f/\partial r) = Q/2\pi; \qquad r^2 = x^2 + (z - d)^2 \qquad (9\text{-}154)$$

Along the interface, from (9-150)

$$\beta_f \partial\Phi_f/\partial t - \beta_s \partial\Phi_s/\partial t - \partial\Phi_f/\partial z = 0$$
$$\beta_f \partial\Phi_f/\partial t - \beta_s \partial\Phi_s/\partial t - \partial\Phi_s/\partial z = 0 \qquad\qquad (9\text{-}155)$$

where

$$\beta_i = n\alpha_i; \qquad i = f, s$$

The elevations of the interface are given by

$$\zeta = \alpha_s\Phi_s - \alpha_f\Phi_f$$

Hence the initial condition is

$$t = 0, \qquad z = 0, \qquad \zeta = \alpha_s\Phi_s - \alpha_f\Phi_f = 0 \qquad (9\text{-}156)$$

At infinity, the interface remains undisturbed. Hence we have there

$$z = 0, \qquad x = \pm\infty, \qquad \zeta = \alpha_s\Phi_s - \alpha_f\Phi_f = 0 \qquad (9\text{-}157)$$

Because of the singularity introduced by the drain, let the required solution for Φ_f be decomposed into two parts, Φ_{f1} and Φ_{f2} (both satisfying the Laplace equation)

$$\Phi_f(x, z, t) = \Phi_{f1}(x, z) + \Phi_{f2}(x, z, t) \qquad (9\text{-}158)$$

and such that Φ_{f1} represents the singular character of Φ_f at $x = 0$, $z = d$, whereas Φ_{f2} is an unsteady state solution which is regular everywhere in the region R_f. Hence

$$\Phi_{f1} = \frac{Q}{4\pi} \ln[x^2 + (z - d)^2] - \frac{Q}{4\pi} \ln[x^2 + (z + d)^2] \qquad (9\text{-}159)$$

The boundary conditions for Φ_{f2} and Φ_s are

$$z = 0, \ \partial\Phi_{f2}/\partial z = \partial\Phi_f/\partial z - \partial\Phi_{f1}/\partial z = \frac{Q}{2\pi}\left[\frac{a - d}{x^2 + (a - d)^2} - \frac{a + d}{x^2 + (a + d)^2}\right]$$
$$(9\text{-}160)$$

$$z = -b, \qquad \partial\Phi_s/\partial z = 0 \qquad (9\text{-}161)$$

$$z = 0, \ \Phi_{f1} = 0; \quad \beta_f \partial\Phi_{f2}/\partial t - \beta_s \partial\Phi_s/\partial t - \partial\Phi_{f1}/\partial z - \partial\Phi_{f2}/\partial z = 0$$
$$(9\text{-}162)$$
$$\beta_f \partial\Phi_{f2}/\partial t - \beta_s \partial\Phi_s/\partial t - \partial\Phi_s/\partial z = 0$$

or

$$\beta_f \partial\Phi_{f2}/\partial t - \beta_s \partial\Phi_s/\Phi_s/\partial t - \partial\Phi_{f2}/\partial z = -Qd/(x^2 + d^2)$$
$$(9\text{-}163)$$
$$\beta_f \partial\Phi_{f2}/\partial t - \beta_s \partial\Phi_s/\partial t - \partial\Phi_s/\partial z = 0$$

The initial conditions become

$$t = 0, \qquad z = 0, \qquad \zeta = \alpha_s\Phi_s - \alpha_f\Phi_f = 0 \qquad (9\text{-}164)$$

Since Φ_{f2} and Φ_s are everywhere regular, let us now assume that they may be represented by Fourier integral transforms, enabling separation of variables

$$\Phi_{f2} = \frac{1}{\pi}\int_0^\infty [A_1(\lambda, t)\exp(-\lambda z) + A_2(\lambda, t)\exp(\lambda z)]\cos(\lambda x)\,d\lambda \qquad (9\text{-}165)$$

$$\Phi_s = \frac{1}{\pi}\int_0^\infty A_3(\lambda, t)\cosh\lambda(z + b)\cos(\lambda x)\,d\lambda \qquad (9\text{-}166)$$

In this form both Φ_{f2} and Φ_s satisfy the Laplace equation; A_1, A_2, and A_3 are functions to be determined by the boundary and the initial conditions (9-163) and (9-164). The boundary condition (9-161) is satisfied by (9-166).

Using tables, the right-hand side of (9-160) and the first equation of (9-163) are now represented in the form of a Fourier integral, similar to (9-165)

$$\frac{Q}{2\pi}\left[\frac{a - d}{x^2 + (a - d)^2} - \frac{a + d}{x^2 + (a + d)^2}\right]$$
$$= \frac{Q}{2\pi}\int_0^\infty \{\exp[-\lambda(a - d)] - \exp[-\lambda(a + d)]\}\cos(\lambda x)\,d\lambda \qquad (9\text{-}167)$$

$$\frac{Q}{2\pi}\frac{d}{x^2 + d^2} = \frac{Q}{\pi}\int_0^\infty \exp(-\lambda d)\cos(\lambda x)\,d\lambda \qquad (9\text{-}168)$$

By inserting (9-165) through (9-168) into (9-160), (9-163), and (9-164), we obtain three equations in A_1, A_2, A_3, and λ. We also obtain for ζ

$$\zeta = \frac{1}{\pi} \int_0^\infty [\alpha_s A_3 \cos(\lambda b) - \beta_f(A_1 + A_2)] \cos(\lambda x) \, d\lambda \qquad (9\text{-}169)$$

Denoting

$$\bar{\zeta} = \alpha_s A_3 \cos(\lambda b) - \beta_f(A_1 + A_2) \qquad (9\text{-}170)$$

we observe, in view of (9-169) that $\bar{\zeta}$ is the inverse transform of ζ, such that

$$\zeta = \frac{1}{\pi} \int_0^\infty \bar{\zeta}(\lambda, t) \cos(\lambda x) \, d\lambda \qquad (9\text{-}171)$$

After some algebraic manipulations, we obtain from (9-170) a partial differential equation for $\bar{\zeta}$

$$[\beta_f \coth(\lambda a) + \beta_s \coth(\lambda b)] \, \partial\bar{\zeta}/\partial t + \lambda\bar{\zeta} = \beta_f Q \cosh[\lambda(a - d)]/\sinh(\lambda a) \quad (9\text{-}172)$$

By solving this equation and inserting the result in (9-171), we obtain

$$\zeta = \frac{\alpha_f Q}{\pi} \int_0^\infty \frac{1}{\lambda} \frac{\cosh[\lambda(a-d)]}{\sinh(\lambda a)} \left\{ 1 - \exp\left(\frac{-\lambda}{\beta_f \coth(\lambda a) + \beta_s \coth(\lambda b)} \right) t \right\}$$
$$\times \cos(\lambda x) \, d\lambda \quad (9\text{-}173)$$

With: $L = \lambda a$, $Z = \zeta/\alpha_f Q$; $T = t/\beta_f a$, $B = b/a$, $X = x/a$; $D = d/a$, and $M = \mu_s/\mu_f$, (9-173) becomes

$$Z = \frac{1}{\pi} \int_0^\infty \frac{\cosh L(1 - D)}{L \sinh L} \{1 - \exp[-LT/(\coth L + M \coth LB)]\} \cos(LX) \, dL$$
$$(9\text{-}174)$$

Because of symmetry, the crest of the upconing interface is at $X = 0$. There, (9-174) becomes

$$Z_{\text{crest}} = \frac{1}{\pi} \int_0^\infty \frac{\cosh L(1 - D)}{L \sinh L} \{1 - \exp[-LT/(\coth L + M \coth LB)]\} \, dL$$
$$(9\text{-}175)$$

For a very thick aquifer, $a \to \infty$, $b \to \infty$, (9-174) and (9-175) become

$$Z = \frac{1}{2\pi} \ln \frac{(X/D)^2 + (T' + 1)^2}{(X/D)^2 + 1}; \qquad T' = t(\beta_f + \beta_s)d \qquad (9\text{-}176)$$

$$Z_{\text{crest}} = \frac{1}{\pi} \ln(T' + 1) \qquad (9\text{-}177)$$

When the porous medium is anisotropic with permeabilities $k_x \neq k_z$ (x, z assumed to be the principal directions of anisotropy), the relationships presented in Sec. 5-9 may be applied to (9-173) leading to

$$\zeta = \frac{\mu_f Q}{\pi \, \Delta\gamma (k_x k_z)^{1/2}} \int_0^\infty \frac{1}{\lambda} \frac{\cosh[\lambda(a-d)]}{\sinh(\lambda a)}$$

$$\times \left\{ 1 - \exp\left(\frac{-\lambda k_z (\Delta\gamma)}{n[\mu_f \cotgh(\lambda a) + \mu_s \cotgh(\lambda b)]} \right) \right\} \cos\left[\lambda \left(\frac{k_z}{k_x} \right)^{1/2} x \right] d\lambda$$

$$(9\text{-}178)$$

Equation (9-176) becomes

$$\zeta = \frac{\mu_f Q}{2\pi \, \Delta\gamma (k_x k_z)^{1/2}} \ln\left\{ \left[\left(\frac{k_z}{k_x} \right)^2 x^2 + \left(\frac{k_z \Delta\gamma}{n(\mu_f + \mu_s)} t + d \right)^2 \right] \Big/ \left[\left(\frac{k_z}{k_x} \right)^2 x^2 + d^2 \right] \right\}$$

$$(9\text{-}179)$$

The position of ζ_{crest} is derived from (9-179) by setting $x = 0$. Figure 9-34 shows ζ and ζ_{crest} for an infinitely thick aquifer in a graphical form.

The above development is presented in detail in order to illustrate the method. Bear and Dagan (1964a, and Dagan and Bear, 1968) apply this method to additional cases:

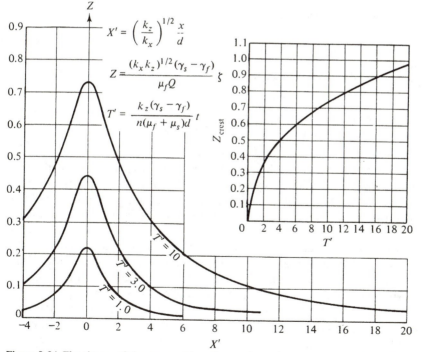

Figure 9-34 The shapes of the upconed interface under a drain in unsteady flow *(Bear and Dagan, 1964a)*.

(a) *Decay of the upconed interface* considered above. For example, for a very thick aquifer they obtain

$$\zeta = (\alpha_f Q/2\pi) \ln\{(x^2 + [t/(\beta_f + \beta_s) + d]^2)/(x^2 + [(t - t_1)/(\beta_d + \beta_s) + d]^2)\} \quad (9\text{-}180)$$

where pumping stops at $t = t_1$. To obtain this result, they assume, in view of the linearity of the problem, that superposition is permitted. A second solution, without superposition, in the form of an infinite integral is also given.

(b) *Upconing and decay of the upconed interface with an initially inclined interface* (i.e., with seaward flow, as it is in reality). Here the upconed interface is no more symmetric, although this effect is negligible for interface slopes less than 30°. The upconed interface, while undergoing decay, moves seaward.

(c) *Upconing and decay of the upconed mound (when pumping stops) under a point sink in three-dimensional flow above initially horizontal and inclined interfaces.*
   For the elevations of the interface above an initially horizontal one, they obtain

$$\zeta(r, t) = \frac{\alpha_f Q}{2\pi} \left\{ (r^2 + d^2)^{-1/2} - \left[\left(d + \frac{t}{\beta_s + \beta_f} \right)^2 + r^2 \right]^{-1/2} \right\} \quad (9\text{-}181)$$

As $t \to \infty$, (9-181) reduces to

$$\zeta(r, t) = \frac{\alpha_f Q}{2\pi(r^2 + d^2)^{1/2}} \quad (9\text{-}182)$$

For the crest ($r = 0$), (9-181) reduces to

$$\zeta_{\text{crest}}(t) = \frac{\alpha_f Q}{2\pi} \left[\frac{1}{d} - \frac{1}{d + t/(\beta_f + \beta_s)} \right] \quad (9\text{-}183)$$

In all these cases, the analytical solutions are compared with results of scaled experiments carried out in a Hele–Shaw analog, an electrolytic tank analog, and a sand box model. Figure 9-35 shows typical experimental results for the rise and decay of an interface under a drain, with seaward fresh water flow.

Based on their investigations, Bear and Dagan (1964a) reach the following conclusions.

For the two dimensional case, the approximate solutions are valid for interface rises of up to 1/2 the initial distance from the interface to the drain, with a recommendation to use ratios of 1/4–1/3 for safety purposes. This compares well with Muskat's conclusions presented earlier in this section. For the small interface slopes encountered in the field, the seaward flow may be neglected during the upconing period. During the decay period (no pumping), the interface crest is displaced towards the sea at a velocity approximately equal to 1/2 the fresh water velocity at the distance of the drain from the sea. The sea water body under the interface acts as a velocity attenuator.

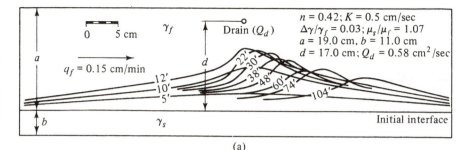

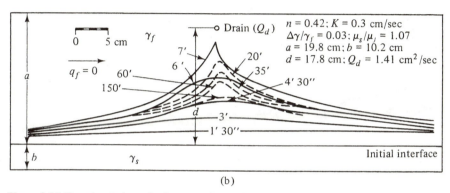

Figure 9-35 Experimental results for upconing and decay of upconed interface *(Bear and Dagan. 1964a)*.

Similar conclusions are reached for the three-dimensional case.

Schmorak and Mercado (1969) carry out field investigations in wells in the Ashkelon region in the coastal plane of Israel in order to check the validity of the results obtained by Bear and Dagan (1964a). They conclude that the theoretical formulae are in agreement with field results up to a critical rise which seems to be approximately half the distance between the initial undisturbed interface and the well.

9-8 TRANSITION ZONE

Before leaving this subject of the coastal interface, it seems worthwhile to remind the reader again that the *interface* as defined and discussed in this chapter is an *approximate concept*. It is a most useful one wherever the transition from fresh water to sea water occurs, indeed, over a short distance, relative to the thickness of the aquifer. Figure 9-36 gives an example of the situation along the coast in Israel. showing a rapid transition. On the other hand, Fig. 9-37 shows a case along the coast of Florida (Kohout, 1960) where a sharp interface approximation is not justified. In the latter case, one should study the problem as one of variable salinity (in this case, also of variable density), using the theory presented in Chap.

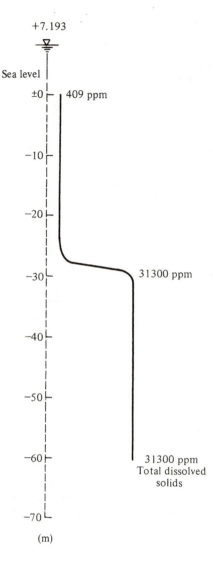

Figure 9-36 Total dissolved solids in Ashkelon Well 602 *(October, 1963; Courtesy of Hydrological Service, Israel).*

7. The example given at the end of Sec. 7-6 and Fig. 7-5 correspond to this case.

One should also recall that even when the assumption of a sharp interface is justifiable, a transition zone always exists. It may cause the contamination by sea water of shallow wells pumping above the interface, whenever the upconed (assumed) sharp interface gets too close to the pumping well. Some authors (e.g., Schmorak and Mercado, 1969) attempt to estimate the width of the transition zone (using the 50% concentration surface as the assumed sharp interface) by assuming that we have here a one-dimensional dispersion problem, with flow taking place in the vertical direction only.

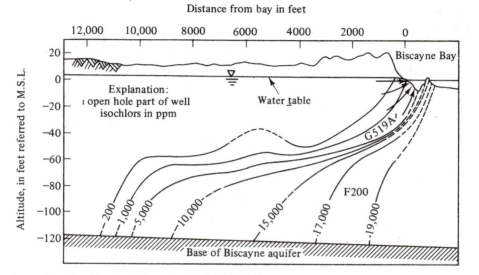

Figure 9-37 Isochlors in Cutler area, Florida *(Sept. 1958: after Kohout, 1960)*.

TEN

MODELING OF AQUIFER SYSTEMS

So far, we have been discussing the *forecasting problem* for both water quantity and water quality. In both cases the required groundwater regime (related both to water quantity and quality) was shown to be the solution of balance equations. We have shown what is involved in order to state in each case a well-posed problem.

However, the main objective of engineers and planners of water resource systems is to make the decisions which formulate a management policy. Among such decisions we may list those related to the quantity, location, and time of pumping from an aquifer and/or artificially recharging it with imported water. When making these decisions, we have to consider the need for additional pumping and recharge installations, the quality of pumped water, the dangers of quality deterioration (e.g., by sea water intrusion or encroachment of water of inferior quality from adjacent aquifers) and the level and quality of water to be maintained in water bodies (e.g., lakes) connected to an aquifer. Making the best decisions, according to some specified criteria and subject to specified constraints, of how to operate a groundwater system, constitutes the management (of groundwater) problem. In Chap. 12 we shall present a brief introduction to the management problem, showing how every management problem is based on the knowledge of the excitation–response relationship of the aquifer system as discussed in the first nine chapters of this book.

Aquifer models are used for studying proposed management policies. Each such policy, comprised, for example, of a list of timing and location of new installations and specification of the temporal and spatial distribution of pumping and artificial recharge, can be tested for physical feasibility by using a model. For each policy which is found to be physically and technologically feasible, we can compute the values of some criteria selected for evaluating the different policies. In this way, the "best" policy can be found. This is the *simulation* approach, it is rather straightforward, but does not ensure an optimal solution.

Another approach is to seek an optimal policy by using a management model. In this case the aquifer model becomes a set of constraints in the management model. To enable a solution of the latter, the aquifer model should be sufficiently simple, yet not too approximate so as to allow solutions which are not feasible in reality.

In these two ways, the aquifer model serves management purposes.

In the present chapter we shall see how models of aquifers and their behavior serve as tools in the solution of the forecasting, and hence also of the management problem. Obviously, additional models (e.g., of the economic and/or social structure and processes) are needed in order to state and solve a management problem. Some such models are introduced in Chap. 12.

10-1 THE NEED FOR AQUIFER MODELS

Obviously it is impossible to carry out experiments and tests in the aquifer itself in order to determine its response (e.g., in the form of water levels, water quality, or spring discharge) to activities proposed in the future and to make comparisons among responses to different possible activities in order to determine the most desirable one, according to some specified criteria, or to incorporate the responses in some decision making procedure. Like in all branches of science and engineering, whenever the treatment of real systems or phenomena is impossible (or the cost of such treatment is prohibitive), models of the considered systems or phenomena are introduced. Instead of treating the real system, we manipulate its model and use the results of these manipulations in order to make decisions regarding the operation of the real system.

The term model is introduced here in its most general sense. Often the terms *conceptual model* and *mathematical model* are also used. A model is a simplification, or abstraction, of the complex physical reality and the processes in it.

There is no need to elaborate on the fact that most real systems, and certainly the aquifer system considered here, are indeed complicated beyond our capability to describe them and to treat them exactly as they really are. The very passage from the microscopic level of treating flow through porous media to the macroscopic level of treating it as a continuum involves already a certain simplification of the real world. The porous medium continuum is inhomogeneous, anisotropic, etc., and further simplifications are necessary. These take the form of a set of assumptions which should not be forgotten whenever the model is being employed in the course of investigations. Examples of assumptions are that the flow is essentially horizontal, that in a phreatic aquifer water is released from storage immediately upon a decline of the water table, or that the water table is a surface which separates between a fully saturated region and a region with no moisture at all.

On the basis of these simplifying assumptions, a model of an investigated groundwater system is constructed. It is always presented in the form of a set of (mathematical) equations, the solution of which yields the behavior of the considered system. Hence, the term *mathematical model*. In almost all cases, the equations are balance equations (of water or of some constituent describing water quality).

Often the considered balances are written only for groundwater in the saturated zone. Sometimes, however, the unsaturated zone is also included (see examples in Sec. 10-6). In the former case, no attempt is usually made to relate the natural replenishment in one period to any specific earlier period, taking into account the lag of time which exists between the two events. If an appropriate balance period is chosen, this lag may be neglected.

Obviously it is possible to construct models of water resource systems including, for example, surface runoff, lakes, etc., in which the aquifer is just one element, and to include in them also processes in the unsaturated zone. The various components of a groundwater balance are discussed in Chap. 3. Each component is also characterized by the concentration of some chemical constituents of interest. The latter are used as a measure of water quality.

In systems engineering, a system is often defined as an ordered assembly of objects which interact in a well-defined interdependent manner. Accordingly, an aquifer model is a system, and the entire methodology of systems engineering, or systems analysis, can be applied to it.

For example, using the terminology of systems analysis, we speak of:

(a) *system states*, e.g., piezometric head (ϕ), water level (h), moisture content (θ), solute concentration (c), interface elevation (ζ),
(b) *system inputs*, e.g., controlled inputs, such as pumping, artificial recharge, specified boundary conditions, or uncontrolled ones, such as natural replenishment. Inputs are given in the form of functions of space and time which excite (or *stress* or *force*) the system,
(c) *state transition functions* which control the transition of the system from one state into another, at a later time. All the balance equations are actually state transition functions.

It is obvious that the same physical reality, i.e., an aquifer, part of an aquifer, or a number of interconnected aquifers, can be modeled in a great number of ways, depending on the assumptions we make in order to simplify the real physical system. The choice of the most appropriate conceptual model for a given aquifer system and for a given management problem is dictated not only by the features of the aquifer itself (e.g., its geological properties), but also by the following criteria:

(a) it should be sufficiently simple so as to be amenable to mathematical treatment,
(b) it should not be too simple so as to exclude those feature which are of interest to the investigation on hand.

As the range of possible models between these two limits is still wide, we should add two more important criteria, namely, that:

(c) information should be available for calibrating the model (this point is further discussed below), and
(d) the model should be the most economic one for solving the problem on hand. It is wasteful to select a very sophisticated model which may give very accurate results, whose construction and solution are costly and time consuming, when

satisfactory results for the problem on hand can be obtained by a simpler model, the operation of which is much cheaper. It is also unwise to construct a complicated model, which may describe the behavior of the real system more closely, when information is not available to calibrate it at that level of accuracy, and give the user the confidence that the model indeed simulates the behavior of the real system. It is meaningless to seek a model which gives very accurate results when the input data is much less accurate, sometimes by far. Similarly it is useless to choose a model which yields very detailed results when these cannot be verified by observing the behavior of the real system in the future.

One should realize that every considered system is usually only part of a larger system, with interactions between the subsystem and the rest of the system across the boundaries which delineate the considered subsystem. In previous chapters we referred to these interactions as boundary conditions. Accordingly, we may always model any part of an aquifer system, provided we specify the conditions on the boundaries of the considered subsystem.

Sometimes a planner wishes to determine at first only average values (say, of pumpage) for a large region. At a later stage of the investigations (often, as more data become available) he may wish to go into details and determine actual pumping rates and pumping schedules for individual wells. This example demonstrates the need for a hierarchy of models for the same real system, where at every stage of the investigation the appropriate model is used, to be replaced in subsequent stages by more detailed and sophisticated models for the same region. Often the results of one model are used as input information to the subsequent, more refined, models.

The emphasis in the above comments is that there is no unique model for a given region (or a given problem) and that one should not seek the most sophisticated (or accurate) model, but that a hierarchy of models of increased sophistication, refinement, and accuracy may be constructed for the same region. In each case the most appropriate one, from the points of view of needs, cost, and availability of data for calibration, should be selected.

10-2 MODEL CALIBRATION

The selected model must be well defined. The definition should be based on the detailed geometry of the aquifer, information about its physical parameters, boundaries, inputs, and outputs, etc. All this information is derived from geological studies and from observations in the real aquifer system. Whenever information is not available, it must be assumed on the basis of experience (or even guessed) and then verified during the calibration process.

The *calibration*, or *identification*, of a model is the process in which the various model parameters (and that may also include its geometry, inputs, etc.) are determined, if no previous knowledge of them is available, or verified (if such information is available). The calibration is based on data obtained from observations of the behavior of the aquifer in the past. Such data usually include water levels, pumping and recharge rates and volumes, water quality, interface positions, spring discharges, etc. The calibration, or identification, procedure is often referred to as the inverse problem. Methods of solving this problem are discussed in Chap. 11. Here we shall only present some general comments and emphasize the importance of always using a calibrated model.

In principle, the calibration involves the simulation on the aquifer model of a period in the past for which data are available on the behavior of the real aquifer system (e.g., water levels). When the model is excited in accordance with recorded input data for that period (e.g., pumpage and natural replenishment), its response (e.g., in the form of water levels) is compared with the recorded past response observed in the aquifer. The model is said to be calibrated when the difference between these two responses is less than some value specified by the planner. At the end of the calibration phase, we have a well-defined model of the aquifer system under consideration. All its parameters are well defined and it can now be used with confidence for forecasting the response of the aquifer in the future to the planned operations.

Actually there is an important flaw in this last statement, which should not be overlooked. We have here two systems, the (real) aquifer and the (conceptual) model. We take data from the first system, say, on water level changes as a result of natural replenishment and pumping, in order to calibrate the latter. Eventually, we end up with a calibrated model with identified model parameters, which we have established to describe the behavior of the aquifer system. Then we investigate the response of this model to future excitations, say, in the form of future pumpage and replenishment. Having obtained these *model responses*, we suggest that they represent also those of the *real aquifer* system. This line of thought explains why certain deviations should be expected in the behavior of the real system in the future when compared with that which was predicted on the basis of model behavior. We say that a noise is introduced because of the various *assumptions* which underlie the passage from the real system to the model one. The noise stems also from the fact that we compare responses of two different systems, where one only approximately reproduces the behavior of the other. In order to reduce these deviations, one should (a) construct a model which simulates the considered aquifer as close as possible, and (b) use as much data as is available for its calibration. As more data become available (and this is always the case in a developing area), the calibration process should be repeated leading to improved models and model parameters.

10-3 CLASSIFICATION OF AQUIFER MODELS

The discussion presented above is applicable to models of all kinds. Aquifer models can be classified in several ways. First we may distinguish between *physical models* and *mathematical models*. In the latter, the aquifer system and its behavior are represented in the form of a set of mathematical expressions, e.g., partial differential equations or linear algebraic equations. Among the former we may mention the sand box model, the Hele–Shaw analog, the RC-network, the electrolytic tank analog, and the membrane analog (see Bear, Chap. 11). They are discussed among the methods of solving the forecasting problem in Sec. 5-6.

Another possibility is to distinguish between continuous models, or models with *distributed parameters*, and those with a discrete distribution of parameters (*lumped-parameter models*). The first type includes mathematical models in the form of partial differential equations, and physical models and analogs such as the electrolytic tank or the Hele–Shaw analog (except when boundary conditions are discretized). The second group includes the numerical models and the RC-network, where the behavior of the system is defined only at specified points in space. In the numerical models, time is also discretized. From this point of view, the physical model is regarded as a simulator; the flow regime in the aquifer is simulated in the model.

Another attitude may be taken towards the physical laboratory models and analogs mentioned above. We prefer to view laboratory models and analogs as being special purpose computers, each designed to solve a specific problem or a limited group of problems. The problem to be solved is presented by the mathematical model. The use of such models and analogs is justified by showing the analogy between the set of mathematical equations which describe the behavior of the real aquifer system (actually its mathematical model!) and those describing the behavior of the (physical) model.

From this point of view, once a mathematical model has been stated, the decision as to whether a general purpose computer should be used for its solution or whether to resort to a special purpose one (e.g., an analog) should be based on economic considerations, availability of trained manpower, facilities, etc. Except for rare cases, there seems to be no advantage nowadays to the use of analogs. Some attempts have been made in recent years to combine an RC-electric analog and a digital computer and to form a *hybrid* computer. The interface equipment (e.g., analog to digital converter) is very expensive and there seems to be no advantage in doing so (Hefez *et al.*, 1975a).

Against this background of ideas regarding the use of mathematical models in investigating groundwater systems, in both the development and the management phases, the objective of the following paragraphs is to present and discuss several useful mathematical models. In most cases of practical interest, the actual solutions are obtained by means of digital computers. Some of the models are treated elsewhere in the book within other contexts. They will only be mentioned here briefly to complete the picture; the reader is referred to other sections in the book for more details.

Following is a list of some of the more commonly used models.

(a) *Single and multicell models.* These are discussed, for water quantity and for water quality, in Secs 10-4 and 10-5.

(b) *Partial differential equations.* These were developed in Chaps 5 and 6 for water quantity and in Chap. 7 for water quality. Equation (5-59), with appropriate initial and boundary conditions, may serve as an example. To develop this equation, we have assumed: (1) essentially horizontal flow in a confined aquifer, (2) the aquifer is isotropic, but is inhomogeneous; $T = T(x, y)$ is a continuous function having continuous first spatial derivatives, (3) the storativity S is due to the elastic properties of the water and the aquifer, yet the water is assumed to be almost incompressible, and (4) the flow obeys Darcy's law.

Because of all these assumptions, a statement of the flow problem in terms of a partial differential equation and boundary and initial conditions is a mathematical model of the real world. In order to solve (5-59), for example, analytically, we need the information on boundaries, and boundary conditions, on initial conditions and on $T(x, y)$, $S(x, y)$ and $N(x, y, t)$ in analytic forms. This is seldom, if ever, possible in cases of practical interest. It is also most difficult to calibrate this model, i.e., to determine these functions on the basis of observed value of ϕ. Therefore, these continuous models, which forecast what will happen in the aquifer at every point and every instant of time, although very refined, are usually impractical for regional studies of practical interest.

(c) *Finite difference* and *finite element* (or, in general, *numerical*) *models*. These models are briefly discussed in Sec. 5-6. However, the numerical schemes and their solutions may be viewed either as tools for solving, numerically, the partial differential equation, or as mathematical models on their own merits, as they require certain additional assumptions with respect to those listed for developing the partial differential equations.

Following the first point of view, the finite difference equations, for example, are derived mathematically as a discrete approximation of the partial differential equation. Following the second point of view, they are derived directly by considering the water balance for a control volume, assuming, for example, among other assumptions, a uniform gradient between nodal points (Sec. 10-4).

(d) *Abrupt fronts and interfaces*. These are also models (or elements of models) as they involve a simplified description of the real world. A discussion of the movement of abrupt fronts is presented in Sec. 7-10. The coastal interface—another abrupt front model—is discussed in Chap. 9.

10-4 SINGLE CELL MODELS

The simplest model is, perhaps, the one which visualizes an entire basin as a single cell. One assumes that average conditions (e.g., an average water table elevation) suffice to describe the behavior of this aquifer cell. The groundwater quantity balance for a cell of horizontal area A, bounded by impervious boundaries (Fig. 10-1), takes the form (see Chap. 3)

$$\Delta t [A\{N + R - P\} - Q] = A \times S \times \left(\bar{h} \Big|_{t+\Delta t} - \bar{h} \Big|_{t} \right) \tag{10-1}$$

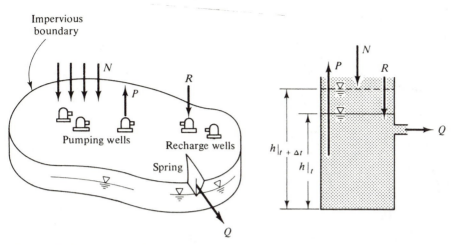

Figure 10-1 A single cell aquifer model.

where the period of balance is from t to $t + \Delta t$, S is aquifer storativity, N is natural replenishment, R is artificial recharge, P is rate of pumping, Q is rate of spring discharge, and $\bar{h}|_t$ is average water table elevation in the cell at time t. In (10-1), N, R, P are in terms of volume per unit area per unit time and Q is in volume per unit time. It is assumed that N, R, P, and Q remain constant during Δt; otherwise we have to use their average values over Δt. Sometimes Q is also considered a function of $\bar{h}$.

The unknown aquifer parameters which have to be determined during the calibration phase are usually S and N. Sometimes $Q = Q(\bar{h})$ has to be determined also. The calibration is based on known values of $\bar{h}$, R, and P for a number of periods in the past. The average values $\bar{h}$ of h over the cell are obtained from contour maps at t and $t + \Delta t$. If the area A is large, it is convenient to subdivide it into areas $(\Delta A)_i$ for which the h_i's are easily determined. Then $\bar{h} = \sum_{(i)} h_i (\Delta A)_i / A$.

Another single cell model can be constructed for a case where groundwater inflow and outflow through cell boundaries constitute components of the water balance and contour maps are available for determining them (Fig. 10-2). This corresponds, for example, to a case where the investigated region is a part of a larger aquifer. The volume of inflow into the cell minus that of outflow from it during Δt is then added to the left-hand side of (10-1). By dividing the boundary of the cell into N segments, the net inflow into the cell can be expressed, in principle, by $\Delta t \sum_{i=1}^{N} W_i T_i J_i$, where W_i, T_i, and J_i are the length, average transmissivity and average normal gradient (positive when inward) of the ith segment, respectively. Since the groundwater contours vary with time, while we maintain the boundaries of the cell fixed, we also have to use an average value $\bar{J}_i$ of J_i taken from the contour maps over the period Δt. It is sometimes convenient to use streamlines as parts of the cell boundary, as no flow takes place through them (Fig. 10-2). The balance period Δt, is usually chosen as 2 months, 6 months, or a year. If the period is too long, we have to be careful to obtain the correct time-averaged inflows and outflows. Altogether the balance equation for a cell with groundwater inflow and

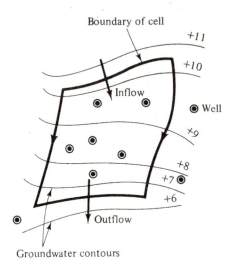

Figure 10-2 A single cell aquifer model with groundwater flow through boundaries.

outflow through cell boundaries is

$$\Delta t \left[A\{N + R - P\} - Q + \sum_{(i)} W_i \overline{T}_i \overline{J}_i \right] = A \times S \times \left(\overline{h} \bigg|_{t + \Delta t} - \overline{h} \bigg|_t \right) \quad (10\text{-}2)$$

where the average $\overline{J}_i$ is over time and length of boundary segment, and $\overline{h}$ is the average over the area A.

When warranted, other terms expressing inflows or outflows (Chap. 3) may be added to the water balance of a single cell aquifer model: (1) evapotranspiration, (which may be a function of h when the water table is sufficiently close to the surface), (2) infiltration from influent streams passing through the region, and drainage into effluent streams, (3) drainage, (4) leakage from a water supply system, (5) return flow from irrigation (which may be a function of the volume of water pumped from the aquifer), from septic tanks, etc., and (6) leakage into or out of the considered aquifer through overlying and/or underlying semipervious layers.

Obviously, in a confined aquifer, we do not have those components of the balance which involve percolation of water to the aquifer through the unsaturated zone.

As more components are included in the water balance of a cell, the magnitude of some of them may be a priori unknown and have to be determined as part of the calibration procedure.

In spite of its simplicity, the single cell model is a very useful one, especially in the early stages of development and management when data are scarce and the planner is interested in an overall picture (average water levels) of the investigated area, and in making decisions (e.g., on pumping) related to the area as a whole.

A single cell model can also be constructed for *water quality* as measured by some dissolved constituent of interest or by Total Dissolved Solids (TDS). As discussed in Chap. 7, all chemical constituents are carried with the flowing water. However, as constituents are being transported with the water, hydrodynamic dispersion as well as decay (of radioactive substances) and/or chemical reactions take place (among dissolved constituents and with the solid skeleton of the porous medium). To every element (inflow and outflow) in the water balance discussed above (see also Chap. 3), we may assign a water quality, expressed in terms of some concentration of the constituent of interest. For the sake of simplicity we shall consider a *solute balance* (or a *salt balance*), where solute ($\equiv$salt) means a constituent of interest.

As in the balance of water quantity, here also we consider a balance of solute (= some constituent describing water quality) only of groundwater in the saturated zone. Unless we specifically desire otherwise, we consider only inflows and outflows of salinity into or out of the saturated zone. The difference between the two is stored also in that zone. When the movement of solutes originates at the groundsurface (e.g. salinity of natural replenishment, return flow from septic tanks or irrigation or leachate from landfills), there exists also movement through and storage in the unsaturated zone, possibly accompanied by various chemical reactions and decay. We may also construct a model for this phase of the solute movement and accumulation. The outflow from such a model will become the inflow into the groundwater quality model. Mercado (1976) and Lyons (1976) present examples of such a model (see Sec. 10-6).

Accordingly, for groundwater in the saturated zone only, the following components, when present, are incorporated in a balance for a single cell quality model.

(*a*) Solute transported with the moving groundwater entering or leaving the cell. One should note that in view of the discussion in Chap. 7, we recognize two modes of movement: convection and hydrodynamic dispersion. However, in the model we usually include only the former (although it is also possible to incorporate the latter). We may also include here leakage of solutes through underlying or overlying semipervious layers. Sometimes salt filtering takes place upon passage through such layers.

(*b*) Dissolved salt contained in the natural replenishment (i.e., precipitation reaching the aquifer, Sec. 3-2). As in the case of the water balance, we take into account only the actual amount of salt reaching the aquifer during the balance period. This amount need not be the same as that actually dissolved in the rainwater, as changes may take place (depending on which component we consider) upon passage through the unsaturated zone. For example, the infiltrating rainwater may dissolve salts from point, or distributed, surface sources such as landfills, fertilizers, herbicides, pesticides, etc., and transport them, subject to possible modifications, through the unsaturated zone to the groundwater table. A lag of time may also exist between the time of infiltration and the time of arrival of the salts at the water table.

(*c*) Solutes contained in irrigation water. These include those initially contained in the irrigation water and those added by leaching. They are carried through the unsaturated zone, again with possible modifications, to the water table by the return flow from irrigation (Sec. 3-3). One should note that irrigation water may come from groundwater pumped within the balance area and/or imported into the area from outside.

(d) Solutes carried by water infiltrating from septic tanks.

(e) Solutes introduced into the aquifer with artificial recharge water (Sec. 3-4). This may also include artificial recharge with reclaimed municipal wastewater (whose salinity with respect to the water supplied has been increased by a few hundreds p.p.m. TDS due to domestic and commercial uses and ion-exchange water softeners).

(f) Solutes leaving the aquifer with pumped water and/or with spring water.

(g) Solutes leaving the aquifer with drainage water (in situations of high water table).

For example, referring to Fig. 10-1 and to the water balance (10-1), we may write the following salt balance for the components (b), (e), and f) mentioned above

$$\Delta t \left[A \left\{ N c_N + R c_R - P c \Big|_t \right\} - Q c_t \right] = U \Big|_{t+\Delta t} c \Big|_{t+\Delta t} - U \Big|_t c \Big|_t \qquad (10\text{-}3)$$

where c_N, c_R are the average (over space and time) concentrations in the natural replenishment and in the artificial recharge water during Δt, respectively; $c|_t$ is the average concentration in the aquifer at time t and $U|_t$ is the volume of water in the aquifer at time t. All concentrations are in p.p.m. (i.e., parts per million). In a phreatic aquifer, changes in U are related to changes in h and to the storativity (= specific yield). In a confined aquifer, U remains practically unchanged with time, and we do not have balance components which percolate to the water table through the unsaturated zone (e.g., natural replenishment).

If we wish to incorporate component (a) above, we should add on the left-hand side of (10-3) $(Q'_{in} c_G - Q'_{out} c|_t)$, where Q'_{in} and Q'_{out} are (average) groundwater inflow and outflow through the basin boundary, and c_G is the concentration of the inflowing groundwater. It is possible to write a salt balance which will include more of the components listed above.

It is important to reemphasize that in writing (10-1) through (10-3), we have considered only the arrival of water and salt to the saturated zone. This means that we have disregarded processes (including transport) which take place in the unsaturated zone. However, models may be constructed which account for such processes (see example in Sec. 10-6).

Furthermore, in (10-3) we have assumed complete continuous mixing in the cell. This means that as the various sources contribute solutes to the aquifer, complete mixing in the entire volume of water present in the aquifer cell takes place. This is obviously a very crude assumption, especially in deep aquifers. Natural replenishment, artificial recharge by spreading and return flow from irrigation introduce solutes at the water table; artificial recharge through wells and pumping produce flows mainly in the upper parts of a deep aquifer.

One way to overcome this difficulty is to assume that the aquifer is composed of two layers: an upper, active one, in which all the mixing takes place, say, from the water table down to some depth below the wells' screens, and a lower one which contains only native water, the quality of which remains practically unaffected by recharge and pumping operations. In this case, U in (10-3) is the volume of water in the upper layer only; it is assumed that complete mixing takes place in this volume and that the concentration is uniform throughout this volume. In this way we actually introduce another (model) parameter which has to be determined during the calibration process (see examples in Sec. 10-6).

We have also assumed no interaction between the water and the solid matrix. If such interactions occur, additional terms may have to be added to the salt balance. For example, solutes may be added to the water in the aquifer as a result of solution of the rock. Solutes (or ions) may also be extracted from the water or added to it as a result of adsorption and exchange phenomena. Sometimes, lenses of material of a permeability much lower than that of the aquifer, saturated with highly mineralized water, are imbedded in the aquifer. This may happen if the entire aquifer was originally occupied by this kind of water, and leaching from the more pervious parts of the aquifer was much faster. The process of leaching continues by molecular diffusion and by dispersion due to the (very small) flow passing through these lenses, thus contributing salt to the salt balance in the aquifer. An appropriate term expressing this additional source of salt should then be added to (10-3).

10-5 MULTIPLE CELL MODELS

The basic ideas underlying the single cell model, also underlie the multicell one. In this case the investigated aquifer is divided into a relatively small number of cells, usually of rectangular shape. We emphasize here that the number of cells into which the aquifer is subdivided may be small in order to make a clear distinction between this approach and that reflected in the numerical (finite difference and finite element) methods described in Sec. 5-6, where we also have cells. In the latter approach we think of the numerical, say, finite difference, representation as an approximation of the partial differential equation, which we would have liked to solve analytically if we could. Hence we aim at as large a number of cells as possible (subject, of course, to such constraints as available computer facilities, costs, etc.) in order to get better accuracy, say, in piezometric head at nodes.

In the *multicell approach*, where the number of cells modeling the behavior of a large aquifer may be as small as two, we take a completely different point of view. We actually write water balances similar to (10-1) or (10-2), for quantity-type models, for each of the cells in the model. Like in the single cell model discussed above, we assume that no flow takes place within each cell, but we take into account flow through the common boundaries of adjacent cells. For each cell we assume average properties, and an average water level. Figure 10-3 shows an aquifer in the shape of an elongated strip visualized as a four-cell model. The flow between adjacent cells is visualized as flow through a linear resistance between reservoirs.

Thus, although the numerical schemes which approximate the partial differential equation and the multicell model may both take the form of similar algebraic equations, the ideas and the way of thinking underlying the two approaches are different.

Figure 10-4 shows a number of rectangular cells in a multicell aquifer model.

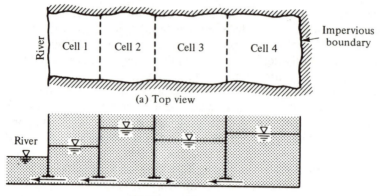

(a) Top view

(b) Cross section

Figure 10-3 A 4-cell model of a strip-shaped aquifer.

The water balance for the i,j cell can be stated for a balance period between t and $t + \Delta t$ in the following way (compare with (5-55))

$$\Delta t \left(Q_x \bigg|_{i-1/2,j} - Q_x \bigg|_{i+1/2,j} + Q_y \bigg|_{i,j-1/2} - Q_y \bigg|_{i,j+1/2} + R_{i,j} - P_{i,j} + N_{i,j} \right)$$
$$= S_{i,j} \Delta x_i \Delta y_j (\phi_{i,j}^{t+\Delta t} - \phi_{i,j}^t) \qquad (10\text{-}4)$$

where Q_x; Q_y are total rates of flow at time t through cell boundaries, positive when in positive direction of the coordinate axes (dims. L^3/T),

$R_{i,j} + N_{i,j}$ is the total recharge rate (artificial and natural, respectively) in cell i,j during Δt (dims. L^3/T),
$P_{i,j}$ is the total pumping rate in cell i,j during Δt, (dims. L^3/T),
$S_{i,j}$ is the average aquifer storativity in cell i,j (dimensionless), and
$\phi_{i,j}^t$ is the piezometric head in cell i,j at time t (dims. L).

Other balance components (e.g., return flow from irrigation and septic tanks, spring discharge, effluent and influent streams, etc.) may also be added to the balance.

In the multicell approach, aquifer properties are usually related to cells (i.e. we have transmissivities $T_{i,j}$, $T_{i,j-1}$, $T_{i,j+1}$, $T_{i-1,j}$, etc.). However, in order to express the Q's in terms of the ϕ's we need the values of T on the boundary between cells. These values may either be given or calculated by one of the following methods.

(i) $$T_{i+1/2,j} = (T_{i+1,j} + T_{i,j})/2 \qquad (10\text{-}5)$$
i.e., arithmetic average, or

(ii) $$T_{i+1/2,j} = (\Delta x_i + \Delta x_{i+1})/(\Delta x_i/T_{i,j} + \Delta x_{i+1}/T_{i+1,j}) \qquad (10\text{-}6)$$

i.e., a hormonic average. Similar expressions can be written for T on the other boundaries. The latter case is more accurate as it takes into account the fact that we have flow through two different transmissivities (flow normal to layers, Bear, 1972, p. 153).

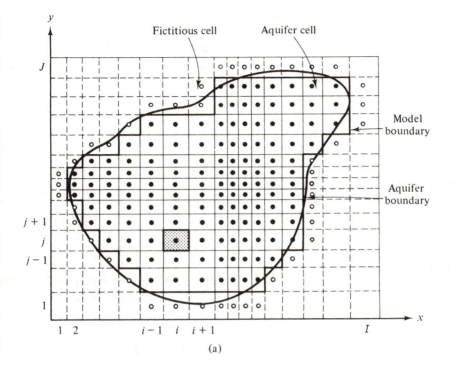

(a)

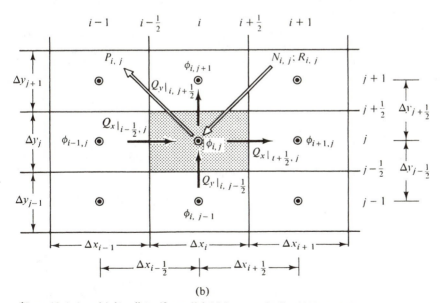

(b)

Figure 10-4 A multiple-cell aquifer model. (a) Layout of cells. (b) Water balance components for cell i, j.

We may rewrite (10-4) in the form

$$T_{i-1/2,j}\,\Delta y_j\frac{\phi_{i-1,j}^t - \phi_{i,j}^t}{(\Delta x_i + \Delta x_{i-1})/2} + T_{i+1/2,j}\,\Delta y_j\frac{\phi_{i+1,j}^t - \phi_{i,j}^t}{(\Delta x_i + \Delta x_{i+1})/2}$$

$$+\, T_{i,j-1/2}\,\Delta x_i\frac{\phi_{i,j-1}^t - \phi_{i,j}^t}{(\Delta y_j + \Delta y_{j-1})/2} + T_{i,j+1/2}\,\Delta x_i\frac{\phi_{i,j+1}^t - \phi_{i,j}^t}{(\Delta y_j + \Delta y_{j+1})/2}$$

$$+\, N_{i,j} + R_{i,j} - P_{i,j} = S_{i,j}\,\Delta x_i\,\Delta y_j\frac{\phi_{i,j}^{t+\Delta t} - \phi_{i,j}^t}{\Delta t} \tag{10-7}$$

where we have expressed the Q's by Darcy's law written for time t and the various T's are expressed by either (10-5) or (10-6).

An equation similar to (10-7) is written for every cell. Special attention should be given to boundary cells (see Sec. 5-6). The mathematical model for an N cells model is thus made up of N equations similar to (10-7).

The selection of Δx_i and Δy_j is not independent of the selection of Δt; they are related to each other by a *stability criterion* similar to (5-104). To illustrate this point, using a non-rigorous physical approach, let us rewrite (10-7) for the case: Δx_i = const. = Δx, Δy_j = const. = Δy, $S_{i,j}$ = const. = S $T_{i,j}$ = const. = T, $N_{i,j} = R_{i,j} = P_{i,j} = 0$, in the form

$$\Delta t\left\{\frac{T\,\Delta y}{\Delta x}(\phi_{i-1,j}^t - 2\,\phi_{i,j}^t + \phi_{i+1,j}^t) + \frac{T\,\Delta x}{\Delta y}(\phi_{i,j-1}^t - 2\,\phi_{i,j}^t + \phi_{i,j+1}^t)\right\}$$

$$= S\,\Delta x\,\Delta y(\phi_{i,j}^{t+\Delta t} - \phi_{i,j}^t) \tag{10-8}$$

The left-hand side of (10-8) gives excess of inflow over outflow in the cell $\Delta x\,\Delta y$; the right-hand side gives the resulting increase in storage.

Consider the special case where $\phi_{i-1,j}^t = \phi_{i+1,j}^t = \phi_{i,j-1}^t = \phi_{i,j+1}^t = 0$, while $\phi_{i,j}^t = -H$. Since the maximum possible increase in storage during Δt is $HS\,\Delta x\,\Delta y$, we have $-H \le \phi_{i,j}^{t+\Delta t} \le 0$. Hence, $0 \le (\phi_{i,j}^{t+\Delta t} + H)/H \le 1$, and we obtain from (10-8)

$$\frac{T}{S}\left[\frac{\Delta t}{(\Delta x)^2} + \frac{\Delta t}{(\Delta y)^2}\right] \le \frac{1}{2}$$

which is identical to the stability criterion (5-104).

For a phreatic aquifer we may use (10-7), with S representing specific yield, h replacing ϕ, and $T = Kh$. The resulting equation is

$$K_{i-1/2,j}h_{i-1/2,j}^t\,\Delta y_j\frac{h_{i-1,j}^t - h_{i,j}^t}{(\Delta x_i + \Delta x_{i-1})/2} + K_{i+1/2,j}h_{i+1/2,j}^t\,\Delta y_j\frac{h_{i+1,j}^t - h_{i,j}^t}{(\Delta x_i + \Delta x_{i+1})/2}$$

$$+\, K_{i,j-1/2}h_{i,j-1/2}^t\,\Delta x_i\frac{h_{i,j-1}^t - h_{i,j}^t}{(\Delta y_j + \Delta y_{j-1})/2} + K_{i,j+1/2}h_{i,j+1/2}^t\,\Delta x_i\frac{h_{i,j+1}^t - h_{i,j}^t}{(\Delta y_j + \Delta y_{j+1})/2}$$

$$+\, N_{i,j} + R_{i,j} - P_{i,j} = S_{i,j}\,\Delta x_i\,\Delta y_j\frac{h_{i,j}^{t+\Delta t} - h_{i,j}^t}{\Delta t} \tag{10-9}$$

We can express K or Kh^t on the boundaries by (10-5) or (10-6). Other schemes for writing the balance are also possible. For example, we may use $Q^{t+\Delta t/2} = (Q^t + Q^{t+\Delta t})/2$ to express the flow through cell boundaries (see (11-23)). However, similar to the forward scheme in the finite difference method (Sec. 5-6), the resulting algebraic equations will include five unknown values of ϕ at time $t + \Delta t$, as compared with one in (10-7). We see here the similarity between writing a water balance for a cell and the numerical approximation of the partial differential equation, except that the balance equation is derived on the basis of physical (rather than mathematical) considerations only.

In a leaky aquifer we have also to add a term representing the leakage into or out of the cell through the area $\Delta x_i \Delta y_j$.

In principle, given the values of ϕ at time t for all cells, we should be able to solve the set of linear equations for all the cells in the aquifer and to obtain the values of ϕ at $t + \Delta t$.

The choice of size and number of cells for a given aquifer depends on the following factors

(a) The use of the model. If the model is to be used for making management decisions (say, on rates of pumping and recharge), the cells should conform to the nature of the decisions. For example, if the decisions are to be related to certain specified (e.g., administrative) subregions, we may wish to use them as model cells. The same consideration is true for the choice of Δt. In areas where decisions have to be more accurate or detailed, we may introduce smaller cells.

(b) Availability of information on aquifer parameters and on their spatial variability. In regions of large changes, we may get more accurate results by using smaller cells. The same is true for water table elevations; where slopes are large we should use smaller cells. Needless to say that geological information (e.g., presence of faults) will influence the choice of cell boundaries.

(c) Recalling that no model should be used before it is calibrated, the availability of information on past water levels may dictate the choice of cell size. It is meaningless to introduce a large number of small cells if there is no information on past water levels in them for determining their coefficients in the calibration process.

For this last reason, polygonal cells are sometimes introduced, based on a Thiessen (1911) polygon network (e.g., Tyson and Weber, 1964).

Consider, for example, an aquifer with a number of wells with recorded data on water levels (Fig. 10-5). The Thiessen polygon method assumes a linear variation of a measured quantity (here water levels) between each pair of two close observation points (here wells). Perpendicular bisectors of the lines connecting adjacent observation points (shorter distance where two possible lines may be drawn) form polygons around each observation point (or partial polygons with the aquifer's boundary). Values measured at an observation point are assumed to represent the entire polygon corresponding to that point.

We may write the water balance for a typical cell, using the nomenclature of Fig. 10-5, where N_i, P_i, R_i are in volume per unit time over entire cell, in the form

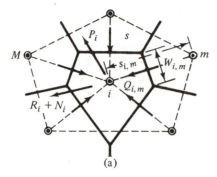

(a)

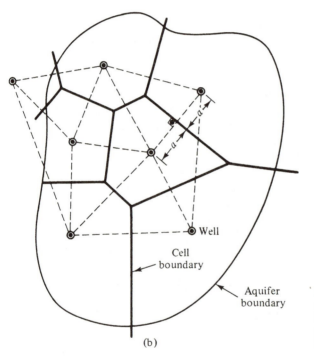

Well

Cell
boundary

Aquifer
boundary

(b)

Figure 10-5 A multicell model
using Thiessen polygons as cell
boundaries. (a) A typical polygonal cell. (b) Aquifer divided
into 6 cells.

$$\sum_{m=1}^{M} T_{i,m} W_{i,m} \frac{\phi_m^t - \phi_i^t}{s_{i,m}} + N_i + R_i - P_i = S_i \frac{\phi_i^{t+\Delta t} - \phi_i^t}{\Delta t} \text{ Area} \quad (10\text{-}10)$$

where S_i represents the average storativity of the cell and $T_{i,m}$ is the equivalent
transmissivity along the line $s_{i,m}$ connecting the two cells. Similar equations may
be written for phreatic aquifers.

(d) The technique to be used for solving the set of balance equations written
for the cells (e.g.. size of digital computer).
(e) Cost of solution which depends on the number of cells.

Quality balances (i.e., balances related to some constituent representing water quality) can also be written for a multicell model. For the multicell model shown in Fig. 10-4, we can write

$$Q_x|_{i-1/2,j} \left\{ \frac{1}{2} \left(1 + \frac{Q_x|_{i-1/2,j}}{|Q_x|_{i-1/2,j}|} \right) c_{i-1,j}^t + \frac{1}{2} \left(1 - \frac{Q_x|_{i-1/2j}}{|Q_x|_{i-1/2,j}|} \right) c_{i,j}^t \right\}$$

$$- Q_x|_{i+1/2,j} \left\{ \frac{1}{2} \left(1 + \frac{Q_x|_{i+1/2,j}}{|Q_x|_{i+1/2,j}|} \right) c_{i,j}^t + \frac{1}{2} \left(1 - \frac{Q_x|_{i+1/2,j}}{|Q_x|_{i+1/2,j}|} \right) c_{i+1,j}^t \right\}$$

$$+ Q_y|_{i,j-1/2} \left\{ \frac{1}{2} \left(1 + \frac{Q_y|_{i,j-1/2}}{|Q_y|_{i,j-1/2}|} \right) c_{i,j-1}^t + \frac{1}{2} \left(1 - \frac{Q_y|_{i,j-1/2}}{|Q_y|_{i,j-1/2}|} \right) c_{i,j}^t \right\}$$

$$- Q_x|_{i,j+1/2} \left\{ \frac{1}{2} \left(1 + \frac{Q_y|_{i,j+1/2}}{|Q_y|_{i,j+1/2}|} \right) c_{i,j}^t + \frac{1}{2} \left(1 - \frac{Q_y|_{i,j+1/2}}{|Q_y|_{i,j+1/2}|} \right) c_{i,j+1}^t \right\}$$

$$+ R_{i,j} c_R + N_{i,j} c_N - P_{i,j} c_{i,j}^t = \frac{1}{\Delta t} (U_{i,j}^{t+\Delta t} c_{i,j}^{t+\Delta t} - U_{i,j}^t c_{i,j}^t) \quad (10\text{-}11)$$

where U is the volume of water in the cell and the Q's, taken at time t, are obtained by solving the water balance model (10-4).

In writing (10-11) we took care of the direction of the flows. We have also neglected dispersion and assumed complete mixing in the cell within each time interval Δt. Obviously this will be far from true in thick aquifers. We may then use some part of U as the effective volume with respect to salinity changes.

As in writing the water balance for a cell, other ways of expressing the salt balance are also possible. For example, (Fig. 10-6), we may express flows of salt at the midpoint of the boundary as a product of rate of flow and salinity at that point, where the latter is taken as the arithmetic (or weighted) average concentration in the adjacent cells

$$c_{i-1/2,j} = \delta c_{i-1,j} + (1 - \delta) c_{i,j},$$

$$c_{i+1/2,j} = \delta c_{i,j} + (1 - \delta) c_{i+1,j}, \qquad \text{etc.} \quad (10\text{-}12)$$

In (10-12), $\delta = \{1 + Q/|Q|\}/2$. Water Resources Engineers Inc. have suggested $\delta = 0.75$ for the study of the Upper Santa Ana River Basin in California. However, $\delta = 0.5$ can be used only for flow which does not reverse its direction. Otherwise we have to keep track of the flow direction, as there should be a difference between the salt transported by $Q_x > 0$ and $Q_x < 0$.

Cells of polygonal shapes may also be used.

It is important to emphasize again the fact (underlying the idea of all cell models) that the model cannot predict water quality at individual pumping wells. It is only the overall regional picture which the model will provide. This fact

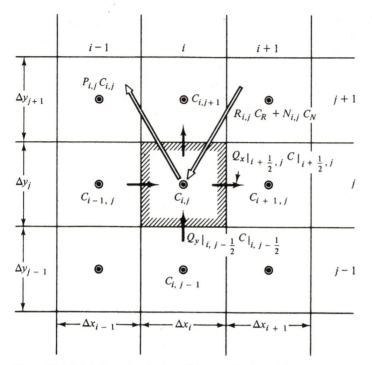

Figure 10-6 Salt balance for the $i. j$ cell in a multicell model.

should also be kept in mind when calibrating the model. Whereas the water level in a well may indeed represent water levels in an area around it, the quality of a water sample from a pumping well is indicative only of the quality of the mixed water from the section of the aquifer supplying water to that particular well. Also, a measure of water quality in a well (even if the sample is taken from pumped water) does not always provide information of the possible variation in water quality along a vertical. The aquifer may be stratified with respect to quality, with saline water underlying the fresh water layer tapped by the well. This problem of quality stratification may happen also in an aquifer with imbedded clay lenses. Above these lenses the quality of water may deteriorate by direct percolation of water of inferior quality from the surface, without being detected by wells whose screens are underneath these lenses.

Simpson and Duckstein (1975) propose a multicell model, which they call Finite State Mixing Cell Model, suitable for the modeling of karst water resources in a systems framework. The model consists of a set of interconnected cells of any desired size through which the transport of water and dissolved matter is represented by a sequence of finite states. Each elementary cell may be either a pure mixing cell or a cell that simulates partial or complete piston flow (i.e., no mixing).

10-6 TWO EXAMPLES OF WATER QUALITY MODELS

Two models are reviewed below to illustrate some of the ideas discussed in Secs. 10-4 and 10-5 above. Both examples show how the unsaturated zone can be incorporated in a multicell and in a single cell models.

Example 1 Lyons (1976) uses a groundwater basin water quality model to study alternate management plans. His model (Fig. 10-7) is composed of polygonal cells. Each cell, j, is divided into two parts: an unsaturated part and an underlying saturated one. The latter is again divided into two parts: an upper one,

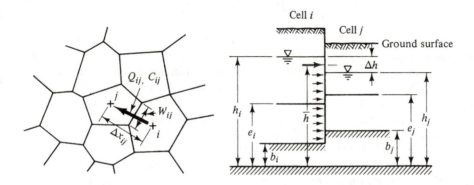

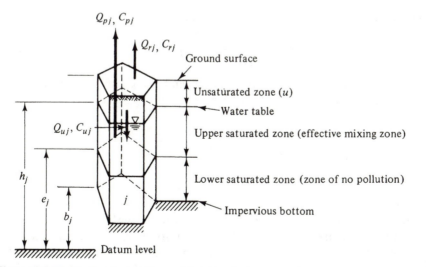

Figure 10-7 Multicell water quality model, taking into account the unsaturated zone *(after Lyons. 1976)*.

in which effective mixing of the considered constituent takes place (continuous, complete mixing), and a lower one, which is never penetrated by the considered constituent. The polluting dissolved constituent is transported with the water from the unsaturated part of the j th cell to its saturated part. It is also carried with the water in the saturated zone by convection only (neglecting hydrodynamic dispersion) from one cell to the next.

The balance of mass (of a considered constituent) during Δt for the j th cell can be described by the following equation

$$\Delta t \left\{ Q_{rj}\bar{c}_{rj} - Q_{pj}\bar{c}_{pj} + \underset{(i)}{\sum} Q_{ij}\bar{c}_{ij} \right\} =$$

$$(U_{uj}c_{uj}) \Big|_{t+\Delta t} - (U_{uj}c_{uj}) \Big|_t + (U_{sj}c_{sj}) \Big|_{t+\Delta t} - (U_{sj}c_{sj}) \Big|_t \quad (10\text{-}13)$$

where: Q_{ij} is the rate of flow (L^3/T) from cell i to adjacent cell j, Q_{rj} is the rate of recharge (L^3/T) in the j th cell during Δt, Q_{pj} is the rate of pumping (L^3/T) in the j th cell during Δt, c_{ij} is the constituent concentration in the flowing water between adjacent cells i and j, c_{pj}, c_{rj} is the constituent concentration in the water pumped from and in the water recharging node j, respectively, during Δt, $\bar{c}_{rj} = (c_{rj}|_{t+\Delta t} + c_{rj}|_t)/2$, $\bar{c}_{pj} = (c_{pj}|_{t+\Delta t} + c_{pj}|_t)/2$, c_{uj} is the constituent concentration in the unsaturated part of the j th cell, and U_{sj}, U_{uj} are the volumes of water in the j th cell in the saturated and unsaturated parts, respectively.

Separate balances can also be written for the unsaturated and for the saturated zones. For the unsaturated zone, the mass balance may be stated in the form

$$(U_{uj}c_{uj}) \Big|_{t+\Delta t} - (U_{uj}c_{uj}) \Big|_t = \Delta t \left\{ Q_{rj}\bar{c}_{rj} - Q_{uj}\bar{c}_{uj} \right\} \quad (10\text{-}14)$$

where Q_{uj} is the flow leaving the unsaturated zone, and

$$\bar{c}_{uj} = \left(c_{uj} \Big|_{t+\Delta t} + c_{uj} \Big|_t \right) \Big/ 2$$

Given conditions at the beginning of Δt, and rates and concentrations of recharge, one can determine c_{uj} from (10-14)

$$c_{uj} \Big|_{t+\Delta t} = \frac{Q_{rj}\bar{c}_{rj}\,\Delta t + c_{uj}|_t \left(U_{uj}|_t - \dfrac{\Delta t}{2} Q_{uj} \right)}{U_{uj} + \dfrac{\Delta t}{2} Q_{uj}} \quad (10\text{-}15)$$

It is obvious that in this simplified model, Lyons (1976) assumes that both the moisture level and the concentration are uniform throughout the unsaturated zone, and that Q_{uj} is always downward. The validity of the first assumption depends on the depth of the water table, the permeability of the unsaturated zone, and the chosen time interval Δt. In a more sophisticated model, the unsaturated zone itself may be divided into several cells.

The saturated zone is divided into two parts. The upper part, designated as the *zone of effective mixing*, represents the portion of the saturated zone from which water is pumped. It also receives the water and solutes that leave the unsaturated zone. The lower zone is not directly affected by pumping and recharge. The two zones are denoted by subscripts 1 and 2, respectively.

For a *conservative* quality constituent, i.e., one which does not interact with the soil or with other constituents in the water, the mass balance for a typical cell (or node) j is

$$(U_{1j}c_{1j})\Big|_{t+\Delta t} - (U_{1j}c_{1j})\Big|_t + (U_{2j}c_{2j})\Big|_{t+\Delta t} - (U_{2j}c_{2j})\Big|_t =$$

$$\Delta t \left\{ Q_{uj}\bar{c}_{uj} - Q_{pj}\bar{c}_{pj} + \sum_{(i)} Q_{ij}\bar{c}_{ij} \right\} \quad (10\text{-}16)$$

Direct artificial recharge of the saturated zone (either subzone 1 or 2, or both) as well as pumping from the lower subzone, can be added. In (10-16) it is assumed that complete mixing in the upper zone takes place continuously, so that the concentration c_{pj} of the pumped water is equal to the time average concentration $\bar{c}_{pj}$ in the upper subzone.

The groundwater flow term $\sum_{(i)} Q_{ij}\bar{c}_{ij}$ appearing in (10-13) and (10-16) requires special attention. Lyons (1976) divides Q_{ij} into several parts, as shown in Fig. 10-7c

$$Q_{ij} = Q_{ij}^{(1)} + Q_{ij}^{(2)} + Q_{ij}^{(3)}$$

for $e_i \geq e_j$

for $e_i < e_j$

$$Q_{ij}^{(1)} = Q_{ij}(\bar{h} - e_i)/(\bar{h} - B_{ij}) \qquad Q_{ij}^{(1)} = Q_{ij}(\bar{h} - e_j)/(\bar{h} - B_{ij})$$

$$Q_{ij}^{(2)} = Q_{ij}(e_i - e_j)/(\bar{h} - B_{ij}) \qquad Q_{ij}^{(2)} = Q_{ij}(e_j - e_i)/(\bar{h} - B_{ij})$$

$$Q_{ij}^{(3)} = Q_{ij}(e_j - B_{ij})/(\bar{h} - B_{ij}) \qquad Q_{ij}^{(3)} = Q_{ij}(e_i - B_{ij})/(\bar{h} - B_{ij})$$

$$(10\text{-}17)$$

where: $B_{ij} = (b_i + b_j)/2$ and $\bar{h} = (h_i + h_j)/2$. Equation (10-17) is written for flow from cell j to cell i. A similar set of equations can be written for flow in the opposite direction.

The volume of water in the lower subzone is assumed constant. Hence, if horizontal inflows $(Q_{ij}^{(2)} + Q_{ij}^{(3)})$ into this subzone exceed outflows, the excess flows vertically upward into the upper subzone, and vice versa.

As for the concentration, Lyons expresses c_{ij} by

$$c_{ij} = \delta c_i + (1 - \delta) c_j \quad (10\text{-}18)$$

(compare with (10-12)), where δ is an interpolation factor ($0.5 \leq \delta < 1.0$). He calculates separately the transfer of mass of the considered constituent by the various partial flows $Q_{ij}^{(k)}$, $k = 1, 2, 3$, at corresponding interpolated concentrations $c_{ij}^{(k)}$ as defined by (10-18).

The constituent is also transported vertically from the upper subzone into the unsaturated zone and vice versa with the flowing water at the average (or interpolated) concentration at the boundary between the two zones.

The flows Q_{ij} are expressed by (Fig. 10-7)

$$Q_{ij} = W_{ij}T_{ij}\frac{h_i - h_j}{\Delta x_{ij}} \tag{10-19}$$

where T_{ij} is some average transmissivity between the two cells.

Although the model is rather simple, it proved a powerful tool for water quality management studies. Improvements can be introduced by adding dispersion phenomena and especially by considering nonconservative solutes (e.g. exhibiting adsorption on the solid matrix, decay, or chemical reactions with other dissolved constituents).

Example 2 Mercado (1976) also presents a groundwater quality manage-model which includes the unsaturated zone. Figure 10-8 shows Mercado's groundwater quality system which integrates pollution sources on the land surface, and approximately represents the hydrological and physicochemical parameters of the aquifer and the overlying unsaturated zone as well as contaminants' concentration in the pumped water. Mercado (1976) applied his model to examine alternative measures for protecting the quality of groundwater in the coastal aquifer of Israel.

The modeled groundwater quality system, described schematically in Fig. 10-8 consists of three major parts.

(*a*) The land surface, where most human activities associated with the release of contaminants occur.
(*b*) The unsaturated zone, including the root zone, which is viewed as the "chemical reactor" of the whole system and which is also responsible for the time lag between the release of contaminants at the ground surface and the arrival at the groundwater table.
(*c*) The aquifer which dilutes the contaminants and transports them to the outlets (e.g., wells and springs).

The concentrations of the considered contaminants serve as decision variables. The selection of contaminants to be included in the model depends on quality standards and the planned use of the water. Decision variables include the location and timing of pumping and artificial recharge, maximum tolerable loads of various surface contaminants, concentration of toxic substances in sewage effluents used for artificial recharge, etc. As in all other models, the basis of Mercado's model is also the conservation of mass of any considered species dissolved in the water

$$\begin{Bmatrix} \text{Rate of mass} \\ \text{accumulation} \end{Bmatrix} = \begin{Bmatrix} \text{Net rate of mass} \\ \text{transfer} \end{Bmatrix} + \begin{Bmatrix} \text{Net rate of production} \\ \text{by chemical reactions} \\ \text{and surface phenomena} \end{Bmatrix}$$

$$+ \begin{Bmatrix} \text{Rate of mass contribution} \\ \text{by external sources} \end{Bmatrix}$$

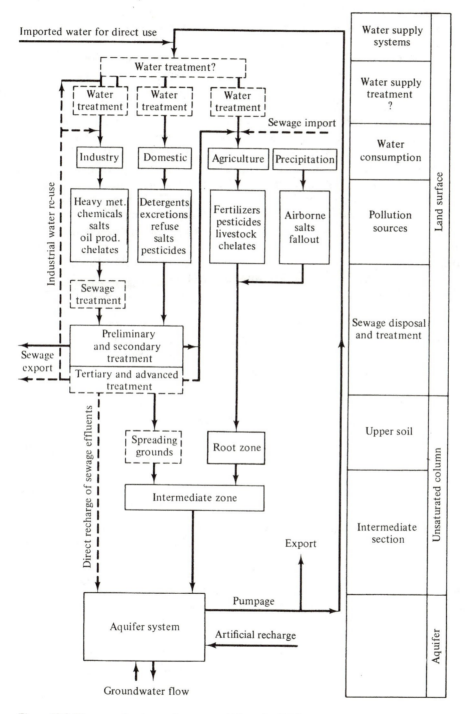

Figure 10-8 The groundwater quality system *(Mercado, 1976)*.

This balance is valid and can be written separately for every section of the model (Fig. 10-8). One should recall that the output of the unsaturated zone becomes the input to the aquifer section.

The considered system can be represented as a single cell model or as a multicell one. In the latter case, contaminants are transported with the groundwater between adjacent cells. Figure 10-9 shows a single cell model as applied to pollution by nitrates and chlorides in the coastal aquifer of Israel. Complicated hydrological and biochemical processes in the unsaturated zone are simplified to the extent they can be represented by two basic parameters: transit time of pollutants from land surface to the aquifer and nitrogen losses in the soil column. Mercado (1976) uses a working hypothesis which states that linear relationships exist between quantities of nitrogen released at the ground surface and those reaching the water table. Chloride and nitrate balances have been studied simultaneously, and a zone of effective mixing in the aquifer was introduced.

An interesting feature of Mercado's model is the way he selects the values of model parameters. In view of the uncertainty involved, the values of model parameters are chosen at random from within their predetermined ranges using one of the simplest verions of the Monte Carlo technique. Altogether, his model involves nine parameters which are relevant to the future concentration of nitrate.

Let X_{min}, X_{max}, and X_{exp} denote the minimum, maximum, and probable values, respectively, of a parameter X, and let F_X denote the cumulative frequency function, with $F_{X\,min} = 0$, $F_{X\,max} = 1$. A computer program which generates random numbers between 0 and 1, representing values of F_X, draws random values of the parameters.

For $F > F_{Xexp}$ $[= (X_{exp} - X_{min})/(X_{max} - X_{min})]$

$$X = X_{max} - [(1 - F)(X_{max} - X_{min})(X_{max} - X_{min})]^{1/2} \qquad (10\text{-}20)$$

For $F < F_{Xexp}$

$$X = X_{min} + [F(X_{max} - X_{min})(X_{max} - X_{min})]^{1/2} \qquad (10\text{-}21)$$

The same computer program (SCMON) performs both the random choice of parameters and the integration of the nitrate and chloride balance equations. For each set of the nine parameters, thus selected, the model is run leading to a resultant future concentration distribution. The procedure is repeated until both the parameter distribution and the resultant concentration distribution converge to the normal Gauss distribution (checked by the T-test for a given confidence limit). Figure 10-10 gives an example of the resulting distribution of average groundwater predicted concentration.

The model is used among other usages for determining the outcome of a large number of proposed protection measures, such as the elimination of most of the major nitrogen sources (sewage and fertilizers) in the area.

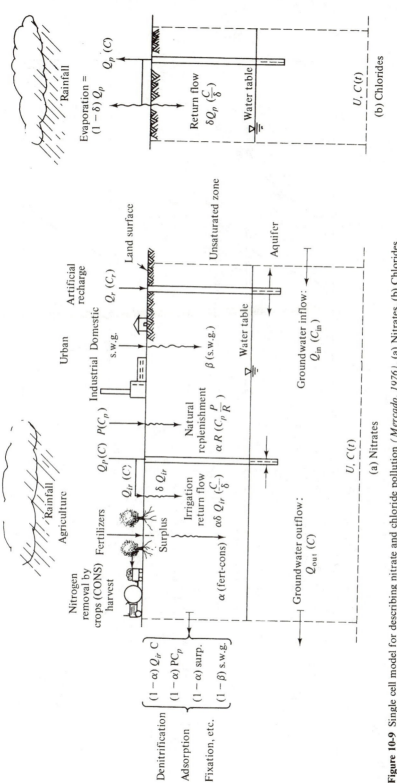

Figure 10-9 Single cell model for describing nitrate and chloride pollution (*Mercado, 1976*). (a) Nitrates. (b) Chlorides.

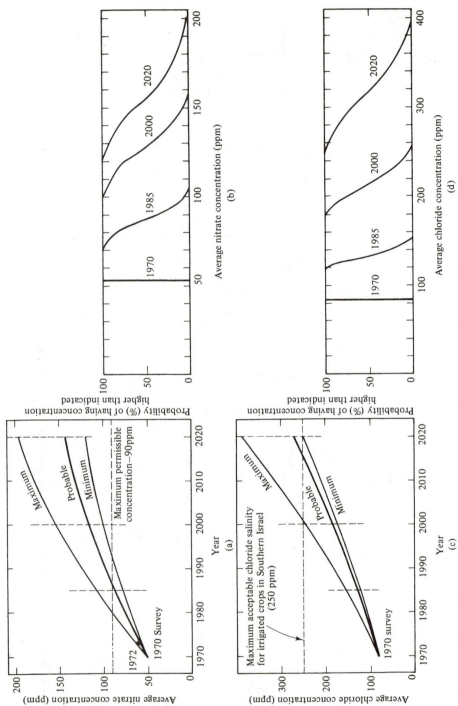

Figure 10-10 Predicted average nitrate and chloride concentration as a function of time (*Mercado, 1976*).

ELEVEN

IDENTIFICATION OF AQUIFER PARAMETERS

The solution of any forecasting problem requires: (1) a mathematical model and (2) information on the numerical values of the parameters appearing in the model. In Sec. 10-4, we have referred to the problem of determining the parameters appearing in an aquifer model as *model calibration*. The problem is also referred to as the *inverse problem* or the *problem of identifying model parameters*.

In the discussion on model calibration presented in Sec. 10-4, we have emphasized that when aquifer models are constructed on the basis of physical considerations, the parameters appearing in the model equations have physical interpretations, i.e., they represent physical properties of the aquifer, averaged over some volume, or area. Nevertheless, one should always remember that these are *model parameters*. Hence, for a given point in an aquifer, we may obtain different values of *presumably* the same parameters (say, transmissivity) in different models. The reader is referred to Sec. 10-4 for a further discussion on the identification of parameters of aquifer models.

In addition to the transport and storage parameters appearing in the various aquifer models, one may wish to identify also boundary conditions and inputs to the model (e.g. from natural replenishment).

The problem of regional parameter identification, although long recognized as essential to the use of models, has not yet become a matter of routine in which well-established procedures are used (except for pumping tests for determining local values of T and S, but not dispersion coefficients). Hence, only some general considerations are brought in this chapter.

11-1 STATEMENT OF THE INVERSE PROBLEM

The basic idea underlying the identification problem is rather simple. It always involves considering a period in the past for which information is available *on both aquifer excitations* (e.g., pumping) *and responses* (e.g., water levels) and determining the values of the parameters which will cause the model equations, relating the two sets of data to each other, to be satisfied.

We may exemplify this idea by referring to the two models—(5-58) and (10-7). Referring to (5-58), let N be replaced by $N + R - P$, where N, R, and P are natural replenishment, artificial recharge, and pumping, respectively. In a forecasting problem $T_x(x, y)$, $T_y(x, y)$, $S(x, y)$ and $N(x, y, t)$ are known parameters and we seek a solution of (5-58) for the unknown response $\phi = \phi(x, y, t)$ to the known operations $R(x, y, t)$ and $P(x, y, t)$. In the inverse problem, $\phi(x, y, t)$, $R(x, y, t)$, and $P(x, y, t)$ are known for a period of time in the past, and we seek a solution of (5-58) for the values of $T_x(x, y)$, $T_y(x, y)$, $S(x, y)$, and $N(x, y, t)$. In a similar way, in the inverse problem of (10-7), we have to solve the set of equations (10-7) for the unknowns $T_{i-1/2,j}$, $T_{i+1/2,j}$, $\ldots$, $N_{i,j}$, and $S_{i,j}$ for all points i, j in the model. However, a basic difficulty inherent in solving the inverse problem as stated above stems from the observation that while the forecasting problem is well posed and we are therefore assured that a unique and stable solution exists, the inverse one is not always well posed and a unique stable solution is not assured, even when parameters of a specific model are being considered. Several factors contribute to the non-uniqueness.

First, we have to realize that the knowledge of heads, ϕ, alone is never sufficient as long as we do not have an independent knowledge of the discharge rate. We can demonstrate this statement by considering the simple case of steady one-dimensional flow in the column shown in Fig. 4-2 and described by (4-3). Even when ϕ_1 and ϕ_2 (and L and A) are known, we cannot determine K unless we know Q. In a regional problem (in the xy plane), let us consider a streamtube in an aquifer (e.g., Fig. 5-36). For that streamtube, $\Delta Q/\Delta \phi = T_1 \Delta n_1/\Delta s_1 = T_2 \Delta n_2/\Delta s_2 = \ldots$. It is obvious that we must know the value of T at least at one point along the streamtube in order to determine it at all other points along the same streamtube. In a regional steady flow problem, we must therefore know at least a value of T for each streamtube (as we have no way of determining the Q's). In mathematical terms, we wish, for example, to determine $T(x, y)$ by solving the equation $\nabla \cdot (T \cdot \nabla \phi) + N(x, y) = 0$, or $\nabla T \cdot \nabla \phi + T \nabla^2 \phi = 0$, where $N(x, y)$, $\nabla^2 \phi$, and $\nabla \phi$ are known. We have also conditions on the boundaries, with those of the second kind (Sec. 5-3) including the unknown T. This is a first order partial (hyperbolic) differential equation. A necessary condition for a unique solution of this equation is that $T(x, y)$ be known on a given curve within the flow domain (*Cauchy data*). The reader is referred to texts on partial differential equations for more details (e.g., Garabedian, 1964, p. 102).

Secondly, there exists inaccuracy in field measurements of ϕ. This error in itself is usually small. When determining S, using changes in ϕ expressed by $\partial\phi/\partial t$, we may neglect this inaccuracy. Also, when comparing field results with those predicted by solving the forecasting problem, these errors are insignificant. However, because of the rather small gradients occurring in groundwater flow, (e.g., 0.001–0.01) a small error in ϕ may result in a very large error in the gradient of ϕ which appears as a coefficient in the equation to be solved for $T(x, y)$.

Then, we should mention the contribution that stems from the fact that we never have all the data; we usually have information at selected points in time and space. Hence it is possible that different combinations of the parameters will give rise to the same set of responses at the selected points.

Finally, since our model never completely represents reality, some unknown factors which are not represented in the model, may always affect the observed data. When the observed data is plugged into the model, this noise is imbedded in an unknown manner in the values of the resulting parameters. Obviously, different partial sets of past data will lead to different values of the sought parameters. This may be the major source for the non-uniqueness of the solution to the inverse problem.

Although we have referred above to the partial differential equation as a model of the aquifer's behavior, with $T = T(x, y)$, most works on the identification, or inverse, problem until now have been related to discrete numerical models.

The identification methods to be described in Sec. 11-3 below are related to regional aquifer models. For example, we seek the variations of aquifer storativity and transmissivity over the aquifer domain. A pumping test is also a parameter identification technique, except that in this case we focus our attention on the vicinity of a point in the aquifer. The considered model is one of radially converging flow to a single pumping well in a homogeneous, usually isotropic, aquifer, and our objective is to determine aquifer parameters at the well and its vicinity. Section 11-2 presents a summary of selected pumping test techniques. In the absence of any other data, the results of pumping tests can be used in regional models for forecasting aquifer behavior. Otherwise, the results of a pumping test should be limited to the prediction of aquifer behavior in the same locality and under conditions similar to those prevailing during the test.

In general, it is relatively easy, fast, and inexpensive to perform pumping tests. Hence their importance as a tool for determining aquifer parameters, or at least obtaining a good first estimate of them.

11-2 PUMPING TESTS

A pumping test is a controlled field experiment aimed at determining the basic aquifer parameters in the vicinity of a pumping well. The controlled experiment replaces in this case the use of past hydrological history in the regional identification problem. For example, for a confined aquifer, the sought parameters are T and S; for a leaky aquifer they are T and S of the main aquifer, and the leakage factor λ and the storativity S' of the semipervious formation. For an anisotropic aquifer, we may wish to determine T_x and T_y. In some cases we also seek information on aquifer boundaries, making use of the ideas presented in Sec. 8-10. No prediction of drawdown produced by pumping wells (and/or build-up produced by recharging ones) is possible without information on these parameters.

During a pumping test, a well is pumping at a constant rate Q_w and variations of drawdown with time are observed in it and/or in one or more observation wells in its vicinity. This information on $s = s(r, t)$ and Q_w is inserted in the drawdown formula corresponding to the considered type of aquifer and kind of flow (e.g., unsteady flow in a leaky aquifer, taking into account delayed storage), in order to obtain the values of the various aquifer parameters appearing in the formula. The formula, which as we have seen in Chap. 8, is based on a set of simplifying assumptions, serves as the model in this case.

The selection of the appropriate formula is, perhaps, the most difficult part of a pumping test. It should be obvious, especially in view of the discussion presented above and in Sec. 10-4, that when we insert data on $s(r, t)$ and Q_w in any formula, we shall always obtain, within some accuracy stemming from the matching of observed and calculated drawdown, $s(r, t)$, numerical values for the parameters appearing in the equation, even if the formula is the wrong one for the considered aquifer. Here, the hydrologist has to use both geological information and his experience related to the nature of the $s = s(r, t)$ relationship and to values of the parameters to be expected for different aquifers, in order to select the most appropriate drawdown formula for each case. No methodology is available for selecting the most appropriate formula in a systematic way. A knowledge of the geology, the boundaries, the type of aquifer, etc., in the neighborhood of the test site prior to a test, is therefore essential, as it serves the only guide for selecting the appropriate formula.

One should keep in mind that each formula, in addition to being correspondent to a specific type of aquifer, also corresponds to a specific set of assumptions ($\equiv$ the model of the aquifer). These include assumptions related to homogeneity and isotropy of aquifer, radius and partial penetration of well, essentially horizontal flow, storage in the semipermeable layer, relative magnitude of drawdown, etc. The flow pattern produced during the test should correspond, as much as possible, to that described by the assumptions underlying the development of the formula selected for use in the considered case.

Several standard types of tests exist.

(*a*) A drawdown test, in which the drawdown is observed only in the pumping well itself.

(*b*) A recovery test, in which a pumping well is shut-off after pumping at a constant rate for a long period of time, and the recovery of water level elevations in the pumped well is observed.

(*c*) An interference test, in which variations of drawdown with time are observed at one, or more observation wells in the vicinity of the pumped well.

(*d*) A step-drawdown test in which pumping is increased in steps.

In this section, the drawdown s is the difference between the initial water table, or piezometric surface, usually assumed horizontal, and the one observed during a pumping test. Whenever the initial water table is not a steady one, but a declining, rising, or a fluctuating one (as in response to changes of barometric pressure), drawdown observations should be corrected to eliminate these effects and obtain the drawdown caused by the pumping alone. These corrections can be done by extrapolation of observed trends in undisturbed (by pumping) water levels for a certain period of time prior to the test and by recording barometric pressures during the test.

Only the principles of pumping test analysis are discussed in the following paragraphs through several examples. With these principles, the reader should be able to use additional information available in the literature in order to analyze test data in the various cases which are not mentioned here. For additional information, especially on the actual execution of tests in the field, the reader is referred, among many others, to such publications as Wenzel (1942), Brown (1953), Bruin and Hudson (1955), Ferris et al. (1962), Stallman (1962), Walton (1960a, 1962), Bentall (1963, 1963a), Hantush (1964), Prickett (1965), and Bear, Zaslavsky, and Irmay (1968, Sec. 13-12).

In addition to these summaries, numerous papers have been published on specific cases. Recently, computer methods have been developed for the analysis of pumping tests (e.g., Saleem, 1970, Labadie and Helweg, 1975, Rushton and Chan, 1977).

Steady Flow in a Confined Aquifer

For steady flow in a confined aquifer, use is made of Thiem's equilibrium equation (8-5), from which we have

$$T = \frac{Q_w}{2\pi(s_1 - s_2)} \ln \frac{r_2}{r_1} = \frac{2.30 \, Q_w}{2\pi(s_1 - s_2)} \log \frac{r_2}{r_1} \qquad (11\text{-}1)$$

where Q_w is the constant discharge rate of the pumping well, and s_1, s_2 are the drawdowns at observation wells located at distances r_1 and r_2 from the pumping well, respectively. When more observation wells are available, a plot of $s(r)$ versus r on a semilog paper (with r on the log scale) should give a straight line from the slope of which we obtain T. By picking any two points on the line, we obtain $\Delta s = s_1 - s_2$, $\Delta(\log r) = \log r_2 - \log r_1$. Then

$$T = \frac{2.30 \, Q_w}{2\pi} \frac{\Delta(\log r)}{\Delta s} \qquad (11\text{-}2)$$

The data points will seldom, if ever, fall exactly on a straight line. We have to draw the line which will best fit the data (usually "by eye" or by some curve fitting technique).

Unsteady Flow in a Confined Aquifer; $Q_w(t) = \text{const.}$

The drawdown formula used in this case is (8-61) which contains the two aquifer characteristics T and S. Because one cannot eliminate S and T from this formula, a graphical method was suggested by Theis. It is called the *Theis method*, or the *type curve method*.

Equation (8-61) written in the form

$$s = \frac{Q_w}{4\pi T} W(u), \qquad \frac{r^2}{t} = \frac{4T}{S} u \qquad (11\text{-}3)$$

can be written as

$$\log s = \log (Q_w/4\pi T) + \log W(u) \qquad (11\text{-}4)$$

$$\log(r^2/t) = \log(4T/S) + \log u \tag{11-5}$$

Since Q_w, T, and S are constants, the relationships between s and r^2/t and between $W(u)$ and u are similar in form when both are plotted on a logarithmic paper, except for a certain displacement of the curves with respect to each other, caused by the first term on the right-hand side of (11-4) and (11-5).

This means that a plot of $W(u)$ versus u on a logarithmic paper is similar in form to that of s versus r^2/t plotted on the same paper. The former plot is called the *type curve* of $W(u)$, whereas the latter is called the *data curve*. If we have only one observation well, i.e., only one value of r, one can plot the data curve as s versus $1/t$.

The similarity between the two curves, as expressed by (11-4) and (11-5), indicates that if we plot the two curves on separate sheets of paper and super-impose the data curve on the segment of the type curve corresponding to the data curve, we shall find that the s and the $W(u)$ scales are displaced with respect to one another by the constant amount $\log(Q_w/4\pi T)$, while the r^2/t and u scales are displaced by the amount $\log(4T/S)$. This observation makes it possible to solve (8-61) graphically for the unknown values S and T, using observed values of $s(r, t)$ at one or more observation wells. Accordingly, the procedure for analyzing a pumping test is as follows.

(a) A type curve $W(u)$ versus u is drawn on a logarithmic paper.

(b) The observed values of s are plotted versus r^2/t (or against $1/t$ if only one observation well is used) on a transparent logarithmic paper of the same scale.

(c) The data sheet is superimposed on the type curve sheet and translated vertically and/or horizontally, keeping the coordinate axes of the two sheets parallel, until a position of "best fit" between the two curves is reached.

(d) An arbitrary point—called the match point—is chosen on the overlapping sheets and the coordinates of this point on the two sheets are recorded: s, r^2/t, $W(u)$, u.

(e) The four coordinates are substituted in (11-3) to yield S and T.

Figure (11-1) gives an example of the application of the type curve method. Table 8-2 gives values of the well function $W(u)$ for a confined aquifer.

Chow (1952) developed a method which avoids curve fitting. He showed that a function $F(u) = e^u W(u)/2.3$ is equal to the ratio of drawdown to slope for the straight line portion of the graph of drawdown plotted against $\log t$ for a single observation point within the cone of depression. Hence, for a given value of u, values of $F(u)$ and $W(u)$ can be determined, or we can determine u and $W(u)$ if $F(u)$ is known. Chow presents a graph which facilitates this procedure.

Jacob (1946, 1950) suggested a method based on the modified nonequilibrium formula (8-62), valid for $u < 0.01$. These equations may be rewritten as

$$s = (2.3Q_w/2\pi T)\left[\tfrac{1}{2}\log(2.25Tt/S) - \log r\right] \tag{11-6}$$

$$s = (2.3Q_w/4\pi T)\left[\log t - \log(r^2 S/2.25T)\right] \tag{11-7}$$

$$s = (2.3Q_w/4\pi T)\left[\log(2.25T/S) - \log r^2/t\right] \tag{11-8}$$

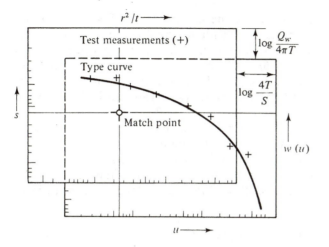

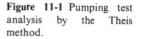

Figure 11-1 Pumping test analysis by the Theis method.

Plotting on semilog paper the test results s versus $\log r$ (at a given t), or s versus $\log t$ (for a given r), or s versus $\log r^2/t$ (for an interference test with a number of observation wells), straight lines described by (11-6) through (11-8), respectively, are obtained (Fig. 11-2). Actually, the data lie along straight lines only for sufficiently large t, or sufficiently small r (i.e., say, for $u < 0.01$). The value of the transmissivity T can be derived from the slope of the straight line portion of any of these graphs, using the following relationships.

In Fig. 11-2a: $\Delta s/\Delta(\log r) = 2.3Q_w/2\pi T$; $T = (2.3Q_w/2\pi)\,\Delta(\log r)\,/\Delta s$

In Fig. 11-2b: $\Delta s/\Delta(\log t) = 2.3Q_w/4\pi T$, $T = (2.3Q_w/4\pi)\,\Delta(\log t)\,/\Delta s$

In Fig. 11-2c: $\Delta s/\Delta(\log r^2/t) = 2.3Q_w/4\pi T$, $T = (2.3Q_w/4\pi)\,\Delta(\log r^2/t)\,/\Delta s$

It is convenient to choose Δs such that $\Delta(\log t)$, $\Delta(\log r)$, or $\Delta(\log r^2/t)$ equals one.

In each case, only one type of straight line is possible. The case $s = s(r)$ at a constant t is rather theoretical. For a drawdown test we obtain $s = s(t)$ for a given $r = r_w$, while for an interference test we obtain $s = s(r^2/t)$.

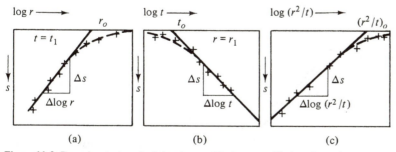

Figure 11-2 Pumping test analysis by the modified nonequilibrium formulae.

To obtain the storativity S, we extend the straight lines of Figs 11-2a, b, and c, and determine their points of intersection with the axis $s = 0$, at r_0, t_0, or $(r^2/t)_0$, respectively. By inserting $s = 0$ together with the already known value of T and r_0, t_0, or $(r^2/t)_0$, respectively, in (11-6), (11-7), or (11-8), the value of S is obtained.

It should be noted that in the case of a drawdown test, we must know the value of r_w (= radius of well) in order to determine S (which appears in the combined form Sr^2).

Obviously, the straight line method is applicable only if one is sure that a straight line is indeed indicated by the plotted test data.

Although both the type curve method and the straight line method were originally suggested for tests of unsteady flow in confined aquifers, they are applicable also to many other types of flow. This is so as long as the drawdown formula has the general form

$$s(r, t) = \frac{Q_w}{4\pi T} F(r, t ; \; S, T, \ldots)$$

with an asymptotic behavior described by a straight line; F should be a function of S/T and at most one more unknown aquifer parameter.

Lohman (1957) describes a method which does not require extrapolation of the straight lines.

In a recovery test, the well pumping at a constant rate Q_w since $t = 0$ is suddenly shut off at $t = t_1$ and the *residual drawdown*, s', is observed in it, or in a neighboring observation well. When u and u' are sufficiently small, say <0.01, or according to Hantush (1964), when $(t - t_1) > 5Sr^2/T$, (8-69) is applicable. The value of T can be obtained from a plot of s' versus $\log[t/(t - t_1)]$. Figure 11-3 shows the straight line obtained by plotting the test data. From the slope of this line we have

$$\frac{\Delta s'}{\Delta[\log t/ \; (t - t_1)]} = \frac{2.30 \; Q_w}{4\pi T} \tag{11-9}$$

One should note that S cannot be derived from a pumping test of this type.

When a pumping well only partially penetrates an aquifer, but observation wells are located at distances $r > 1.5B$ from it, the procedures described above for a fully penetrating pumping well are applicable as a very good approximation.

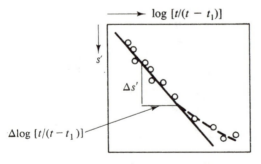

$$\longrightarrow \; \log [t/(t - t_1)]$$

s'

$\Delta s'$

$\Delta \log [t/(t - t_1)]$

(s' = residual drawdown)

Figure 11-3 Analysis of a recovery test.

If such observation wells are not available, formulas accounting for partial penetration (e.g., Hantush, 1961a, 1964; Stallman, 1965) should be used. Sternberg (1968) suggested a pumping test technique for variable discharge, $Q_w = Q_w(t)$, based on (8-59).

Aron and Scott (1966) present a method for analyzing pumping test data in which a drawdown curve, obtained during a pumping test with abrupt changes in well discharge, can be transformed into a synthetic curve for constant discharge by a simple reversal of the Theis procedure. From this generated constant discharge curve, T and S can be computed by the Theis type curve method.

Unsteady Flow to a Flowing Well in a Confined Aquifer; $s_w = $ const.

In an artesian ($=$ flowing) well, the well's discharge, Q_w, varies with time, while the boundary condition at the well is $s_w = $ const. The drawdown formula for this case is (8-74). From this equation we can write

$$Q_w = 2\pi T s_w G(\alpha); \qquad \log Q_w = \log 2\pi T s_w + \log G(\alpha)$$

$$t = (r_w^2 S/T)\alpha; \qquad \log t = \log r_w^2 S/T + \log \alpha \tag{11-10}$$

where $\alpha = 1/4u_w$.

Following the Theis method, we use here the type curve $G(\alpha)$ on a logarithmic paper, plotting the test data Q_w versus t on a transparent logarithmic paper of a similar scale. We bring the two curves—the type curve and the data curve—to the "best fit" position as explained above, and select an arbitrary matching point on the overlapping portion of the two sheets.

From this matching point we obtain four coordinates: Q_w, t, $G(\alpha)$, and α. These are inserted into (10-10) in order to obtain T and S.

For large values of t, $G(\alpha)$ may be replaced by $2/W(u)$ which, in turn, can be approximated by $2/2.3 \log 2.25 Tt/Sr_w^2$. We obtain

$$s_w/Q_w = \frac{2.3}{4\pi T}[\log t/r_w^2 + \log 2.25 T/S] \tag{11-11}$$

which gives a straight line when s_w/Q_w is plotted versus $\log t/r_w^2$. The value of T can be determined from the slope of this line

$$T = \frac{2.3\,\Delta(\log t/r_w^2)}{4\pi\,\Delta(s_w/Q_w)} \tag{11-12}$$

The storativity S is obtained from (11-11) using the value of t/r_w^2 at $s_w = 0$. A difficulty encountered in this and in other methods is that it is difficult to determine (or estimate) the effective radius, r_w, of the pumping well.

Steady and Unsteady Flow to a Well in a Phreatic Aquifer

Using (8-30) for steady flow with relatively small drawdowns (say, $s < 0.25H$) in a phreatic aquifer, a straight line is obtained by plotting the corrected drawdown

$s'(=s-s^2/2H)$ versus $\log r$, or $\Delta s'$ versus $\log(r_2/r_1)$. We obtain

$$T = KH = 2.3 Q_w (\log r_2/r_1)/2\pi(s'_1 - s'_2) \qquad (11\text{-}13)$$

or

$$T = 2.3 Q_w/2\pi \,\Delta s', \quad \text{with} \quad \Delta s' = s'_1 - s'_2 \quad \text{and} \quad \log r_2/r_1 = 1 \qquad (11\text{-}14)$$

where H is the initial thickness of the saturated flow below the phreatic surface.

In unsteady flow with small drawdowns relative to the initial (or ultimate, or average) thickness of the saturated domain below the phreatic surface, (8-89) is applicable. Following Jacob (1944), we may therefore use the Theis type curve method described above to determine T and S'. Once S' is known, we may determine n_e by

$$n_e = [(H - \bar{s})/H]\,S' \qquad (11\text{-}15)$$

where $\bar{s}$ is the drawdown averaged over the distances of the observation wells, i.e., the drawdown at the geometric mean radius of all observation wells at the end of pumping. However, Jacob's correction procedure is applicable only to essentially horizontal, radially converging flow. It should not be used where vertical flow components are appreciable (e.g., close to a partially penetrating well).

According to Jacob (1944), when the observed drawdowns are adjusted by the factor $s^2/2H$, the nonequilibrium formula gives satisfactory results even when the dewatering is as much as 25% of the initial thickness of flow. Jacob (1945) also suggests a procedure for adjusting observed drawdowns when the pumping well only partially penetrates the aquifer. One should, however, recall that data collected during the early part of a test may not conform to the theory, since the theory neglects vertical flow components and time lag in the release of water from storage. The effect of these factors may be appreciable during the first period of pumping.

Prickett (1965), following Boulton (1963), presents a type curve technique for analyzing a pumping test in a phreatic aquifer. The well is assumed to fully penetrate the aquifer and $Q_w = $ const. during the test. Values of the well function $W(u_{\alpha y}, r/D_t)$ appearing in (8-101) are plotted against values of $1/u_\alpha$ and $1/u_y$ on a logarithmic paper and two families of time-drawdown type curves are drawn as shown in Fig. 11-4. The type curves which lie to the left of the values r/D_t are referred to as Type A curves; they are essentially the same as those corresponding to a leaky confined aquifer (Walton, 1960). They are used for analyzing time-drawdown data collected during the early stages of a pumping test. The type curves shown to the right of the values of r/D_t in Fig. 11-4 are called Type Y curves. They are used for analyzing late time-drawdown data. The right-hand portion of each Type A curve and the left-hand portion of each Type Y curve in Fig. 11-4 approach a horizontal asymptote given by $W(u_{\alpha y}, r/D_t) = 2K_0(r/D_t)$, where K_0 is the modified Bessel function of the second kind and order zero.

Prickett (1965; see also Walton, 1970, p. 222) outlines the details of this method and demonstrates the type curve technique he proposes for determining T, S, S', and α which appear in (8-101).

Kriz, Scott, and Burgy (1966) use the dimensionless curves shown in Fig. 11-5

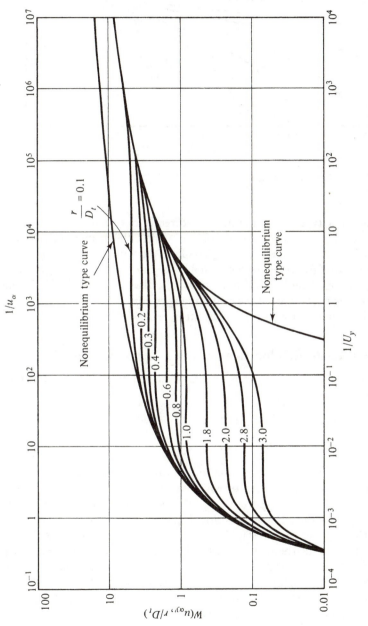

Figure 11-4 Water table, fully penetrating, constant discharge, time-drawdown type curves *(after Prickett, 1965)*.

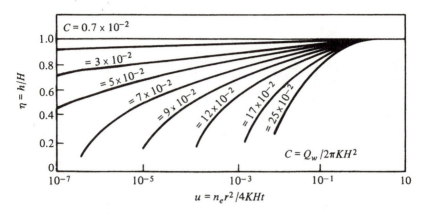

Figure 11-5 Unconfined aquifer type curve *(Kriz et al., 1966)*.

as a basis for a type curve method for directly determining the hydraulic conductivity K of an unconfined aquifer in which the transmissivity varies with variations in water table elevations. The field data plot is that of $\eta = h/H$ versus r^2/t. Once one of the type curves matches the data curve (in the usual procedure of superposition, keeping the axes parallel), the value of K is obtained from the value of the parameter of the matching type curve. Then from any match point, the coordinates u and r^2/t yield n_e from $n_e = 4KHu/(r^2/t)$.

Steady and Unsteady Flow to a Well in a Leaky Aquifer

Steady flow to a well in a leaky aquifer is described by (8-41), which can be written in the form

$$\log s = \log Q_w/2\pi T + \log K_0(x)$$

$$\log r = \log \lambda + \log x \qquad (11\text{-}16)$$

Following the type curve method outlined above for a confined aquifer, we plot the test results as drawdown, s, versus distance, r, for three or more observation wells on a (transparent) logarithmic paper. The type curve in this case is that of the modified Bessel function $K_0(x)$; it is plotted also on a logarithmic paper of the same scale. We then superimpose the sheet of test results on the type curve sheet and shift the former until a best fit is obtained between the points of the test results and the type curve (Fig. 11-6a). Picking an arbitrary match point in the overlapping portion of the two sheets, we obtain four coordinates on the two sheets: s, r, K_0, and x. When inserted in (11-16), they yield the values of T and λ. From the latter, the value of the resistance $\sigma'(=\lambda^2/T)$ of the semipervious layer is obtained.

As in the case of a confined aquifer, a *straight line method* based on (8-42) is also applicable. In this case we plot s versus r on a semilogarithmic paper (Fig. 11-6b). Because

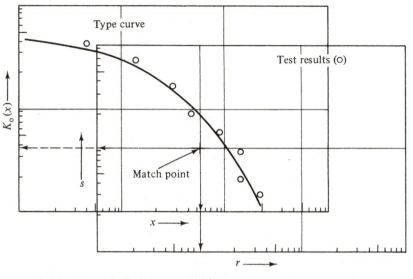

(a)

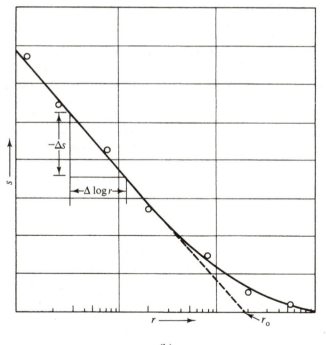

(b)

Figure 11-6 Determination of T and λ for a leaky aquifer. (a) The type curve method (log–log paper). (b) The straight line method (semilog paper).

$$s = - \frac{2.3 Q_w}{2\pi T} (\log r - \log 1.123) \tag{11-17}$$

we may derive the value of T from the slope of this straight line

$$\frac{-\Delta s}{\Delta(\log r)} = \frac{2.3 Q_w}{2\pi T}; \qquad T = \frac{2.3 Q_w}{2\pi [(-\Delta s)/\Delta(\log r)]} \tag{11-18}$$

The value of λ is obtained from the intercept $r = r_0$ of the straight line with the $s = 0$ axis: $\sigma' = (r_0/1.123)^2/T$. One should note that (8-42) is a good approximation for $r/\lambda < 0.05$; the test points fall on a curve which approaches zero drawdown as $r \to \infty$.

For unsteady flow in a leaky aquifer, (8-105), or (8-106), developed by Hantush and Jacob (1955b), is applicable. For $a > 2r/\lambda$, (8-106) may, for all practical purposes, be approximated by (Hantush, 1964)

$$s_m - s = (Q_w/4\pi T)\, W(a); \qquad s_m = (Q_w/2\pi T)\, K_0(r/\lambda) \tag{11-19}$$

where s_m is the maximum, or steady state, drawdown.

Walton (1960) suggests a pumping test method based on a family of type curves constructed by plotting the function $W(u, r/\lambda)$ versus $1/u$, on a logarithmic paper, with r/λ as the running parameter of the family (Fig. 8-17). The curve $r/\lambda = 0$ is the Theis type curve (with abscissa $1/u$ instead of u). The test results are plotted on a logarithmic paper of the same scale as s versus t/r^2 (or versus t if only one observation well is used). The solution procedure is that of the Theis type curve method. When the best fit is obtained, the data curve will follow one of the family of type curves. Because the leakage effects may be insignificant during the early period of pumping, the early points of the data curve fall on, or, at least approach asymptotically, the Theis type curve. This type curve will, therefore, serve as a guide in obtaining the best fit position (Hantush, 1964). Having obtained the best fit position, the value of r/λ is read off the appropriate matching curve (sometimes by interpolation). Denoting this value by $(r/\lambda)_0$, the value of λ of the leaky aquifer is obtained by taking the mean of values computed from $\lambda = r/(r/\lambda)_0$. Using the four coordinates of a match point: $W(u, r/\lambda)$, $1/u$, s and t/r^2 (or s and t), together with the specific value of r/λ obtained above, the values of T and S can be obtained from

$$T = (Q_w/4\pi s)\, W(u, r/\lambda); \qquad S = 4T(t/r^2)/(1/u) \tag{11-20}$$

Hantush (1964) indicates that in this procedure, the data points that fall on the Theis type curve are generally those which correspond to the period $t < 0.25\, t_i$ on the semilogarithmic data plot, where t_i is the time at which an inflection point occurs on this curve. The position of the inflection point may be reasonably estimated by inspection; if the maximum drawdown s_m can be extrapolated from the semilogarithmic data curve, the inflection occurs at a value $s_i = s_m/2$.

When the test duration is sufficiently long, so that a sufficient number of observed data points fall in the period $t > 4t_i$, and the distribution of these points on a semilogarithmic plot is such that the maximum drawdown s_m can be reasonably extrapolated, Hantush (1964) suggests the use of a type curve method

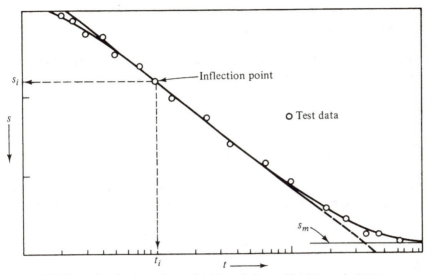

Figure 11-7 Hantush inflection point method for a leaky aquifer *(Hantush, 1956)*.

based on the second expression of s in (8-106)

$$s_m - s = \frac{Q_w}{4\pi T} W(a); \qquad a = Tt/S\lambda^2; \qquad s_m = \frac{Q_w}{2\pi T} K_0\left(\frac{r}{\lambda}\right) \qquad (11\text{-}21)$$

for the determination of T, S, and λ. In this case the type curve is a logarithmic plot of $W(a)$ versus a (an adequate range of a is $10^{-3} < a < 5$) and the data plot is that of $(s_m - s)$ versus t from all the available observation wells. The values of s_m are extrapolated from the semilogarithmic plot of each well, and the corresponding t_i's are obtained from the position of the curve's inflection point, using the value of s_i obtained from $s_i = s_m/2$. The matching should be based on a point (for each well) for which $t > 4t_i$. For the period $t < 4t_i$ the procedure described above is to be used. The four coordinates of the match point yield T and $S\lambda^2$. The value of λ for each well is then obtained from the expression for s_m in (11-19) and a table of $K_0(r/\lambda)$. The average of these values will be the average value of λ. Consequently, the average values of S and σ' are obtained.

Another method suggested by Hantush (1956) is based on the inflection point which occurs in a plot of the test data, $s = s(t)$, on a semilogarithmic paper, with s on the linear scale (Fig. 11-7).

Step-Drawdown Test

The objective of this type of test is to determine the well characteristic r_w, C, and n (when the latter is assumed different from 2) of (8-182). In a step-drawdown test (Jacob, 1946a, 1950; Rorabough, 1953; Bierschenk, 1963; Harrill, 1970; Lennox, 1966; Sheahan, 1971, 1972; Labadie and Helweg, 1975) the pumping rate is increased in a stepwise manner during successive periods of time. Essentially this is an interference test, as values of drawdown in observation wells are needed in order to determine the storativity S which appears as a factor in the group Sr_w^2, together with the well's radius r_w.

By applying the principle of superposition (Sec. 5-7 and (8-68)), the total drawdown s_{tw} in a well during the jth step may be written as

$$s_{tw,j}(t) = C(Q_{w,j})^n + \sum_{i=1}^{j} (Q_{w,i} - Q_{w,i-1}) B(r_w, t - t_i) \qquad (11\text{-}22)$$

where $Q_{w,j}$ is the constant well's discharge during the jth period, and t_i is the time at which the ith step begins, with $t_1 = 0$; (11-22) follows from (8-182), with n replacing the exponential 2.

The sum of the increments of drawdown referred to a fixed time interval $t^*(= t - t_i)$, taken from the beginning of each step may be obtained from (11-22) as

$$\sum_{i=1}^{j} \delta S_{tw,i}(t^*) = s_{tw}(t^*)_j = B(r_w, t^*) Q_{w,j} + C(Q_{w,j})^n \qquad (11\text{-}23)$$

where $\delta S_{tw,i}(t^*)$ is the drawdown increment between the ith step and the one preceding it, taken at time t^* from the beginning of the ith step; $s_{tw}(t^*)_j$ is the drawdown in the well that would have taken place at $t = t^*$, had the well been pumped at the constant rate $Q_{w,j}$; $B(r_w, t^*)$ is the head loss in the formation per unit discharge at time t^* since the commencement of pumping.

Figure 11-8a shows (on a semilogarithmic paper) the drawdown s_{tw} versus time during a step-drawdown test. The following procedure leads to the determination of B and C for $n = 2$ (Hantush, 1964)

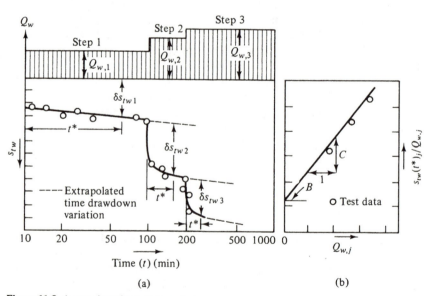

Figure 11-8 A step-drawdown test analysis.

1. Obtain the increments of drawdown $\delta s_{tw,i}(t^*)$, with t^* usually taken as 1 or 2 hours (Fig. 11-8a).
2. Obtain $s_{tw}(t^*)_j$ corresponding to the discharge $Q_{w,j}$ from $s_{tw}(t^*)_j = \delta s_{tw,1}(t^*)_j + s_{tw,2}(t^*) + \ldots + s_{tw,j}(t^*)$, and compute the specific drawdown $s_{tw}(t^*)_j/Q_{w,j}$.
3. Plot $Q_{w,j}$ versus the specific drawdown $s_{tw}(t^*)_j/Q_{w,j}$ on a linear scale paper and obtain the best fitting straight line.
4. From (8-185) it follows that the slope of this line, $\Delta\big(s_{tw}(t^*)_j/Q_{w,j}\big)/\Delta Q_{w,j}$, is equal to C (Fig. 11-8b).
5. Equation (11-23) and the coordinates of any point on this line will solve for $B(r_w, t^*)$.

For $n \neq 2$, (e.g., for high discharge rates) the procedure of analysis is essentially the same, except that the values of $s_{tw}(t^*)_j/(Q_w - B)$ will be plotted versus Q_w on a logarithmic paper for assumed values of B. The value of B which gives the straightest line on this plot is the required one. The intercept of this line on the axis $Q_w = 1$ is equal to the value of C, while the slope $\Delta[s_{tw,j}/(Q_{w,j} - B)]/\Delta Q_{w,j}$ is equal to $n - 1$.

Once B is obtained from the analysis described above, and T and S are known from a regular pumping test analysis carried out simultaneously during the first step of a step-drawdown test, or as a separate test, the value of r_w may be obtained by substituting these data in the appropriate expression for $B(r_w, t^*)$ and solving for r_w. The appropriate expression for B depends on the type of aquifer and the type of flow regime in the aquifer. For example, for unsteady flow in a confined aquifer $B = (1/4\pi T)\, W(Sr_w^2/4Tt^*)$.

Pumping Test in a Bounded Aquifer

In the various pumping tests described above, it was assumed that the aquifer has an infinite areal extent; accordingly the solutions for a single well in an infinite aquifer were employed in the analysis of these tests. However, pumping tests are often performed in aquifers of limited areal extent, i.e., with boundaries of the types considered in Sec. 8-10. In general, the location of an aquifer's impervious boundary is unknown. However, it can be determined by the pumping test itself. The modified nonequilibrium formula is particularly advantageous for the analysis of boundary effects, because it is easier to recognize changes in the slope of straight lines than to detect changes in the curvature of plotted test data on a logarithmic paper.

To illustrate the method of pumping test analysis, consider the case of unsteady flow to a well of constant discharge, Q_w, in a confined aquifer which has the shape of an infinite strip between two parallel impermeable boundaries. Figure 11-9 shows the time-drawdown curve plotted on a semilogarithmic paper at an observation well placed between the pumping well and the nearest boundary. At first, no drawdown is observed at the observation well. Then the propagating cone of depression reaches the well and drawdown begins. The first points of the time-drawdown curve lie on a transition curve. Then, as u becomes smaller (say, smaller than 0.01), a straight line is obtained for $s = s(\log t)$. From this straight

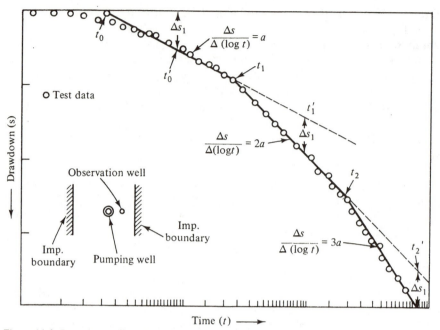

Figure 11-9 Drawdown affected by impervious boundaries.

line (called first limb), or by the type curve method for this initial period, we determine the aquifer characteristics T and S. As the test continues, the observed drawdown curve starts to bend downwards again, approaching the straight line of the second limb. This is caused by the cone of depression being reflected back to the observation well from the nearest boundary. The observed drawdown is the net effect of the pumping well and the boundary, the influence of which is equivalent to that of the image wells replacing it (Sec. 8-10). Because Q_w (of the real pumping well) and T are constant, the (pumping) image well operates at the same rate as the real well. Hence, the slope of the straight line of s versus $\log t$ is doubled when the effect of the first image well is felt at the observation well. When the influence of another image well, the one replacing the other impervious boundary, begins to be felt at the observation well, the rate of drawdown is tripled (as if the total pumping rate is $3Q_w$). The instants t_0, t_1, t_2, etc. (Fig. 11-9) are obtained from the intersection of the straight line segments. From (8-65) and Fig. 11-9 it follows that

$$t_0/r_0^2 = t_1/r_1^2 = t_2/r_2^2 = \ldots = \text{const.} \tag{11-24}$$

where r_0, r_1, r_2, etc. are the distances from the observation well to the pumping well and to the image wells. If we consider an increment of drawdown Δs_1 (Fig. 11-9) we obtain

$$t_0'/r_0^2 = t_1'/r_1^2 = t_2'/r_2^2 = \ldots = \text{const.} \tag{11-25}$$

Since the distance r_0 (to the real well) is known and t_0, t_1, and t_2 (or t_0', t_1', t_2') can be obtained from Fig. 11-9, it is possible to solve (11-24) or (11-25) for the distances r_1 and r_2 from the observation well to the image wells, respectively. Actually these are the first two of an infinite sequence of image wells. If more than one observation well is available, (11-24) may be applied to each of them and the location of the image wells may be determined more accurately. Once the location of image wells is known, the position of the boundary itself can be easily traced.

Stallman (1952) suggests a matching curve method based on Theis' method. For nonsteady flow to a well in a confined aquifer near a straight line recharge or impervious boundary, we have

$$ s = \frac{Q_w}{4\pi T}\left[W(u_p) \pm W(u_i)\right] = \frac{Q_w}{4\pi T}\sum W(u); \qquad u_p = Sr_p^2/4Tt; \qquad u_i = Sr_i^2/4Tt $$

$$ (11\text{-}26) $$

where u_p and u_i are computed from the distances r_p and r_i of the observation well to the real (pumping) and image wells, respectively. He writes $k = r_i/r_p$, $u_i = ku_p$ and plots on a logarithmic paper a family of type curves $\sum W(u)$ versus $1/u_p$ for various values of k and u_p. The plot of the observed data is superimposed on the family of type curves, keeping the $\sum W(u)$ and s axes parallel (and so are the $1/u_p$ and the t axes). Having matched the data to one of the curves, the four coordinates $\sum W(u)$, $1/u_p$, s, and t of a matching point are substituted into (11-26) to yield T and S. From the value of k of the particular curve, the value of r_i is computed. Van Poolen (1965) applies a similar procedure to determine the angle between intersecting (impervious) faults.

Hantush (1959a) uses a special type curve, $M(u, \beta)$, as a basis for determining T and S for a well near a straight line stream. Here, and in other methods, when a "distance to a stream" is mentioned, we mean the (effective) distance to some line in the stream bed (or in a lake bounding the confined aquifer) which is taken as the line of contact between the stream and the aquifer (i.e., where the water enters the aquifer).

Injection-Pumping Test for Determining Dispersion Coefficients

So far we have been discussing in this section only pumping tests for determining aquifer parameters related to flow and storage of water. Another important aquifer coefficient is that of hydrodynamic dispersion, D_h or $\tilde{D}_h$ (see Sec. 7-7) which appears in the equations used for forecasting the movement and accumulation of solutes in aquifers.

The controlled single well experiment in this case is made up of two parts. First we inject through a well into the aquifer a certain volume of tracer labeled water. If observation wells are present in the vicinity, within the area invaded by the injected water, the tracer concentration, $C(r, t)$, in these wells is recorded. At the end of the injection period, pumping in the same well commences, and $C(r, t)$ in the pumped water is recorded. By inserting this information into the solution for $C(r, t)$ or $C(r_w, t)$ obtained by solving the appropriate equation of hydrodynamic dispersion, say (7-65), we obtain the longitudinal dispersivity, a_L. In order to obtain both longitudinal and transversal dispersivities, we have to produce a situation which is nonsymmetrical with respect to the pumping well. This can be done either by using two different close by wells—one for pumping and one for injection, with both operations taking place either simultaneously or in sequence—or by using a single well, but with a delay time between the periods of injection and pumping. During this period, the body of injected water is displaced by the flow in the aquifer. A test of this kind can also be used for determining the velocity of this flow. It is also possible to inject a certain volume of labeled water, continue with unlabeled water and then pump.

The labeling can be done by a radioactive tracer, by a dye, or by any detectable, nontoxic solute.

The analysis of the field tests in order to derive the coefficients is done by trial and error, curve fitting, or numerical techniques. The methods have not yet crystallized to well-established routines.

Discussions and examples on the use of tracer experiments for determining aquifer dispersion coefficients are given by Harpaz and Bear (1963), Mercado and Bear (1965), Nir (1968), Hoopes and Harleman (1967a), and Grove and Beetem (1971).

Halevy and Nir (1962), Mercado and Halevy (1966), and Nir (1968) also discuss the use of tracer tests as described above for determining the porosity and permeability of aquifers.

The discussion of the various pumping test procedures presented above, as is the discussion on hydraulics of wells presented in the previous sections, is based on idealized aquifer models. Each model consists of a set of assumptions. The main assumptions are given in Sec. 8-1. One should keep these in mind when deriving parameters from a pumping test and applying them to field conditions. In the practice, wells never fully penetrate the entire thickness of an aquifer (see, for example, Hantush, 1961b, 1964, for a pumping test procedure for partially penetrating wells), the well's discharge Q_w is not maintained constant during a pumping test because of pressure variations in the water supply system (see, for example, Sternberg, 1968), the specific yield of a phreatic aquifer should be handled with care as water is not instantaneously released from storage upon a decline in water table elevations (see, for example, Boulton, 1963), aquifers are anisotropic both in the horizontal plane (see, for example, Hantush, 1966) or in the vertical direction (see, for example, Dagan, 1967) and a pumped well may penetrate and pump from several aquifers simultaneously (see, for example, Bennett and Patten, 1962). To some of these modified models, special analysis techniques have already been developed.

These, and many other deviations from the idealized model, emphasize the importance of a careful examination of the field conditions, the geological data, etc. against the assumptions underlying any formula before using it.

11-3 INVERSE METHODS FOR DETERMINING REGIONAL AQUIFER PARAMETERS

Perhaps the simplest method of parameter identification is *trial and error*. In this method, a subjective set of parameters is first related and inserted into the aquifer model, together with the known past history of excitation (pumping, recharge of both water and tracers, etc.). Usually the selection of values of aquifer parameters is based on geological information and on our past experience as modelers in the same or in similar aquifers. We then compare the calculated response of the aquifer model (say, in the form of water levels) with the actual historical data observed in the field. The procedure is repeated for different sets of values of the considered parameters until the two responses—the calculated and the observed ones—match. Obviously, a perfect match should never be expected, and attempts are made to reach a best fit between these responses, using some criterion which will indicate an optimal estimate of the parameters.

Bear and Schwarz (1966) calibrated a large limestone aquifer model by trial and error, using an RC-electric analog. The work of Pinder and Bredehoeft (1968) may serve as an example of the trial and error technique, implemented by using a digital computer. They identified the parameters of a model composed of 1150 cells, obtaining a satisfactory fit after 37 runs. Another example is presented by Hughes and Robson (1973).

The main disadvantage of the trial and error technique is that it does not involve any algorithm for approaching the "best" solution in a systematic way. A lot (including required computer time) depends on the skill and experience of the modeller. Because of the non-uniqueness discussed above, we never really know if the optimal estimate has really been attained.

These considerations led researchers to seek methods for obtaining, in a systematic and rational way, the parameters which will lead to the best fit between observed and predicted aquifer responses. One should note that the requirement for certain initial information on transmissivity, T, to ensure uniqueness is not eliminated by using these methods.

Some methods attempt to improve the basic trial and error technique by proposing some policy for correcting the values of the parameters after each trial run, and criteria for examining the agreement between observed and calculated responses. Examples for such procedures are given by Coats *et al.* (1970), Slater and Durrer (1971), Sage and Melsa (1971), Addison *et al.* (1972), and Knowles *et al.* (1972).

In other methods, an optimal set of aquifer parameters is directly obtained by minimizing some error criterion. The works of Kleinecke (1971) and Hefez *et al.* (1973, 1975), are examples of this approach. Emsellem and de Marsily (1969, 1971) use more than a single error criterion. While trying to minimize a functional related to the difference (= error) between observed and calculated response, they also require that the resulting parameter distributions be rather smooth and uniform.

Neuman (1973a) emphasizes the strong element of uncertainty that invariably enters into the inverse problem. He shows that because of this uncertainty, any approach that is based on the minimization of a single error functional does not, in general, lead to satisfactory results. He proposes a multiple objective decision process which takes into account all the available information on the aquifer flow system, as well as the range of environmental conditions under which the system is expected to operate in the future. Accordingly, he first generates a set of alternative solutions to the identification problem by means of some mathematical programming technique and then lets the decision maker apply his own value judgement in order to select a particular model structure. The reliability of each parameter is ascertained with the aid of a post optimal sensitivity analysis.

Let us illustrate the principles of deriving an optimal set of parameters, using a single error criterion, by reviewing the work of Hefez *et al.* (1973, 1975).

Use of Linear and Quadratic Programming for Solving the Regional Identification Problem

We shall demonstrate the method by applying it to the multicell aquifer model shown in Fig. 10-4. The balance equation for a typical cell is

$$
T_{i-1/2,j} \Delta y_j \frac{\phi_{i-1,j}^{k+1/2} - \phi_{i,j}^{k+1/2}}{\Delta x_{i-1/2}} + T_{i+1/2,j} \Delta y_j \frac{\phi_{i+1,j}^{k+1/2} - \phi_{i,j}^{k+1/2}}{\Delta x_{i+1/2}}
$$

$$
+ T_{i,j-1/2} \Delta x_j \frac{\phi_{i,j-1}^{k+1/2} - \phi_{i,j}^{k+1/2}}{\Delta y_{j-1/2}} + T_{i,j+1/2} \Delta x_j \frac{\phi_{i,j+1}^{k+1/2} - \phi_{i,j}^{k+1/2}}{\Delta y_{j+1/2}}
$$

$$
+ N_{i,j}^{k+1/2} + R_{i,j}^{k+1/2} - P_{i,j}^{k+1/2} = S_{i,j} \frac{\phi_{i,j}^{k+1} - \phi_{i,j}^k}{(\Delta t)^{k+1/2}} \tag{11-27}
$$

where the transmissivities at the cells' boundaries are expressed by (10-5), and the gradients of ϕ are computed at the midpoint of the time interval, that is, $(\Delta t)^{k+1/2} = t^{k+1} - t^k$, $t^k = \sum_{(k)} (\Delta t)^k$, $\phi_{i,j}^{k+1/2} = (\phi_{i,j}^k + \phi_{i,j}^{k+1})/2$.

In a forecasting problem, $T_{i,j}$, $S_{i,j}$, $N_{i,j}$, $R_{i,j}$, and $P_{i,j}$ are known and we seek the values of $\phi_{i,j}^k$. The identification problem can be stated as follows. Determine $T_{i,j}$ and $S_{i,j}$ for each cell of the model, using the equations obtained by writing (11-27) for every cell and for every time interval during the calibration period. The following characteristics have to be specified for any particular problem: shape of model, its division into cells, boundary conditions in terms of values of ϕ in the fictitious cells (Fig. 10-4), initial values of ϕ, values of ϕ in all cells at the beginning and end of each time interval, and inputs and outputs $N_{i,j}$, $R_{i,j}$, $P_{i,j}$ for all cells and for all time intervals.

The method of solution proposed here permits the relaxation of some of these requirements: (1) when information is missing in some cells, one can identify the parameters for only part of the cells; (2) it is also possible to identify unknown inputs; (3) when boundary conditions are unknown, one may identify parameters for all cells except the storativity in cells adjacent to the unknown boundary, and (4) it is possible to identify all parameters when water levels are known for only part of the historical period in some cells.

For the sake of simplicity, let us first assume that data are available for all cells at all times, that is $P_{i,j}^k$, $N_{i,j}^k$, and $R_{i,j}^k$ are known for all i,j, and k. Denoting the number of cells by C, the number of sought parameters is $2C$, while the number of available equations is $N \times C$, where N is the number of time intervals included in the calibration period. The equations are linear in the unknowns T and S and can be solved as a system of linear equations provided that the number of equations is equal to the number of unknowns and provided that all equations are independent of each other. Obviously, the first condition is satisfied when

$N = 2$, that is, for two time intervals. In cases of parameter identification of practical interest, however, we usually have information for more than two time intervals, that is, $N > 2$, which means that the number of equations is larger than the number of unknowns.

If we wish to solve the problem as one of solving $2C$ equations in $2C$ unknowns, we have to pick two specific time intervals, which brings up the question of which ones to choose and whether each pair of chosen intervals will yield the same solution. The answer to the last question is, in general, negative. This stems from the fact that in the equations, water levels and parameter values are those of a model, whereas the information we use is that of water levels in the field. This discrepancy, together with other errors discussed in Sec. 11-1, introduces a certain noise in each equation. Each equation becomes an approximation, and hence a different solution will be derived from any subset of equations. Obviously, we should be interested in making use of all the information available, *assuming* that as we use more information, a better solution will be obtained.

Hence the problem becomes one of obtaining the best set of parameters, i.e., that set which will satisfy the equations most closely, closeness being defined in some exact sense. In other words, we are faced here with the problem of obtaining the optimal solution of the given set of equations. In what follows we shall show that by choosing the criterion according to which we determine how closely the equations are satisfied, we determine the type of optimization problem we have on hand.

The number of equations for which we seek an optimal solution is reduced when information is missing for some cells during some time intervals.

We shall now be more specific and state the problem on hand as one of *linear* or *quadratic optimization*. We shall show several alternatives for defining the *objective function* and discuss the differences among them.

Let us first rewrite the model equations (11-27) in a form which will emphasize that T and S are the unknowns to be identified

$$a_{i,j}^{k+1/2} T_{i,j-1} + b_{i,j}^{k+1/2} T_{i-1,j} + c_{i,j}^{k+1/2} T_{i,j} + d_{i,j}^{k+1/2} T_{i+1,j} + e_{i,j}^{k+1/2} T_{i,j+1}$$
$$+ f_{i,j}^{k+1/2} S_{i,j} = p_{i,j}^{k+1/2} - r_{i,j}^{k+1/2} \quad (11\text{-}28)$$

where: $p_{i,j} = P_{i,j}/\Delta x_i \Delta y_j$; $r_{i,j} = (N_{i,j} + R_{i,j})/\Delta x_i \Delta y_j$,

$$a_{i,j}^{k+1/2} = \frac{-1}{2\,\Delta y_i} \frac{\Delta \phi_{i,j-1/2}^{k+1/2}}{\Delta y_{i-1/2}}; \qquad b_{i,j}^{k+1/2} = \frac{-1}{2\,\Delta x_i} \frac{\Delta \phi_{i-1/2,j}^{k+1/2}}{\Delta x_{i-1/2}};$$

$$c_{i,j}^{k+1/2} = \frac{1}{2\,\Delta x_i} \left(\frac{\Delta \phi_{i+1/2,j}^{k+1/2}}{\Delta x_{i+1/2}} - \frac{\Delta \phi_{i-1/2,j}^{k+1/2}}{\Delta x_{i-1/2}} \right) + \frac{1}{2\,\Delta y_j} \left(\frac{\Delta \phi_{i,j+1/2}^{k+1/2}}{\Delta y_{j+1/2}} - \frac{\Delta \phi_{i,j-1/2}^{k+1/2}}{\Delta y_{j-1/2}} \right)$$

$$d_{i,j}^{k+1/2} = \frac{1}{2\,\Delta x_i} \frac{\Delta \phi_{i+(1/2),j}^{k+1/2}}{\Delta x_{i+1/2}}; \qquad e_{i,j}^{k+1/2} = \frac{1}{2\,\Delta y_j} \frac{\Delta \phi_{i,j+1/2}^{k+1/2}}{\Delta y_{j+1/2}};$$

$$f_{i,j}^{k+1/2} = - \frac{\phi_{i,j}^{k+1} - \phi_{i,j}^{k}}{\Delta t^{k+1/2}}; \qquad \Delta \phi_{i+1/2,j}^{k+1/2} = \phi_{i+1,j}^{k+1/2} - \phi_{i,j}^{k+1/2}, \text{ etc.}$$

In order to simplify the presentation, while emphasizing the various optimization criteria, (11-28) may be rewritten in the following compact form

$$\sum_{m=1}^{5} g_m T_m + f_{i,j}^{k+1/2} S_{i,j} = p_{i,j}^{k+1/2} - r_{i,j}^{k+1/2}, \qquad \text{for all } i, j, k \qquad (11\text{-}29)$$

where the sum is taken over the five terms which include transmissivity, i.e.,

$$\sum_{m=1}^{5} g_m T_m = a_{i,j}^{k+1/2} T_{i,j-1} + b_{i,j}^{k+1/2} T_{i-1,j} + c_{i,j}^{k+1/2} T_{i,j} + d_{i,j}^{k+1/2} T_{i+1,j} + e_{i,j}^{k+1/2} T_{i,j+1}$$

$$(11\text{-}30)$$

Even when the optimal solution is obtained and inserted back into the equations, only in part of the set of equations (11-29) will equality hold between the right and left-hand sides. In the remaining equations, a difference, or deviation, will be observed between the two sides of the equation. We shall use the absolute values of these deviations and their squares in order to define five different possible criteria for approaching an equality between the two sides of the equations.

Criterion A: the maximum absolute deviation. Let X denote the maximum absolute deviation occurring in one equation (11-29), i.e., for all cells and all time intervals. We may therefore state that all other deviations are less than or equal to X

$$\left| \sum_{m=1}^{5} g_m T_m + f_{i,j}^{k+1/2} S_{i,j} - p_{i,j}^{k+1/2} + r_{i,j}^{k+1/2} \right| \le X, \qquad \text{for all } i, j, k \qquad (11\text{-}31)$$

This inequality may be replaced by two equivalent inequalities

$$\left.\begin{array}{l} \displaystyle\sum_{m=1}^{5} g_m T_m + f_{i,j}^{k+1/2} S_{i,j} - p_{i,j}^{k+1/2} + r_{i,j}^{k+1/2} \le X, \qquad \text{for all } i, j, k \\[2em] \displaystyle\sum_{m=1}^{5} g_m T_m + f_{i,j}^{k+1/2} S_{i,j} - p_{i,j}^{k+1/2} + r_{i,j}^{k+1/2} \ge -X, \qquad \text{for all } i, j, k \end{array}\right\} \qquad (11\text{-}32)$$

Using this criterion, the identification problem may now be stated as the following optimization problem: determine the unknown values of $S_{i,j}$, $T_{i,j}$, and X from

$$\min F = X \qquad (11\text{-}33)$$

subject to the constraints

$$\left.\begin{array}{l} \displaystyle\sum_{m=1}^{5} g_m T_m + f_{i,j}^{k+1/2} S_{i,j} - X \le p_{i,j}^{k+1/2} - r_{i,j}^{k+1/2}, \qquad \text{for all } i, j, k \\[2em] \displaystyle\sum_{m=1}^{5} g_m T_m + f_{i,j}^{k+1/2} S_{i,j} + X \ge p_{i,j}^{k+1/2} - r_{i,j}^{k+1/2}, \qquad \text{for all } i, j, k \end{array}\right\} \qquad (11\text{-}34)$$

and the nonnegativity restrictions

$$T_{i,j}, S_{i,j}, X \ge 0, \qquad \text{for all } i, j \qquad (11\text{-}35)$$

This is a typical statement of a *linear programming* problem in which there are $(2C + 1)$ decision variables and $2CN$ constraints (not counting the physical restrictions expressed by (11-35)). One should note that X is regarded here as a decision variable which appears alone in the objective function.

Criterion B: the sum of the absolute values of the maximum deviations in all time intervals. For each time interval, we have C (equal to number of cells) balance equations, with different deviations between the two sides. Let $Y^{k+1/2}$ denote the maximum absolute value of all deviations corresponding to the $(k + \frac{1}{2})$ time interval. We may then state that all other deviations for the time interval are smaller than or equal to $Y^{k+1/2}$, i.e., for each $k = 0, ..., N - 1$, we have

$$\left| \sum_{m=1}^{5} g_m T_m + f_{i,j}^{k+1/2} S_{i,j} - p_{i,j}^{k+1/2} + r_{i,j}^{k+1/2} \right| \leq Y^{k+1/2}, \quad \text{for all } i,j \quad (11\text{-}36)$$

The number of maximum deviations $Y^{k+1/2}$ is equal to the number of time intervals in the calibration period.

The second linear identification problem can therefore be stated as: determine $T_{i,j}$, $S_{i,j}$ and $Y^{k+1/2}$, such that the sum of the absolute values of the maximum deviations in all time intervals will reach a minimum

$$\min F = \sum_{k=0}^{N-1} Y^{k+1/2} \quad (11\text{-}37)$$

subject to the constraints

$$\left. \begin{array}{l} \displaystyle\sum_{m=1}^{5} g_m T_m + f_{i,j}^{k+1/2} S_{i,j} - Y^{k+1/2} \leq p_{i,j}^{k+1/2} - r_{i,j}^{k+1/2} \\[18pt] \displaystyle\sum_{m=1}^{5} g_m T_m + f_{i,j}^{k+1/2} S_{i,j} + Y^{k+1/2} \geq p_{i,j}^{k+1/2} - r_{i,j}^{k+1/2}, \quad \text{for all } i, j, k \end{array} \right\} \quad (11\text{-}38)$$

and the nonnegativity restrictions

$$T_{i,j}, S_{i,j}, Y^{k+1/2} \geq 0, \quad \text{for all } i, j, k \quad (11\text{-}39)$$

Here we have $(2C + N)$ decision variables and $2CN$ constraints of type (11-38). Again, T and S do not appear in the objective function.

Criterion C: the sum of the absolute values of the maximum deviations of all cells. Denote by $Z_{i,j}$ the maximum of the absolute values of the deviations occurring in the N equations of a cell. We may then state that for each cell the absolute values of the deviations in all the balance equations will be smaller than or at most equal to $Z_{i,j}$, that is, for each i,j we have

$$\left| \sum_{m=1}^{5} g_m T_m + f_{i,j}^{k+1/2} S_{i,j} - p_{i,j}^{k+1/2} + r_{i,j}^{k+1/2} \right| \leq Z_{i,j}, \quad \text{for all } k \quad (11\text{-}40)$$

Accordingly, the third linear optimization problem may now be stated as: determine $T_{i,j}$, $S_{i,j}$, and $Z_{i,j}$ so as to minimize the objective function F

$$\min F = \sum_{(i)} \sum_{(j)} Z_{i,j} \quad (11\text{-}41)$$

subject to the constraints

$$\sum_{m=1}^{5} g_m T_m + f_{i,j}^{k+1/2} S_{i,j} - Z_{i,j} \leq p_{i,j}^{k+1/2} - r_{i,j}^{k+1/2}, \qquad \text{for all } i, j, k$$

$$\sum_{m=1}^{5} g_m T_m + f_{i,j}^{k+1/2} S_{i,j} + Z_{i,j} \geq p_{i,j}^{k+1/2} - r_{i,j}^{k+1/2}, \qquad \text{for all } i, j, k$$

$$(11\text{-}42)$$

and the non-negativity restrictions

$$T_{i,j}, S_{i,j}, Z_{i,j} \geq 0 \qquad \text{for all } i, j \qquad (11\text{-}43)$$

The number of decision variables here is $3C$ and the number of constraints of type (11-42) is $2CN$.

Criterion D: the sum of absolute values of all deviations. Denote by $U_{i,j}^{k+1/2}$ the absolute value of the deviation for each equation, that is, for each i, j, and k

$$\left| \sum_{m=1}^{5} g_m T_m + f_{i,j}^{k+1/2} S_{i,j} - p_{i,j}^{k+1/2} + r_{i,j}^{k+1/2} \right| = U_{i,j}^{k+1/2} \qquad (11\text{-}44)$$

The identification can now be stated as a linear programming problem in the following way: determine the values of $T_{i,j}, S_{i,j}$, and $U_{i,j}^{k+1/2}$, such that the objective function F will attain a minimum

$$\min F = \sum_{(k)} \sum_{(j)} \sum_{(i)} U_{i,j}^{k+1/2} \qquad (11\text{-}45)$$

subject to the constraints

$$\sum_{m=1}^{5} g_m T_m + f_{i,j}^{k+1/2} S_{i,j} - U_{i,j}^{k+1/2} \leq p_{i,j}^{k+1/2} - r_{i,j}^{k+1/2}, \quad \text{for all } i, j, k,$$

$$\sum_{m=1}^{5} g_m T_m + f_{i,j}^{k+1/2} S_{i,j} + U_{i,j}^{k+1/2} \geq p_{i,j}^{k+1/2} - r_{i,j}^{k+1/2}, \quad \text{for all } i, j, k,$$

$$(11\text{-}46)$$

and the nonnegativity restrictions

$$T_{i,j}, S_{i,j}, U_{i,j}^{k+1/2} \geq 0 \qquad \text{for all } i, j, k \qquad (11\text{-}47)$$

Here the number of unknowns is $C(N + 2)$ and the number of constraints of type (11-46) is $2CN$. Since the absolute value of the deviation is defined by (11-44), only one of each pair of constraints will be binding in the optimal solution, while the other will actually be redundant. In order to avoid this unnecessary redundancy which will only burden the solution, let us modify the criteria defined above, and define alternative criteria.

Let the deviation in each equation be expressed as a difference between the values of two positive variables V and W

$$\sum_{m=1}^{5} g_m T_m + f_{i,j}^{k+1/2} S_{i,j} - p_{i,j}^{k+1/2} + r_{i,j}^{k+1/2} = V_{i,j}^{k+1/2} - W_{i,j}^{k+1/2} \qquad (11\text{-}48)$$

The deviation itself may be either positive or negative, depending on the magnitudes of V and W. The linear optimization problem may now be restated as: determine the values of $T_{i,j}$, $S_{i,j}$, $V_{i,j}^{k+1/2}$, and $W_{i,j}^{k+1/2}$ such that the objective function F will attain a minimum

$$\min F = \sum_{(k)} \sum_{(j)} \sum_{(i)} (V_{i,j}^{k+1/2} + W_{i,j}^{k+1/2}) \tag{11-49}$$

subject to the constraints

$$\sum_{m=1}^{5} g_m T_m + f_{i,j}^{k+1/2} S_{i,j} - V_{i,j}^{k+1/2} + W_{i,j}^{k+1/2} = p_{i,j}^{k+1/2} - r_{i,j}^{k+1/2}, \qquad \text{for all } i, j, k \tag{11-50}$$

and the nonnegativity restrictions

$$T_{i,j}, S_{i,j}, V_{i,j}^{k+1/2}, W_{i,j}^{k+1/2} \geq 0, \qquad \text{for all } i, j, k \tag{11-51}$$

In this form there are $2C(N+1)$ decision variables and NC constraints of type (11-50). In the optimal solution, at least one of each pair, V and W, for a cell will vanish. The advantage of introducing this modified form of the criterion is that the number of constraints has been cut in half, resulting in a significant reduction in computational effort.

Criterion E: the sum of the squares of the deviations. A *quadratic objective function* may be defined by taking the sum of the squares of all the deviations between the two sides of the equations for all cells and all time intervals. The optimization problem will then be stated as: determine the values of $T_{i,j}$ and $S_{i,j}$ such that the objective function F will attain a minimum

$$\min F = \sum_{(k)} \sum_{(j)} \sum_{(i)} \left(\sum_{m=1}^{5} g_m T_m + f_{i,j}^{k+1/2} S_{i,j} - p_{i,j}^{k+1/2} + r_{i,j}^{k+1/2} \right)^2 \tag{11-52}$$

subject to the nonnegativity restrictions

$$T_{i,j}, S_{i,j} \geq 0 \tag{11-53}$$

This is a typical *quadratic programming* problem, with a quadratic objective function and linear constraints.

The main difference between this and the previous optimization problems is that here the balance equations appear as part of the objective function, whereas the constraints are only the physical (nonnegativity) ones.

The solution of linear programming problems as stated above, can be obtained by using ready made computer programs, which are nowadays supplied as a routine by computer manufacturers.

Hefez, Shamir, and Bear (1973, 1975) use a synthetic 24-cell model to study and compare the different criteria presented above. They conclude, empirically, that the best results were obtained when the sum of absolute deviations of all balance equations was used as the objective function to be minimized. The criteria of sum of absolute values of deviations was also preferred by Klinecke (1971). They also present an example of identifying the natural replenishment in a single cell aquifer model.

Caspi (1975) continued the work of Hefez, Shamir, and Bear (1973, 1975) by analyzing the effect of the input data on the identification results and by developing tools for a sensitivity analysis of the results. He concluded that the value of the objective function does not necessarily indicate the quality of the solution and should not be considered a good criterion for this purpose. He also showed that the solution is very sensitive to variations in water levels and to the addition or removal of constraints. It thus appears that although linear programming as described above seems to be a natural approach to to the identification of model parameters, it has not yet been developed to the point of becoming a reliable routine method which yields a unique result (see discussion by Neuman, 1973a).

Among additional published works on parameter identification we may mention those of Eykhoff (1968), Nelson (1968), Haimes *et al.* (1968), Vemuri and Karplus (1969), Deninger (1969), Korganoff (1970), Lovell (1971), Labadie (1972, 1975), Lovel *et al.* (1972), Slater and Durrer (1971), Yeh and Tauxe (1971), Frind and Pinder (1973), Bruch *et al.* (1974), Lin and Yeh (1974), Sagar *et al.* (1975), Yeh (1975a,b), Kisiel and Duckstein (1976), and Cooley (1977).

There is no doubt that the problem of parameter identification is of paramount importance. Without a satisfactory solution of this problem, the solution of any forecasting problem, and hence also of any management problem which is based on it, remains in doubt.

Because of the uncertainty inherent in the parameter determination, methods will be developed in the future which will produce probability distributions of the values of the various parameters rather than single values. These, in turn, when inserted in the forecasting models will yield probability distributions of the responses.

TWELVE

USE OF LINEAR PROGRAMMING IN AQUIFER MANAGEMENT

In Chap. 1 we have defined the problem of aquifer management and referred to it as our primary objective. We have also emphasized that in order to manage an aquifer system, we must be able to define it, by solving the *identification problem*, and to predict its response to our implemented decisions, by solving the *forecasting problem*. The first eleven chapters of the book are devoted to these two problems. In the present chapter, we wish to demonstate how information derived from a forecasting problem indeed serves as an essential input to a management one.

We have selected linear programming (LP) as an example of a decision making technique to be used in our demonstration, assuming that this very useful tool is better known to the reader. This should not be interpreted as an indication of the merits or superiority of this technique with respect to other ones used in the management of water resources. The basic manner in which the forecasting problem and the management one are interrelated is the same also for other techniques.

Furthermore, no attempt is made here to present this technique in any depth beyond that required for achieving our objective as stated above. The reader is referred to the literature for further information related to the management of water resources in general and to optimization techniques in particular; e.g., Maass *et al.* (1962), Hall and Dracup (1970), Buras (1972), Biswas (1976), and many articles in professional journals. A large number of publications also exists on the application of linear programming to the management of water resources, including some, e.g., Lynn *et al.* (1962), Deininger (1965), and Futagami *et al.* (1976), that focus on the management of water quality.

Of special interest as far as the present chapter is concerned should be publications which apply linear programming to groundwater systems, or to water resource systems in which aquifers play an important role, e.g., Dracup (1966) and Schwartz (1971).

At this point, the reader should return to Sec. 1-2 for a brief review of goals and objectives of water resources management, objective functions and constraints, especially those related to the management of a groundwater basin.

12-1 BRIEF REVIEW OF LINEAR PROGRAMMING

Linear programming is one of the methods used for solving the problem of allocating limited resources among competing users in an optimal manner. This means that among the various possible ways of allocating the given resources, we are to select the one which will minimize or maximize a specified objective function (Sec. 12-2).

The principles and techniques of solving linear programming problems are well described in the literature, e.g., Hadley (1962); Dantzing (1963); and Gass (1969). The brief review presented here is intended as a refresher for those who are already familiar with linear programming as an optimization technique, and as a brief introduction for those who are not familiar with it, indicating that here is another tool that can be used for making management decisions.

Hadley (1962) describes the general linear programming problem as follows: given a set of m linear inequalities or equations in r variables, we wish to find nonnegative values of these variables which will maximize or minimize some linear function of the variables, while satisfying the linear constraints. Mathematically, the problem may be stated as follows.

Determine the values of the r decision variables x_i ($i = 1, ..., r$) which will maximize (or minimize) the objective function z

$$z = c_1 x_1 + c_2 x_2 + \cdots + c_r x_r \tag{12-1}$$

subject to the m constraints

$$a_{i1} x_1 + a_{i2} x_2 + \cdots + a_{ir} x_r \ \{\geq, =, \leq\} \ b_i, \qquad i = 1, ..., m \tag{12-2}$$

where for each constraint only one of the signs $\leq$, $=$, $\geq$ holds, but the sign may vary from one constraint to another. We also require that all decision variables be nonnegative, i.e., that the nonnegativity restriction

$$x_j \geq 0, \qquad j = 1, ..., r \tag{12-3}$$

be satisfied; a_{ij}, b_i, and c_j are assumed to be known constants.

Any set of x_j's which satisfies the constraints (12-2) is a solution of the linear programming problem. If, also, all the nonnegativity requirements (12-3) are satisfied, it is called a feasible solution. The feasible solution which yields the optimal value of the objective function is called the optimal solution (Hadley, 1962). Each set of decisions, x_j, is also called a policy. Our objective is to select that particular policy which will optimize the objective function, subject to the nonnegativity restrictions and the specified constraints.

When the x_j's are interim or final outputs of a production system, and the

a_{ij}'s are the use that is made of a certain resource for producing a unit of x_j, then the values b_i on the right-hand side of (12-2), with a "smaller or equal" sign, are referred to as the limited available quantities of the resources.

The x_j's are the decision variables of the problem; they may represent, for example, *activities*. When the c_j's are costs or prices associated with the x_j's, then z represents the *total cost* (or *benefit*) from operating the system at the activities x_j. The matrix composed of the coefficients a_{ij} is often referred to as the *technological matrix*.

Although the problem as stated by (12-1) through (12-3) may at first sight seem complicated to the planner who is not familiar with the solution of linear programming problems, he need not worry. His primary task should be to analyze his groundwater (or water resources) system, and state his management problem as an optimization problem (i.e., involving a minimization, or maximization of an objective function—or several objective functions in a multi-objective problem—and constraints which have to be satisfied). He should then be able to recognize a linear programming problem from the fact that the objective function as well as the constraints are linear expressions of the decision variables (that is, the x_j's).

Once a problem has been cast into the standard linear programming form, say (12-1) through (12-3), the common algebraic procedure for solving it is the Simplex method, developed by Dantzig in the late 40's (Dantzig, 1963). This procedure is well suited for solution by digital computers. The MPSX program (available in most IBM library programs) is an example of a standard procedure for solving the linear programming problem. Most computer manufacturers supply packages for solving linear programming by modified and advanced Simplex methods. Problems with hundreds of constraints can be solved by a medium-size computer. Large computers can handle thousands of constraints. When larger problems are encountered, decomposition techniques (see, for example, Lasdon, 1970; Hadley, 1962) permit a solution, although they usually require more computer time.

There is no difficulty in changing a problem from minimizing an objective function to maximizing one, noting that Min z = Max$(-z)$. If a decision variable, x_j, is unrestricted in sign, we may replace it by the difference between two (new) nonnegative decision variables, say, x_j^+ and x_j^- such that $x_j = x_j^+ - x_j^-$.

Each "less than or equal" constraint, e.g.

$$a_{k1}x_1 + a_{k2}x_2 + \cdots + a_{kr}x_r \leq b_k \qquad (12\text{-}4)$$

can be written as an equality constraint

$$a_{k1}x_1 + a_{k2}x_2 + \cdots + a_{kr}x_r + x_{r+1} = b_k \qquad (12\text{-}5)$$

by adding a slack variable x_{r+1}. In a similar way, a "greater than or equal" constraint can be changed to an equality constraint by substracting an excess variable.

Hence any linear programming problem can be stated in the following form.

$$\text{Minimize } z = c_1x_1 + c_2x_2 + \cdots + c_nx_n$$

subject to the m constraints

$$
\left.
\begin{aligned}
a_{11}x_1 + a_{12}x_2 + \cdots + a_{1n}x_n &= b_1 \\
a_{21}x_1 + a_{22}x_2 + \cdots + a_{2n}x_n &= b_2 \\
a_{m1}x_1 + a_{m2}x_2 + \cdots + a_{mn}x_n &= b_m
\end{aligned}
\right\}
\tag{12-6}
$$

and

$$x_j \geq 0, \quad j = 1, 2, \ldots, n$$

which is called the standard linear programming form.

Before leaving these general introductory remarks, it may be of interest to present also the so called dual problem.

Associated with the (primal) problem as stated by (12-6), there exists another problem, referred to as the dual of the original (primal) problem. For a primal stated as

$$\text{Maximize } z = c_1x_1 + c_2x_2 + \cdots + c_rx_r$$

subject to the m constraints

$$
\left.
\begin{aligned}
a_{11}x_1 + a_{12}x_2 + \cdots + a_{1r}x_r &\leq b_1 \\
a_{21}x_1 + a_{22}x_2 + \cdots + a_{2r}x_r &\leq b_2 \\
a_{m1}x_1 + a_{m2}x_2 + \cdots + a_{mr}x_r &\leq b_m
\end{aligned}
\right\}
\tag{12-7}
$$

and

$$x_j \geq 0, \quad j = 1, \ldots, r$$

the dual problem is stated as

$$\text{Minimize } w = b_1y_1 + b_2y_2 + \cdots + b_my_m$$

subject to the r constraints

$$
\left.
\begin{aligned}
a_{11}y_1 + a_{21}y_2 + \cdots + a_{m1}y_m &\geq c_1 \\
a_{12}y_1 + a_{22}y_2 + \cdots + a_{m2}y_m &\geq c_2 \\
a_{1r}y_1 + a_{2r}y_2 + \cdots + a_{mr}y_m &\geq c_r
\end{aligned}
\right\}
\tag{12-8}
$$

and

$$y_j \geq 0, \quad j = 1, \ldots, m$$

It is of interest to note (a) that the dual problem (12-8) contains the same constants as the primal (12-7), but in a rearranged (transposed) order, (b) that the dual of the dual problem is the primal, (c) that if the optimal values of x_j ($= x_j^*$) and of y_j ($= y_j^*$) are used to compute the optimal values of the objective functions z^* of z and w^* of w, then we shall find that $z^* = w^*$ (we may think of z^* as the maximum profit, and of w^* as the minimum value of the resources), and (d) that if $r < m$ then the formulation (12-7) involves more constraints than the formulation (12-8), so that the latter may be more convenient for computer solution. Actually, once one solution is obtained, the other can be obtained from it.

Finally, it may be of interest to note that the dual decision variables y_i are interpreted as marginal values of the resources b_i, or their shadow prices. When z is being maximized, y_i is a measure of the rate of increase of z with respect to b_i

$$y_i = \partial z / \partial b_i \bigg|_{z = z^*} \tag{12-9}$$

12-2 APPLICATION OF LINEAR PROGRAMMING TO AQUIFER MANAGEMENT

Before stating the management problem, we have to establish the link between the management problem and the forecasting one, as presented in Chaps 5–9. This link is the balance equations which have always and everywhere to be satisfied. They act as (equality) constraints in the management problem (obviously in addition to any other constraints of the problem, such as economic and technical ones). Constraints may also be imposed on water levels, gradients, subsidence and solute concentration at selected points and times, spring discharge, etc.

In certain management procedures, we speak of the transition which a given system undergoes from one state to the next as a result of executing operational decisions. We may then visualize the balance equations as state transition equations.

In this and the following paragraph, for the sake of simplicity, we shall deal with flow in confined aquifers and/or phreatic ones with relatively small water level fluctuations. The fundamental equations governing the flow of groundwater in these aquifers are presented in Sec. 5-4. These are linear equations, in which the dependent variable, whether $\phi(x, y, t)$ or $h(x, y, t)$, varies in space and time in a continuous fashion. In Sec. 5-6 the numerical approximations of these equations are presented. Discrete aquifer models are presented in Chap. 10.

It is rather difficult to use a partial differential equation as a constraint (or as an equation of state transition), since this would require the solution of a partial differential equation each time we wish to check whether a constraint is violated (or to determine the new state of the system). It is much easier to make use of the linear algebraic equations which approximate the partial differential equations in the numerical methods of solution. As we shall see below, in many cases it is more convenient to use solutions of the equation in conjunction with superposition (Sec. 5-7), which is permitted as we deal with linear flow problems (or approximated as such), instead of the differential equations themselves, or their numerical representation. This approach will be discussed in details below.

To simplify the discussion, we shall consider henceforth pumping as the only controlled activity (= decision variable), regarding artificial recharge as negative pumping.

Use of Technological Functions

Let us demonstrate this method by applying it to a multicell model of an aquifer (Schwarz, 1971; Maddock, 1972). The following notation will be used: h_k^t is the water level or piezometric head at point (or aquifer cell) k and time t resulting from all activities, both controlled and uncontrolled; $k = 1, 2, ..., N$. h_{k0}^t is the water level at point (or aquifer cell) k and time t resulting only from the uncontrolled factors (including pumping which is beyond the control of the planner). P_j^i is the controlled pumping rate at a well (or cell) located at point j (position vector x_j), $j = 1, 2, ..., M$; duration of pumping is Δt from time $(i - 1)\Delta t$ to time $i\,\Delta t$ (see Fig. 8-15c, with Δt = constant, independent of i). Artificial recharge is visualized as negative pumping. $a_{k,j}^{n-i,\Delta t}$ is the net change of water level (= residual drawdown) produced at point (or cell) k by pumping at a rate of one unit of discharge from a well (or cell) j; duration of pumping is Δt, starting at time $(i - 1)\Delta t$; the drawdown is observed at time $t = n\,\Delta t$: $n - i \geq 0$.

The drawdown $a_{k,j}^{n-i,\Delta t}$ is obtained for the considered aquifer with homogeneous initial and boundary conditions; that is, $h = 0$ everywhere at $t = 0$, and $h = 0$ at $t \geq 0$ on the aquifer's boundaries.

By employing the principle of superposition (Sec. 5-7, especially Ex. 4), we obtain the combined drawdown in cell k at time $t = n\,\Delta t$ produced by M wells (or cells) in the aquifer, each with its pumping schedule P_j^i

$$s(k, t) = h_{k0}^t - \sum_{i=1}^{n} \sum_{j=1}^{M} P_j^i a_{k,j}^{n-i,\Delta t} \tag{12-10}$$

The factor $a_{k,j}^{n-i,\Delta t}$ is called the *influence function* (or *coefficient*). It is also called the *technological function* (e.g., Maddock, 1972). In (12-10) we have summed for all M wells (or cells) and n time intervals (see Fig. 8-15c). If $k = 1, 2, ..., N$ (that is, N points at which changes in water level are measured) and M wells are operating in the considered aquifer, the array of $N \times M$ coefficients form the influence matrix (or technological matrix) $a^{n-1,\Delta t}$. For the selected Δt, an $N \times M$ matrix has to be determined for different times $(n - i)$, Δt.

Let h_{k0}^s denote the steady state water levels in an aquifer with steady boundary conditions and natural replenishment. Pumping takes place at points (or cells) j at constant rates P_j^s and a new steady state is established at points (or cells) k, with the new water levels denoted by h_k^s. Water is drawn only from boundaries or natural replenishment, but not from storage within the aquifer. We then have

$$h_k^s = h_{k0}^s - \sum_{j=1}^{M} P_j^s a_{k,j}^s \tag{12-11}$$

where $a_{k,j}^s$ is the steady state influence (or technological) function. The values of h_{k0}^s and h_{k0}^t already include the effects of boundaries, natural replenishment, and all uncontrollable pumping or recharge activities.

We have to emphasize again that we deal here with situations in which superposition is permitted because of the linearity of the system and the homogeneity of the boundary conditions. If, for example, during the operations, boundary conditions vary, the technological function also varies and superposition as employed above does not describe the resulting water levels.

Thus, instead of using the balance equations as constraints, we use their solutions in the forms of the technological functions, $a_{k,j}^s$ and $a_{k,j}^{n-i,\Delta t}$, which give drawdowns produced by pumping. Since these functions are related to unit pumping rates, they are independent of the decision variables P_j^s or P_j^i.

If minimal or maximal water levels, which may be time dependent, are now introduced as constraints of the problem, e.g.

$$h_{k,min} \leq h_k^t \leq h_{k,max} \quad \text{or} \quad h_{k,min}^t \leq h_k^t \leq h_{k,max}^t$$
$$h_{k,min}^s \leq h_k^s \leq h_{k,max}^s \tag{12-12}$$

we may use (12-10) and (12-11) to express them in the forms

$$\sum_{i=1}^{n} \sum_{j=1}^{M} P_j^i a_{k,j}^{n-i,\Delta t} \leq b_k^t; \qquad b_k^t = h_{k0}^t - h_{k,min}^t; \qquad k = 1, 2, ..., N \tag{12-13}$$

$$\sum_{j=1}^{M} P_j^s a_{k,j}^s \leq b_k^s; \qquad b_k^s = h_{k0}^s - h_{k,min}^s; \qquad k = 1, 2, ..., N \tag{12-14}$$

and similar expressions for maximal water levels. The b_k's represent permissible drawdowns at points k resulting from implementing the planned activities; we may think of water levels as the limited resources at our disposal.

In (12-13), or in (12-14), we have N linear constraints, one for each critical point at which a minimal water level is specified during the planned operation of the groundwater system. The constraints (12-13) and (12-14) are written in terms of the known technological functions and the set of activities, P_j^s, or P_j^i, which constitute the decision variables of the problem. This is the standard form of the constraints in (12-2). Obviously, all P's are positive (or can easily be made so in the case of artificial recharge).

The determination of the limiting values $h_{k,min}$ and $h_{k,max}$ is affected by various considerations. Among them, we may mention the following.

(a) Water levels cannot drop to below the physical bottom of an aquifer.
(b) When artificial recharge is implemented, it is undesirable to let water levels rise to the ground surface (and cause inundation) or above some maximum depth below ground surface (to prevent damage to foundations).
(c) Rising water levels may cause subsidence due to compaction by wetting of certain unconsolidated, moisture deficient soils (e.g., loess and other eolian deposits).
(d) Subsidence may be caused by dewatering certain formations. In general, subsidence is a function of the drop in water levels.
(e) Changes in water levels (and gradients) may produce an undesirable flow pattern (e.g., movement of water bodies of inferior quality towards pumping wells).
(f) Water levels may not drop below the elevations of the screens of pumping wells. Pump characteristics may also limit drawdown.
(g) Considerations of energy (availability and/or cost) required to lift the water to the ground surface.

Obviously, when values of $h^t_{k,min}$ and $h^t_{k,max}$ are not determined by natural conditions they may be considered as decision variables to be determined as part of the optimization anlysis.

In addition to legal, hydrologic, physical, or economic constraints imposed on water levels we have those imposed on the decision variables P^i_j (e.g., due to size of available pumping installations).

Sometimes constraints are imposed on the hydraulic gradients, in order to control the flow. We may then express the gradients in terms of water levels at points (or cells). For example

$$L_{mk} J_{mk,min} \geq h^t_m - h^t_k \geq L_{mk} J_{mk,max} \qquad (12\text{-}15)$$

where m and k denote a pair of adjacent points (or cells) in the aquifer, with a distance L_{mk} and gradient J_{mk} between them.

The above constraints may be imposed at all points in a considered aquifer, or only at certain critical points.

We have shown above how, rather than introduce the balance equations as constraints (in addition to minimum and maximum water levels), we have introduced their solution through the technological functions. We have still to discuss how these functions can be determined. Let us review several procedures for achieving this goal.

(a) Procedures based on well drawdown equations These equations were developed in Chap. 8 for aquifers for various types. For example, for steady flows in a confined aquifer, we have from (8-6)

$$a^s_{k,j} \equiv s^s_{k,j}/Q_j = \frac{1}{2\pi T} \ln(R/r_{kj}), \qquad r_{kj} \leq R \qquad (12\text{-}16)$$

where R is the radius of influence of the well.

For unsteady flow in an infinite confined aquifer, we have from (8-68b)

$$a^{n-i,\Delta t}_{k,j} = s^{n-i,\Delta t}_{k,j}/Q^i_j = \frac{1}{4\pi T}\left[W\left\{ \frac{Sr^2_{kj}}{4T(n-i+1)\,\Delta t} \right\} - W\left\{ \frac{Sr^2_{kj}}{4T(n-i)\,\Delta t} \right\} \right]$$
$$\equiv \beta(k,j,n-i) \qquad (12\text{-}17)$$

where $\beta(k,j,n-1)$ is the value of $\beta(n-i)$ as defined by (8-68b) corresponding to the distance $r_{kj} = [(x_k - x_j)^2 + (y_k - y_j)^2]^{1/2}$ from the well at $\mathbf{x}_j$ to the point $\mathbf{x}_k$, for $k \neq j$ and $r_{kj} \equiv (r_w)_j$ for $k = j$: Q^i_j is the pumping at well j during Δt between $(i-1)\,\Delta t$ and $i\,\Delta t$, $s^{n-i,\Delta t}_{k,j}$ is the drawdown at point k produced by Q^i_j.

When Q^i_j ($\equiv P^i_j$ in (12-10)) varies in a stepwise fashion, say, as represented by Fig. 8-15c with $\Delta t = $ constant, the resultant drawdown at point k at time $t = n\,\Delta t$ is obtained from (8-68b)

$$s(\mathbf{x}_k, t) \equiv s^n_k \equiv \sum_{i=1}^{n} \sum_{j=1}^{M} Q^i_j \beta(k,j,n-i) \equiv a^{n-i,\Delta t}_{k,j} \qquad (12\text{-}18)$$

The conditions under which (12-16) and 12-17) are applicable are given in

Secs 8-2 and 8-5, respectively. In a similar way, the technological function can be obtained for other types of aquifers (e.g., leaky), by introducing the appropriate well function in the definition of β.

In most cases, however, we are interested in a bounded aquifer which is also inhomogeneous. If the region influenced by the pumping is homogeneous, isotropic, and sufficiently removed from the boundaries, we may still use (12-17) as a good approximation. Sometimes the geometry of the boundaries is such that we can modify (12-17) using the method of images as discussed in Sec. 8-10. If all this is not possible, we have to derive the technological function by solving the appropriate partial differential equation.

(b) Procedures based on the solution of the partial differential equation The technological function, for cases of practical interest, may be obtained by solving the forecasting problem for the considered aquifer with its boundaries by any of the methods discussed in Sec. 5-6. The more common method nowadays is a numerical solution, using a digital computer, which can be applied to inhomogeneous and anisotropic aquifers having boundaries of any shape. With homogeneous boundary and initial conditions, we introduce pumping during Δt at one unit rate of discharge at a point j and determine the resulting response ($=$ drawdowns) at all points k of interest at a times $t = (n - i) \Delta t$ and obtain $a_{k,j}^{n-i,\Delta t}$. The procedure is repeated for all points and times of interest.

The technological function $a_{k,j}^{n-i,\Delta t}$ can also be obtained by solving the forecasting problem using an RC-network analog (Bear, 1972, Chap. 11) instead of a digital computer.

Maddock (1974) derived nonlinear technological functions for aquifers whose transmissivities vary with drawdown.

The procedure described above is also applicable to the case where we use a multicell model to represent the aquifer. The mathematical statement is similar in this case to that of the finite difference technique.

We may now return to the linear programming problem of determining the optimal utilization of an aquifer. A typical problem may be stated in the following way.

Given an aquifer with a known influence matrix ($a^{n-1,\Delta t}$ or a^s), determine the pumping (and/or artificial recharge) rates (P_j^i, or P_j^s) at M wells (or M aquifer cells) so as to: (1) to maximize the net benefits (or minimize costs), (2) not to produce at N critical points drawdowns (and/or water level rises) in excess of certain permissible values, (3) meet specified demands, or supply at least a specified demand, if the latter is given, and (4) do not exceed the capacities of pumping (and/or recharge) installations, if these already exist.

Other objective functions and additional constraints may also be used. Following are several examples of possible additional constraints.

(a) Seasonally varied demand In a quasi-steady state problem, the annual pumpage P_j^s may be made up of seasonal demands P_j^m for the mth season. The additional constraint is then $\sum_{(m)} P_j^m \geq P_j^s$ for all wells $j = 1, 2, \ldots, M$.

(b) Maximum installed pumping (or artificial recharge) capacities If MP_j^m denotes the maximum pumping capacity available in the mth season, then, $P_j^m \leq MP_j^m$ for all seasons and all wells.

(c) Meeting demand in a cell from different sources The demand in a typical cell, k, can be supplied by water pumped within the same cell, by water pumped in another cell and transported to the considered cell, or by water (both groundwater and surface water) imported into the area from a source external to the considered aquifer.

(d) Staged development Planned pumping rates and/or installed capacities may vary with the stage of development of a regional water resource system.

Coupling the Management and the Forecasting Problems

By using the technological (or influence) functions, we separate the forecasting problem from the management one. First, the forecasting problem is solved a number of times, yielding, as results, the technological functions. These, in turn, are used in the constraints of the management problem. However, the two problems may be solved simultaneously as a single management problem, with the balance equations serving as constraints. This approach is especially advantageous when the balance equation takes the form of a set of linear algebraic equations such as (10-7), representing simple cell balances in a multicell model of an aquifer, or any of the numerical representation schemes (finite difference or finite element) of the partial differential equation as discussed in Sec. 5-6. A general matrix form of the algebraic equations for solving the continuity equation, say (5-58), is

$$[A]\{h\} + [B]\left\{\frac{\partial h}{\partial t}\right\} = \{N\} \tag{12-19}$$

where $[A]$ is a conductance (or stiffness) matrix depending on aquifer transmissivities (and leakage factors, if leakage exists) and element configuration, $[B]$ is a capacitance matrix depending on aquifer storativity and element configuration, $\{h\}$ is a vector made up of piezometric heads at the nodes, and $\{N\}$ is a vector representing the discharge rates at the nodes. If h is taken as the average piezometric head at time t and $t - \Delta t$, (12-19) becomes

$$[C]\{h^t\} = [D]\{h^{t-\Delta t}\} + 2\{N\} \tag{12-20}$$

where: $[C] = [A] + (2/\Delta t)[B]$, $[D] = -[A] + (2/\Delta t)[B]$. It is also possible to use a certain weight $\frac{1}{2} \leq \theta \leq 1$ between h^t and $h^{t-\Delta t}$ (see Sec. 5-6). Our interest here is only in the end result, namely, that the continuity equation can be expressed as a set of linear algebraic equations which, in turn, can serve as linear equality constraints in the management problem.

　　If the drawdowns are limited by some minimum water levels and we add the constraints $h_k^t \geq h_{k,\min}$, it is sometimes convenient to introduce a new variable $h_k'^t = h^t - h_{k,\min}$ (rather than the water level h_k^t) as decision variable. Each constraint is then expressed as $h_k'' \geq 0$, which is an intrinsic nonnegativity constraint of the linear programming model. Substituting h_k'' for h_k' in the algebraic equations results in constant terms which are transferred to the right-hand side of the equations.

12-3 EXAMPLES

Let us consider several examples of applying the linear programming technique to the management problem.

Example 1: Pumping in a two-cell aquifer This example serves as a demonstration of a graphical method for solving LP problems when only two decision variables are involved. All other problems have to be solved numerically, usually by computers.

Consider a rectangular aquifer represented by a two-cell model (Fig. 12-1). Steady state conditions prevail.

Given the geometrical dimension $(2L, W)$ and area $(A = 2LW)$ of the aquifer, its transmissivity (T) and the rate of its natural replenishment $(N;$ dims. $L/T)$, it is required to determine the pumping rates (dims. L^3/T), P_1 in cell 1 and P_2 in cell 2, so as to supply a specified total demand D at the lowest possible total cost. The pumping rates should be such that water levels in the cells do not drop below certain specified minimum levels, $h_{1,min}$ and $h_{2,min}$ in cells 1 and 2, respectively.

The costs of pumping and supply per unit of P are C_1 and C_2 for water pumped from cells 1 and 2, respectively. In the discussion here we do not specify any particular unit for $T, P, N, L, W, C,$ and h. However, one should be careful to employ a consistent system of units. For example: P in cubic meters per year, N in meters per year, T in square meters per year, $W, L,$ and h in meters, and C in monetary units per cubic meter per year.

The undisturbed water levels are h_{10} and h_{20} in cells 1 and 2, respectively. The drawdown in cell k produced by pumping one unit in cell j is denoted by a_{kj}.

The influence matrix a (elements a_{kj}) is obtained by solving the balance equations for the two cells for unit rates of pumping.

(i)   *Cell 1, $P_1 = 0, P_2 = 0$*

$$NA + (h_{20} - h_{10})\frac{WT}{L} - (h_{10} - 0)\frac{WT}{L/2} = 0 \qquad (12\text{-}21)$$

(ii)   *Cell 2, $P_1 = 0, P_2 = 0$*

$$NA - (h_{20} - h_{10})\frac{WT}{L} = 0 \qquad (12\text{-}22)$$

(iii) *Cell 1, $P_1 = 1, P_2 = 0$*

$$NA - 1 + [(h_{20} - a_{21}) - (h_{10} - a_{11})]\frac{WT}{L} - (h_{10} - a_{11})\frac{WT}{L/2} = 0$$

$$(12\text{-}23)$$

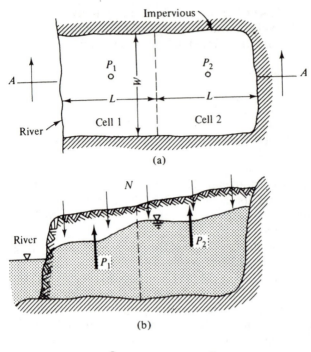

(a)

(b)

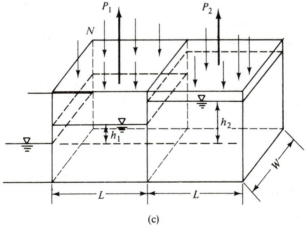

(c)

Figure 12-1 Management of an aquifer represented as a two-cell model. (a) Map. (b) Cross section $A-A$. (c) Two-cell aquifer model.

(iv) *Cell 2, $P_1 = 1$, $P_2 = 0$*

$$NA - [(h_{20} - a_{21}) - (h_{10} - a_{11})]\frac{WT}{L} = 0 \qquad (12\text{-}24)$$

(v) *Cell 1, $P_1 = 0$, $P_2 = 1$*

$$NA + [(h_{20} - a_{22}) - (h_{10} - a_{12})]\frac{WT}{L} - (h_{10} - a_{12})\frac{WT}{L/2} = 0 \qquad (12\text{-}25)$$

(vi) *Cell 2, $P_1 = 0$, $P_2 = 1$*

$$NA - 1 - [(h_{20} - a_{22}) - (h_{10} - a_{12})]\frac{WT}{L} = 0 \qquad (12\text{-}26)$$

The first two equations yield the steady, undisturbed water levels

$$h_{10} = NL^2/T \qquad h_{20} = 2NL^2/T \qquad (12\text{-}27)$$

By solving (12-23) through (12-26), we obtain

$$a_{11} = a_{12} = a_{21} = L/2WT; \qquad a_{22} = 3L/2WT \qquad (12\text{-}28)$$

The water levels with pumping taking place in the two cells are therefore

$$\left.\begin{array}{l} h_1 = h_{10} - (P_1a_{11} + P_2a_{12}) \\ h_2 = h_{20} - (P_1a_{21} + P_2a_{22}) \end{array}\right\} \qquad (12\text{-}29)$$

The objective function is the total cost of supply

$$z = C_1P_1 + C_2P_2 \qquad (12\text{-}30)$$

Here the costs are independent of the water levels, h_1 and h_2, but could be made a function of them if necessary. In the latter case $C_2 = C_2(h)$, as cost of pumping depends on the depth to water table and the problem is nonlinear. It can be linearized by assuming some average depth $\bar{h}$, neglecting the effect of changes in h due to drawdown.

We can now state the complete LP problem: Determine P_1 and P_2 such that

$$\text{Minimize } z = C_1P_1 + C_2P_2 \qquad (12\text{-}31)$$

subject to the constraints:

$$P_1 + P_2 \geq D \qquad (12\text{-}32)$$

$$P_1a_{11} + P_2a_{12} \leq h_{10} - h_{1,\min}(\equiv b_1) \qquad (12\text{-}33)$$

$$P_1a_{21} + P_2a_{22} \leq h_{20} - h_{2,\min}(\equiv b_2) \qquad (12\text{-}34)$$

$$P_1 \geq 0, \qquad P_2 \geq 0 \qquad (12\text{-}35)$$

Equations (12-32) through (12-35) define the domain of feasible solutions (Fig. 12-2). For $C_1 < C_2$, we draw the lines $z = $ const. and seek the line of smallest z which still intersects the domain of feasible solution, at least at one point. We find this point to be B. This means that the optimal solution of our problem in this case is $P_1^* = D$ (that is, to pump the entire demand in cell 1), $P_2^* = 0$.

For $C_1 > C_2$ the solution is given by point A in Fig. 12-2, with optimal values P_1^* and P_2^*.

Note that Fig. 12-2 is drawn for the case $b_1/a_{12} \leq D$ and $b_1/a_{21} \leq D$.

For example, for $D = 45$ Mm^3/yr (1 Mm$^3 = 10^6$ m^3), $C_1 = 0.2$ MU/m^3/ yr (MU $\equiv$ monetary unit), $C_2 = 0.1$ MU/m^3/yr, $W = 10$ km, $L = 10$ km, $T = 10{,}000$ m^2/d, $N = 360$ mm/yr, $h_{1,\min} = 2.5$ m, $h_{2,\min} = 5$ m, we obtain: $a_{11} = 0.139$ m/Mm^3/yr $= a_{12} = a_{21}$, $a_{22} = 0.417$ m/Mm^3/yr, $h_{10} = 10.0$ m,

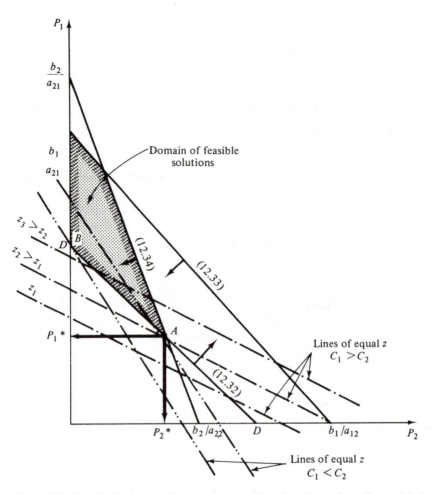

Figure 12-2 Graphical solution of management problem for a two-cell aquifer model (Example 1).

$h_{20} = 20.0$ m, $b_1 = 7.5$ m, $b_2 = 15$ m, $P_1^* = 12.5$ Mm^3/yr, $P_2^* = 32.5$ Mm^3/yr, $z^* = 5.75 \times 10^6$ MU/yr. It may be of interest to note that in this case, the constraint (12-33) does not affect the result.

Since only two decision variables are involved, we could obtain the solution graphically. When a larger number of variables is involved, the *Simplex method* of solution is employed, using a digital computer (see any text on Linear Programming).

Instead of determining the influence matrix a prior to solving the LP problem, let us combine the two problems. Assuming, as is usually the case, that h_{10} and h_{20} are known, the balance equations to be satisfied are

$$NA - P_1 + (h_2 - h_1)\frac{WT}{L} - (h_1 - h_0)\frac{WT}{L/2} = 0$$

$$\text{(12-36)}$$

$$NA - P_2 - (h_2 - h_1)\frac{WT}{L} = 0$$

or

$$P_1 + \frac{3WT}{L}h_1 - \frac{WT}{L}h_2 = NA + h_0\frac{WT}{L/2}$$

$$\text{(12-37)}$$

$$P_2 - \frac{WT}{L}h_1 + \frac{WT}{L}h_2 = NA$$

which serve as equality constraints of the management problem. These equations are actually identical to (12-23) and (12-26) written for P_1 instead of $P_1 = 1$. We can always select a reference level such that

$$h_1 \geq 0, \qquad h_2 \geq 0 \qquad \text{(12-38)}$$

In addition, we have to satisfy

$$h_i \geq h_{1,\min}, \qquad h_2 \geq h_{2,\min} \qquad \text{(12-39)}$$

The constraints (12-32) and (12-35) remain unchanged. We may thus visualize the problem as having four decision variables P_1, P_2, h_1, h_2, where h_1 and h_2 have to satisfy (12-37). The LP problem now consists of minimizing the objective function, subject to the constraints (12-32), (12-35), and (12-37) through (12-39).

It is also possible to have an objective function affected also by the values of h, for example, when we take into account the cost of lifting the water. We may thus have in general

$$z = C_1 P_1 + C_2 P_2 + C_3 h_1 + C_4 h_2 \qquad \text{(12-40)}$$

In (12-31), $C_3 = C_4 = 0$. Note that if C_3 and C_4 are functions of h_1 and h_4, respectively, the problem becomes nonlinear.

Example 2: Allocation of pumping in a 25-cell aquifer (Schwarz, 1971) The investigated aquifer has the shape of a square 10 km × 10 km (Fig. 12-3). On three sides the aquifer's boundaries are impervious. Its fourth side is a lake into which the aquifer is draining. However, because of the poor quality of the water in the lake, it is important to keep certain minimum water levels in the aquifer along the lake so as to prevent lake water from encroaching into the aquifer. Let these water levels be +0.64 m (above lake level used as datum level) and +0.95 m at distances of 1 km and 3 km from the lake, respectively.

The aquifer is replenished from precipitation at a rate $N = 100$ mm/yr, uniformly distributed over the entire area. Under natural conditions the entire replenishment is drained through the aquifer to the lake.

The aquifer is homogeneous with transmissivity $T = 1000$ m²/d. Since we assume steady flow, the knowledge of storativity is not required.

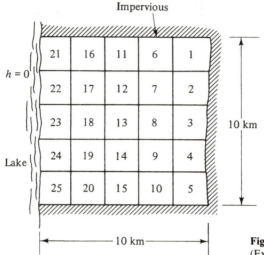

Impervious

$h = 0$

Lake

10 km

10 km

Figure 12-3 Layout of a 25-cell aquifer (Example 2).

After analyzing the areal distribution of pumping, the aquifer is divided into 25 cells of size 2 km × 2 km. Each such cell is considered as a single point at which pumping takes place. Pumping is planned in 15 cells (6 through 20) only.

A digital computer is used to derive the components of the steady state influence matrix $\boldsymbol{a}^s$ for all the 15 cells in which pumping is planned ($j = 15$, $16, \dots, 25$, assuming pumping to be concentrated at the center of each cell). The components of $\boldsymbol{a}^s$ are obtained by solving (10-7) written for all 25 cells. However, since we are interested in the water levels in only 10 critical cells ($k = 16, 17, \dots, 25$), the matrix $\boldsymbol{a}^s$ in this case consists only of 10 rows and columns. The components $a^s_{k,j}$, $k = 16, 17, \dots, 25$, $j = 6, 7, \dots, 20$, are given in Table 12-1.

Initial values h_{i0} (with no pumping anywhere) are computed to be

$$h_{i0} = 13.88 \text{ m}, \quad i = 1, \dots, 5$$
$$h_{i0} = 12.77 \text{ m}, \quad i = 6, \dots, 10$$
$$h_{i0} = 10.55 \text{ m}, \quad i = 11, \dots, 15$$
$$h_{i0} = 7.22 \text{ m}, \quad i = 16, \dots, 20$$
$$h_{i0} = 2.79 \text{ m}, \quad i = 21, \dots, 25$$

In this example we have only a single consumer (or center of consumption) in the region, located in cell 18. The total demand of this consumer is $D = 7.0 \text{ Mm}^3/\text{yr}$. The cost of pumping and conveyance of water from cell 18 to this consumer is 1.0 MU/m³ (MU = monetary unit); the cost increases with distance from the pumping area to this consumer at a rate of 0.5 MU/m³/1000 m distance.

Table 12-1 contains all this information, including the permissible draw-

Table 12-1 LP model for steady state allocation of pumping in a 25-cell aquifer (Example 2)

	Activity															Constraint		
	P_6	P_7	P_8	P_9	P_{10}	P_{11}	P_{12}	P_{13}	P_{14}	P_{15}	P_{16}	P_{17}	P_{18}	P_{19}	P_{20}	Type	Level	Designation
	1.10	0.98	0.82	0.72	0.68	1.32	1.00	0.78	0.60	0.55	1.88	1.00	0.60	0.42	0.38	≤	6.27	Level-16
	0.97	0.92	0.85	0.78	0.72	1.01	1.10	0.88	0.70	0.62	0.98	1.52	0.82	0.53	0.43	≤	6.27	Level-17
	0.82	0.87	0.95	0.87	0.81	0.78	0.93	1.01	0.93	0.78	0.60	0.81	1.43	0.81	0.60	≤	6.27	Level-18
	0.72	0.78	0.85	0.92	0.97	0.62	0.70	0.88	1.10	1.01	0.43	0.53	0.82	1.52	0.98	≤	6.27	Level-19
	0.68	0.72	0.82	0.98	1.10	0.55	0.60	0.78	1.00	1.32	0.38	0.42	0.60	1.00	1.88	≤	6.27	Level-20
	0.37	0.32	0.28	0.24	0.21	0.42	0.35	0.27	0.21	0.20	0.60	0.47	0.22	0.18	0.12	≤	2.15	Level-21
	0.32	0.30	0.29	0.27	0.25	0.35	0.35	0.30	0.23	0.21	0.37	0.45	0.30	0.20	0.15	≤	2.15	Level-22
	0.28	0.30	0.30	0.30	0.28	0.25	0.30	0.31	0.30	0.25	0.21	0.29	0.41	0.29	0.21	≤	2.15	Level-23
	0.25	0.27	0.29	0.30	0.32	0.21	0.23	0.30	0.35	0.35	0.15	0.20	0.30	0.45	0.37	≤	2.15	Level-24
	0.21	0.24	0.28	0.32	0.37	0.20	0.21	0.27	0.35	0.42	0.12	0.18	0.22	0.47	0.60	≤	2.15	Level-25
	1.00	1.00	1.00	1.00	1.00	1.00	1.00	1.00	1.00	1.00	1.00	1.00	1.00	1.00	1.00	=	7.0	Demand
	3.9	3.2	3.0	3.2	3.9	3.2	2.4	2.0	2.4	3.2	3.0	2.0	1.0	2.0	3.0	Free	Min.	Cost

down in cells 16 through 25, the components of the influence matrix, the total to be supplied (row before last), and the costs associated with each P_j (last row).

The first row gives the vector of activities, or decision variables (here the P_j's). Then come the water level constraints and the demand constraint. The last row specifies the costs associated with the P_j's. It is very convenient to summarize the information in such a table.

The objective function is the total costs associated with supplying the specified demand

$$z = C_6 P_6 + C_7 P_7 + \cdots + C_{20} P_{20}; \qquad P_j\text{'s in Mm}^3/\text{yr} \quad (12\text{-}41)$$

It is required to determine the values of $P_j, j = 6, 7, \ldots, 20$, which will supply the required demand at a minimum cost. The constraints are

$$
\left.
\begin{array}{l}
P_6 + P_7 + P_8 + \cdots + P_{20} = 7.0 \\
1.10\,P_6 + 0.98\,P_7 + 0.82\,P_8 + \cdots + 0.38\,P_{20} \leq 6.27 \\
\vdots \\
\vdots \\
\vdots \\
0.21\,P_6 + 0.24\,P_7 + 0.28\,P_8 + \cdots + 0.60\,P_{20} \leq 2.15 \\
\qquad P_j \geq 0, \qquad j = 6, 7, \ldots, 20
\end{array}
\right\} \quad (12\text{-}42)
$$

Table 12-2 presents the solution obtained by using one of the available library programs for solving linear programming problems.

As in Example 1, we can use algebraic balance equations written for the cells as constraints, rather than expressing the latter in terms of components of the influence matrix which have to be determined separately.

Table 12-2 Results of Example 2

Cell No.	Optimal pumping (Mm³/yr)	Drawdown (m)	Permissible drawdown (m)
13	1.06		
14	—		
15	—		
16	0.80	5.59	6.27
17	1.59	6.27	6.27
18	1.16	6.27	6.27
19	1.59	6.27	6.27
20	0.80	5.59	6.27
21	—	2.15	6.27
22	—	2.11	2.15
23	—	2.06	2.15
24	—	2.11	2.15
25	—	2.15	2.15
Total	7.00	—	—

No pumping in cells 6–12.

The total cost $z^* = 14.43 \times 10^6$ MU/yr.

Example 3: Staged development in a two-cell aquifer (Schwarz, 1971) This example is presented to illustrate the effect of time. The studied region is the two-cell aquifer described in Ex. 1 (Fig. 12-1). The water supply system is to be based on both groundwater pumped from within the two cells and on water imported from outside into the region. The project is to be built in four stages, at $t = 0$, 5, 10, and 15 years. The total available budget is limited and serves as a constraint. The investments at the various stages constitute the decision variables. The planning horizon of the project is 20 years.

Investments in the pumping installations are smaller than those required for the conveyance system for the imported water. Initially, the aquifer contains a relatively large volume of water in storage; this may serve as a temporary source of water if we wish to postpone investments in constructing the import installations.

In the last stage, pumping rates may not exceed the steady state (or long term average) yield of the aquifer. At the end of the planning horizon, the drawdowns in the aquifer may not exceed the specified (tolerable) maximum values of 7.5 m in cell 1 and 15.0 m in cell 2. These constraints on ultimate drawdowns will ensure a state of equilibrium at the end of the planning horizon, and prevent the tendency to mine water from the aquifer, leaving it empty for future generations. There is no limit, however, on temporary mining during the first three stages of the project.

The minimum demand which has to be supplied in each cell is 30 Mm^3/yr. Water may be supplied at a higher rate if this will produce an increased net benefit. However, a reduction in supply from one stage to the next is not permitted. Furthermore, in the final stage not more than 55 Mm^3/yr from groundwater and 45 Mm^3/yr of imported water may be supplied.

The interest rate on all costs and benefits is 8 %.

With an existing installed pumping capacity of 5 Mm^3/yr in each cell, it is required to plan the project by determining: (1) rates of supply to each cell and during each stage from groundwater and from imported water, and (2) investment in pumping and conveyance installations (for supplying the imported water) at each stage. The objective is to maximize the present worth of the flow of net benefits to be derived from the operation of the project. All decisions are subject to the hydrologic, economic, and demand constraints listed above. Table 12-3 summarizes costs, benefits and present worth of net benefits.

Table 12-4 summarizes the LP model

Activities are divided into two main groups: Development (columns 1–16) and Operations (columns 17–32). Each of these is further subdivided into those related to pumping (columns 1–8, 17–24) and those related to import (columns 9–16, 25–32). Finally, each activity is considered separately for each cell.

Rows 1–2 give the constraints on groundwater levels in the cells at the end of stage 4 (i.e., at the end of 20 years). The elements of these two rows

Table 12-3 Costs, benefits and present worth of net benefits (MU/m³) for Ex. 3

Cell		1				2			
Stage	1	2	3	4	1	2	3	4
Present worth of investments								
In pumping installations	20.00	13.62	9.26	6.30	20.00	13.62	9.26	6.30
In import installations†	10.00	6.81	4.63	3.15	40.00	27.24	18.52	12.60
Benefits and costs								
Annual benefits from water supplied	9.0	9.0	9.0	9.0	10.0	10.0	10.0	10.0
Annual costs of pumping	2.0	2.0	2.0	2.0	2.0	2.0	2.0	2.0
Annual net benefits from pumping	7.0	7.0	7.0	7.0	8.0	8.0	8.0	8.0
Present worth of flow of net benefits from pumping	27.95	19.03	12.94	8.80	31.95	21.75	14.79	10.06
Annual costs of import††	0.4	0.4	0.4	0.4	0.4	0.4	0.4	0.4
Annual net benefits from import	8.6	8.6	8.6	8.6	9.6	9.6	9.6	9.6
Present worth of flow of net benefits from import	34.33	23.38	15.90	10.82	38.33	26.10	17.75	12.07

† Import to cell 1 is possible only through cell 2. The cost is only that required for transporting the water from cell 2 to cell 1.

are components of the influence matrices $a^{5,20}$ (columns 17–18), $a^{5,15}$ (columns 19–20), $a^{5,10}$ (columns 21–22) and $a^{5,5}$ (columns 23–24), which give the drawdowns (in cm/Mm³/yr) due to 5 years of pumping, observed after 20, 15, 10, and 5 years from the beginning of pumping, respectively.

The constraints given by rows 3, 4 limit the pumping in the last stage to the long term sustained yield by limiting the drawdowns to the maximum steady state ones.

The constraints which limit the total import to the cells to the total available import of 45 Mm³/yr are given by rows 5–8.

Rows 9–16 show the constraints on water demands. During the first stage, water supply may not drop below 30 Mm³/yr to each cell. In each of the following stages the supply during each stage should not be reduced with respect to the preceding one.

The next two groups of constraints (rows 17–24 and 25–32) specify the availability of pumping and import installations in each stage and in each cell. Pumping and import in each cell and in each stage cannot exceed (i.e., should be equal to or smaller than) the capacity installed in that cell in the current stage and in the previous ones. The initial capacity of the pumping installations is 5.0 Mm³/yr. The import capacity to cell 2 should be capable of handling also conveyance to cell 1.

Rows 33–34 specify the budget constraints. The total budget in all stages cannot exceed 3000 MU, while in the first stage the available budget amounts to 1500 MU only.

Table 12-4 Linear programming model for Ex. 3

Col	1	2	3	4	5	6	7	8	9	10	11	12	13	14	15	16	17	18	19	20	21	22	23	24	25	26	27	28	29	30	31	32	33	34
Group	Development — Pumping								Development — Import								Operations — Pumping								Operations — Import								Constraints	
Level	1		2		3		4		1		2		3		4		1		2		3		4		1		2		3		4		Type	Level
Type	2	1	2	1	2	1	2	1	2	1	2	1	2	1	2	1	2	1	2	1	2	1	2	1	2	1	2	1	2	1	2	1		

Constraint rows (Type / Level)

Constraint group	Type	Level
Groundwater levels — 20 yrs.	V	7.50
Groundwater levels — Steady state	V	15.00
Groundwater levels — 20 yrs.	V	7.50
Groundwater levels — Steady state	V	15.00
Import	V	45.00
Import	V	45.00
Import	V	45.00
Water demands — Re-sent	∧	30.00
Water demands — Re-sent	∧	30.00
Water demands — Future increases	∧	0
Water demands — Future increases	∧	0
Water demands — Future increases	∧	0
Water demands — Future increases	∧	0
Water demands — Future increases	∧	0
Water demands — Future increases	∧	0
Available installations — Pumping	V	5.00
Available installations — Pumping	V	5.00
Available installations — Pumping	V	5.00
Available installations — Pumping	V	5.00
Available installations — Pumping	V	5.00
Available installations — Pumping	V	5.00
Available installations — Pumping	V	5.00
Available installations — Import	V	0
Available installations — Import	V	0
Available installations — Import	V	0
Available installations — Import	V	0
Available installations — Import	V	0
Available installations — Import	V	0
Available installations — Import	V	0
Budget — Stage 1	VI	1500
Budget — Total	VI	3000
Total net benefit	Free	Max

Groundwater level coefficients (Operations — Pumping, cols 17–24)

Level	17	18	19	20	21	22	23	24
7.50	0.72	1.08	2.44	2.09	3.27	2.45	8.33	2.45
15.00	1.72	2.44	5.96	3.27	8.62	2.45	13.24	13.24
7.50							13.90	13.90
15.00							13.90	41.70

Budget row coefficients

	1	2	3	4	5	6	7	8	9	10	11	12	13	14	15	16
Stage 1	20	20	20	20	20	20	20	20	10	10	10	10	10	10	10	40
Total	20	20	20	20	20	20	20	20	10	10	10	10	10	10	10	40

Total net benefit (objective) coefficients

Col	1	2	3	4	5	6	7	8	9	10	11	12	13	14	15	16
Net benefit	−20.0	−13.6	−9.3	−9.3	−9.3	−6.3	−6.3	−6.3	−10.8	−6.8	−27.2	−4.6	−18.5	−3.1	−12.6	

Col	17	18	19	20	21	22	23	24	25	26	27	28	29	30	31	32
Net benefit	27.95	31.95	19.03	21.75	12.94	14.79	8.80	10.06	34.33	38.33	23.38	26.10	15.90	17.75	10.82	12.07

The remaining interior cells of the tableau contain unit coefficients ±1.0 linking the development and operations variables to the Import, Water demands, and Available installations constraints.

511

Table 12-5 Results of Ex. 3

Cell	1				2			
Stage	1	2	3	4	1	2	3	4
Planned supply (Mm³/yr)	30.0	30.0	30.0	30.0	43.5	69.0	69.0	69.0
Additional pumping installations (Mm³/yr)	25	0	0	0	27	0	0	0
Additional import installations (Mm³/yr)	0	0	0	0	11.5	25.5	0	8.0
Pumping (Mm³/yr)	30.0	30.0	30.0	30.0	32.0	32.0	32.0	24.0
Import (Mm³/yr)	0	0	0	0	11.5	37.0	37.0	45.0
Water level drawdown at end of stage (m)				6.78				13.0

Present worth of total net benefits : 4787 MU
Total investments: 2840 MU
Investments in stage 1 :   1500 MU

The objective function is the present worth of all net benefits. The elements of this function are shown in row 35.

Table 12-5 summarizes the results obtained by using a standard library program for solving the linear programming problem. From this table it follows that the present demand in cell 1 of 30 Mm³/yr should continue to be supplied from groundwater. The ultimate supply to cell 2 is 69 Mm³/yr; part of it is supplied by temporary mining at a rate of 8 Mm³/yr. This temporary overdraft is replaced by import in the last stage. The import to cell 2 is developed only partly in stage 1 due to the budget constraint.

DERIVATION OF THE BASIC TRANSPORT EQUATION BY AVERAGING

A.1 MICROSCOPIC AND MACROSCOPIC SPACES

Consider the domain (U_0) of volume U_0 which is a Representative Elementary Volume of a porous medium comprised of several phases α. Each phase α occupies a volume $U_{0\alpha}$ within U_0; its volumetric fraction is

$$\theta_\alpha = U_{0\alpha}/U_0; \qquad \sum_{(\alpha)} \theta_\alpha = 1 \tag{A-1}$$

For any tensorial field $G_\alpha(\mathbf{x}', t)$, the spatial average $\bar{G}_\alpha(\mathbf{x}, t)$ of G_α at time t and point $\mathbf{x}$, over the domain $(U_{0\alpha}) \subset (U_0)$ centered at $\mathbf{x}$ is defined by

$$\bar{G}_\alpha(\mathbf{x}, t) = \frac{1}{U_{0\alpha}} \int_{(U_{0\alpha})} G_\alpha(\mathbf{x}', t; \mathbf{x}) \, dU_\alpha(\mathbf{x}') \tag{A-2}$$

The average $\bar{G}_\alpha$ is also called *intrinsic phase average* of G_α in $U_{0\alpha}$.
   The quantity

$$\mathring{G}_\alpha(\mathbf{x}', t; \mathbf{x}) = G_\alpha(\mathbf{x}', t; \mathbf{x}) - \bar{G}_\alpha(\mathbf{x}, t) \tag{A-3}$$

is the deviation of G_α at a point $\mathbf{x}'$ inside U_0 (centered at $\mathbf{x}$) from its average over $U_{0\alpha}$. Also

$$\bar{\mathring{G}}_\alpha = 0; \qquad (\mathring{\bar{G}}_\alpha) = \mathring{G}_\alpha \tag{A-4}$$

The space $\mathbf{x}$ is the *macroscopic space*, whereas $\mathbf{x}'$ is the *microscopic space*. From (A-1) and (A-2), it follows that

$$\theta_\alpha \bar{G}_\alpha(\mathbf{x}, t) = \frac{1}{U_0} \int_{(U_{0\alpha})} G_\alpha(\mathbf{x}', t; \mathbf{x}) \, dU_\alpha(\mathbf{x}') \tag{A-5}$$

is another average, called the *phase average* of G_α in $U_{0\alpha}$. It is the *macroscopic value* of G_α in the α phase.

As an example, let $G_\alpha \equiv V_\alpha$ be the microscopic velocity vector in the α phase. Then $\bar{V}_\alpha$ is the average velocity of the α phase

$$\bar{V}_\alpha(\mathbf{x}, t) = \frac{1}{U_{0\alpha}} \int_{(U_{0\alpha})} V_\alpha(\mathbf{x}', t; \mathbf{x}) \, dU_\alpha(\mathbf{x}') \tag{A-6}$$

and

$$\theta_\alpha \bar{V}_\alpha = \frac{1}{U_0} \int_{(U_{0\alpha})} V_\alpha(\mathbf{x}', t; \mathbf{x}) \, dU_\alpha(\mathbf{x}') \equiv \frac{1}{U_0} \int_{(U_0)} V_\alpha(\mathbf{x}', t; \mathbf{x}) \, dU(\mathbf{x}') \tag{A-7}$$

is the macroscopic velocity field of α (recall that $V_\alpha = 0$ outside $U_{0\alpha}$).

A-2 SPECIFIC DISCHARGE

Let $P(\mathbf{x})$ be a point in a multiphase porous medium domain through which flow takes place. Similar to the definition of a Representative Elementary Volume (Sec. 2-5), we may also define at every point $\mathbf{x}$ a Representative Elementary Area (REA) A_0, such that an average of a property over it will represent a meaningful value of that property at its centroid. The area $A_{0j} \equiv \sum_{(\alpha)} A_{0\alpha j}$ at P is facing the direction $\mathbf{1j}$.

In order to interpret the meaning of the product $\theta_\alpha \bar{V}_\alpha$ at point $P(\mathbf{x})$, let U_0 have the shape of a straight cylinder having a cross-sectional area A_{0j} facing the direction $\mathbf{1j}$ and length L_j. Let $V_{\alpha j}$ be the component in the direction $\mathbf{1j}$ of the local velocity at all points in the α phase, that is, $V_{\alpha j} = 0$ outside the α phase. Then, from (A-7)

$$\theta_\alpha \bar{V}_{\alpha j} = \frac{1}{U_0} \int_{(U_{0\alpha})} V_{\alpha j}(\mathbf{x}', t; \mathbf{x}) \, dU_\alpha(\mathbf{x}') = \frac{1}{L_j} \int_{(L_j)} \left[\frac{1}{A_{0j}} \int_{(A_{0\alpha j})} V_{\alpha j}(\mathbf{x}', t; \mathbf{x}) \, dA_{0\alpha j} \right] ds \tag{A-8}$$

Now, the expression in the square brackets in (A-8) is the specific discharge $q''_{0\alpha j}$, that is, the total discharge per unit area of A_{0j}

$$q''_{0\alpha j} = \frac{1}{A_{0j}} \int_{(A_{0\alpha j})} V_{\alpha j}(\mathbf{x}', t; \mathbf{x}) \, dA_{0\alpha j} \tag{A-9}$$

so that $\theta_\alpha \bar{V}_{\alpha j}$ is the average of $q''_{0\alpha j}$ over parallel cross-sectional areas of the REV (here from $\mathbf{x} - (L_j/2) \mathbf{1j}$ to $\mathbf{x} + (L_j/2) \mathbf{1j}$). For the sake of simplicity, we have assumed that the REV has a cylindrical shape, with $U_0 = L_j A_{0j}$.

We have thus shown that the macroscopic velocity $\theta_\alpha \bar{V}_\alpha$ defined by (A-7) has the meaning of the specific discharge vector $\mathbf{q}_\alpha$ of the α phase at that point. It is of interest to note that $\mathbf{q}_\alpha$ is a function of $(\mathbf{x}, t)$ only, independent of the direction chosen for $\mathbf{1j}$.

A-3 SOME AVERAGING RULES (BACHMAT, 1972; GRAY AND LEE, 1977)

(*a*) Let $\sigma_{0\alpha}$ denote the total surface area between the α phase and the other phases within U_0. We define the specific surface of the α phase by

$$\sigma'_\alpha = \sigma_{0\alpha}/U_0 \tag{A-10}$$

and the average of $G(\mathbf{x}', t)$ over $\sigma_{0\alpha}$ for a given $(U_{0\alpha})$ by

$$\langle G_\alpha \rangle (\mathbf{x}, t) = \frac{1}{\sigma_{0\alpha}} \int_{(\sigma_{0\alpha})} G_\alpha(\mathbf{x}', t; \mathbf{x}) \, dS_\alpha(\mathbf{x}') \tag{A-11}$$

Also

$$\langle G_\alpha \rangle \, \sigma'_\alpha = \frac{1}{U_0} \int_{(\sigma_{0\alpha})} G_\alpha(\mathbf{x}', t; \mathbf{x}) \, dS_\alpha(\mathbf{x}') \tag{A-12}$$

(*b*) Average of a product

$$\overline{G_1 G_2} = \bar{G}_1 \bar{G}_2 + \overline{\mathring{G}_1 \mathring{G}_2} \tag{A-13}$$

Also

$$(G_1^\circ G_2) = \mathring{G}_1 \bar{G}_2 + \bar{G}_1 \mathring{G}_2 + \mathring{G}_1 \mathring{G}_2 - \overline{\mathring{G}_1 \mathring{G}_2}$$

Hence

$$\overline{(G_1^\circ G_2) \, \mathring{G}_3} = \overline{\mathring{G}_1 \mathring{G}_3 \bar{G}_2} + \overline{\bar{G}_1 \mathring{G}_2 \mathring{G}_3} + \overline{\mathring{G}_1 \mathring{G}_2 \mathring{G}_3} \tag{A-14}$$

(*c*) Average of a time derivative

$$\overline{\left(\frac{\partial G_\alpha}{\partial t} \right)_\alpha} = \frac{1}{\theta_\alpha} \left[\frac{\partial}{\partial t} (\theta_\alpha \bar{G}_\alpha) - \langle G_\alpha u_n \rangle \, \sigma'_\alpha \right] \tag{A-15}$$

where u_n is the speed of displacement, along the outward normal, of a point on the interface between the phase α and the other phases within U_0. Cases of interest:

1. $G_\alpha \equiv 1$. Then $\bar{G}_\alpha = 1$ and

$$\partial \theta_\alpha / \partial t = \langle u_n \rangle \, \sigma'_\alpha$$

2. $u_n = 0$, that is, the configuration of the α phase is nondeformable. Then

$$(\overline{\partial G_\alpha / \partial t})_\alpha = \partial \bar{G}_\alpha / \partial t$$

(*d*) Average of a spatial derivative

$$\overline{\left(\frac{\partial G_\alpha}{\partial x_i} \right)_\alpha} = \frac{1}{\theta_\alpha} \left[\frac{\partial}{\partial x_i} (\theta_\alpha \bar{G}_\alpha) + \langle G_\alpha n_i \rangle \, \sigma'_\alpha \right] \tag{A-16}$$

where n_i is the i-component of a unit vector normal to σ_α and pointing outwards. In (A-16) we have to consider only the multiply connected portion of the α domain.

Case of special interest:

$G = V_\alpha^m$. Then

$$\overline{\partial V_\alpha^m / \partial x_i} = \frac{1}{\theta_\alpha} \left[\overline{\partial q_\alpha^m / \partial x_i} + \langle V_\alpha^m n_i \rangle \, \sigma_\alpha' \right]$$

A-4 THE MICROSCOPIC AND MACROSCOPIC BALANCE EQUATIONS FOR A SINGLE PHASE

Let e_α denote the specific value (that is, per unit mass) of an extensive property E in the α phase, and let ρ_α denote the mean density (mass per unit volume) of the α phase

$$e_\alpha = dE/dm_\alpha, \qquad \rho_\alpha e_\alpha = dE/dU_\alpha \tag{A-17}$$

The microscopic differential balance of E within the α phase is given by (Truesdell and Toupin, 1960)

$$\partial(\rho_\alpha e_\alpha)/\partial t = -\operatorname{div}\left[\rho_\alpha e_\alpha \mathbf{V}_\alpha + \mathbf{J}(e_\alpha)\right] + \rho_\alpha \Gamma(e_\alpha) \tag{A-18}$$

where $\mathbf{J}(e_\alpha) = \rho_\alpha e_\alpha [\mathbf{V}(E) - \mathbf{V}_\alpha]$ is the difference between the total flux of E and its part convected by the α phase, $\mathbf{V}_\alpha$ is the velocity of the α-phase and $\Gamma(e_\alpha)$ is the rate of production of E per unit mass of α.

In order to arrive at a macroscopic balance of E, we apply the averaging rules given above to (A-18), obtaining the general macroscopic balance equation of an extensive property of a phase in a porous medium (Bachmat and Bear, 1972)

$$\frac{\partial}{\partial t}(\theta_\alpha \overline{\rho_\alpha e_\alpha}) = -\operatorname{div}\left[\overline{\rho_\alpha e_\alpha}\mathbf{q}_\alpha + \theta_\alpha \overline{(\rho_\alpha^\circ e_\alpha)} \, \overset{\circ}{\mathbf{V}}_\alpha + \theta_\alpha \overline{\mathbf{J}(e_\alpha)}\right]$$

$$- \left[\langle J_n(e_\alpha)\rangle + \langle \rho_\alpha e_\alpha(V_{n\alpha} - u_{n\alpha})\rangle\right]\sigma_\alpha' + \theta_\alpha \overline{\rho_\alpha \Gamma(e_\alpha)} \tag{A-19}$$

It is of interest to note, by comparing the microscopic balance (A-18) with the macroscopic one (A-19), that the latter contains two additional terms introduced as a result of the averaging process: $\theta_\alpha \overline{(\rho_\alpha^\circ e_\alpha)} \, \overset{\circ}{\mathbf{V}}_\alpha$, which is the *macroscopic dispersive flux* of E in excess of average convection by the α phase, and $\left[\langle J_n(e_\alpha)\rangle + \langle \rho_\alpha e_\alpha(V_{n\alpha} - u_{n\alpha})\rangle\right]\sigma_\alpha'$, which describes the transfer of E across the interface between the α-phase and all other phases.

A-5 BALANCE EQUATION FOR THE VOLUME OF α PHASE

In this case, $E = U_\alpha$ = volume of α phase, $e_\alpha = 1/\rho_\alpha$ = specific volume of α phase; $\mathbf{J}(1/\rho_\alpha) = \mathbf{V}_\alpha - \mathbf{V}_\alpha = 0$. Since from (A-18) it follows that $\rho_\alpha \Gamma(1/\rho_\alpha) = \operatorname{div} \mathbf{V}_\alpha$, which is the equation of volumetric dilatation of the α phase, we obtain from (A-19)

$$\partial \theta_\alpha / \partial t = \operatorname{div} \mathbf{q}_\alpha - \langle V_{n\alpha} - u_{n\alpha} \rangle \, \sigma'_\alpha + \theta_\alpha \overline{(\operatorname{div} \mathbf{V}_\alpha)} \tag{A-20}$$

where $\mathbf{q}_\alpha = \theta_\alpha \overline{\mathbf{V}}_\alpha$ is the specific discharge of the α phase.

In incompressible flow $\operatorname{div} \mathbf{V}_\alpha = 0$. If also there exists no transfer of volume across the surface bounding that α phase (that is, the bounding surface is a material surface of α), the middle term on the right-hand side of (A-20) also vanishes and (A-20) reduces to

$$\partial \theta_\alpha / \partial t = -\operatorname{div} \mathbf{q}_\alpha \tag{A-21}$$

This equation is (6-34), used to describe unsaturated flow of water in porous media. In saturated flow of an incompressible fluid through a nondeformable porous medium, $\theta_\alpha = n = \text{const.}$ Hence

$$\operatorname{div} \mathbf{q} = 0 \tag{A-22}$$

A-6 MASS BALANCE OF A PHASE

In this case $E = m_\alpha$; $e_\alpha = 1$; $\Gamma(1_\alpha) = 0$ and $\mathbf{J}(1_\alpha) = \rho_\alpha [\mathbf{V}(m_\alpha) - \mathbf{V}_\alpha]$ is the diffusive flux of the mass of the α phase which can be expressed by Fick's law of diffusion. Equation (A-19) becomes

$$\frac{\partial(\theta_\alpha \bar\rho_\alpha)}{\partial t} = -\operatorname{div} \left[\bar\rho_\alpha \mathbf{q}_\alpha + \theta_\alpha \overline{\mathring{\rho}_\alpha \mathring{\mathbf{V}}_\alpha} + \theta_\alpha \overline{\mathbf{J}(1_\alpha)} \right] - \langle J_n(1_\alpha) + \rho_\alpha (V_{n\alpha} - u_{n\alpha}) \rangle \, \sigma'_\alpha \tag{A-23}$$

In saturated flow, we have $\theta_\alpha = n$ and $\rho_\alpha = p$. Assuming the effect of grad ρ (producing $\bar{\mathbf{J}}$) to be negligible, $\overline{\mathring{\rho}\mathring{\mathbf{V}}} \ll \bar\rho \overline{\mathbf{V}}$ and $V_n = u_n$, that is, no mass transfer across the solid-fluid interface, (A-23) reduces to

$$\frac{\partial(\bar\rho n)}{\partial t} = -\operatorname{div}(\bar\rho \mathbf{q}) \tag{A-24}$$

used for saturated flow of a compressible fluid in a deformable porous medium (see (5-13)).

A-7 MASS BALANCE FOR A SOLUTE γ IN THE α PHASE

In this case, the solution is the α phase, $c_\alpha = dm_{\alpha\gamma}/dU_\alpha$, $e_\alpha = c_\alpha/\rho_\alpha$ and (A-19) reduces to:

$$\partial(\theta_\alpha \bar c_\alpha)/\partial t = -\operatorname{div} \left[\underset{①}{\bar c_\alpha \mathbf{q}_\alpha} + \underset{②}{\theta_\alpha \overline{\mathring{c}_\alpha \mathring{\mathbf{V}}_\alpha}} + \underset{③}{\theta_\alpha \overline{\mathbf{J}(c_\alpha/\rho_\alpha)}} \right]$$

$$- \underset{④}{\langle J_n(c_\alpha/\rho_\alpha) + c_\alpha(V_{n\alpha} - u_n) \rangle \, \sigma'_\alpha} + \underset{⑤}{\theta_\alpha \overline{\rho_\alpha \Gamma(c_\alpha/\rho_\alpha)}} \tag{A-25}$$

which is the *generalized equation of hydrodynamic dispersion*, (7-26), for a solute in a multiphase system. In this equation, we identify the convective flux of the α phase ①, the dispersive flux of γ ②, which is an extra flux resulting from variations of V_α at the microscopic level, the flux of γ by molecular diffusion ③, transfer across phase boundaries, e.g., due to adsorption, desorption, precipitation, ion exchange, etc., ④ and production (or decay) of the considered solute ⑤.

A-8 MACROSCOPIC BOUNDARIES AND BOUNDARY CONDITIONS (BACHMAT AND BEAR, 1972)

The boundaries of a porous medium domain may be of two types

(*a*) A boundary across which an abrupt change ($=$ jump) takes place in the volumetric fraction θ_α of any of the α phases present. Examples are the boundary between two porous media of different porosities, an impervious boundary, and a boundary between a porous medium and an external space free of the solid matrix.

(*b*) An arbitrary (mathematical) surface within a porous medium.

Referring to the first type, in view of the averaging approach presented above, leading to macroscopic variables and parameters, no discontinuity in θ_α can exist. This is illustrated in Fig. A-1. The dashed line shows how the porosity (which is a macroscopic parameter) varies along x. However, an idealized boundary is introduced as an abrupt surface separating the two regions. Usually, when we speak of a boundary of a flow domain, we mean the idealized boundary in the sense of Fig. A-1.

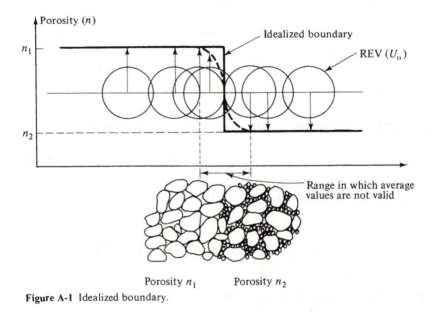

Figure A-1 Idealized boundary.

The idealized boundary is thus a hypothetical surface of discontinuity lying within a narrow region (width of U_0) in which the macroscopization is invalid. One should be careful when comparing observations taken close to such an idealized boundary with results of computations based on such boundaries.

The second type of boundary is already one in the macroscopic sense.

In general, both types may be moving and nonmaterial with respect to a considered phase. Here, a material surface is one on which the velocity of the phase ($V_{n\alpha}$) normal to the surface is equal to the (macroscopic) speed of displacement ($u_{n\alpha}$) of the surface.

Consider a finite domain U (Fig. A-2) fixed in space, composed of two parts, $U = U_1 + U_2$, separated from each other by a persistent (*macroscopic*) surface S^* which may be nonmaterial with respect to the α phase.

For the sake of generality, S^* may contain portions with a jump in θ_α defined by

$$[\theta_\alpha]_{1,2} \equiv \theta_\alpha \Big|_{\text{on } S^* \text{ on } U_1 \text{ side}} - \theta_\alpha \Big|_{\text{on } S^* \text{ on } U_2 \text{ side}} \tag{A-26}$$

The existence of the jump $[\theta_\alpha]_{1,2}$ across S^* implies that at least on some portion of S^*, the α phase on one side faces another phase (say, β) on the other side.

With the differential balance equation (A-19), the integral balance equation for E over each U_i, $i = 1, 2$, is

$$\int_{(U_i)} \frac{\partial}{\partial t} (\theta_\alpha \overline{\rho_\alpha e_\alpha}) \, dU + \int_{(S_i)} \overline{\rho_\alpha e_\alpha} q_{n_i} \, dS + \int_{(S^*)} \theta_\alpha \overline{\rho_\alpha e_\alpha} u_{n_i} \, dS =$$

$$- \int_{(S^*)} \{ \overline{\rho_\alpha e_\alpha} (q_{n_i} - \theta_\alpha u_{n_i}) + \theta_\alpha J^*_{n_i}(e_\alpha) \} \, dS - \int_{(S_i)} \theta_\alpha J^*_{n_i}(e_\alpha) \, dS + \int_{(U_i)} \tau^*(e_\alpha) \, dU$$

$$\tag{A-27}$$

where we have made use of the Gauss divergence theorem, and $\mathbf{J}^*(e_\alpha) = \overline{(\rho^\circ_\alpha e_\alpha) \mathring{\mathbf{V}}_\alpha} + \overline{\mathbf{J}(e_\alpha)}$ is the total flux in excess of the macroscopic convection by the α phase; $\tau^*(e_\alpha) = \theta_\alpha \overline{\rho_\alpha \Gamma(e_\alpha)} - \langle \rho_\alpha e_\alpha (V_{n\alpha} - u_{n\alpha}) + J_n(e_\alpha) \rangle \sigma'_\alpha$ is the total (macroscopic) rate of supply of E in the α phase within U_0; it includes transfer across the microscopic

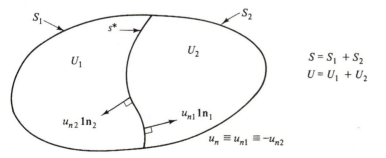

Figure A-2 Definition sketch for determining the generalized boundary conditions.

interface surface of the α phase in U_0. We now write (a) Eq. (A-27) separately for U_1 and U_2 and add the two equations and (b) a balance equation for U as a whole, taking into account that it contains a surface of singularity S^*. By comparing the results of (a) and (b), we obtain

$$\int_{(S^*)} [\overline{\rho_\alpha e_\alpha}(q_{n\alpha} - \theta_\alpha u_n) + \theta_\alpha J_n^*(e_\alpha)]_{1.2} = 0 \tag{A-28}$$

Hence

$$[\overline{\rho_\alpha e_\alpha}(q_{n\alpha} - \theta_\alpha u_n) + \theta_\alpha J_n^*(e_\alpha)]_{1.2} = 0 \tag{A-29}$$

is the general macroscopic boundary condition. It states that the total flux of E across any macroscopic boundary is continuous.

Bachmat and Bear (1972) consider the possibility of transfer of E (for example heat) across S^* from any phase to any other phase.

Raats (1972) considers the possibility that a source, Γ''', of E exists on the surface of discontinuity. We then have to write Γ''' instead of zero on the right-hand side of (A-29).

One should note that all values in (A-29) are *macroscopic* values.

AVERAGING ALONG THE VERTICAL

B-1

Let $F(x, y, z, t) = 0$ denote the equation of an upper or lower bound of an aquifer, and $b(x, y, t)$ denote the elevations, z, of points along the surface $F = 0$, that is, $z = b(x, y, t)$. Hence

$$F(x, y, z, t) \equiv z - b(x, y, t) = 0 \qquad \text{(B-1)}$$

As this boundary is displaced at a velocity $\mathbf{u}$, points along it satisfy the condition

$$\frac{dF}{dt} \equiv \frac{\partial F}{\partial t} + \mathbf{u} \cdot \nabla F = 0, \quad \text{or} \quad \mathbf{u} \cdot \nabla F = -\frac{\partial F}{\partial t} \qquad \text{(B-2)}$$

On any such boundary, continuity requires that

$$[\rho(\mathbf{q} - \theta\mathbf{u})]_{1,2} \cdot \mathbf{1n} = 0 \qquad \text{(B-3)}$$

where $[A]_{1,2} = A|_1 - A|_2$ denotes a jump in A across the boundary, θ is the volumetric fraction of water in the porous medium, and $\mathbf{1n}$ is a unit outward normal to the boundary. Since by definition $\mathbf{1n} = \nabla F / |\nabla F|$, (B-3) reduces to

$$[\rho(\mathbf{q} - \theta\mathbf{u})]_{1,2} \cdot \nabla F = 0, \qquad [\rho(\mathbf{q} - \theta\mathbf{u})]_{1,2} \cdot \nabla(z - b) = 0;$$

$$\rho(\mathbf{q} - \theta\mathbf{u})\Big|_{\text{aquifer}} \cdot \nabla(z - b) - \rho(\mathbf{q} - \theta\mathbf{u})\Big|_{\text{external}} \cdot \nabla(z - b) = 0 \qquad \text{(B-4)}$$

When combined with (B-2), this yields

$$\rho q\Big|_{\text{aquifer}} \cdot \nabla(z - b) - \rho q\Big|_{\text{external}} \cdot \nabla(z - b) = \left(\rho\theta\Big|_{\text{external}} - \rho\theta\Big|_{\text{aquifer}} \right)\frac{\partial F}{\partial t}$$

$$= \left(\rho\theta\Big|_{\text{aquifer}} - \rho\theta\Big|_{\text{external}} \right)\frac{\partial b}{\partial t} \quad \text{(B-5)}$$

For the different types of bounding surfaces, we obtain

(a) *For an impervious boundary,* $q_{\text{external}} \cdot \nabla F = 0$, *and* $\rho\theta|_{\text{external}} = 0$. Equation (B-5) reduces to

$$\rho q\Big|_{\text{aquifer}} \cdot \nabla(z - b) = \rho\theta\Big|_{\text{aquifer}} \frac{\partial b}{\partial t}$$

(b) *For a semipervious boundary,* $q|_{\text{external}} = q_{\text{leakage}}$.
(c) *For a phreatic surface with accretion,* $q|_{\text{external}} = N =$ *rate of accretion, and* $b \equiv h$. *Also,* $\theta|_{\text{aquifer}} = n$ *and* $\theta|_{\text{external}} = \theta_r =$ *specific retention.*

B-2

For an aquifer of thickness $B = b_2(x, y, t) - b_1(x, y, t)$ and any vector $A(x, y, z, t)$, we obtain by Leibnitz rule

$$\int_{b_1(x,y,t)}^{b_2(x,y,t)} \nabla \cdot A \, dz = \int_{b_1}^{b_2} \left(\nabla' \cdot A' + \frac{\partial A_z}{\partial z} \right) dz = \nabla' \cdot \int_{b_1}^{b_2} A' \, dz - A'\Big|_{b_2} \cdot \nabla b_2$$

$$+ A'\Big|_{b_1} \cdot \nabla b_1 + A_z\Big|_{b_2} - A_z\Big|_{b_1}$$

$$= \nabla' \cdot B\tilde{A}' + A\Big|_{b_2} \cdot \nabla(z - b_2)$$

$$- A\Big|_{b_1} \cdot \nabla(z - b_1) \quad \text{(B-6)}$$

where

$$\tilde{A}' = \frac{1}{B}\int_{b_1}^{b_2} A' \, dz; \quad A' = A_x \mathbf{1x} + A_y \mathbf{1y}; \quad \nabla' \cdot A' = (\partial A_x/\partial x)\,\mathbf{1x} + (\partial Ay/\partial y)\,\mathbf{1y}$$

For any scalar $\phi(x, y, z, t)$

$$\int_{b_1(x,y,t)}^{b_2(x,y,t)} \frac{\partial \phi}{\partial t} \, dz = \frac{\partial}{\partial t}\int_{b_1}^{b_2} \phi \, dz - \phi\Big|_{b_2}\frac{\partial b_2}{\partial t} + \phi\Big|_{b_1}\frac{\partial b_1}{\partial t}$$

$$= \frac{\partial}{\partial t} B\tilde{\phi} - \phi\Big|_{b_2}\frac{\partial b_2}{\partial t} + \phi\Big|_{b_1}\frac{\partial b_1}{\partial t} \quad \text{(B-7)}$$

where

$$\tilde{\phi} = \frac{1}{B}\int_{b_1}^{b_2} \phi \, dz.$$

Bear (1977) presents more details on the averaging along the vertical.

Chapter 4

4-1 Water at 10°C flows through a vertical sand (grain diameter = 0.5 mm) column (Fig. 4-1) of length $L = 120$ cm and cross-sectional area $A = 200$ cm². The difference between water levels at the inflow and the outflow reservoirs is 120 cm. The porosity of the sand is $n = 0.36$. The hydraulic conductivity $K = 20$ m/day.

Required: (a) Is Darcy's law applicable? (b) What is the total discharge Q? (c) What is the specific discharge q? (d) What is the average flow velocity in the column? (e) What is the hydraulic gradient along the column? (f) How will the answer to the above questions be changed if the water temperature changes from 10°C to 30°C.

4-2 Using a radioactive tracer, the average flow velocity at a point in an aquifer was found to be $V = 0.75$ m/day. The slope of the piezometric surface at that point is $J = 0.002$.

Required: Determine the hydraulic conductivity at that point for a porosity $n = 0.2$.

4-3 Given a vertical column of length $L = 100$ cm filled with homogeneous sand (Fig. 4-1). The constant water levels in the inflow and outflow reservoirs are at elevations h_1 and h_2, respectively, above some datum level.

Required: (a) For $h_1 = 115$ cm and $h_2 = 15$ cm, draw the piezometric head, ϕ, and pressure, p, distributions along the column. (b) Repeat (a) for $h_1 = 125$ cm, $h_2 = 15$ cm. (c) Repeat (a) for $h_1 = 115$ cm, $h_2 = 25$ cm. (d) Discuss the difference between cases (a), (b), and (c). (e) Will the ϕ and p distributions change if the hydraulic conductivity and/or the porosity of the sand were doubled?

4-4 The equation of state for a slightly compressible fluid is given by (1) $\rho = \rho_0 \exp[\beta(p - p_0)]$, or (2) $\rho = \rho_0[1 + \beta(p - p_0)]$, where ρ_0 is the density corresponding to p_0, and β is the constant coefficient of compressibility.

Required: Determine (a) Hubbert's potential, ϕ^*, (b) its gradient, $\nabla\phi^*$, and (c) $\partial\phi^*/\partial t$.

4-5 Given the following observations of the piezometric heads in three observation wells:

Well	A	B	C
Coordinate x	0	300 m	0
Coordinate y	0	0	200 m
Piezometric head (m)	+10 m	+11.5 m	+8.4 m

Assume that the wells penetrate a homogeneous, isotropic confined aquifer of constant thickness $B = 20$ m, porosity $n = 0.2$ and hydraulic conductivity $K = 15$ m/day, and that the piezometric surface between the wells can be approximated as a plane.

Required: Determine (a) the hydraulic gradient (magnitude and direction), (b) the total discharge in the aquifer, per unit width, and (c) the velocity of water at point P (100, 100).

4-6 Given the cross section of a confined aquifer shown in Fig. P-1. The piezometric heads at points A and B are:

Case (1): $\phi_A = +5.2$ m $\qquad \phi_B = +10.2$ m
Case (2): $\phi_A = +10.2$ m $\qquad \phi_B = +5.2$ m

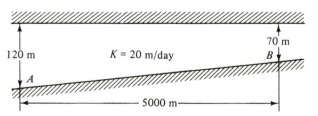

120 m

$K = 20$ m/day

70 m

B

A

5000 m

Figure P-1.

Assume that the flow in the aquifer is essentially horizontal (can you justify this assumption?).

Required: Determine (a) the rate of flow, Q', (b) the shape of the piezometric surface between A and B, (c) the piezometric head at point C, midway between A and B. (d) Repeat (a), (b), and (c) for an inhomogeneous aquifer with $K = 0.004x + 20$; K in meters per day, x in meters.

4-7 A fully penetrating well of radius $r_w = 0.3$ m pumps at a constant rate $Q_w = 600$ m³/h from a confined sandy aquifer: thickness $B = 50$ m, porosity $n = 0.25$.

Required: (a) Considering only macroscopic flow velocity, how long will it take for a pollutant to reach the well, if introduced into the aquifer at 1000 m from it? (b) Will your answer to (a) be affected by doubling the hydraulic conductivity of the aquifer? (c) Discuss your answer to (b).

4-8 The hydraulic conductivity of an inhomogeneous isotropic aquifer of constant thickness $B = 50$ m varies with distance and height according to the relationship $K = 100 + 0.004x - 1.5z$, where $0 \le x \le 5000$ m; z is measured from the bottom of the aquifer; x, z are in meters and K in meters per day.

Required: Determine the aquifer's transmissivity.

4-9 Let the aquifer of Prob. 4-5 be anisotropic, with $[K] = \begin{bmatrix} 20 & 6 \\ 6 & 10 \end{bmatrix}$; K in meters per day.

Determine: (a) The specific discharge $\mathbf{q}$, (b) the angle α between $\mathbf{q}$ and $\mathbf{J}$.

4-10 Let $K_x = 36$ m/day and $K_y = 9$ m/day be the principal values of K in an anistropic aquifer, in the x and y directions, respectively, in two-dimensional flow in the xy plane. Let the hydraulic gradient be 0.005 in a direction making an angle of 30 with the $+x$ axis.

Required: Determine $\mathbf{q}$.

4-11 Given the cross section of a confined aquifer with two observation wells as shown in Fig. P-2.

Required: (a) Determine Q' = (flow in aquifer per unit width). (b) Draw the piezometric surface. (c) Repeat (a) and (b) when K_2 changes to 5 m/day. (d) Discuss the results.

4-12 A layered soil is made up of N layers as shown in Fig. P-3; $N/2$ layers of thickness B_1 and hydraulic conductivity K_1 and $N/2$ layers of thickness B_2 and hydraulic conductivity K_2.

Required: (a) Determine the equivalent hydraulic conductivities K^H and K^V in the horizontal and vertical directions, respectively. (b) Show that $K^H > K^V$.

4-13 A well of radius r_w is located at the center of a confined circular aquifer of radius r_e and constant thickness B. The hydraulic conductivity, K, varies: $K = K_1$ for $r_w \le r < r_1$, and $K = K_2$ for $r_1 \le r \le r_e$.

Required: Determine the well's discharge Q_w if the piezometric heads ϕ_w and ϕ_e are maintained constant at r_w and r_e, respectively.

4-14 A horizontal stratified aquifer of constant thickness, $B = 60$ m, is made up of three layers: 15 m of $K = 30$ m/day, 25 m of $K = 10$ m/day and 20 m of $K = 40$ m/day. The corresponding porosities are 27 %, 20 % and 32 %. Two observation wells, 2000 m apart in the direction of the flow, show water levels at $+109$ m and $+105$ m, respectively, assumed the same for all layers.

Required: (a) Determine the rate flow through the aquifer. (b) If a pollution front is located in all three strata at some point, determine the position of this front three months later. (c) Discuss and compare the movements of the water and of the pollution.

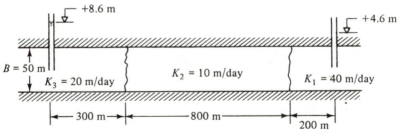

Figure P-2.

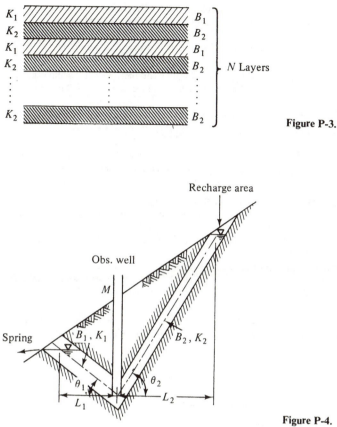

Figure P-3.

Figure P-4.

4-15 Water enters the confined aquifer shown in Fig. P-4, and leaves in the form of a spring.

Required: (a) Determine the piezometric head at a well located at M. (b) Specify the conditions for the well to become a flowing (artesian) one.

4-16 Piezometric heads are measured simultaneously in thirteen wells penetrating a confined aquifer of thickness $B = 50$ m, hydraulic conductivity $K = 20$ m/day and porosity $n = 0.27$ (see Table P4-16).

Table P4-16

Observation well	1	2	3	4	5	6	7	8	9	10	11	12	13
Coordinate x	4.3	16.5	7.0	3.0	11.0	22.0	8.0	3.2	18.1	13.5	4.0	8.7	19.5
Coordinate y	1.0	3.5	5.1	6.5	7.0	6.5	9.0	11.8	10.0	12.9	15.5	16.1	16.3
Head (m)	34.6	35.1	32.8	32.1	31.5	34.5	33.3	34.4	34.3	35.2	35.2	37.3	36.3

Each coordinate unit $= 200$ m

Required: (a) Draw equipotentials ($\Delta\phi = 1$ m) and streamlines. (b) Determine the specific discharge (magnitude and direction) at points A (10, 4) and B (16, 11). (c) Determine the total flow through the aquifer between observation wells No. 10 and No. 9. (d) What is the (average) time of travel for a pollutant introduced into the aquifer in the vicinity of well No. 12 to reach a pumping well in the vicinity of well No. 5?

4-17 Given a cross section of a leaky aquifer (Fig. P-5). The hydraulic conductivity of the semipermeable layer is $K' = 0.1$ m/day. The piezometric head in the main leaky confined aquifer is $+9$ m. The water table in the overlying leaky phreatic aquifer is at $+10$ m.

Required: Determine the rate and direction of leakage through the semipervious layer.

4-18 The piezometric head in an aquifer in the vicinity of point P (1, 1) is given by $\phi = 3x^2 + 2xy + 3y^2 + 7$.

Required: Determine the specific discharge (magnitude and direction) at P if $K = 30$ m/day.

4-19 Repeat Prob. 4-18, assuming that the aquifer is anisotropic with $T_x = 4000$ m²/day, $T_y = 1000$ m²/day (x, y are principal directions).

4-20 Given a cross section of two confined aquifers as shown in Fig. P-6.

Required: (a) Determine the flows Q_1 and Q_2 in the two aquifers. (b) Discuss the ratio Q_1/Q_2 and the factors affecting it.

4-21 A uniform flow in the horizontal $+y$ direction takes place in a homogeneous confined aquifer of constant thickness.

Required: (a) Determine the direction of the equipotentials if the aquifer's transmissivity is given by $T_\xi = 35$ m²/day, $T_\eta = 10$ m²/day; $+\xi$ makes an angle of 30° with the $+x$ axis, and ξ and η are principal directions. (b) Repeat (a) when the aquifer's transmissivity is given by $T_{\xi\xi} = 35$ m²/day, $T_{\xi\eta} = 10$ m²/day $T_{\eta\eta} = 5$ m²/day.

4-22 Two observation wells in a phreatic aquifer show water levels at $+75$ m and $+83$ m above mean sea level. The impervious horizontal bottom of the aquifer is at $+25$ m. The wells are 2000 m apart in the direction of the flow.

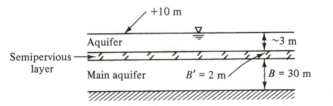

Figure P-5.

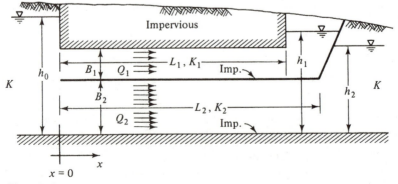

Figure P-6.

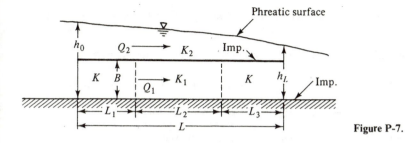

Figure P-7.

Required: (a) Determine the rate of flow per unit width of aquifer for a hydraulic conductivity of of 20 m/day. (b) Repeat (a) using the formula for a confined aquifer and taking the average thickness of flow to calculate the aquifer's transmissivity. (c) Compare and discuss the results.

4-23 Given the cross section shown in Fig. P-7 through a homogeneous phreatic aquifer (K_2) and a nonhomogeneous confined one (K, K_1, K). The two aquifers are separated by a thin impervious layer.

Required: (a) Determine Q_1 and Q_2 for $K = K_1 = K_2$. (b) Repeat (a) for $K = K_2 \neq K_1$. (c) Determine the ratio Q_1/Q_2 for cases (a) and (b) and discuss the results. (d) Draw the piezometric surface along L in each of the two aquifers for cases (a) and (b).

4-24 Repeat Prob. 4-23 for $K = K_1 \neq K_2$. Compare the result with (4-67) obtained in the absence of the impervious layer.

Chapter 5

5-1 A volume of water of 40×10^6 m^3 has been pumped from a phreatic aquifer through wells that are, more of less, uniformly distributed over the area, $A = 100$ km^2, of the aquifer. The aquifer's specific yield is $S = 20\%$.

Required: Determine the average drawdown of the water table over the area.

5-2 Figure P-8 shows two contour maps (scale 1:20000) of a portion $(EFGH)$ of a phreatic aquifer. The aquifer's storativity is $S = 0.1$.

Required: Determine the volume of excess total inflow into $ABCD$ (from all sources) over total outflow during the specified time interval.

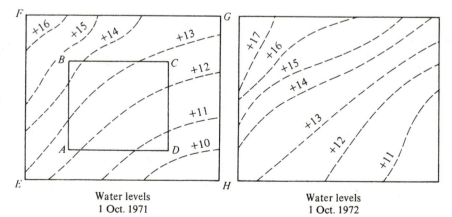

Water levels
1 Oct. 1971

Water levels
1 Oct. 1972

Figure P-8.

5-3 In order to prevent sea water intrusion into a confined coastal aquifer, the water level (= piezo-metric head) has to be raised by 5 m over an area of 2 km × 20 km along the coast. The aquifer's storativity is $S = 5 \times 10^{-5}$.

Required: Determine the volume of water which should be injected for this purpose in an array of wells located along this aquifer strip.

5-4 Discuss and compare the possibility of storage of water by artificial recharge in a confined aquifer and in a phreatic one.

5-5 A formula often used for estimating natural replenishment (N) from precipitation (P) is: $N = \alpha(P - P_0)$, where α is a coefficient and P_0 is another coefficient (interpreted as threshold precipitation).

Required: For $\alpha = 0.9$ and $P_0 = 200$ mm, what is the rise in the water table produced by a seasonal precipitation of 600 mm, if the specific yield of the aquifer is 25%?

5-6 The water table of a phreatic aquifer drops 4.5 m over an area of 1 km². The aquifer's porosity is 30%. Its specific specific retention is 0.10.

Required: (a) Determine the aquifer's specific yield. (b) What is the change in the volume of water in the aquifer?

5-7 Given the maps of Prob. 5-2.

Required: Prepare a groundwater balance for subdomain $ABCD$, and use it to estimate the net volume of water pumped from or recharged into the area between the specified dates. Assume $S_y = 10\%$; $T = 1000$ m²/day (assumed constant, although the elevations of the water table vary). Also assume that the change in water levels throughout the area is monotonous.

5-8 Rewrite the general mass conservation equation (5-13), assuming: (a) that the solid matrix is rigid and nondeformable, (b) that the fluid is homogeneous and incompressible, (c) that both (a) and (b) are valid, (d) that the flow is steady, and (e) that $\mathbf{q} \cdot \nabla \rho \ll n \partial \rho / \partial t$; $n =$ const.

5-9 Start from (5-23), assuming that the fluid is homogeneous and S_0 is a constant, and rewrite it for the following: (a) Inhomogeneous anisotropic porous medium; use x, y, z coordinates which are not principal directions of K. (b) Inhomogeneous anisotropic porous medium; use x, y, z coordinates which are principal directions of K. (c) Inhomogeneous but isotropic aquifer, using cylindrical co-ordinates r, θ, z. (d) Same as (c) but the medium is anisotropic with $K_x = K_y \neq K_z$. (e) Homogeneous isotropic porous medium using x, y, z coordinates. In each case, express $\mathbf{q}$ and its components by the appropriate motion equation.

5-10 Rewrite the Laplace equation (5-30) (a) in cylindrical coordinates, and (b) in spherical coordinates.

5-11 An aquifer domain in two-dimensional horizontal flow has the shape of an ellipse defined by $x^2/a^2 + y^2/b^2 = 1$; a, b are constants. The transmissivity is anisotropic; x and y are not principal directions.

Required: Express the boundary condition on the elliptical impervious boundary.

5-12 A segment of an equipotential boundary $\phi = 0$ in two-dimensional flow in the xy plane is de-scribed by the ellipse $x^2/a^2 + y^2/b^2 = 1$.

Required: Determine the direction of $\mathbf{q}$ at points along the boundary when (a) the medium is isotropic, (b) the medium is nonisotropic, with x, y not principal directions.

5-13 Determine the angle of refraction of a streamline upon passage through an interface between two isotropic media, K_1, K_2, when the angle of incidence is 30° for (a) $K_1 = 10K_2$, (b) $K_1 = 0.1K_2$.

5-14 In two-dimensional flow in the xy plane, the line $y = 0$ separates two zones of different trans-missivities; T_1 in $y > 0$ and T_2 in $y < 0$. Both zones are anisotropic, with x and y principal directions.

Required: Determine the law of refraction of streamlines and of equipotentials at the interface between the two zones.

5-15 Given the cross section of an inhomogeneous anisotropic aquifer strip between two streams shown in Fig. P-9. Assume that x and z are principal directions for both media.

Required: Give the complete mathematical statement of the exact problem of groundwater flow in the cross section. List all the assumptions underlying your statement.

5-16 Transform the flow domain of Prob. 5-15 into an equivalent one with two isotropic subdomains but still with $K_1 \neq K_2$, and state the flow problems for the transformed domains.

5-17 Show that for an aquifer with a thickness $B = B(t)$, the storativity S is given by $S = B(\alpha + \beta n) \rho g$.

5-18 Derive (5-69) by integrating (5-26) over the vertical.

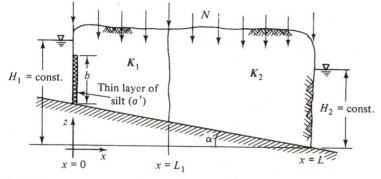

Figure P-9.

5-19 Use integration along the vertical to derive the continuity equation for flow in a leaky phreatic, homogeneous, isotropic aquifer.

5-20 Show that the right-hand side of (5-64) reduces to approximately $S_0 \beta \partial \tilde{\phi}/\partial t$ when both b_1 and b_2 are time dependent, but we assume vertical equipotentials.

5-21 Figure 5-40 represents a cross section of an aquifer between two streams.

Required: For $L = 3000$ m, $K = 20$ m/day, $h_0 = +30$ m, $h_L = +20$ m and $N = 500$ mm/yr, determine the rates of flow to the two streams and the shape of the phreatic surface. Justify and use the Dupuit assumption.

5-22 Determine the rates of flow to the two streams and the shape of the phreatic surface for the cross section of Prob. 5-21, except that the region $0 < x < 1000$ has the hydraulic conductivity of 10m/day, while in the remaining part the hydraulic conductivity is 30 m/day.

5-23 Determine the rates of flow to the two streams and the shape of the phreatic surface for the cross section of Prob. 5-21, except that $N = Ax + B$, $A = 10^{-4}$ yr^{-1}, $B = 0.3$ m/yr, x in meters.

5-24 In Prob. 5-21, add a horizontal gallery parallel to stream B at a distance of 200 m from it.

Required: Determine the minimum rate of pumping from the gallery required to produce induced recharge from stream B.

5-25 Figure P-10 shows an aquifer in the form of a rectangle bounded by an impervious rock outcrop and by a river in contact with the aquifer. Water levels in the river vary linearly, with elevations at A, B, C at $+6.5$ m, $+7.0$ m, and $+5.0$ m, respectively. The sandy aquifer is underlain by a thin layer of loam (semipervious) under which a very highly permeable layer of gravel is present, with (approximately) a constant head at $\phi_0 = +6.0$ m.

Required: (a) Draw (schematically) two cross sections showing the phreatic surface, equipotentials, and streamlines. (b) Present the problem of predicting water levels for a variable natural replenishment (no wells) in an exact mathematical form. (c) Repeat (b) assuming essentially horizontal flow in the aquifer. (d) With the assumption introduced in (c), add several wells scattered over the area and give a numerical scheme for solving the forecasting problem.

5-26 Figure P-11 shows a cross section between a lake and a channel parallel to the lake. At $t = 0$, the water level in the channel is instantaneously lowered to h_0 and thereafter maintained there.

Required: (a) Give the exact mathematical statement of the flow problem and describe how the flow through each of the aquifers and the shape of the (unsteady) phreatic surface can be determined. (b) Assume that a steady flow has been established, use the integrated aquifer equations to determine the shape of the phreatic surface and the rates of flow through the two aquifers. (c) What is the leakage in case (b)? (d) Give a numerical scheme for solving (b) for $h_0 = h_0(t)$.

5-27 Figure P-12 shows a map and a cross section of an aquifer system. Two well fields operate in the area; well field A pumping 10×10^6 m^3/yr and well field B pumping 16×10^6 m^3/yr. All screens in well field B are below the semipervious layer. A new well field, C, is planned, with an annual withdrawal of 15×10^6 m^3/yr and, if possible, up to 25×10^6 m^3/yr.

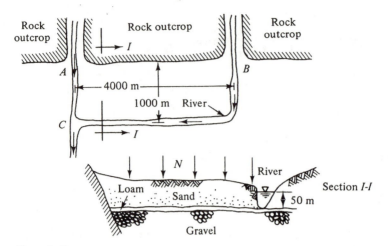

Figure P-10.

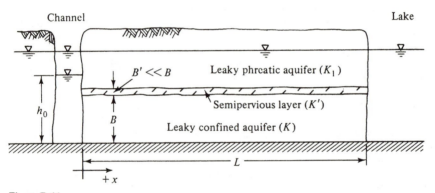

Figure P-11.

Required: How would you determine the feasibility of pumping at these rates from the planned well field C? The river itself cannot serve as a source of water because it is heavily polluted. Present equations and discuss methods of solution.

The figure gives all the known information. If you need more information—assume. List all the assumptions you make.

5-28 The solution of a certain steady, two-dimensional flow problem in an isotropic leaky aquifer is

$$Q = \frac{4KH}{\pi} \sum_{m=1,3,5\ldots} \frac{1}{m} \left\{ \coth\left[\frac{m\pi D}{2B} \left(1 + \frac{\sigma'K}{D} \right) \right] \right\}^{-1}$$

where H and D are lengths in the vertical direction, and B is a length in the horizontal, x, direction. The flow is in the vertical xz plane.

Required: What will the corresponding solution be for an anisotropic case with $K_x \neq K_z$ (x, z principal directions)?

5-29 An artificial recharge pond is located in a sandy area ($K = 10$ m/day, $n = 0.30$). The pond covers a large area. A flood wave fills the pond within a very short time up to a water level of 5.5 m above the

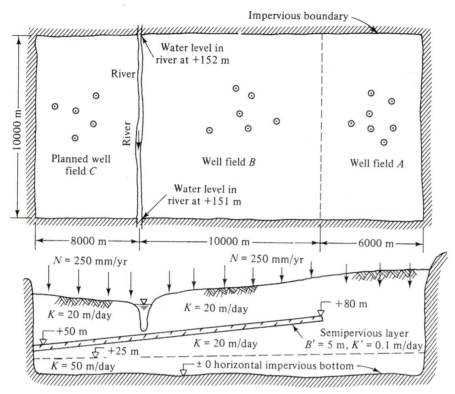

Figure P-12.

sandy bottom. By controlling the inflow, the water level is maintained at this elevation. During previous operations, the bottom of the pond was covered by a layer of fines (semipervious) of thickness 0.5 m, $K = 0.1$ m/day, $n = 0.5$.

Required: Assuming that the infiltration is only vertically downward, and under practically saturated flow conditions, show the rate of advance of the wetting front from the recharge pond into an assumed dry soil as a function of time. Assume a very deep water table. What will be the rate of infiltration after a very long time?

5-30 A recharge pond has the shape of a circle of diameter 100 m. The pond is fed such that a steady state is established. A groundwater mound is established under the pond. At a distance of approximately 500 m from the center of the pond, practically no rise in the water levels in the aquifer can be observed.

Required: Determine the shape of the mound (especially its peak) for a rate of infiltration of 1 m³/ day/m².

Chapter 6

6-1 Derive a relationship between the capillary pressure across the oil-water interface in the capillary tube and the height h shown in Fig. P-13.

6-2 Determine the height z above the phreatic surface from which a sample was taken, if a laboratory test of the sample gives $S_w = 0.35$. Using the apparatus shown in Fig. 6-6a, we obtain a retention curve on which we have the following point

$$S_w = 0.35; \qquad h_c = 108 \text{ cm}$$

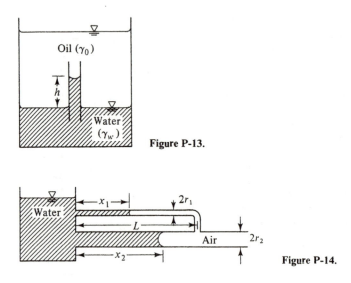

Figure P-13.

Figure P-14.

6-3 Determine the distance of the advancing front in each of the two horizontal capillary tubes shown in Fig. P-14, assuming air to be at atmospheric pressure. Under what conditions will air be entrapped in one of the tubes?

6-4 The following expression has been suggested for relative permeability

$$k_{rw} = \frac{\int_0^{S_w} dS_w/p_c^3(S_w)}{\int_0^1 dS_w/p_c^3(S_w)}$$

Use this expression, together with the data of one of the curves in Fig. 6-9b (drainage) to obtain a curve $k_{rw} = k_{rw}(S_w)$.

6-5 Write the continuity equation for one-dimensional horizontal unsaturated flow in terms of (a) S_w, (b) ψ, and (c) p_w. The soil is homogeneous, isotropic, and nondeformable.

6-6 The downward specific discharge is a constant 0.001 cm/sec, and the unsaturated hydraulic conductivity is given by $K = K_0 \exp(-a\psi)$, where $K_0 = 10^{-2}$ cm/sec, $a = 10^{-2} (\text{cm H}_2\text{O})^{-1}$.

   Required: Determine the distribution of tension, ψ, along the vertical, if the water table is at a depth of 1500 cm.

6-7 Use one of the curves in Fig. 6-9b ($n = 0.35$) together with Fig. 6-16 ($K_0 = 10^{-2}$ cm/sec), to determine the diffusivity $D(\theta_w)$.

6-8 Develop a single continuity equation which governs the flow in both the saturated and the unsaturated zones.

   Discuss how this equation can be applied to a flow domain composed of two such zones. What information on parameters is required for the solution of such a problem? List all the assumptions which serve as the basis for the development of the equation.

Chapter 7

7-1 Fresh water at relative concentration $c/c_0 = 0$ is introduced at $x = 0$, $t = 0$, into a semi-infinite sand column saturated with saline water at relative concentration $c/c_0 = 1.0$. As the saline water is being displaced, the following salinity measurements were made at some time t

Distance along column (cm)	48.2	49.7	51.5	53.6	55.4	57.2	59.3	61.3	63.2	65.4	68.2	73.4
c/c_0 (%)	0.8	2.2	3.5	9.5	21.7	50.3	78.0	94.5	94.5	98.7	98.7	100

The average flow velocity in the column is 1.6 cm/min.

Required: (a) At what time t were these measurements taken? (b) Determine the coefficient of dispersion D_h and the dispersivity a_L.

7-2 Estimate the width of the transition zone between contaminated water and noncontaminated aquifer water, after an initially sharp front between them has advanced (a) 100 m, (b) 500 m, and (c) 2 km, from its initial position. Use (7-117) and define the transition zone as the distance between the relative concentrations of $\varepsilon = 0.1$ and 0.9 (or: 0.15 and 0.85). Use $a_L = 5$ m.

7-3 Given an infinite homogeneous confined aquifer with a number of recharge and pumping wells. The recharge water has 200 ppm Cl⁻ more than the native water of the aquifer.

Required: Give the complete mathematical statement needed in order to determine the variation of salinity in the pumped water. List the parameters which appear in the statement and suggest a method of solution.

7-4 Pumping from a phreatic aquifer takes place at a certain distance from a point proposed as a solid waste disposal site (Fig. P-15).

Required: Discuss the investigations which should be undertaken by the planners in order to determine the danger of contamination of the pumped water.

7-5 (a) Write the equation of hydrodynamic dispersion for uniform flow ($V_x = V$, $V_y = V_z = 0$) in a three-dimensional flow domain. (b) Relate the components of the coefficient of dispersion to the components of the dispersivity and the velocity. (c) Rewrite the equation for the case of a radioactive tracer and adsorption, assuming a linear equilibrium adsorption isotherm as described by (7-28), with $m = 1$. (d) Define a retardation factor for (c) and explain how it affects the spreading of a pollutant.

7-6 A fully penetrating well is injecting tracer labelled water at a rate of 1500 m³/h into an infinite homogeneous confined aquifer of thickness $B = 50$ m. Steady uniform flow at a velocity of 8 m/day exists in the aquifer. The longitudinal and lateral dispersivities of the aquifer are 20 m and 3 m, respectively. Porosity is 0.2.

Required: (a) Determine the steady state areal distribution of tracer concentration in the aquifer, neglecting the velocity produced by the injection. (b) Give the complete mathematical statement of the problem without neglecting the velocity produced by the injection.

7-7 A well pumping at a constant rate $Q_w = 300$ m³/h is located a distance $L = 1000$ m downstream of a recharging well operating at the same rate. The aquifer is homogeneous isotropic and confined, with a constant thickness $B = 50$ m and porosity $n = 0.2$.

Required: (a) If both wells start operating at $t = 0$, when will the first trace of recharged water

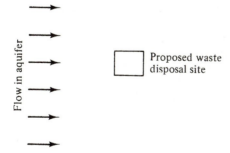

Proposed waste disposal site

⊙ Well

⊙ Well

Figure P-15.

reach the pumping well? (b) Repeat (a) if a uniform flow is present in the aquifer at a constant velocity $V_0 = 0.3$ m/day. The recharging well is directly upstream of the pumping one.

7-8 Let the recharging well of Prob. 7-7 be located downstream of the pumping one.

Required: (a) Will recharge water reach the pumping well? (b) What are the conditions for no recirculation between the two wells? (c) Indicate the area eventually occupied by recharged water and that from which water will eventually reach the pumping well.

7-9 A fully penetrating well injects a volume of 6×10^6 m^3 into a confined aquifer of thickness 35 m and porosity $n = 15\%$. Assume that a sharp front separates the injected water from the native water in the aquifer (but a distinguishable difference exists between the two types of water as the injected water has a concentration of 300 ppm Cl$^-$, while the native water has only 100 ppm Cl$^-$). At the end of the recharge period, a nearby pumping well starts to pump from the aquifer.

Required: Draw the Cl$^-$ concentration in the pumped water as a function of the pumped volume. The distance between the two wells is (a) 200 m, (b) 600 m. Neglect natural flow in the aquifer.

7-10 Repeat Prob. 7-9, but pumping starts twelve months after the termination of recharge. The hydraulic gradient of the uniform flow in the aquifer is 0.004 and the hydraulic conductivity is 80 m/day, $Q_P = Q_R = 3000$ m^3/day. (Use Fig. 7-18.)

Chapter 8

8-1 A fully penetrating well is pumping at a constant rate $Q_w = 400$ m^3/h from a homogeneous isotropic confined aquifer of constant thickness $B = 40$ m. Two observation wells, one at a distance $r_1 = 25$ m shows a steady water level at $+85.3$ m and the other at $r_2 = 75$ m shows a steady water level at $+89.6$ m.

Required: Determine the aquifer's transmissivity and hydraulic conductivity.

8-2 A well is pumping from an infinite leaky confined aquifer until steady flow is reached.

Required: Determine the drawdown at a distance of 50 m from the well (the initial water level in the upper phreatic aquifer remains unchanged), for $Q_w = 200$ m^3/h, $B = 50$ m, $B' = 3$ m, $K' = 0.1$ m/day, $K = 10$ m/day.

8-3 For the well of Prob. 8-2, what percentage of the well's discharge is derived from leakage within a radius of 30 m from the well?

8-4 A well is pumping at a constant rate $Q_w = 300$ m^3/h from a homogeneous phreatic aquifer with a horizontal bottom at $+17$ m. Two observation wells, at $r_1 = 50$ m and $r_2 = 100$ m, show water levels at $+57$ m and $+60$ m, respectively.

Required: Estimate the hydraulic conductivity of the aquifer.

8-5 A phreatic aquifer is made of two layers above an impervious bottom at $+30$ m. Between $+30$ m and $+45$ m, $K_1 = 20$ m/day, between $+45$ m and $+90$ m ($=$ ground surface), $K_2 = 40$ m/day. A well of radius 0.3 m pumps at a constant rate Q_w, producing a drawdown of 3 m at the well itself. No drawdown is observed at a distance of 300 m, where the initially horizontal water level at $+75$ m remains unchanged.

Required: Determine Q_w.

8-6 Develop the equation for $h(r)$ for steady flow to a well in a homogeneous phreatic aquifer with accretion at a constant rate N.

8-7 A well of a radius $r_w = 0.25$ m pumps at a constant rate $Q_w = 360$ m^3/h from a confined aquifer (transmissivity $= 2000$ m^2/day) at a distance of $L = 200$ m from a stream. A good hydraulic contact exists between the aquifer and the stream. Assume that steady flow has been established.

Required: (a) Draw the flow net in the vicinity of the well and up to the stream. (b) What is the drawdown at an observation well midway between the well and the stream? (c) What percentage of pumped water is coming from the stretch of stream between 100 m upstream of the well and 100 m downstream of it? (d) Determine the area within which drawdown exceeds 0.1 m.

8-8 Repeat Prob. 8-7, assuming that prior to the pumping, uniform flow from the aquifer to the stream took place, with the water level at the well at 0.5 m above the stream. Discuss the effect of changing Q_w on the shape of the flow net.

8-9 Repeat Prob. 8-8 for a phreatic aquifer, with $K = 20$ m/day, when initially the thickness of the saturated layer is 30 m below stream level.

8-10 A well pumps at a constant rate $Q_w = 240 \text{ m}^3/\text{h}$ from a homogeneous confined aquifer. The well's radius of influence is 300 m and it is located at a distance of 200 m from a straight impervious boundary.

Required: Map the drawdown, $s = s(x, y)$ for steady flow, in the vicinity of the well and up to the boundary. $T = 1000 \text{ m}^2/\text{day}$.

8-11 Given four pumping wells spaced 100 m apart along a straight line which is parallel to a stream (in contact with the aquifer) at a distance of 150 from it. Assume a radius of influence of 600 m, equal for all the wells. Assume $T = 2400 \text{ m}^2/\text{day}$ and $r_w = 0.25$ m.

Required: If we wish the drawdown to be 1 m, in all wells, determine the discharge rate in the wells.

8-12 Four 16 inch wells are located at the corners of a square 100 m $\times$ 100 m in a confined aquifer of constant transmissivity $T = 2000 \text{ m}^2/\text{day}$. Each well pumps at a constant rate of 120 m^3/h and steady flow has been established. The radius of influence of the wells is $R = 300$ m.

Required: (a) Determine the drawdown in each of the wells. (b) What should be the diameter of a single well located at the center of the square if the drawdown in it is the same as in each of the wells of (a), when its pumping rate is 480 m^3/h, and it has the same R?

8-13 Determine the relationship between pumping rate and drawdown in steady flow to a well located inside the corner between two perpendicular streams which fully penetrate a phreatic aquifer.

8-14 The drawdown caused by a pumping well in a confined aquifer is 0.5 m at an observation well located 50 m from the pumping well after two hours of pumping.

Required: When will the same drawdown occur at an observation well located at 150 m from the pumping well?

8-15 A well is pumping 250 m^3/h from a confined aquifer ($T = 1000 \text{ m}^2/\text{day}$, $S = 3 \times 10^{-5}$).

Required: Determine the drawdown at a point of observation located 200 m from the pumping well after (a) one hour, (b) one day, (c) seven days.

8-16 Let the well of Prob. 8-15 be shut off at the end of seven days of pumping.

Required: Determine the residual drawdown (a) one hour, (b) one day after shut-off.

8-17 Let the well of Prob. 8-15 be located at a distance of 100 m from a stream (hydraulically connected to the aquifer) and let the point of observation be located such that the line connecting the observation well and the pumping one is parallel to the stream.

Required: Calculate the drawdown as required in Prob. 8-15 and the residual drawdown as required in Prob. 8-16.

8-18 Given a leaky confined aquifer overlain by a leaky phreatic one. A single well penetrates the two aquifers which are homogeneous and of infinite areal extent. The well pumps at a constant rate Q_w.

Required: Write the complete mathematical statement of the problem, the solution of which will yield the drawdowns, $s_i(r, t)$, $i = 1, 2$, in each of the two aquifers. List all the assumptions underlying your equations.

8-19 Repeat Prob. 8-18 for the case of three aquifers, two leaky confined ones and an upper phreatic one, with a well screened only in the middle aquifer.

8-20 Two additional pumping wells are installed in the area of Prob. 4-16: one near observation well No. 5 and the other near observation well No. 7.

Required: Draw the piezometric surface at the end of a pumping season of six months, if the rates of pumping in the two wells were 600 m^3/h and 750 m^3/h, respectively. The aquifer's storativity is 7×10^{-5}.

8-21 Repeat Probs 8-15 through 8-17 for the leaky confined aquifer of Prob. 8-2.

8-22 Repeat Prob. 8-15, but assuming that the aquifer is anisotropic, with $T_x = 1000 \text{ m}^2/\text{day}$ and $T_y = 5000 \text{ m}^2/\text{day}$. Place the observation well (a) in the x direction, (b) in the y direction, (c) along the bisector between them. (d) What is the shape of lines of equal drawdown?

8-23 A well in a homogeneous confined aquifer pumps at a constant rate of 300 m^3/h for one week, stops for one week, and then starts again and continues one week at 200 m^3/h.

Required: What will the residual drawdown in the well ($r_w = 0.4$ m) be three days later if the aquifer's transmissivity and storativity are 5000 m^2/day and 4×10^{-5}, respectively.

8-24 A pumping well is located at a distance of 100 m from an impervious straight line boundary. The semi-infinite aquifer is homogeneous and isotropic ($T = 2000 \text{ m}^2/\text{day}$, $S = 3 \times 10^{-5}$). The well's constant pumping rate is $Q_w = 400 \text{ m}^3/\text{h}$; its radius is 0.3 m.

Required: Draw $s = s(t)$ at the well.

Chapter 9

9-1 Give the complete mathematical statement of the problem of determining the shape and position of the moving interface shown in Fig. 9-2. The time-dependent pumping rates and natural replenishment are given.

9-2 A point on a steady interface in a phreatic aquifer is located 30 m below sea level.
Required: Determine the elevation of the water table above this point.

9-3 Figure P-16 shows a cross section perpendicular to the coast of a coastal aquifer. The aquifer is overlain by impervious material above $+3.5$ m and extends 12 km from the coast. The piezometric surface at a distance of 10 km from the coast is at $+12$ m.
Required: (a) Determine the amount of fresh water flowing to the sea, assuming steady flow that is everywhere perpendicular to the coast. (b) Show the shape and position of the interface.

9-4 Repeat Prob. 9-3 for a phreatic aquifer, i.e., without the impervious layer, but with natural replenishment from above at a constant rate. The aquifer ends at 12 km.

9-5 Let an array of closely spaced shallow wells be installed 600 m from the coast and parallel to it.
Required: Determine the shape and position of the interface if the wells intercept 300,000 m³/yr/km. Use the data of Prob. 9-4.

9-6 Figure P-17 shows a cross section perpendicular to the coast through a coastal aquifer.
Required: Determine the shape and position of the interface, assuming steady flow. The pumping rate is 2×10^6 m³/yr/km.

9-7 Given a homogeneous isotropic phreatic coastal aquifer with a horizontal impervious base 120 m below sea level. The aquifer's hydraulic conductivity is $K = 15$ m/day. Uniform seaward fresh water flow takes place at a rate of 6.0 m³/m/day. No replenishment occurs close to the coast.
Required: (a) Determine the position of the toe of the interface in the xy plane. (b) When a single pumping well is located at 800 m from the coast, draw the positions of the interface toe as a function of the well's discharge rate, Q_w.

9-8 Develop the equation describing the shape of the fresh water lens under a circular island, as shown in Fig. 9-1c.

9-9 A well at an elevation $+60$ m pumps fresh water above an initially horizontal interface at an elevation of $+37$ m. The aquifer is homogeneous, isotropic, confined, and of infinite areal extent. The

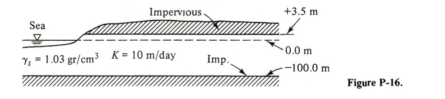

Figure P-16.

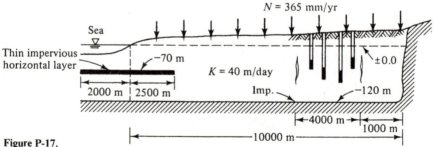

Figure P-17.

impervious top and bottom of the aquifer are at elevations $+72$ m and $+17$ m, respectively. Initially, the piezometric head in the stationary fresh water is at $+100$ m. The aquifer's porosity is 0.25 and its hydraulic conductivity is $K = 20$ m/day.

Required: Determine the approximate shapes of the upconing interface as a function of the well's discharge rate and of time.

Chapter 10

10-1 Let Fig. 10-1 represent a single cell aquifer. The total area is 100 km². The natural replenishment over the area is 400 mm/yr. The aquifer's hydraulic conductivity is $K = 40$ m/day. In the absence of pumping and artificial recharge, the entire natural replenishment is drained through a spring at elevation $+50$ m. We assume that the spring's discharge is related to the average elevation, $\bar{h}$, in the aquifer by $Q = \alpha K (\bar{h} - 50)$, where $\alpha = 140$ m. The aquifer's specific yield is 0.20.

Required: (a) Assume steady flow and determine $\bar{h}$. (b) Repeat (a) when pumping at a rate of 30×10^6 m³/yr and artificial recharge at a rate of 5×10^6 m³/yr (both assumed evenly distributed over the aquifer) take place. (c) Estimate the transition period from the initial to the new average water level in the aquifer.

10-2 Let the initial concentration of groundwater in the aquifer of Prob. 10-1 be 20 ppm Cl⁻ and assume that the new equilibrium with respect to water levels is established practically at $t = 0$. The chloride concentration of the natural replenishment is 8 ppm and that of the artificial recharge is 100 ppm. Assume that the aquifer's bottom is at $+10$ m.

Required: (a) Estimate the rate of rise or drop of chloride concentration in the pumped water, assuming that complete mixing takes place in the aquifer. (b) Repeat (a) assuming that, practically, changes in concentration take place only in the aquifer volume above $+25$ m.

10-3 Figure 10-3 shows a four-cell aquifer model. The width of the aquifer strip is 3000 m. The length of each cell is 5000 m. The natural replenishment of the aquifer strip is 400 mm/yr, 300 mm/yr, 300 mm/yr, and 200 mm/yr in Cells 1, 2, 3, and 4, respectively. The water level in the river is maintained at a constant elevation of 160 m above the horizontal impervious bottom. The hydraulic conductivity in Cells 1 and 2 is 30 m/day, while in Cells 3 and 4 it is 60 m/day.

Required: (a) Use this model to determine (obviously approximately) the elevations of water levels in the four cells, without pumping. (b) Repeat (a) when pumping takes place in Cells 2 and 3 at the rates of 4×10^6 m³/yr and 7×10^6 m³/yr, respectively.

10-4 Figure P-18 shows a four-cell aquifer model. The data for the model is summarized in Table P10-4.

Table P10-4

Pumping (10^6 m³/yr)		Nat. replenishment (mm/yr)	K (m/day)
Cell 1	17.8	200	40
Cell 2	0	420	40
Cell 3	6.3	300	30
Cell 4	2.7	300	30

Required: Determine the steady average water levels in the four cells.

10-5 For the aquifer of Prob. 10-4, let the initial chloride (Cl⁻) concentration in the aquifer be 100 ppm. The Cl⁻ concentrations of the natural replenishment and of the river water are 5 ppm and 150 ppm, respectively. Assume complete mixing in the cells, with an effective volume equal to 50% of the cell's volume.

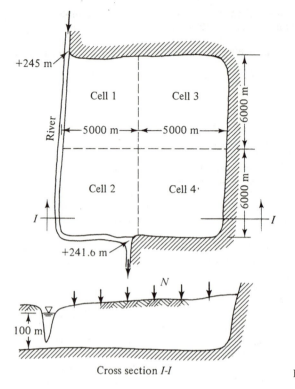

Cross section *I-I*   **Figure P-18.**

Required: (a) Assuming that a new equilibrium has been established, what is the concentration of the pumped water in the different cells? (b) Set up a simple scheme for determining annual changes in concentration during the transition period from the initial concentration distribution to the new one.

10-6 A multicell aquifer model is proposed for studying aquifer behavior. The aquifer's transmissivity is 500 m^2/day. It's storativity is (a) 0.2, (b) 2 × 10^{-5}

Required: Can you find cell dimensions and a time interval to obtain a stable solution in each case?

10-7 For Mercado's (1976) single-cell model, shown in Fig. 10-9a, write all the equations necessary in order to determine $c(t)$.

10-8 Expand the model considered in Prob. 10-7 to a multicell model to be employed for determining $c(x, y, t)$.

Chapter 11

11-1 The data in Table P11-1 were recorded during a pumping test in an aquifer. Pumping rate is $Q_w = 40$ m^3/h. The pumping well is taken as the origin of the coordinate system.

Required: Assuming that the aquifer is a confined one and that steady flow has been established, analyze the data to determine the aquifer's transmissivity and the well's radius of influence.

11-2 Determine the aquifer's transmissivity for the data of Prob. 11-1, when the well is located at a distance of $L = 50$ m from a river which is in contact with the aquifer.

11-3 Figure P-19 shows data that were recorded during the drilling of a well. The water table in the well was first encountered at 12 m below ground surface and stayed there until the completion of drilling, Accordingly, the well was equipped with a screen between 17 and 27 m below ground surface.

A pumping test was then performed in the well. The constant pumping rate was 11 m^3/h. Water

Table P11-1

Observation Well No.	Coordinates (m)		Drawdown (m)
	x	y	
1	10	12	0.85
2	− 200	230	0.20
3	− 72	− 70	0.55
4	10	45	0.72

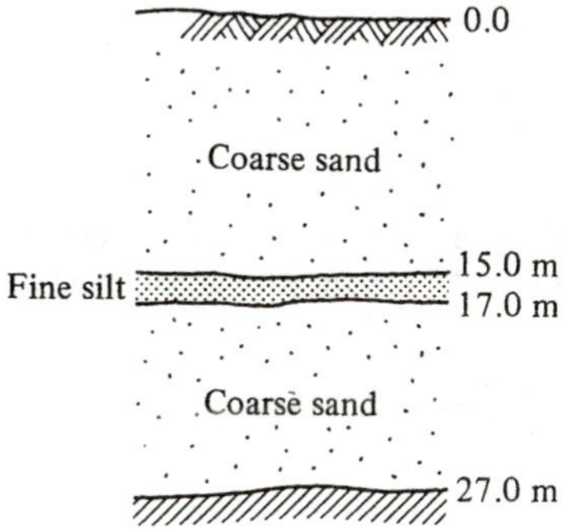

Figure P-19.

levels in four wells in the vicinity dropped during a short transition period and then remained steady. Table P11-3 summarizes the pumping test data.

Table P11-3

Well No.	1	2	3	4	5
Screen from (m)	17	18	20	18	14
to (m)	19	20	22	21	15
Distance from pumping well (m)	20	300	50	100	50
Steady drawdown (m)	0.65	0.05	0.40	0.23	0
Direction	N	N	W	E	S

Required: Analyze the pumping test to determine aquifer parameters.

11-4 A pumping test was performed in a well pumping from a sandy aquifer. Originally, the water level in the well was at + 20 m above mean sea level. The observed water levels during the test were

t (min)	0	3	4	5	6	7	8	10	12	15	20	25
Water level (m)	20.00	19.92	19.85	19.78	19.70	19.64	19.57	19.45	19.32	19.16	18.95	18.78

t (min)	30	35	40	45	50	60	70	80	90	100	125	150
Water level (m)	18.62	18.49	18.38	18.27	18.18	18.00	17.87	17.72	17.61	17.50	17.30	17.21

t (min)	200	300	400	500	600	700	800	1000	1500	2000	3000	4000
Water level (m)	16.83	16.41	16.15	15.92	15.75	15.60	15.45	15.25	14.92	14.58	14.15	13.80

The constant discharge during the test was $Q_w = 1000$ m^3/h. The observation well is at a distance of 1000 m from the pumping well.

Required: (a) Use the type curve method to determine aquifer parameters. (b) Use the straight line (Jacob's) method to determine aquifer parameters. (c) Plot residual drawdown as a function of time for the recovery period, if the well stopped pumping after 4000 min. (d) List *all* the assumptions underlying your analysis.

11-5 A well pumps at a constant rate of $Q_w = 100$ m^3/h for 100 minutes from a homogeneous, isotropic confined aquifer of infinite areal extent. Then pumping stops and the following residual water levels have been observed:

Time (min)	0	20	50	80	140
Residual drawdown (m)	2.38	1.01	0.62	0.48	0.30

Required: Determine the aquifer's transmissivity.

11-6 A pumping test was performed in a leaky confined homogeneous, isotropic aquifer of infinite areal extent. The constant pumping rate was 200 m^3/h. The following drawdowns were observed in an observation well located 500 m from the pumping one:

Time (h)	0.1	0.6	1.6	5	16	100
Drawdown (m)	0.04	0.50	0.90	1.35	1.66	1.71

Required: Analyze the test to determine aquifer parameters.

Chapter 12

12-1 For the four-cell aquifer model of Prob. 10-3 (Fig. 10-3), determine the optimal values of pumping P_2 in Cell 2 and P_3 in Cell 3, so as to maximize the benefits from operating the system. The total pumpage must at least supply the demand which amounts to 14×10^6 m^3/yr. Because the river water is polluted, no induced recharge from the river is permitted. Also, the water levels in Cells 2 and 3 may not be lowered below 150 m and 155 m, respectively, above the horizontal impervious base, as well screens in these cells are located just below these elevations. The benefits from pumping 10^6 m^3/yr from Cells 2 and 3 are 3 MU (Monetary Units) and 5 MU, respectively.

12-2 Repeat Prob. 12-1 if 3 MU and 5 MU represent costs and we wish to minimize the costs.

12-3 For the four-cell aquifer model of Prob. 10-4, determine the optimal *additional* pumping rates P_2 and P_3 in Cells 2 and 3, respectively. The following constraints have to be satisfied: (1) Total additional pumping rate should supply at least the increased demand of 6×10^6 m^3/yr in the area. (2) Drawdown in Cell 1 should not exceed 1 m. (3) Drawdown in Cell 3 should not exceed 2 m. The criterion for optimization is: (a) Minimize total costs, when costs of pumping in Cells 2 and 3 are 3 MU/10^6 m^3/yr and 5 MU/10^6 m^3/yr, respectively, or (b) minimize total costs, when pumping costs in Cells 2 and 3 are 5 MU/10^6 m^3/yr and 3 MU/10^6 m^3/yr. Repeat criteria (a) and (b) with the figures representing benefits which we wish to maximize. Use the method of influence functions in solving this problem.

12-4 For the four-cell model of Prob. 12-3 it is required to determine *additional* pumping and/or artificial recharge $P_1, P_2, P_3, P_4, R_1, R_2, R_3$, and R_4 in the four cells. The costs of pumping, C_{pi}, and of artificial recharge, C_{ri}, ($i = 1, 2, 3, 4$) are given. The criterion for optimization is minimum costs. The optimal operation should be such that (1) total pumping should supply at least the increased demands, D, (2) the maximum available annual total volume of recharge is R, (3) induced recharge from the (polluted) river should be eliminated.

Required: Give the complete statement of the optimization problem, using water balances as constraints.

BIBLIOGRAPHY

Abu-Zied, M. A. and V. H. Scott, Modified solutions for decreasing discharge wells, *Proc. Am. Soc. civ. Engrs,* **90**(HY6), 145–160, 1963.

Adamson, A. W., *Physical Chemistry of Surfaces,* 2nd edn, Interscience, New York, 1967.

Addison, L. E., D. R. Friedrichs, and K. L. Kipp, "Transmissivity iterative program, User's Manual," Report BNWL-1708-UC-70, Battelle Pacific Northwest Labs., Richland, Washington, 1972.

American Water Works Assoc. (AWWA), "Standards for deep wells," American Water Works Assoc. Inc., New York, revised edn., Publ. no. A100–58, 51 pp., 1958.

Aravin, V. I. and S. N. Numerov, *Theory of Motion of Liquids and Gases in Undeformable Porous Media* (in Russian), Gostekhizdat, Moscow, 616 pp. (English translation by A. Moscona, Israel Prog. for Scientific Trans., 511 pp., 1965), 1953.

Aris, R., *Vectors, Tensors and the Basic Equations of Fluid Mechanics,* Prentice Hall, Englewood Cliffs, N.J., 286 pp., 1962.

Aron, G. and V. H. Scott, Data analysis from stepwise throttled pump test, *Proc. Am. Soc. civ. Engrs,* **92**(HY6), 95–99, 1966.

Bachmat, Y., Spatial macroscopization of processes in heterogeneous systems, *Israel J. of Tech.,* **10**(5), 391–403, 1972.

Bachmat, Y. and J. Bear, The general equations of hydrodynamic dispersion, *J. geophys. Res.,* **69**(12), 2561–2567, 1964.

Bachmat, Y. and J. Bear, Mathematical formulation of transport phenomena in porous media, Proc. Int. Symp. of IAHR on the Fundamentals of Transport Phenomena in Porous Media, Guelph, Canada, pp. 174–193, 1972.

Badon-Ghyben, W., Nota in Verband met de Voorgenomen Putboring Nabij Amsterdam (Notes on the probable results of well drilling near Amsterdam), *Tijdschr. Kon. Inst. Ing.,* The Hague, 1888/9, pp. 8–22, 1888.

Barak, A. and J. Bear, "Flow of a newtonian fluid in an anisotropic porous medium at large Re" (in Hebrew), Technion–Israel Institute of Technology, Dept. of Civil Eng., 265 pp., 1973.

Bastian, W. C. and L. Lapidus, Longitudinal diffusion in ion exchange and chromatographic column, finite column, *J. phys. Chem.,* **60**, 816–817, 1956.

Baumann, P., Technical development in groundwater recharge, in *Advances in Hydroscience* (V. T. Chow, Ed.), **2**, 209–279, 1965.

Bear, J., "The transition zone between fresh and salt waters in coastal aquifers," Ph.D. dissertation, Univ. of California, Berkeley, California, 1960.

Bear, J., On the tensor form of dispersion, *J. geophys. Res.,* **66**(4), 1185–1197, 1961.

Bear, J., Some experiments on dispersion, *J. geophys. Res.,* **66**(8), 2455–2467, 1961a.

Bear, J., Hydrodynamic dispersion, in *Flow through Porous Media* (R. J. M. De Wiest, Ed.), Chap. 4, pp. 109–200, Academic Press, New York, 1969.

Bear, J., *Dynamics of Fluids in Porous Media*, American Elsevier, New York, 764 pp., 1972.

Bear, J., On the aquifer's integrated balance equations, *Advances in Wat. Resour.*, **1**(1), 15–23, 1977.

Bear, J. and Y. Bachmat, A generalized theory on hydrodynamic dispersion in porous media, I.A.S.H. Symposium on Artificial Recharge and Management of Aquifers, Haifa, Israel, I.A.S.H. Publ. No. 72, pp. 7–16, 1967 (Also as Hydraulics Lab. Report in Hebrew, 1966).

Bear, J. and G. Dagan, "The transition zone between fresh water and salt water in a coastal aquifer," Hydraulics Lab., Technion, Haifa, Israel, Prog. Rep. 1: "The steady interface between two immiscible fluids in a two-dimensional field of flow," 170 pp., 1962; Prog. Rep. 2: "A steady flow to an array of wells above the interface; approximate solution for a moving interface," 45 pp., 1963; Prog. Rep. 3: "The unsteady interface below a coastal collector," 122 pp., 1964a; Prog. Rep. 4: "Increasing the yield of a coastal collector by means of special operation techniques," 81 pp., 1966.

Bear, J. and G. Dagan, Some exact solutions of interface problems by means of the hodograph method, *J. Geophys. Res.*, **69**(2), 1563–1572, 1964.

Bear, J. and G. Dagan, Moving interface in coastal aquifers, *Proc. Am. Soc. civ. Engrs*, **99**(HY4), 193–215, 1964b.

Bear, J. and G. Dagan, The relationship between solutions of flow problems in isotropic and anisotropic soils, *J. Hydrol.*, **3**, 88–96, 1965.

Bear, J. and M. Jacobs, On the movement of water bodies injected into aquifer, *J. Hydrol.*, **3**(1), 37–57, 1965.

Bear, J. and Y. Schwartz, "Electric analog for regional groundwater studies," Water Planning for Israel, Ltd., Tel Aviv, Israel, P.N. 609, 1966.

Bear, J. and D. Zaslavsky, "Underground storage and mixing of water," Prog. Report 1, Technion–Israel Inst. of Technology, Hydraulics Lab. P.N. 2/62, 102 pp., 1962.

Bear, J., D. Zaslavsky, and S. Irmay, *Physical Principles of Water Percolation and Seepage*, UNESCO, Paris, 465 pp., 1968.

Bender, S. J., G. F. Pinder, and W. G. Gray, "A comparison of numerical approximations to the one-dimensional convective-diffusion equation," Internal Report, Water Resources Program, Dept. of Civil Eng., Princeton Univ., Princeton, N.J., 52 pp., 1975.

Ben Zvi, A. and Y. Goldstoff, "Calculation and updating of the natural replenishment of the Yarkon Taninim Basin" (in Hebrew), Water Planning for Israel, Tel Aviv, Israel, 14 pp., P.N.HR/72/056, 1972.

Bennett, G. D. and E. P. Patten Jr., "Constant head pumping test of a multiaquifer well to determine characteristics of individual aquifers," U.S. Geol. Survey, Water Supply Paper 1536-G, pp. 181–203, 1962.

Bentall, R., (Ed.), "Methods for determining permeability, transmissivity and drawdown," U.S. Geol. Survey, Water Supply Paper 1536-I, pp., 1963.

Bentall, R., (Ed.), "Shortcuts and special problems in aquifer tests," U.S. Geol. Survey, Water Supply Paper 1545-C, 1963a.

Bhuijan, S. E., E. A. Hiler, C. H. M. van Bavel, and A. R. Aston, Dynamic simulation of vertical infiltration into unsaturated soils, *Wat. Resour. Res.*, **7**(6), 1597–1606, 1971.

Bierschenk, W. H., "Determining well efficiency by multiple step-drawdown tests," Intern. Assoc. Scientific Hydrology, Publ. no. 64, Symposium at Berkeley, Calif., pp. 493–507, 1963.

Biot, M. A., General theory of three-dimensional consolidation, *J. appl. Phys.*, **12**, 155–164, 1941.

Biswas, A. K. (Ed.), *Systems Approach to Water Resources*, McGraw-Hill, New York, 429 pp., 1976.

Bize, J., L. Bourguet, and J. Lemoine, *L'alimentation artificielle des nappes souterraines*, Mason et Cie., Paris, 1972.

Blaney, H. F. and W. D. Criddle, "Determining water requirements in irrigation areas from climatological and irrigation data," USDA Soil Conserv. Service, Tech. Paper no. 96, 1950.

Boast, C. W., Modeling the movement of chemicals in soils by water, *Soil Sci.*, **115**(3), 224–230, 1973.

Boen, D. F., J. H. Bunts and R. J. Currie, "Study of reutilization of wastewater recycled through groundwater," Office of Research and Monotoring, Environmental Protection Agency, 1971.

Bondarev, E. A. and V. N. Nikolaevskij, Convective diffusion in porous media with the influence of absorption phenomena, The Acad. Sci. U.S.S.R., *Inl. Appl. Mech. & Tech. Phys.*, no. 5, pp. 128–134, 1962.

Boreli, M., Free surface flow toward partially penetrating wells, *Trans. Amer. Geophys. Un.*, **36**, 664–672, 1955.

Boulton, N. S., The flow pattern near a gravity well in a uniform water bearing medium, *J. Instn civ. Engrs* (London), **36**, 534, 1951.

Boulton, N. S., The drawdown of the water table under non-steady conditions near a pumped well in an unconfined formation, *Proc. Instn civ. Engrs* (London), **3**(3), 564–579, 1954.

Boulton, N. S., Unsteady radial flow to a pumped well allowing for delayed yield from storage, *Internat. Assoc. Hydrology*, Publ. 37., pp. 472–477, 1954a.

Boulton, N. S., Analysis of data from non-equilibrium pumping tests allowing for delayed yield from storage, *Proc. Instn civ. Engrs* (London), **26**, 469–482, (and further discussion pp. 603–610), 1963.

Boulton, N. S., The discharge to a well in an extensive unconfined aquifer with constant pumping, *J. Hydrol.* **3**(2), 124–130, 1965.

Braester, C., Moisture variation at the soil surface and the advance of the wetting front during infiltration at constant flux, *Wat. Resour. Res.*, **9**(3), 687–694, 1973.

Braester, C., G. Dagan, S. P. Neuman, and D. Zaslavsky, "A survey of the equations and solutions of unsaturated flow in porous media," Tech. Rept. Technion—Israel Inst. of Technology, Haifa, Israel, 176 pp., 1971.

Brebbia, C. A. and J. J. Connor, *Fundamentals of Finite Element Techniques*, Butterworth, London, 269 pp., 1974.

Bredehoeft, J. and B. Hanshaw, On the maintenance of anomalous pressures, I: Thick sedimentary sequences, *Bull. geol. Soc. Am.*, **79**, 1097–1106, 1968.

Bredehoeft, J. D. and G. F. Pinder, Digital analysis of areal flow in multiaquifer groundwater systems: A quasi three-dimensional model, *Wat. Resour. Res.*, **6**(3), 883–888, 1970.

Brooks, R. H. and A. T. Corey, "Hydraulic properties of porous media," Colorado State Univ., Hydrology Papers no. 3, Fort Collins, Colorado, 27 pp., 1964.

Brooks, R. H. and A. T. Corey, Properties of porous media affecting fluid flow, *Proc. Am. Soc. civ. Engrs*, **92**(IR2), 61–87, 1966.

Brown, R. H., Selected procedures for analyzing aquifer test data, *J. Am. Wat. Wks Ass.*, **45**(8), 844–866, 1953.

Bruch, J. C. Jr., Two-dimensional unsteady water flow in unsaturated porous media, Proc. Inter. Symp. on Finite Elements in Water Resources, 17 pp., Princeton Univ., Princeton, N.J., 1976.

Bruch, J. C. Jr., C. M. Lam, and T. H. Simundich, Parameter identification in field problems, *Wat. Resour. Res.*, **10**(1), 73–79, 1974.

Bruin, J. and H. Hudson, "Selected methods for pumping test analysis," Illinois State Water Survey, Report Invest. 25, 1955.

Brutsaert, W., The adaptability of an exact solution to horizontal infiltration, *Wat. Resour. Res.*, **4**(4), 785–789, 1968a.

Brutsaert, W., A solution of vertical infiltration into a dry porous medium, *Wat. Resour. Res.*, **4**(5), 1031–1038, 1968b.

Brutsaert, W. and M. Y. Corapcioglu, Pumping of aquifer with viscoelastic properties, *Proc. Am. Soc. civ. Engrs*, **102**(HY11), 1663–1675, 1976.

Brutsaert, W. and R. N. Weisman, Comparison of solutions of a nonlinear diffusion equation, *Wat. Resour. Res.*, **6**(2), 642–644, 1970.

Buras, N., *Scientific Allocation of Water Resources*, American Elsevier, New York, 208 pp., 1972.

Campbell, M. D. and J. H. Lehr, *Water Well Technology*, McGraw-Hill, New York, 681 pp., 1973.

Carman, P. C., Fluid flow through a granular bed, *Trans. Instn chem. Engrs* (London), **15**, 150–156, 1937.

Carslaw, H. S. and J. C. Jaeger, *Conduction of Heat in Solids*, 2nd edn, Oxford Univ. Press, London, 510 pp., 1959.

Caspi, I., "Linear programming as a method for parameter identification of an aquifer model" (in Hebrew), M.Sc. Thesis, Technion—Israel Inst. of Technology, Haifa, Israel, 166 pp., 1975.

Castany, G., *Prospection et Exploitation des Eaux Souterraines*, Dunod, Paris, 717 pp., 1966.

Charni, I. A., Rigorous proof of the Dupuit formula for unconfined flow during seepage, *Dokl. Akad. Nauk SSSR* (in Russian), Moscow, **79**(6), 1951.

Chertousov, M. D., *A Specialized Hydraulics Course* (in Russian), Gosenergouzdat, Moscow, 1949.

Chertousov, M. D., *Hydraulics* (in Russian), Gosenergouzdat, Moscow, 630 pp., 1962.

Childs, E. C., Soil moisture theory, in *Advances in Hydroscience* (Ven T. Chow, Ed.), **4**, 73–117, Academic Press, New York, 1967.

Childs, E. C., *An Introduction to the Physical Basis of Soil Water Phenomena*, Wiley, New York, 1969.

Childs, E. C. and N. Collis-George, The permeability of porous materials, *Proc. Roy. Soc. (London) A*, **201**, 392–405, 1950.

Chow, V. T., On the determination of transmissibility and storage coefficients from pumping test data, *Trans. Am. Geophys. Un.*, **33**, 397–404, 1952.

Chow, V. T., Sequential generation of hydrologic information, in *Handbook of Applied Hydrology* (V. T. Chow, Ed.), Chap. 8 pt. IV, pp. 8.91–8.97, McGraw-Hill, New York, 1964.

Coats, K. H., J. R. Dempsey, and H. H. Henderson, A new technique for determining reservoir description from field performance data, *Soc. Petrol Eng. J.*, **10**(1), 66, 1970.

Coats, K. H. and B. D. Smith, Dead-end pore volume and dispersion in porous media, *Soc. Petrol Eng. J.*, **4**, 73–84, 1964.

Collins, R. E., *Flow of Fluids Through Porous Materials*, Reinhold, New York, 270 pp., 1961.

Columbus, N., Viscous model study of sea water intrusion in water table aquifers, *Wat. Resour. Res.*, **1**(2), 313–323, 1965.

Conkling, H. *et al.*, "Ventura County Investigation," California Div. of *Water Resources Bull.*, **6**, 244 pp., 1934.

Cooley, R. L., A method of estimating parameters and assessing reliability for models of steady state groundwater flow, 1, Theory and numerical properties, *Wat. Resour. Res.*, **13**(2), 318–324, 1977.

Cooper, H. H. Jr., A hypothesis concerning the dynamic balance of fresh water and salt water in a coastal aquifer, *J. geophys. Res.*, **64**, 461–467, 1959.

Cooper, H. H. Jr., J. D. Bredehoeft, and I. S. Papadopulos, Response of a finite diameter well to an instantaneous charge of water, *Wat. Resour. Res.*, **3**(1), 263–269, 1967.

Cooper, H. H. Jr. and C. E. Jacob, A generalized graphical method for evaluating formation constants and summarizing well field history, *Trans. Am. Geophys. Un.*, **27**, 526–534, 1946.

Corapcioglu, M. Y. and W. Brutsaert, Viscoelastic aquifer model applied to subsidence due to pumping, *Wat. Resour. Res.*, **13**(3), 597–604, 1977.

Corey, A. T., Measurement of water and air permeability in unsaturated soils, *Proc. Soil Sci. Soc. Am.*, **21**, 7–10, 1957.

Crank, J., *Mathematics of Diffusion*, Oxford University Press, New York and London, 347 pp., 1956.

Crawford, N. H. and R. K. Linsley, "Digital simulation in hydrology, Stanford watershed model IV," Stanford Univ., Dept. of Civil Eng., Tech. Report no. 39, 210 pp., 1966.

DaCosta, J. A. and R. R. Bennett, "The pattern of flow in the vicinity of a recharging and discharging pair of wells in an aquifer having a real parallel flow," Int. Assoc. Sci. Hydrology, IUGG General Assembly of Helsinki, pp. 524–536; Publ. no. 52, 1960.

Dagan, G., Second order linearized theory of free surface flow in porous media, *La Houille Blanche*, **8**, 901–910, 1964.

Dagan, G., Second order correction of the formula of pumping test in unconfined aquifers, *J. Hydraul. Res.*, **5**(2), 127–133, 1967.

Dagan, G. and J. Bear, Solving the problem of interface upconing in a coastal aquifer by the method of small perturbations, *J. Hydraul. Res.*, **6**(1), 15–44, 1968.

Dantzig, G. B., *Linear Programming and Extensions*, Princeton Univ. Press, Princeton, N.J., 627 pp., 1963.

Darcy, H., *Les Fontaines Publiques de la Ville de Dijon*, Victor Dalmont, Paris, 647 pp., 1856.

Davis, S. N. and R. J. M. De Wiest, *Hydrogeology*, Wiley, New York, 463 pp., 1966.

DeGlee, G. J., *Over Grondwaterstromingen bij Wateronttrekking Door Middel van Putten* (in Dutch), J. Waltman, Delft, 175 pp., 1930.

Deininger, R. A., Water quality management in the planning of economically optimal pollution control systems, Proc. First Ann. Meeting, Amer. Water Resources Assoc., Univ. of Chicago, 1965.

Deininger, R. A., Linear programming for hydrologic analysis, *Wat. Resour. Res.*, **5**(5), 1105, 1969.

De Josselin de Jong, G., Longitudinal and transverse diffusion in granular deposits, *Trans. Am. Geophys. Un.*, **39**, 67–74, 1958.

De Josselin de Jong, G., Consolidatie in drie dimensies (vervolg; in Dutch), *L.G.M.-Mededelingen*, **7**, 57–73, 1963.

Desai, C. S., Finite element method for flow in porous media, Sec. 10 in *Finite Element Methods in Flow Problems* (J. T. Oden, O. C. Zienkiewicz, R. H. Gallagher, and C. Taylor, Eds.), Univ. of Alabama, at Huntsville, pp. 511–515, 1974.

De Wiest, R. J. M., On the theory of leaky aquifers, *J. geophys. Res.*, **66**(12), 4257–4262, 1961.

De Wiest, R. J. M., Flow to an eccentric well in a leaky circular aquifer with varied lateral replenishment, *Geofisica Pura e Aplicata*, **54**, 87–102, 1963.

Dracup, J., "The optimum use of a ground water and surface water system, A parametric linear programming approach," Univ. of California at Berkeley, Water Resources Center, Contr. no. 107, 134 pp., 1966.

Dupuit, J., *Etudes Theoriques et Pratiques sur le Mouvement des Eaux dans les Canaux de Couverts et a Travers les Terrains Permeables*, 2nd edn, Dunod, Paris, 304 pp., 1863.

Emich, V. N., On horizontal drains in layered soil, *PMTF*, **4**, 131–133 (in Russian), 1962.

Emsellem, Y. and G. de Marsily, "An automatic solution for the inverse problem," Ecole Nationale Superieure des Mines de Paris, 1969.

Emsellem, Y. and G. de Marsily, An automatic solution for the inverse problem, *Wat. Resour. Res.*, **7**(5), 1264–1283, 1971.

Ergatoudis, I., B. M. Irons, and O. C. Zienkiewicz, Curved, isoparametric "quadrilateral" elements for finite element analysis, *Int. J. solids Struct.*, **4**, 31–42, 1968.

Ergun, S., Fluid flow through packed columns, *Chem. Engng Prog.*, **48**, 89–94, 1952.

Eykhoff, P., Process parameter and state estimation, *Automatica*, **4**, 205–233, 1968.

Fair, G. M. and L. P. Hatch, Fundamental factors governing the streamline flow of water through sand, *J. Am. Wat. Wks Ass.*, **25**, 1551–1565, 1933.

Ferris, J. G., D. B. Knowles, R. H. Brown, and R. W. Stallman, "Theory of aquifer tests," U.S. Geol. Survey, Water Supply Paper 1536-E, pp. 69–174, 1962.

Fetter, C. W. and R. C. Holzmacher, Groundwater recharge with treated waste water, *J. Water Pollution Control Federation*, **46**(2), 260–270, 1974.

Fiering, M. B. and B. B. Jackson, "Synthetic streamflow," American Geophysical Union, Washington, D.C., 98 pp., 1971.

Fleming, G., *Computer Simulation Techniques in Hydrology*, American Elsevier, New York, Environmental Science Series, 333 pp., 1975.

Forchheimer, P., Wasserbewegung durch Bodem, *Z. Ver. Deutsch. Ing.*, **45**, 1782–1788, 1901.

Forsythe, G. E. and W. R. Wasow, *Finite Difference Methods for Partial Differential Equations*, Wiley, New York, 444 pp., 1960.

Freeze, R. A., Three-dimensional, transient saturated-unsaturated flow in a groundwater basin, *Wat. Resour. Res.*, **7**(2), 347–366, 1971.

Fried, J. J., *Groundwater Pollution*, Elsevier, Amsterdam, 330 pp., 1975.

Frind, E. O. and G. F. Pinder, Galerkin solution of the inverse problems for aquifer transmissivity, *Wat. Resour. Res.*, **9**(5), 1397–1410, 1973.

Futagami, T., N. Tamai, and M. Yatsuzuka, FEM coupled with LP for water pollution control, *Proc. Am. Soc. civ. Engrs*, **102**(HY7), 881–897, 1976.

Gambolati, G. and R. A. Freeze, Mathematical simulation of the subsidence of Venice, 1, Theory, *Wat. Resour. Res.*, **9**, 721–733, 1973.

Gambolati, G., P. Gatto, and R. A. Freeze, Mathematical simulation of the subsidence of Venice, 2, Results, *Wat. Resour. Res.*, **10**, 563–577, 1974.

Garabedian, P. R., *Partial Differential Equations*, John Wiley, New York, 672 pp., 1964.

Gardner, A. O. Jr., D. W. Peaceman, and A. L. Pozzi Jr., Numerical calculation of multidimensional miscible displacement by the method of characteristics, *Soc. Petrol. Eng. J.*, **4**, 26–36, 1964.

Gardner, W. R., Some steady state solutions of the unsaturated moisture flow equation with application to evaporation from a water table, *Soil Sci.*, **85**(4), 228–232, 1958.

Gardner, W. R. and M. S. Mayhugh, Solutions and tests on the diffusion equation for the movement of water in soil, *Proc. Soil Sci. Soc. Am.*, **22**, 197–201, 1958.

Gass, S. I., *Linear Programming*, McGraw-Hill, New York, 358 pp., 1969.

Gelfand, I. M. and S. V. Fomin, *Calculus of Variations* (translated from Russian by R. A. Silverman), Prentice Hall, Englewood Cliffs, N.J., 232 pp., 1963.

Gelhar, L. W., Stochastic analysis of flow in aquifers, AWRA Symp. on Advances in Groundwater Hydrology," Chicago, Ill., 15 pp., 1976.

Gershon, N. D. and A. Nir, Effects of boundary conditions of models on tracer distribution in flow through porous mediums, *Wat. Resour. Res.*, **5**(4), 830–839, 1969.

Glover, R. E., The pattern of fresh water flow in a coastal aquifer, *J. geophys. Res.*, **64**, 439–475, 1959.

Gray, W. G. and P. C. Y. Lee, On the theorems for local volume averaging, *Int. J. of Multiphase Flow*, **3**, 333, 1977.

Gray, W. G. and K. O'Neill, On the general equations for flow in porous media and their reduction to Darcy's law, *Wat. Resour. Res.*, **12**(2), 148–154, 1976.

Green, D. W., H. Dabiri, C. F. Weinaug, and R. Prill, Numerical modelling of unsaturated groundwater flow and comparison of the model to a field experiment, *Wat. Resour. Res.*, **6**, 862–875, 1970.

Grove, B. and W. A. Beetem, Porosity and dispersion calculation for fractured carbonate aquifer, using the 2-well tracer method, *Wat. Resour. Res.*, **7**(1), 128–134, 1971.

Gupta, S. K. and K. K. Tanki, A three-dimensional Galerkin finite element solution of flow through multiaquifers in Sutter Basin, California, *Wat. Resour. Res.*, **12**(2), 155–162, 1976.

Gupta, S. K., K. K. Tanki, and J. N. Luthin, "A three-dimensional finite element ground water model," California Water Resources Center, Univ. of Calif., Davis, California, Contr. no. 152, 119 pp., 1975.

Guymon, G. L., V. H. Scott, and L. R. Hermann, A general numerical solution of the two-dimensional diffusion-convection equation by finite element method, *Wat. Resour. Res.*, **6**(6), 1611–1617, 1970.

Hadley, G., *Linear Programming*, Addison-Wesley, Reading, Mass., 520 pp., 1962.

Haimes, Y. Y., R. L. Perrine, and D. A. Wismer, Identification of aquifer parameters by decomposition and multilevel optimization via decomposition, *Israel J. Tech.*, **6**(5), 322–329, 1968.

Halevy, E. and A. Nir, The determination of aquifer parameters with the aid of radioactive tracers, *J. geophys. Res.*, **67**(6), 2403–2409, 1962.

Hall, W. A. and J. A. Dracup, *Water Resources Systems Engineering*, McGraw-Hill, New York, 372 pp., 1970.

Hanks, R. J. and S. A. Bowers, Numerical solution of the moisture flow equation for infiltration into layered soils, *Proc. Soil Sci. Soc. Am.*, **26**, 530–534, 1962.

Hanks, R. J., A. Klute, and E. Bresler, A numeric method for estimating infiltration, redistribution, drainage and evaporation of water from soil, *Wat. Resour. Res.*, **5**, 1064–1069, 1969.

Hansen, V. E., "Evaluation of unconfined flow to multiple wells by membrane analogy," Thesis, Iowa State Univ., 1949.

Hantush, M. S., "Plain potential flow of ground water with linear leakage," Doctoral dissertation, Univ. of Utah, 1949.

Hantush, M. S., Analysis of data from pumping tests in leaky aquifers, *Trans. Am. Geophys. Un.*, **37**(6), 702–714, 1956.

Hantush, M. S., Nonsteady flow to a well partially penetrating an infinite leaky aquifer, *Proc. Iraq. Sci. Soc.* **1**, 10–19, 1957.

Hantush, M. S., Nonsteady flow to flowing wells in leaky aquifers, *J. geophys. Res.*, **64**(8), 1043–1052, 1959.

Hantush, M. S., Analysis of data from pumping wells near a river, *J. geophys. Res.*, **64**(11), 1921–1932, 1959a.

Hantush, M. S., Modification of the theory of leaky aquifers, *J. geophys. Res.*, **65**(11), 3713–3715, 1960.

Hantush, M. S., "Tables of functions," New Mexico Inst. of Mining and Technology, Socorro, New Mexico, Professional Paper 104, 1961.

Hantush, M. S., Drawdown around a partially penetrating well, *J. Am. Soc. civ. Engrs*, **87**(HY4), 83–98, 1961a.

Hantush, M. S., Aquifer tests on partially penetrating wells, *J. Am. Soc. civ. Engrs*, **87**(HY5), 171–195, 1961b.

Hantush, M. S., Hydraulics of wells, in *Advances in Hydroscience* (V. T. Chow, Ed.), Academic Press, New York, **1**, 281–442, 1964.

Hantush, M. S., Wells near streams with semipervious beds, *J. geophys. Res.*, **70**(12), 2829–2838, 1965.

Hantush, M. S., Analysis of data from pumping tests in anisotropic aquifers, *J. geophys. Res.*, **71**(2), 421–426, 1966.

Hantush, M. S., Flow to wells in aquifers separated by a semipervious layer, *J. geophys. Res.*, **72**(6), 1909–1920, 1967a.

Hantush, M. S., Flow of groundwater in relatively thick leaky aquifers, *Wat. Resour. Res.*, **3**(2), 583–590, 1967b.

Hantush, M. S. and C. E. Jacob, Plane potential flow of ground water with linear leakage, *Trans. Am. Geophys. Un.*, **35**(6), 917–936, 1954.

Hantush, M. S. and C. E. Jacob, Steady three-dimensional flow to a well in a two layered aquifer, *Trans. Am. Geophys. Un.*, **36**, 286–292, 1955a.

Hantush, M. S. and C. E. Jacob, Non-steady radial flow in an infinite leaky aquifer, *Trans. Am. Geophys. Un.*, **36**(1), 95–100, 1955b.

Hantush, M. S. and I. S. Papadopulos, Flow of ground water to collector wells, *Proc. Am. Soc. civ. Engrs*, **88**(HY5), 221–244, 1962.

Harpaz, Y., Artificial ground water recharge by means of wells in Israel, *Proc. Am. Soc. civ. Engrs*, **97**(HY12), 1947–1964, 1971.

Harpaz, Y. and J. Bear, Investigations on mixing of waters in underground storage operations, *Intern. Assoc. Scientific Hydrology*, **64**, 132–153, 1963.

Harr, M. E., *Foundations of Theoretical Soil Mechanics*, McGraw-Hill, New York, 381 pp., 1966.

Harrill, J. R., "Determining transmissivity from water level recovery of a step-drawdown test," U.S. Geol. Survey, Prof. Paper 700-C, 212–213, 1970.

Hefez, E., "The use of digital and hybrid computers for solving ground water problems," D.Sc. Thesis, Technion—Israel Inst. of Technology, Haifa, Israel (Hebrew with English summary), 184 pp., 1972.

Hefez, E., U. Shamir, and J. Bear, "The use of linear and quadratic programming for identifying the parameters of a cell model of an aquifer" (in Hebrew), Technion—Israel Inst. of Technology, Haifa, Israel, Water Resources Lab., P.N. 30/73, 86 pp., 1973.

Hefez, E., U. Shamir, and J. Bear, Identifying the parameters of an aquifer cell model, *Wat. Resour. Res.*, **11**(6), 993–1004, 1975.

Hefez, E., U. Shamir, and J. Bear, Forecasting water levels in aquifers by numerical and semihybrid methods, *Wat. Resour. Res.*, **11**(6), 988–992, 1975a.

Helm, D., One dimensional simulation of aquifer system compaction near Pixley, California, 1. Constant parameters, *Wat. Resour. Res.*, **11**, 465–478, 1975.

Henry, H. R., Salt intrusion into freshwater aquifers, *J. geophys. Res.*, **64**(11), 1911–1919, 1959.

Herzberg, A., Die Wasserversorgung einiger Nordseebaden (The water supply on parts of the North Sea coast in Germany), *Z. Gasbeleucht. Wasserversorg.*, **44**, 815–819; 824–844, 1901.

Hildebrand, F. B., *Advanced Calculus for Applications*, Prentice Hall, Englewood Cliffs, N.J., 646 pp., 1962.

Hoopes, J. A. and D. R. F. Harleman, Waste water recharge and dispersion in porous media, *Proc. Am. Soc. civ. Engrs*, **93**(HY5), 51–72, 1967.

Hoopes, J. A. and D. R. F. Harleman, Dispersion in radial flow from a recharge well, *J. geophys. Res.*, **72**(14), 3595–3607, 1967a.

Hougen, O. A. and W. R. Marshall, Adsorption from fluid stream flowing through a stationary granular bed, *Chem. Engn Prog.*, **43**(4), 197, 1947.

Hubbert, M. K., The theory of ground water motion, *J. Geol.*, **48**, 785–944, 1940.

Hubbert, M. K., Darcy law and the field equations of flow of underground fluids, *Trans. Am. Inst. Min. Metal. Eng.*, **207**, 222–239, 1956.

Huges, J. L. and S. G. Robson, Effects of waste percolation on ground water in alluvium near Barslow, California, Proc. Intern. Symp. on Underground Waste Management and Artificial Recharge, New Orleans, Sept. 1973 (also *Amer. Assoc. Petrol. Geol.*, **1**, 91–129, 1973).

Huisman, L., Artificial replenishment, Internation. Water Supply Assoc., 6th Congress, Stockholm, pp. J1–18, 1964.

Huisman, L., *Groundwater Recovery*, MacMillan, London, 336 pp., 1972.

Huisman, L., "Artificial groundwater recharge," Delft Univ. of Technology, Dept of Civil Eng., Delft, 1975.

Huyakorn, P. S. and C. R. Dudgeon, "Groundwater and well hydraulics," The Univ. of New South Wales, Water Research Lab. Report no. 121, 148 pp., 1972.

Hydrocomp International, Inc., "Hydrocomp simulation programming operations manual," Palo Alto, California, 1968.

Hydrologisch Colloquium, "Steady flow of ground water towards wells," Committee for Hydrological Research, T.N.O., Publ. no. 10, 179 pp., The Hague, 1964.

International Association of Scientific Hydrology, "International survey of existing water recharge facilities," Publ. no. 87, 1970.

Irmay, S., On the hydraulic conductivity of unsaturated soils, *Trans. Am. Geophys. Un.*, **35**, 463–468, 1954.

Irmay, S., Solutions of the non-linear diffusion equation with a gravity term in hydrology, I.A.S.H. Symposium on Water in the Unsaturated Zone, Wageningen, 1966.

Jacob, C. E., On the flow of water in an elastic artesian aquifer, *Trans. Am. Geophys. Un.*, pt. 2, pp. 574–586, 1940.

Jacob, C. E., "Notes on determining permeability by pumping tests under water table conditions," U.S. Geol. Survey, Mimeo. Rept., 1944.

Jacob, C. E., "Partial penetration of pumping wells, adjustments for," U.S. Geol. Survey, Water Resources Bull., 12 pp., 1945.

Jacob, C. E., Radial flow in a leaky artesian aquifer, *Trans. Am. Geophys. Un.*, **27**(2), 198–208, 1946.

Jacob, C. E., Drawdown test to determine effective radius of artesian well, *Proc. Am. Soc. civ. Engrs*, **72**(5), (also *Trans. Am. Soc. civil Eng.*, **112**, 1047–1070, 1947), 1946a.

Jacob, C. E., Flow of groundwater, in *Engineering Hydraulics* (H. Rouse, Ed.), Chap. 5, pp. 321–386, Wiley, New York, 1950.

Jacob, C. E., "Determining the permeability of water table aquifers," U.S. Geol. Survey Water Supply Paper 1536-I, pp. 245–271, 1963.

Jacob, C. E. and S. W. Lohman, Nonsteady flow to a well of constant drawdown in an extensive aquifer, *Trans. Am. Geophys. Un.*, **33**(4), 559–569, 1952.

Jacobs, M. and S. Schmorak, "Sea water intrusion and interface determination along the coastal plane of Israel," State of Israel, Hydrological Service, Hydrological Paper no. 6, 12 pp., 1960.

Jahnke, E. and Emde, F., *Tables of Functions*, 4th edn, Dover Publ., New York, 304 pp., 1945.

Javandel, I. and P. A. Witherspoon, Application of the finite element method to transient flow in porous media, *Soc. Petrol. Eng. J.*, **8**(3), 241–252, 1968.

Johnson, E. E. Inc., *Ground Water and Wells*, Edward E. Johnson, Saint Paul, Minn., 440 pp., 1966.

Kapuler, Y. and J. Bear, "A numerical solution for the movement of the interface in a multilayered coastal aquifer," (in Hebrew), Water Planning for Israel, Ltd., Tel Aviv, Israel, P.N. 01/74/70, 169 pp., 1975.

Karanjac, J., Well losses due to reduced formation permeability, *Ground Water*, **10**(4), 42–46, 1972.

Kipps, K. L. Jr., Unsteady flow to a partially penetrating well, finite radius well in an unconfined aquifer, *Wat. Resour. Res.*, **9**(2), 448–462, 1973.

Kirkham, D., Exact theory of flow into a partially penetrating well, *J. geophys. Res.*, **64**(9), 1317–1327, 1959.

Kirkham, D., Exact theory for the shape of the free water surface about a well in a semi-confined aquifer, *J. geophys. Res.*, **69**(12), 2537–2549, 1964.

Kirkham, D. and W. L. Powers, *Advanced Soil Physics*, Wiley-Interscience, New York, 534 pp., 1972.

Kisiel, C. C. and L. Duckstein, Ground-water models, in *System Approach to Water Management* (A. K. Biswas, Ed.), Chap. 4, pp. 80–155, McGraw-Hill, New York, 1976.

Kleinecke, D., Use of linear programming for estimating geohydrological parameters of ground water basin, *Wat. Resour. Res.*, **7**(2), 367–374, 1971.

Klute, A., A numerical method for solving the flow equation for water in unsaturated materials, *Soil Sci.*, **73**, 105–116, 1952.

Knowles, T. R., B. J. Claborn, and D. M. Wells, "A computerized procedure to determine aquifer characteristics," Rep. WRC-72-5, Water Resources Center, Texas Tech. Univ., Lubbock, Texas, 1972.

Kohout, F. A., Cyclic flow of salt water in the Biscayne aquifer of southeastern Florida, *J. geophys. Res.*, **65**(7), 2133–2141, 1960.

Korganoff, A., Sur la résolution de problèmes "Inverse" en hydrogéologie, *Bull. Int. Assoc. Scientific Hydrology*, vol. XV, pp. 67–78, 1970.

Kozeny, J., Theorie und Berechnung der Brunnen, *Wasserkraft und Wassenwirtschaft*, **28**, 88–92, 101–105, 113–116, 1933.

Kriz, G. J., V. H. Scott, and R. H. Burgy, Analysis of parameters of an unconfined aquifer, *Proc. Am. Soc. civ. Engrs*, **92**(HY5), 49–56, 1966.

Kutilek, M., Non-Darcian flow of water in soils (laminar region), 1st IAHR Symp. Fundamentals of Transport Phenomena in Porous Media, Haifa, Israel (Publ. by Elsevier, 1972), 327–340, 1969.

Labadie, J. W., "Decomposition of a large scale nonconvex parameter identification problem in geohydrology," Report no. ORC 72/23, Operations Research Center, University of California, Berkeley, Calif., 72 pp., 1972.

Labadie, J. W., A surrogate-parameter approach to modelling groundwater basins, *Water Resources Bull.*, **11**(1), 97–114, 1975.

Labadie, J. W. and O. J. Helweg, Step-drawdown test analysis by computer, *Ground Water*, **13**(5), 438–449, 1975.

Laliberte. G. E.. A. T. Corey. and R. H. Brooks, "Properties of unsaturated porous media," Colorado State Univ., Hydrology Paper, no. 17, 40 pp., 1966.

Lambe, T. W. and R. V. Whitman, *Soil Mechanics*, Wiley, New York, 553 pp., 1969.

Lasdon, L. S., *Optimization Theory for Large Systems*, Macmillan, New York, 1970.

Lau, L. K., W. J. Kaufman, and D. K. Todd, "Dispersion of a water tracer in radial laminar flow through homogeneous porous media," Hydraulic Lab., Univ. of California, Berkeley, Calif., 1959.

Laverty, F. B. and H. A. van der Goot, Development of a fresh-water barrier in southern California for the prevention of sea water intrusion, *J. Am. Wat. Wks Ass.*, **47**, 886–908, 1955.

Lembke, K. E., Groundwater flow and the theory of water collectors (in Russian), *The Engineer*, J. of the Ministry of Communications, no. 2, 1886 and nos. 17–19, 1887.

Lennox, D. H., Analysis and application of step-drawdown test, *Proc. Am. Soc. civ. Engrs*, **92**(HY6), 25–48, 1966.

Lin, A. C. and W. W. G. Yeh, Identification of parameters in an inhomogeneous aquifer by use of the maximum principle of optimal control and quasi-linearization, *Wat. Resour. Res.*, **10**(4), 829–838, 1974.

Lohman, S. W., "Method for determining the coefficient of storage from straight line plots without extrapolation," U.S. Geol. Survey, Ground Water Note 33, 1957.

Lovell, R. E., "Collective adjustment of the parameters of the mathematical model of a large aquifer," Technical Report no. 4, The Univ. of Arizona, Tuscon, Ariz., 87 pp., 1971.

Lovell, R. E., L. Duckstein, and C. C. Kisiel, Use of subjective information in estimation of aquifer parameters, *Wat. Resour. Res.*, **8**(3), 680–690, 1972.

Lynn, W. R., J. A. Logan, and A. Charnes, System analysis for planning wastewater treatment plants, *J. Water Pollution Control Federation*, **34**(6), 565–581, 1962.

Lyons, T. C., Groundwater basin water quality simulation to study alternative management plans, WRC Conference on Groundwater Quality, Measurement, Prediction and Protection, 10 pp., Reading, England, 1976.

Maas, A. M., M. M. Hufschmidt, R. Dorfman, H. A. Thomas, and G. M. Fair, *Design of Water Resource Systems*, Harvard University Press, Cambridge, Mass., 620 pp., 1962.

Maddock, T. III, Algebraic technological function from a simulation model, *Wat. Resour. Res.*, **8**(1), 129–134, 1972.

Maddock, T. III, Nonlinear technological functions for aquifers whose transmissivities vary with drawdown, *Wat. Resour. Res.*, **10**(4), 877–881, 1974.

Marino, M. A. and W. W. G. Yeh, Nonsteady flow in a recharge well–unconfined aquifer system, *J. Hydrol.*, **16**, 159–176, 1972.

Marle, C., P. Simandoux, J. Pacsirszky, and C. Gaulier, Etude du déplacement de fluides miscibles en milieu poreux stratifié, *Rev. Inst. Fr. Pétrole*, **22**(2), 272–294, 1967.

Marvin, J., M. Droracek, and S. H. Peterson, Artificial recharge in water resources management, *Proc. Am. Soc. civ. Engrs*, **97**(IR2), 219–232, 1971.

Matalas, N. C. and J. R. Wallis, Generation of synthetic sequences, in *Systems Approach to Water Management* (A. K. Biswas, Ed.), Chap. 3, pp. 54–79, McGraw-Hill, New York, 1976.

Mavis, F. T. and T. P. Tsui, "Percolation and capillary movement of water through sand prisms," Bull. 18, Univ. of Iowa, Studies in Eng., Iowa City, 1939.

McCracken, D. D. and W. S. Dorn, *Numerical Methods and Fortran Programming with Application to Engineering and Science*, Wiley, New York, 457 pp., 1964.

Meinzer, O. E. (Ed.), *Hydrology*, Dover, New York, 712 pp., 1942.

Mercado, A., Nitrate and chloride pollution of aquifers—a regional study with the aid of a single cell model, *Wat. Resour. Res.*, **12**(4), 731–747, 1976.

Mercado, A. and J. Bear, "Mixing of labeled water by injecting and pumping in the same well" (in Hebrew), Water Planning for Israel, Ltd., P.N. 511, 1965.

Mercado, A. and E. Halevy, Determining the average porosity and permeability of a stratified aquifer with the aid of radio-active tracers, *Wat. Resour. Res.*, **2**(2), 525–531, 1966.

Mero, F., Application of the ground water depletion curves in the analysis and forecasting of spring discharges influenced by wells, Symp. Intern. Assoc. for Scientific Hydrology, IUGG. Publ. no. 63, pp. 107–117, Berkeley, California, 1963.

Mero, F., An approach to hydrometeorological water balance computations for surface and ground water basins, Seminar on Integrated Surveys for River Basin Development, Delft, The Netherlands, October, 1969.

Mitchell, A. R., *Computational Methods in Partial Differential Equations*, Wiley, New York, 1971.

Morel-Seytoux, H. J., Two-phase flows in porous media, in *Advances in Hydroscience* (V. T. Chow, Ed.), **9**, 119–202, Academic Press, New York, 1973.

Morse, P. M. and H. Feshbach, *Methods of Theoretical Physics*, McGraw-Hill, New York, 1978 pp., 1953.

Mualem, Y. and J. Bear, The shape of the interface in steady flow in a stratified aquifer, *Wat. Resour. Res.*, **10**(6), 1207–1215, 1974.

Muskat, M., Potential distribution in large cylindrical discs (homogeneous sands) with partially penetrating electrodes (partially penetrating wells), *Physics*, **2**(5), 329–384, 1932.

Muskat, M., *The Flow of Homogeneous Fluids through Porous Media*, McGraw-Hill, New York, 763 pp., 1937.

Nahrgang, G., "Zur Theorie des Volkommenen und Unvolkommenen Brunnen," Springer, Berlin, 54 pp., 1954.

Narasimhan, T. N., "A unified numerical model for saturated-unsaturated ground water flow," Ph.D. dissertation, Dept. of Civil Eng., Univ. of California, Berkeley, California, 1975.

Nelson, R. W., "Steady Darcian transport of fluids in heterogeneous, partially saturated porous media, I., Mathematical and numerical formulation," General Electric, Hanford Atomic Products Operations, HW-72335, Pt. 1, 33 pp., 1962.

Nelson, R. W., In place determination of permeability distribution for heterogeneous porous media, through analysis of energy dissipation, *Soc. Petrol. Eng. J.*, **8**(1), 33–42, 1968.

Nelson, R. W., "Evaluating the environmental consequences of groundwater contamination," Management Summary and Technical Papers, BCS Richland, Inc., 129 pp., 1977.

Neuman, S. P., Theory of flow in unconfined aquifers considering delayed response of the water table, *Wat. Resour. Res.*, **8**(4), 1031–1045, 1972.

Neuman, S. P., "Finite element computer programs for flow in saturated-unsaturated porous media," Hydraulics Lab., Technion—Israel Inst. of Technology, Report on Project no. A10-SWC-77, 87 pp., 1972.

Neuman, S. P., Saturated-unsaturated seepage by finite elements, *Proc. Am. Soc. civ. Engrs*, **99**(HY12), 2233–2250, 1973.

Neuman, S. P., Calibration of distributed parameter groundwater flow models viewed as a multiple objective decision process under uncertainty, *Wat. Resour. Res.*, **9**(4), 1006–1021, 1973a.

Neuman, S. P., Galerkin approach to unsaturated flow in soils, in *Finite Elements in Flow Problems* (J. T. Oden, O. C. Zienkiewicz, R. H. Gallagher, and C. Taylor, Eds.), UAH Press, Huntsville, Alabama, pp. 517–522, 1974.

Neuman, S. P., Galerkin method for analyzing nonsteady flow in saturated-unsaturated porous media, in *Finite Elements in Fluids* (Gallagher *et al.*, Eds), **1**, Chap. 10, Wiley, London, 1975.

Neuman, S. P. and P. A. Witherspoon, Theory of flow in confined two aquifer system, *Wat. Resour. Res.*, **5**(4), 803–816, 1969.

Neuman, S. P. and P. A. Witherspoon, Applicability of current theories of flow in leaky aquifers, *Wat. Resour. Res.*, **5**(4), 817–829, 1969a.

Nir, A., "On the determination of aquifer characteristics," sec. IV B.2 in "Guidebook on Nuclear Techniques in Hydrology," Tech. Rep. Series. no. 91, International Atomic Energy Agency, Vienna, pp. 128–138, 1968.

Noblanc, A. and H. J. Morel-Seytoux, Perturbation analysis of two-phase infiltration, *Proc. Am. Soc. civ. Engrs*, **98**(HY9), 1527–1541, 1972.

Nutting, P. G., Physical analysis of oil sands, *Bull. Am. Ass. petrol. Geol.*, **14**, 1337–1349, 1930.

Ogata, A., "Dispersion in porous media," Ph.D. dissertation, Northwestern Univ., Ill., 1958.

Ogata, A. and R. B. Banks, "A solution of the differential equation of longitudinal dispersion in porous media," U.S. Geol. Survey, Prof. Paper no. 411-A, 1961.

Papadopulos, I. S. and H. H. Cooper, Drawdown in a well of large diameter, *Wat. Resour. Res.*, 3(1), 241–244, 1967.

Parlange, J. Y. and D. K. Babu, A comparison of techniques for solving the diffusion equation with an exponential diffusivity, *Wat. Resour. Res.*, 12(6), 1317–1318, 1976.

Parlange, J. Y. and D. K. Babu, On solving the nonlinear diffusion equation—a comparison of perturbation, iterative and optimal techniques for an arbitrary diffusivity, *Wat. Resour. Res.*, 13(1), 213–214, 1977.

Peaceman, D. W. and H. H. Rachford Jr., The numerical solution of parabolic and elliptic differential equations, *Soc. Indust. appl. Math. J.*, 3(1), 28–41, 1955.

Penman, H. L., Natural evaporation from open water, bare soil and grass, *Proc. Roy. Soc. (London) A*, **193**, 120–145, 1948.

Pfankuch, H. O., Contribution a l'etude des deplacement de fluides miscible dans un milieu poreux, *Rev. Inst. Fr. Petrol*, 18(2), 215–270, 1963.

Philip, J. R., Numerical solution of equations of the diffusion type with diffusivity—concentration dependent, *Trans. Faraday Soc.*, **51**, 885–892, 1955.

Philip, J. R., Numerical solutions of equations of the diffusion type with diffusivity—concentration dependent II, *Australian J. Phys.*, **10**, 29–42, 1957a.

Philip, J. R., The theory of infiltration 1: The infiltration equation and its solution, *Soil Sci.*, **83**, 345–357, 1957b.

Philip, J. R., The theory of infiltration and its solution 2: The profile at infinity, *Soil Sci.*, **83**, 435–448, 1957c.

Philip, J. R., The theory of infiltration and its solution 3: Moisture profile and relation to experiments, *Soil Sci.*, **84**, 163–178, 1957d.

Philip, J. R., Theory of infiltration 4: Sorptivity and algebraic infiltration equations, *Soil Sci.*, **84**, 257–264, 1957e.

Philip, J. R., The theory of infiltration 5: The influence of the initial moisture content, *Soil Sci.*, **84**, 329–339, 1957f.

Philip, J. R., The physical principles of soil water movement during the irrigation cycle, *3rd Cong. Int. Comm. Irrig. Drainage*, **8**, 125–154, 1957g.

Philip, J. R., The theory of infiltration 6: Effect of water depth over soil, *Soil Sci.*, **85**, 278–286, 1958a.

Philip, J. R., The theory of infiltration 7, *Soil Sci.*, **85**, 333–337, 1958b.

Philip, J. R., General method of exact solution of the concentration dependent diffusion equation, *Australian J. Phys.*, **13**, 1–12, 1960a.

Philip, J. R., A very general class of exact solutions in concentration dependent diffusion, *Nature*, **185**, 233, 1960b.

Philip, J. R., Theory of infiltration, in *Advances in Hydroscience* (V. T. Chow, Ed.), **5**, 215–296, Academic Press, New York, 1969.

Philip, J. R., Flow through porous media, *Ann. Rev. Fluid Mechan.*, **2**, 177–204, 1970.

Pinder, G. F., "A digital model for aquifer evaluation," U.S. Geol. Survey, Techniques in Water Resources Investigation, Book 7, Automatic Data Processing and Computations, 18 pp., 1970.

Pinder, P. G., A Galerkin finite element simulation of groundwater contamination in Long Island, New York, *Wat. Resour. Res.*, 9(6), 1657–1669, 1973.

Pinder, G. F., "Galerkin finite element models for aquifer simulation," Water Resources Program, Dept. of Civil Eng., Princeton Univ., Princeton, N.J., 1974.

Pinder, G. F. and J. D. Bredehoeft, Application of the digital computer for aquifer evaluation, *Wat. Resour. Res.*, 4(5), 1069–1093, 1968.

Pinder, G. F. and E. O. Frind, Application of Galerkin's procedure to aquifer analysis, *Wat. Resour. Res.*, 8(1), 108–120, 1972.

Pinder, G. F. and W. G. Gray, Is there a difference in the finite difference method, *Wat. Resour. Res.*, 12(1), 105–107, 1976.

Pinder, G. F. and W. G. Gray, *Finite Element Simulation in Surface and Subsurface Hydrology*, Academic Press, New York, 295 pp., 1977.

Pinder, G. F. and H. Page, Finite element simulation of salt water intrusion on the South Fork of Long

Island, Proc. Intern. Conf. on Finite Elements, Princeton Univ., Princeton, N.J., 19 pp., 1976.

Polubarinova-Kochina, P. Ya., Theory of filtration of liquids in porous media, in *Advanc. Appl. Mech.* (R. V. Mises and Th. U. Karman, Eds), **2**, 153–225, Academic Press, New York, 1951.

Polubarinova-Kochina, P. Ya., *Theory of Ground Water Movement* (in Russian), Gostekhizdat, Moscow, 1952; English transl. by R. J. M. De Wiest, Princeton Univ. Press, Princeton, N.J., 1962.

Prakash, A., Radial dispersion through adsorbing porous media, *Proc. Am. Soc. civ. Engrs*, **102**(HY3), 379–396, 1976.

Prickett, T. A., Type curve solutions to aquifer tests under water table conditions, *Ground Water*, **3**(3), 5–14, 1965.

Prickett, T. A. and C. G. Lonnquist, "Selected digital computer techniques for groundwater resource evaluation," Illinois State Water Survey, Bull. 55, 62 pp., 1971.

Prickett, T. A. and C. G. Lonnquist, "Aquifer simulation for use on disk supported small computer system," Illinois State Water Survey, Circ. 114, 21 pp., 1973.

Reisenaur, A. E., R. W. Nelson, and C. N. Knudsen, "Steady Darcian transport of fluids in heterogeneous partially saturated porous media," General Electric, Hanford Atomic Products Operations, HW-72335, Pt. 2, 88 pp., 1963.

Raats, P. A. C., Jump conditions in the hydrodynamics of porous media, Proc. 2nd IAHR-ISSS Symp. Fundamentals of Transport Phenomena in Porous Media, 151–173, Guelph, 1972.

Remson, I., G. M. Hornberger, and F. J. Moltz, *Numerical Methods in Subsurface Hydrology*, Wiley-Interscience, New York, 389 pp., 1971.

Richards, L. A. and W. Gardner, Tensiometers for measuring the capillary tension and soil water, *J. Am. Soc. Agron.*, **28**, 352–358, 1936.

Richards, L. A. and L. R. Weaver, Moisture retention by some irrigated soils as related to soil moisture tension, *J. Agric. Res.*, **69**, 215–235, 1944.

Richtmeyer, R. D., *Difference Methods for Initial Value Problems*, Interscience, New York, 288 pp., 1957.

Rider, N. E., Water losses from various land surfaces, *Quart. J. Roy. Meteorological Soc.*, **83**, 181–193, 1957.

Rorabaugh, M. I., Graphical and theoretical analysis of step-drawdown test of artesian well, *Proc. Am. Soc. civ. Engrs*, **79**, separte no. 362, 23 pp., 1953.

Rosenberg, D. U. von, *Methods for the Numerical Solution of Partial Differential Equations*, Elsevier, New York, 128 pp., 1969.

Rubin, J., Theoretical analysis of two-dimensional transient flow of water in unsaturated and partly saturated soils, *Proc. Soil Sci. Soc. Am.*, **32**(5), 607–615, 1968.

Rumer, R. R. and D. R. F. Harleman, Intruded salt water wedge in porous media, *Proc. Am. Soc. civ. Engrs*, **89**(6), 193–220, 1963.

Rushton, K. R. and Y. K. Chan, Numerical pumping test analysis, *Proc. Am. Soc. civ. Engrs*, **103**(IR1), 1–12, 1977.

Sage, A. P. and J. L. Melsa, *System Identification*, Academic Press, New York, 221 pp., 1971.

Sagar, B., S. Yakowitz, and L. Duckstein, A direct method for the identification of the parameters of dynamic nonhomogeneous aquifers, *Wat. Resour. Res.*, **11**(4), 563–570, 1975.

Saffman, P. G., A theory of dispersion in a porous medium, *J. Fluid Mech.*, **6**(3), 321–349, 1959.

Saffman, P. G., Dispersion due to molecular diffusion and macroscopic mixing in flow through a network of capillaries, *J. Fluid Mech.*, **7**(2), 194–208, 1960.

Saleem, Z. A., A computer method for pumping test analysis, *Ground Water*, **8**(5), 21–24, 1970.

Scheideger, A. E., *The Physics of Flow through Porous Media*, 2nd edn, Univ. of Toronto Press, 313 pp., 1960.

Scheidegger, A. E., General theory of dispersion in porous media, *J. Geophys. Res.*, **66**(4), 3273–3278, 1961.

Schiff, L., The status of water spreading for groundwater replenishment, *Trans. Am. Geophys. Un.*, **36**, 1009–1020, 1955.

Schiff, L., Ground-water recharge hydrology, *Ground Water*, **2**(3), 1964.

Schmorak, S., "Salt water encroachment in the coastal plain of Israel," International Association of Scientific Hydrology, Symposium on Artificial Recharge and Management of Aquifers, Haifa, Israel, IASH Publ. no. 72, pp. 305–318, 1967.

Schmorak, S. and A. Mercado, Upconing of fresh water—sea water interface below pumping wells, field study, *Wat. Resour. Res.*, **5**(6), 1290–1311, 1969.

Schoeller, H., *Arid Zone Hydrology, Recent Developments*, UNESCO, Paris, 125 pp., 1959.

Schulze, J., *Die Grundwasserabsenkung in Theorie und Praxis*, Springer, Berlin, 140 pp., 1924.

Schwarz, J., "Linear models for groundwater management," Water Planning for Israel, Ltd., Tel Aviv, Israel, P.N. ET/71/062, 1971.

Schwarz, J. and M. Rebhum, Clogging and contamination processes in recharge wells, *Wat. Resour. Res.*, **4**, 1207–1217, 1968.

Scott, R. F., *Principles of Soil Mechanics*, Addison-Wesley, Reading, Mass., 550 pp., 1963.

Scriven, L. E., Dynamics of a fluid interface, *Chem. Engng Sci.*, **12**, 98–108, 1960.

Segol, G., A three-dimensional Galerkin-finite element model for the analysis of contaminant transport in saturated-unsaturated porous media, Intern. Conference on Finite Elements in Water Resources, Princeton Univ., Princeton, N.J., 1976.

Segol, G., G. F. Pinder, and W. G. Gray, A Galerkin-finite element technique for calculating the transition position of the salt-water front, *Wat. Resour. Res.*, **11**(2), 343–347, 1975.

Shamir, U. and D. R. F. Harleman, "Numerical and analytical solutions of the dispersion problems in homogeneous aquifers," Hydrodynamics Lab., M.I.T., Report no. 89, 1966.

Sheahan, N. T., Type-curve solution of a step-drawdown test, *Ground Water*, **9**(1), 25–29, 1971.

Sheahan, N. T., Discussion of well losses due to reduced formation permeability and comments on the step-drawdown test, *Ground Water*, **10**(4), 46–49, 1972.

Shima, M., Hydrodynamics of flow in porous media, Transient characteristics of salt water wedge, *Proc. 13th Congress International Ass. for Hydraulic Research*, **4**, 433–440, Japan, 1969.

Signor, D. C., D. J. Growitz, and W. Kam, "Annotated bibliography on artificial recharge of ground water," U.S. Geol. Survey, Water Supply Paper no. 1990, 1970.

Silin-Bekchurin, A. I., *Dynamics of Ground Water* (in Russian), Moscow Izdat., Moscow Univ., 258 pp., 1958.

Simpson, E. S. and L. Duckstein, Finite state mixing cells models, Proc. U.S.-Yugoslavian Symp. on Karst Hydrology and Water Resources, 489–508, Dubrovnik, June, 1975.

Slater, G. E. and E. J. Durrer, Adjustment of reservoir simulation models to match field performance, *Soc. Petrol. Eng. J.*, **11**(3), 295, 1971.

Slattery, J. C., Fundamentals, *Ind. Eng. Chem.*, **6**, 108, 1967.

Slichter, C. S., "Field measurement of the rate of movement of underground waters," U.S. Geol. Survey Water Supply Paper no. 140, 1905.

Smith, G. O., *Numerical Solution of Partial Differential Equations*, Oxford Univ. Press, London, 179 pp., 1965.

Sonu, J. and H. J. Morel-Seytoux, Water and air movement in a bounded deep homogeneous soil, *J. Hydrol.*, **29**, 23–42, 1976.

Spiegel, M. R., *Theory and Problems of Vector Analysis*, Schaum Dulline Series, McGraw-Hill, New York, 225 pp., 1959.

Stallman, R. W., "Nonequilibrium type curves modified for two wells systems," U.S. Geol. Survey, Ground Water Notes, Hydraulics, no. 3, 1952.

Stallman, R. W., "Boulton's integral for pumping test analysis," in U.S. Geol. Survey Research 1961; U.S. Geol. Survey Prof. Paper 424-C, pp. 24–29, 1962.

Stallman, R. W., "Multiphase fluid flow in porous media—a review of theories pertinent to hydrologic studies," U.S. Geol. Survey, Prof. Papers, no. 411-E, 51 pp., 1964.

Stallman, R. W., Effects of water table conditions on water level changes near pumping wells, *Wat. Resour. Res.*, **1**(2), 295–312, 1965.

Stallman, R. W., Flow in the zone of aeration, in *Advances in Hydroscience* (V. T. Chow, Ed.), **4**, 151–195, Academic Press, New York, 1967.

Steggewentz, J. H. and B. A. Van Nes, Calculating the yield of a well taking account of replenishment of the groundwater from above, *Water and Water Engng*, **41**, 561–563, 1939.

Sternberg, Y. M., Simplified solution for variable rate pumping test. *Proc. Am. Soc. civ. Engrs*, **94**(HY1), 177–180, 1968.

Sternberg, Y. M. and V. H. Scott, Mutual interference of water wells, *Proc. Am. Soc. civ. Engrs*, **93**(HY4), 169–181, 1967.

Strack, O. D. L., "Many valuedness encountered in groundwater flow," Doctoral Thesis, Delft Univ. Press, Delft, 126 pp., 1973.

Strack, O. D. L., Some cases of interface flow towards drains, *J. Eng. Math.* **6**, 175–191, 1972.

Strack, O. D. L., A single-potential solution for regional interface problems in coastal aquifers, *Wat. Resour. Res.*, **12**(6), 1165–1174, 1976.

Streltsova, T. D., Unsteady radial flow in an unconfined aquifer, *Wat. Res.*, **8**, 1059–1066, 1972.

Streltsova, T. D., Flow near a pumped well in an unconfined aquifer under nonsteady conditions, *Wat. Resour. Res.*, **9**(1), 227–235, 1973.

Streltsova, T. D. and K. R. Rushton, Water table drawdown due to a pumped well in an unconfined aquifer, *Wat. Resour. Res.*, **9**(2), 236–242, 1973.

Swartzendruber, D., The flow of water in unsaturated soils, in *Flow Through Porous Media* (R. J. M. De Wiest, Ed.), Chap. 6, pp. 215–292, Academic Press, New York, 1969.

Terzaghi, K., *Erdbaumechanik auf Bodenphysikalische Grundlage*, Franz Deuticke, Leipzig, 390 pp., 1925.

Theis, C. V., The relation between lowering of the piezometric surface and the rate and duration of discharge of a well using ground water storage, *Trans. Am. Geophys. Un.*, 16th annual meeting, Pt. 2, 519–524, 1935.

Thiem, G., *Hydrologische Methoden* (Hydrologic methods), J. M. Gebhardt, Leipzig, 56 pp., 1906.

Thiessen, A. H., Precipitation for large areas, *Montly Weather Bur. Rev.*, **39**, 1082–1084, 1911.

Thomas, H. A. Jr. and M. B. Fiering, Mathematical synthesis of streamflow sequences for the analysis of river basins by simulation, in *Design of Water-Resource Systems* by Dorfman, R., H. A. Thomas Jr., S. A. Marglin, and G. M. Fair, Chap. 12, pp. 459–493, Harvard Univ. Press, Cambridge, Mass., 1962.

Thornthwaite, C. W., A re-examination of the concept and measurement of potential evapotranspiration, *Publ. in Climate*, **7**, 1954.

Thornthwaite, C. W. and F. K. Hare, The loss of water to the air, *Meteorological Monographs*, **6**, 163–180, 1965.

Todd, D. K., "Annotated bibliography on artificial recharge," U.S. Geol. Survey, Water Supply Paper no. 1477, 113 pp., 1959a.

Todd, D. K., *Ground Water Hydrology*, John Wiley, New York, 336 pp., 1959.

Toth, J., A theory of ground water motion in small drainage basins in central Alberta, Canada, *J. geophys. Res.*, **67**, 4375–4387, 1962.

Trescott, P. C., G. F. Pinder, and S. P. Larson, "Finite difference model for aquifer simulation in two dimensions with results of numerical experiments," U.S. Geol. Survey, Techniques of Water Resources Investigations, Book 7, Automated Data Processing and Computations, 116 pp., 1976.

Trescott, P. C. and S. P. Larson, Comparison of iterative methods of solving two dimensional groundwater flow equations, *Wat. Resour. Res.*, **13**(1), 125–136, 1977.

Truesdell, C. and R. A. Toupin, Classical field theories, in *Handbuch der Physik*, Vol. III-1, pp. 226–793, Springer, Berlin, 1960.

Tyson, N. H. and E. M. Weber, Groundwater management for the nation's future-computer simulation of groundwater basins, *Proc. Am. Soc. civ. Engrs*, **90**(HY4), 59–77, 1964.

UNESCO, jointly with WMO, *Hydrological Maps*, UNESCO, Paris, 204 pp., 1977.

Vachaud, G., J. P. Gaudet, and V. Kuraz, Air and water flow during ponded infiltration in a vertical bounded column, *J. Hydrol.*, **22**(1/2), 89–108, 1974.

Van Genuchten, M. Th., On the accuracy and efficiency of several numerical schemes for solving the convective-dispersive equation, Intern. Conference on Finite Elements in Water Resources, Princeton Univ., Princeton, N.J., 1976.

Van Genuchten, M. Th., G. F. Pinder, and E. O. Frind, Simulation of two-dimensional contaminant transport with isoparametric-Hermitian finite elements, *Wat. Resour. Res.*, **13**(2), 451–458, 1977.

Van Poolen, H. K., Radius of drainage and stabilization time equations, *Oil and Gas J.*, 138–146, Sept. 1964.

Van Poolen, H. K., Drawdown curves give angle between intersecting faults, *Oil and Gas J.*, 5 pp., Dec. 1965.

Vappicha, V. N. and S. H. Nagaraja, An approximate solution for the transient interface in a coastal aquifer, *J. Hydrol.* **31**(1/2), 161–173, 1976.

Vauclin, M., G. Vachaud, and J. Khanji, Two-dimensional numerical analysis of transient water transfer in saturated-unsaturated soils, in *Computer Simulation of Water Resources* (G. C. Vanteenkiste, Ed.), p. 103, North Holland Publ., Amsterdam, and American Elsevier, New York, 1975.

Vemuri, V. and W. J. Karplus, Identification of non-linear parameters of ground water basins by hybrid computation, *Wat. Resour. Res.*, **5**, (1), 172–185, 1969.

Vermeer, P. A. and C. Van den Akker, Performance of a recharge and recovery system in an aquifer with uniform flow, *Bull. Int. Assoc. of Hydrological Sciences*, **21**(2), 387–396, 1976.

Verruijt, A., "Elastic storage of aquifers," NSF sponsored Hydrology Institute at Princeton Univ., 1965; also in *Flow Through Porous Media* (R. J. M. De Wiest, Ed.), Academic Press, New York, pp. 331–376, 1969.

Verruijt, A., Steady dispersion across an interface in a porous medium, *J. Hydrol.*, **14**, 337–347, 1971.

Walton, W. C., "Leaky artesian aquifer conditions," Dept. of Registration and Ed., State of Illinois, Report of Investigation, no. 39, 1960.

Walton, W. C., Application and limitation of methods used to analyze pumping test data, *Water Well J.*, 6 pp., Feb.–March, 1960a.

Walton, W. C., "Selected analytical methods for well and aquifer evaluation," Illinois State Water Survey, Bull. 49, 81 pp., 1962.

Walton, W. C., *Groundwater Resource Evaluation*, McGraw-Hill, New York, 664 pp., 1970.

Ward, J. C., Turbulent flow in porous media, *Proc. Am. Soc. civ. Engrs*, **90**(HY5), 1–12, 1964.

Wenzel, L. K., "Methods of determining permeability of water bearing materials, with special reference to discharging well methods," U.S. Geol. Survey Water Supply, Paper 887, Washington, D.C., 192 pp., 1942.

Wesseling, J., Principles of unsaturated flow and their application to the penetration of moisture into soil, *Tech. Bull. Inst. Land and Water Res.*, no. 23, Wageningen, The Netherlands, 1961.

White, W. N., "A method of estimating ground water supplies based on discharge by plants and evaporation from soil," U.S. Geol. Survey Water Supply Paper no. 659, Washington, D.C., 105 pp., 1932.

Wiener, A., *The Role of Water in Development*, McGraw-Hill, New York, 483 pp., 1972.

Wilson, J. L. and L. W. Gelhar, "Dispersive mixing in partially saturated porous medium," Hydro. Lab., Report 191, Mass. Inst. of Tech., Cambridge, Mass., 353 pp., 1974.

Wolff, R. G., Field and laboratory determination of hydraulic diffusivity of a confining bed, *Wat. Resour. Res.*, **6**(1), 194–203, 1970.

Wyckoff, R. D. and H. G. Botset, The flow of gas–liquid mixture through unconsolidated sands, *Physics*, **7**, 325–345, 1936.

Yih, C. S., A transformation for free-surface flow in porous media, *Physics of Fluids*, **1**, 20–24, 1964.

Yeh, W. W. G., Aquifer parameter identification, *Proc. Am. Soc. civ. Engrs*, **101**(HY9), 1197–1209, 1975a.

Yeh, W. W. G., Optimal identification of parameters in an inhomogeneous medium with quadratic programming, *Soc. Petrol. Eng. J.*, **15**(5), 371–375, 1975b.

Yeh, W. and G. W. Tauxe, Quasilinearization and the identification of aquifer parameters, *Wat. Resour. Res.*, **7**(2), 375–381, 1971.

Youngs, E. G., Moisture profiles during vertical infiltration, *Soil Sci.*, **84**(4), 283–290, 1957.

Youngs, E. G., Redistribution of moisture in porous materials after infiltration, I, *Soil Sci.*, **86**, 117–125, II, 202–207, 1958.

Youngs, E. G., The drainage of liquids from porous materials, *J. geophys. Res.*, **65**(12), 4025–5040, 1960.

Zienkiewicz, O. C., *The Finite Element Method in Engineering Science*, McGraw-Hill, London, 521 pp., 1971.

Zienkiewicz, O. C. and C. J. Parekh, Transient field problems: Two-dimensional and three-dimensional analysis by isoparametric finite elements, *Int. J. Numer. Meth. Eng.*, **2**, 61–71, 1970.

AUTHOR INDEX

Abu-Zied, M. A., 320
Adamson, A. W., 240
Addison, L. E., 483
Aravin, V. I., 306
Aris, R., 72
Aron, G., 471
Aston, A. R., 224

Bachmat, Y., 232, 234, 235, 248, 257, 515, 516, 520
Babu, D. K., 221
Badon-Ghyben, W., 384
Banks, R. B., 268
Barak, A., 73
Bastian, W. C., 270
Baumann, P., 50
Bear, J.,* 69, 73, 126, 128, 169, 196, 204, 231, 232, 234, 235, 248, 257, 264, 265, 276, 282, 288, 290, 294, 300, 309, 386, 393, 395, 399, 401, 408–412, 414, 416, 418, 420–422, 425, 427, 431, 433, 441, 482–490, 516, 520
Beetem, W. A., 482
Bender, S. J., 262
Ben Zvi, A., 38, 39
Bennett, G. D., 482
Bennett, R. R., 370

Bentall, R., 467
Bhuijan, S. E., 224
Bierschenk, W. H., 477
Biot, M. A., 184
Biswas, A. K., 491
Bize, J., 51
Blaney, H. F., 57
Boast, C. W., 240
Boen, D. F., 51
Bondarev, E. A., 275
Boreli, M., 344
Botset, H. G., 210
Boulton, N. S., 309, 332–335, 337, 338, 467, 472, 482
Bourguet, L., 51
Bowers, S. A., 224
Braester, C., 221
Brebbia, C. A., 129
Bredehoeft, J. D., 110, 328, 331, 483
Brooks, R. H., 212, 213
Brown, R. H., 300, 467
Bruch, J. C., 223, 224, 490
Bruin, J., 300, 467
Brutsaert, W., 189, 221
Bunts, J. H., 51
Buras, N., 491
Burgy, R. H., 339, 472

* References to Bear (1972) are not included.

Campbell, M. D., 58, 300
Carman, P. C., 67
Carslaw, H. S., 124, 176, 320, 325
Caspi, I., 490
Castany, G., 160
Chan, Y. K., 339, 467
Charnes, A., 491
Charni, I. A., 311
Chertousov, M. D., 306
Childs, E. C., 191, 197, 210, 212, 214
Chow, V. T., 40, 468
Claborn, B. J., 483
Coats, K. H., 242, 483
Collins, R. E., 65
Collis-George, N., 210
Columbus, N., 396
Conkling, H., 88
Connor, J. J., 129
Cooley, R. L., 490
Cooper, H. H. Jr., 321, 327, 328, 331, 381
Corapcioglu, M. Y., 189
Corey, A. T., 211–213
Crank, J., 266
Crawford, N. H., 38
Criddle, W. D., 57
Currie, R. J., 51

DaCosta, J. A., 370
Dagan, G., 169, 321, 386, 395, 408–412, 414, 416, 418, 420–422, 425, 427, 431, 433, 482
Dantzig, G. B., 492, 493
Darcy, H., 60
Davis, S. N., 88
De Glee, G. J., 312, 344, 347
Deininger, R. A., 490, 491
De Josselin de Jong, G., 184, 232
Dempsey, J. R., 483
De Marsily, G., 483
Desai, C. S., 142
De Wiest, R. J. M., 88, 312, 343
Dorfman, R., 491
Dorn, W. S., 129
Dracup, J., 491
Droracek, M., 50
Duckstein, L., 454, 490
Dudgeon, C. R., 300
Dupuit, J., 76
Durrer, E. J., 483, 490

Emde, F., 321
Emich, V. N., 183
Emsellem, Y., 483
Ergatoudis, I., 148
Ergun, S., 66
Eykhoff, P., 490

Fair, G. M., 67, 491
Ferris, J. G., 300, 407
Feshbach, H., 72
Fetter, C. W., 50
Fiering, M. B., 40
Fleming, G., 38
Fomin, S. V., 142
Forchheimer, P., 66
Forsythe, G. E., 129
Freeze, R. A., 189, 224
Fried, J. J., 231, 259
Friedrichs, D. R., 483
Frind, E. O., 147–150, 261, 490
Futagami, T., 491

Gambolati, G., 189
Garabedian, P. R., 95, 464
Gardner, A. O. Jr., 259
Gardner, W., 196
Gardner, W. R., 212, 215, 216, 221
Gass, S. I., 492
Gatto, P., 189
Gaudet, J. P., 218
Gaulier, C., 258
Gelfand, I. M., 142
Gelhar, L. W., 232, 257
Gershon, N. D., 270, 271
Glover, R. E., 391, 396
Goldstoff, Y., 38, 39
Gray, W. G., 64, 129, 142, 151, 224, 251, 259, 261–263, 515
Grove, D. B., 482
Growitz, D. J., 51
Gupta, S. K., 142, 261
Guymon, G. L., 261

Hadley, G., 492, 493
Haimes, Y. Y., 490
Halevy, E., 482
Hall, W. A., 491
Hanks, R. J., 224
Hansen, V. E., 309
Hanshaw, B., 110
Hantush, M. S., 108, 300, 312, 332–335, 339–345, 349, 350, 365, 467, 470, 471, 476–478, 481, 482
Hare, F. K., 57
Harleman, D. R. F., 276, 412, 482
Harpaz, Y., 49, 50, 288, 290, 294, 482
Harr, M. E., 185
Harrill, J. R., 477
Hatch, L. P., 67
Hefez, E., 128, 441, 483, 484–490
Helm, D., 189
Helweg, O. J., 467, 477
Henderson, H. H., 483

Henry, H. R., 391, 396, 413
Herzberg, A., 384
Herrmann, L. R., 261
Hildebrand, F. B., 142
Hiler, E. A., 224
Holzmacher, R. C., 50
Hoopes, J. A., 276, 482
Hornberger, G. M., 129, 142
Hougen, O. A., 240
Hubbert, M. K., 30, 62
Hudson, H., 300, 467
Hufschmidt, M. M., 491
Hughes, J. L., 483
Huisman, L., 50, 300, 318
Huyakorn, P. S., 300

Irmay, S., 69, 126, 196, 204, 210, 221, 300, 306, 467
Irons, B. M., 148

Jacob, C. E., 303, 304, 312, 318, 321, 326, 332, 339, 342, 344, 348, 374, 377, 468, 472, 476, 477
Jacobs, M., 282, 288, 380
Jackson, B. B., 40
Jaeger, J. C., 124, 176, 320, 325
Jahnke, E., 321
Jahvandel, I., 145

Kam, W., 51
Kapuler, Y., 393, 401
Karanjac, J., 378
Karplus, W. J., 490
Kaufman, W. J., 275
Khanji, J., 224
Kipp, K. L., 483
Kipps, K. L. Jr., 345
Kirkham, D., 191, 308, 345, 349
Kisiel, C. C., 490
Kleinecke, D., 483, 490
Klute, A., 209, 221
Knowles, D. B., 300, 467
Knowles, T. R., 483
Knudson, C. N., 224
Kohout, F. A., 433
Korganoff, A., 490
Kozeny, J., 344, 348
Kriz, G. J., 339, 472
Kuraz, V., 218
Kutilek, M., 66

Labadie, J. W., 467, 477, 490
Laliberte, G. E., 213
Lam, C. M., 490
Lambe, T. W., 85
Lapidus, L., 270
Larson, S. P., 130, 136, 138, 140

Lasdon, L. S., 493
Lau, L. K., 275
Laverty, F. B., 43
Lee, P. C. Y., 515
Lehr, J. H., 58, 300
Lembke, K. E., 306
Lemoine, J., 51
Lennox, D. H., 477
Lin, A. C., 490
Linsley, R. K., 38
Logan, J. A., 491
Lohman, S. W., 326, 342, 470
Lonnquist, C. G., 129, 138, 140
Lovell, R. E., 490
Luthin, J. N., 142, 261
Lynn, W. R., 491
Lyons, T. C., 445, 455–458

Maas, A. M., 491
McCracken, D. D., 129
Maddock, T. III., 496, 499
Marino, M. A., 377
Marle, C., 258
Marshal, W. R., 240
Marvin, J., 50
Matalas, N. C., 40
Mavis, F. T., 74
Mayhugh, M. S., 215
Meinzer, O. E., 21
Melsa, J. L., 483
Mercado, A., 276, 433, 434, 445, 458–462, 482
Mero, F., 38, 56
Mitchel, A. R., 129
Moltz, F. J., 129, 142
Morel-Seytoux, H. J., 191, 218, 221
Morse, P. M., 72
Mualem, Y., 399
Muskat, M., 278, 279, 320, 325, 326, 344, 348, 352, 417, 423

Nagaraja, S. H., 396, 412, 414
Nahargang, G., 344
Narasimhan, T. N., 224
Nelson, R. W., 224, 294, 298, 299, 490
Neuman, S. P., 89, 98, 224, 331, 343, 483, 490
Nikolaevskij, V. N., 275
Nir, A., 270, 271, 482
Noblanc, A., 218, 221
Numerov, S. N., 306
Nutting, P. G., 67

Ogata, A., 268, 275, 276
O'Neill, K., 64

Page, H., 390, 393
Papadopulos, I. S., 300, 327, 328, 331

Pacsirszky, J., 258
Parekh, C. J., 150
Parlange, J. Y., 221
Patten, E. P. Jr., 482
Peaceman, D. W., 139, 140, 259
Penman, H. L., 57
Perrine, R. L., 490
Peterson, S. H., 50
Pfankuch, H. O., 238
Philip, J. R., 191, 221
Pinder, G. F., 110, 129, 130, 136, 138–140, 142, 147–151, 224, 251, 259, 261–263, 390, 393, 483, 490
Polubarinova-Kochina, P. Ya., 74, 311, 312, 344, 408
Powers, W. L., 191
Pozzi, A. L. Jr., 259
Prakash, A., 276
Prickett, T. A., 129, 138, 140, 338, 339, 467, 472

Raats, P. A. C., 520
Rachford, H. H. Jr., 139, 140
Rebhun, M., 50
Reisenaur, A. E., 224
Remson, I., 129, 142
Richards, L. A., 196, 199
Richtmeyer, R. D., 129
Rider, N. E., 57
Robson, S. G., 483
Rorabaugh, M. I., 375, 477
Rosenberg, D. V. von, 129, 134
Rubin, J., 221
Rumer, R. R., 412
Rushton, K. R., 339, 467

Saffman, P. G., 232, 233, 257
Sage, A. P., 483
Sagar, B., 490
Saleem, Z. A., 467
Scheidegger, A. E., 66, 234
Schiff, L., 50
Schmorak, S., 380, 433, 434
Schoeller, H., 332, 333, 338
Schulze, J., 306
Schwartz, Y., 50, 128, 483, 491, 496, 505, 509
Scott, R. F., 185
Scott, V. H., 261, 320, 339, 355, 471, 472
Scriven, L. E., 194
Segol, G., 251, 261
Shamir, U., 128, 276, 441, 483, 484–490
Sheahan, N. T., 477
Shima, M., 412
Signor, D. C., 51
Silin-Bekchurin, A. I., 75
Simandoux, P., 258
Simundich, T. M., 490

Simpson, E. S., 454
Slater, G. E., 483, 490
Slattery, J. C., 194
Slichter, C. S., 227
Smith, B. D., 242
Smith, G. O., 129
Sonu, J., 221
Spiegel, M. R., 72
Stallman, R. W., 191, 193, 300, 467, 471, 481
Steggewentz, J. H., 312
Sternberg, Y. M., 355, 471, 482
Strack, O. D. L., 395, 399, 400, 403, 405, 406, 416
Streltsova, T. D., 339
Swartzendruber, D., 191

Tamai, N., 491
Tanji, K. K., 142, 261
Tauxe, G. W., 490
Terzaghi, K., 84, 185, 189
Theis, C. V., 174, 318, 321
Thiem, G., 306
Thiessen, A. H., 451
Thomas, H. A., 40, 491
Thornthwaite, C. W., 57
Todd, D. K., 21, 51, 275
Toth, J., 169
Toupin, R. A., 516
Trescott, P. C., 130, 136, 138, 140
Truesdell, C., 516
Tsui, T. P., 74
Tyson, N. H., 451

Vachaud, G., 218, 224
Van Baveland, C. H. M., 224
Van der Goot, H. A., 43
Van Genuchten, M. Th., 261–263
Van Nes, B. A., 312
Van Poolen, H. K., 323, 481
Vappicha, V. N., 396, 412, 414
Vauclin, M., 224
Vemuri, V., 490
Verruijt, A., 184, 274

Wallis, J. R., 40
Walton, W. C., 331, 338, 374, 467, 472, 476
Ward, J. C., 65, 66
Wasow, W. R., 129
Weaver, L. R., 199
Weber, E. M., 451
Weisman, R. N., 221
Wells, D. M., 483
Wenzel, L. K., 321, 348, 467
Wesseling, J. 191
White, W. N., 58
Whitman, R. V., 85

Witherspoon, P. A., 145, 343
Wiener, A., 4
Wilson, J. L., 232
Wismer, D. A., 490
Wolff, R. G., 110
Wyckoff, R. D., 210

Yakowitz, S., 490
Yatsuzuka, M., 491

Yeh, W. W-G., 377, 490
Yih, C. S., 416
Youngs, E.G., 221

Zaslavsky, D., 69, 126, 196, 204, 288, 300, 306, 467
Zienkiewicz, O. C., 142, 146, 148, 150

SUBJECT INDEX

Abrupt front (*see* Fronts)
ADI (*see* Finite difference methods, ADI)
Adsorption, 240, 266
    isotherms of, 240
Analogs (*see* Models and analogs)
Air entry value (*see* Bubbling pressure)
Anisotropy, 31–33
    transformation to equivalent isotropic medi-
        um, 169–175
Aquiclude, 21
Aquifer:
    artesian, 24
    boundaries of, 119
    classification of, 24
    compressibility of, 85
    confined, 24
    definition of, 21
    dispersion equation for, 252
    functions of, 7
    leaky, 26
    models of, 436–462
        (*see also* Models of aquifers)
    perched, 25, 26
    phreatic, 24
    relations to rivers, 51–53
    storativity of, 14, 59, 86–89

    transmissivity of, 36, 69–71
    unconfined, 24
    yield of, 9, 12
Aquifuge, 21
Aquitard, 21
Arrival distribution, 294–299
Artesian aquifer, definition of, 24
Artificial recharge, 10, 11, 34, 42–51
    clogging of wells, 42–51, 377
    enhancing infiltration, 45
    induced recharge, 49
    methods of, 45–50
    objectives of, 42–45
    by surface spreading, 46
    through wells, 49
Average velocity, 62, 64, 514
Averaged aquifer equations:
    for confined aquifer, 106
    for hydrodynamic dispersion, 254–256
    for leaky aquifer, 109
    for phreatic aquifer, 114
Averages:
    phase average, 31, 514
    intrinsic phase average, 31, 513
Averaging:
    along the vertical, 521
    rules of, 521

Boundary conditions, 94, 518
  Cauchy type, 98, 220
  for coastal interface, 390–393
  Dirichlet type, 96, 219
  at discontinuity in permeability, 100
  of first type, 96, 219
  general, 518
  in hydrodynamic dispersion, 248
  of mixed type, 98
  Neumann type, 98
  in numerical methods, 138
  on phreatic surface, 98, 99
  of prescribed flux, 98, 117, 139
  of prescribed potential, 96, 117, 138
  of second type, 98, 219
  on seepage face, 99
  on semipervious boundary, 98, 117
  at a spring, 117
  in ·unsaturated ·flow, 219–220
  in saturated flow, 219–220
Balance of groundwater (*see* Groundwater balance)
Boussinesq equation, 113
Breakthrough curve, 228
Bubbling pressure, 197, 199, 212
Build-up, 302, 377

Calibration of models (*see* Models of aquifers, calibration of)
Calibration problem (*see* Identification problem)
Capacitance matrix, 500
Capillary diffusivity, 209 (*see also* Diffusivity in unsaturated flow)
Capillary fringe, 23, 74, 202
Capillary pressure, 193–197
Capillary pressure head, 197
  critical, 198
Capillary rise, 202
Cauchy–Riemann conditions, 167, 245
Clogging of wells in artificial recharge, 49, 377
Coefficient of permeability (*see* Permeability)
Collector wells, 300
Compression Index, 186
Cone of depression, 301
Confined aquifer:
  continuity equation for, 104
  definition of, 24
Consolidation:
  (*see also* Land subsidence)
  primary, 189
  secondary, 189
Constraints, 11
  in aquifer management, 497, 499
  in linear programming, 492–494
Contact angle, 191
Continuity equation, 83

  for coastal interface, 386–393
  for confined aquifer, 103
  for flow in aquifers, 103
  for leaky confined aquifer, 109
  for leaky phreatic aquifer, 115
  for phreatic aquifer, 111
  for three-dimensional flow, 92
  for unsaturated flow, 213–217
Continuum approach, 14, 28

Darcy's law, 60–63
  for compressible fluid, 64
  for three-dimensional flow, 63
  range of validity of, 65
Dead end pores, 242
Decision variables, 1, 10, 492
Delayed storage, 338
Diffusivity, in unsaturated flow, 209, 214, 215
Dimensionless variables, 175
Dirac delta function, 104, 112, 217
Dispersion:
  (*see also* Hydrodynamic dispersion)
  coefficient of mechanical dispersion, 232, 239
    in curvilinear coordinates, 235
    longitudinal, 233
    principal axes of, 235
    transversal, 233
  mechanical, 229
  phenomena of, 227–229
Dispersive flux, 232
Dispersivity, 234, 239
  longitudinal, 234, 239
  transversal, 234, 239
Drainage, 6
  in groundwater balance, 34
Drawdown:
  in array of wells, 352
  corrected, 311
  definition of, 301
  in mutilayered aquifer, 317–318
  near partially penetrating well, 344–350
  recovery of, 325
  residual, 323
  in steady flow to a well:
    in confined aquifer, 304–308
    in leaky confined aquifer, 312–318
    in phreatic aquifer, 308–311
  Thiem equation, 306
  in unsteady flow to a well:
    in confined aquifer, 318–330
    in leaky confined aquifer, 339–344
    in phreatic aquifer, 331–339
Dual problem, 494
Duhamel's integral, 158
Dupuit, assumptions, 74–82
  for coastal interface, 386

estimate of error, 78
  for interface upconing, 425–433
  regions of inapplicability, 82
Dupuit–Forchheimer discharge formula, 78
Dupuit–Forchheimer well discharge formula, 310, 335

Effective permeability, 204, 209–213
Effective porosity, 63
Effective stress (*see* Stress, effective)
Effluent stream, 34, 51
Einstein's summation convention, 72
Entrance head loss to wells, 303
Eulerian approach, 90
Evapotranspiration, 34, 39, 57

Fair and Hatch formula, 67
Fick's law, 233
Field capacity, 23, 203
Finite difference methods, 129–141
  ADI, 139
  alternating direction implicit method (*see* Finite difference method, ADI)
  backward difference, 132
  boundary conditions for, 138
  central difference, 132
  confined aquifer equation, 134, 136, 140
  Crank–Nicolson scheme, 135
  for dispersion equation, 259
  explicit scheme, 134
  forward difference, 131
  grid for, 130
  IADI, 139
  implicit scheme, 134
  iterative alternating direction implicit method (*see* IADI)
  for Laplace equation, 132
  LSOR, 140
  for phreatic aquifer equation, 141
  round-off error in, 130
  SIP, 140
  stability of, 135
  truncation error in, 130
Finite difference schemes (*see* Finite difference methods)
Finite difference techniques (*see* Finite difference methods)
Finite element method, 141–152
  calculus of variations approach, 142
  for dispersion equation, 260
  for unsaturated flow equation, 222
  weighted residual method, 146
    basis functions, 147
    Galerkin method, 146–152
    isoparametric quadrilateral elements, 148
Flow at large Reynolds numbers, 66, 67

Flow nets, 165–169
Flowing well, 326, 342
Fokker–Plank equation, 214, 221
Forecasting problem, 2, 14 (*see also* Groundwater flow problem)
  methods of solution, 124
Fractured porous medium, pollution in, 243
Free interfacial energy, 191
Free surface (*see* Phreatic surface)
Fronts (*see* Sharp fronts)
Fundamental tensor of Riemannian space, 236
Funicular saturation, 193

Galerkin method (*see* Finite difference method, Galerkin method)
Ghyben–Herzberg approximation, 384–386
Groundwater:
  balance of, 14, 34–59
  characteristics of, 4
  classification of problems, 15
  definition of, 19
  equations of motion for, 14, 60
  flow problem, 84–116
    methods of solution, 124
  flow through boundaries, 35
  leakage of, 35, 36
    coefficient of, 108
  management of, 1, 10, 15, 491–512
  modeling of, 15, 436–462
  quality of, 5, 15, 225
  role of, 2
  storage of, 59
  vis surface water, 2
Groundwater divide, 165
Groundwater pollution, 225
  mathematical statement of problem, 250–252
  sources of, 226
Gravitational water, 23

Hele–Shaw analog (*see* Models and analogs, Hele–Shaw)
Homogeneity, 31–33
Horizontal flow approach (*see* Hydraulic approach)
Horizontal wells (*see* Collector wells)
Hubbert's potential, 64, 93
Hybrid computer, 441
Hydraulic approach, 14, 26, 103, 114, 252
Hydraulic conductivity, 61, 63, 66–69
  effective, 204, 210
  principal directions of, 72
  tensor of, 72
  units of, 68
Hydraulic diffusivity (*see* Diffusivity in unsaturated flow)
Hydraulic gradient, 61

Hydraulics of wells (*see* Wells, hydraulics of)
Hydrodynamic dispersion, 15, 230
  boundary conditions of, 248–250
  coefficients of, 233, 239
    determination of, 272
  equation of, 239–248, 517
    in cartesian coordinates, 244
    in curvilinear coordinates, 244
    in cylindrical coordinates, 246–247
    in radial coordinates, 246-247
  occurrence of, 231
Hydrologic cycle, 19
Hydrologic map, 160–165
Hysteresis, 192, 200, 209, 211, 214

IADI (*see* Finite difference methods, IADI)
Identification:
  of aquifer parameters, 463–490
    by pumping tests, 465–482
  of model parameters, 439
    by linear programming, 484–489
    by quadratic programming, 489
    by regional methods, 482–490
Identification problem, 14, 15 (*see also* Identification)
Images, method of (*see* Method of images)
Imaginary wells (*see* Method of images)
Impervious formation (*see* Aquiclude)
Influence function, 496 (*see also* Technological matrix)
Influent stream, 34, 51
Inhomogeneity, 31–33
Initial conditions, 94, 96
Ink bottle effect, 200
Insular saturation, 193
Interface in coastal aquifer, 373–435
  approximation involved, 380
  continuity equation for, 386–393
  exact statement of problem, 381–383
  movement of, 406–414
  slope of, 384
  stationary, shape of, 393–406
  transition zone at, 414, 433
  in a two-layered aquifer, 396–399
Interfacial tension, 191
Intergranular stress (*see* Stress, intergranular)
Intermediate zone (*see* Vadose water)
Intrinsic permeability (*see* Permeability, intrinsic)
Intrinsic phase average (*see* Averages, intrinsic phase average)
Inverse problem, 463
  statement of, 463–465 (*see also* Identification)
Irreducible water content, 200, 210
Isotropy, 31–33
Iterative alternating direction implicit method (*see* Finite difference method, IADI)

Lagrangian approach, 90
Laminar flow, 65
Land subsidence, 184–189
Laplace equation, 93, 105
Leaching of irrigated soil, 42
Leakage of groundwater (*see* Groundwater, leakage of)
Leakage factor, 109, 339
Leaky aquifer, 26
  continuity equation for, 109, 115
Leibnitz rule, 70, 105, 522
Linear programming, 491
  application to aquifer management, 495–512
  decision variables, 492
  dual problem, 494
  objective function, 492
  review of, 492–495
Linearization, 114, 332
LSOR (*see* Finite difference methods, LSOR)

Macrodispersion, 257
Macrodispersivity, 257
Macroscopic level, 31
Macroscopic space, 513
Macroscopic velocity, 514 (*see also* Average velocity)
Management problem, 1 (*see also* Constraints in aquifer management)
Management, use of linear programming in, 491–512
Mass conservation equation, 83, 89
Material derivative, 185
Mathematical statement of problems:
  of hydrodynamic dispersion, 250–252
  of saturated flow, 95, 116
  of unsaturated flow, 218
Mechanical dispersion (*see* Dispersion, mechanical)
Method of images, 356–367
  image well, 480
Methods of solving the forecasting problem, 124
  analog methods, 125
  analytical methods, 124
  numerical methods, 128–152
Microscopic level, 31
Microscopic space, 513
Mining of groundwater, 9
Models (*see also* Models and analogs)
  of watershed, 38
Models and analogs, 125–128
  electrolytic tank, 128
  Hele–Shaw analog, 126, 408
  membrane analog, 126
  RC-network, 126
  sand box model, 126, 221

Models of aquifers:
  calibration of, 439 (*see also* Identification of model parameters)
  classification of, 440–442
  conceptual, 437
  criteria for selection, 438
  with distributed parameters, 441
  finite difference, 442
  finite element, 442
  with lumped parameters, 441
  mathematical, 437, 440
  mixing cells, 454
  multicell, 441, 447–455
  stability criterion for, 450
  need for, 437–439
  physical, 440
  polygonal cells, 451
  single cell, 441, 442–447
Moisture diffusivity (*see* Diffusivity in unsaturated flow)
Moisture distribution, 22
Molecular diffusion, 229
  coefficient of, 233, 235, 239
Monte Carlo technique, 460
Multicell aquifer, management of, 501–512

Natural replenishment, 34, 37–41
  synthetic sequence of, 40
Non-Darcy flow (*see* Flow at large Reynolds numbers)
Nonlaminar flow (*see* Flow at large Reynolds numbers)
Numerical dispersion, 261
Numerical methods. 128–152 (*see also* Finite difference methods *and* Finite element methods)

Objective functions, 11
  in linear programming, 492
One time reserve, 9 (*see also* Mining of groundwater)
Overshooting, 261

Partially penetrating well, drawdown in, 344–350
Peclet number, 232
Pellicular water, 22
Pendular ring, 192
Perched groundwater, 25, 26
Permeability, 63, 67
  effective, 204, 209–213
  Fair and Hatch formula, 67
  intrinsic, 67
  Kozeny–Carman equation, 67
  principal directions of, 72
  relative, 204, 209–213
  units of, 68

Phase average (*see* Averages, phase average)
Phreatic aquifer:
  continuity equation for, 110
    linearization of, 114
  definition of, 24
  stratified, 80–84
Phreatic surface, 22, 24
  conditions on, 98, 99
Piezometric head, 62
Policy, optimal, 436
Porosity, 30
  aerated, 204
  change in, 86
  drainable, 206
  effective, 63, 99, 204
Pressure head, 61
Pumping tests, 465–482
  in bounded aquifer, 479–481
  in confined aquifer, 467–471
    Chow's method, 467
    flowing well, 471
    Jacob's method, 468
    Kriz et al. method, 471
    recovery test, 470
    straight line method (*see* Pumping tests, Jacob's method)
    Theis method, 467
    type curve method (*see* Pumping tests, Theis method)
  for dispersion coefficients, 481
  in leaky aquifer, 474–477
  in phreatic aquifer, 471–474
  step-drawdown test, 477–479
  types of, 466

Radioactive decay, 239
Radius of influence, 306
  effective, 303
  empirical relations for, 306
  in leaky aquifer, 315
  in unsteady flow, 322
Raindrop effect, 200
RC-network (*see* Models and analogs, RC-network)
Recovery of water levels (*see* Drawdown, recovery of)
Refraction of streamlines, 102
Relative permeability, 208, 209–213
Representative elementary volume, 29, 30
Residual drawdown (*see* Drawdown, residual)
Resistance-capacitance analog (*see* Models and analogs, RC-network)
Retardation factor, 242, 266
Retention curve, 198
  scanning curves, 200
Return flow, 34, 41

Reynolds number, 65
  large Reynolds number, 66
River-aquifer relations, 51–53, 163

Salt water intrusion (*see* Sea water intrusion)
Scanning curves (*see* Retention curves, scanning curves)
Sea water encroachment (*see* Sea water intrusion)
Sea water intrusion, 15, 381
  control of, 394
Seepage face, 79
  conditions on, 99
  in wells, 309
Semipervious formation (*see* Aquitard)
Sharp front, 276–299
  approximation of, 277
  Muskat's approach, 279
SIP (*see* Finite difference methods, SIP)
Skin effect, 378
Small perturbations, method of, 425
Soil water zone, 23
Specific capacity of well (*see* Specific discharge of well)
Specific discharge, 61, 514
Specific discharge of well, 376
Specific retention, 88
Specific storativity, 14, 84–86
Specific yield, 88, 204 (*see also* Storativity of a phreatic aquifer)
  time dependence, 89
Springs, 1, 13, 53–57
  boundary conditions at, 117
  discharge of, 34
  models of, 55–56
  types of, 54
Stagnation point:
  for well in uniform flow, 283, 368, 373
Step-drawdown test (*see* Pumping tests, step-drawdown test)
Stiffness matrix, 500
Storage in semipervious layer, 109
Storativity, 14, 59, 86–89
  of a confined aquifer, 86, 107
  of a phreatic aquifer, 86
  specific, 14, 84–86
Stream function, 166
Stream tube, 165
Streamline, 165
Stress:
  effective, 84
  intergranular, 84
  total, 84
Subsidence, 6 (*see also* Land subsidence)
Subsurface water, 19
Successive approximations, method of 408
Suction, 194 (*see also* Capillary pressure head)

Superposition, 152–159
  in multiple well systems, 350–356
  principle of, 152
Surface tension, 191
Synthetic sequence, 40

Technological functions (*see* Technological matrix)
Technological matrix, 493, 496–499
Tensiometer, 196
Tension, 194 (*see also* Capillary pressure)
Theis solution, 320
Thiem equation, 306
Threshold pressure (*see* Bubbling pressure)
Tortuosity, 234, 235, 239
Tracer, 227
  ideal, 230
Transition zone at coastal interface, 433–435
Transmissivity of aquifer, 36, 69–70
Transport equation, derivation by averaging, 513–518

Uncertainty, 12
Unconfined aquifer (*see* Phreatic aquifer)
Unsaturated hydraulic conductivity (*see* Effective permeability)
Unsaturated zone, flow in, 15
Upconing, 414–425
  by Dupuit assumption, 421, 425
  by Muskat's approximation, 417–421
  by small pertubations, 425–433

Vadose water, 22
Void ratio, 186
Void space, 21
Volumetric fraction, 31

Water capacity, 209, 214
Water content, 191
  irreducible, 200, 210
Water divide, 3, 81, 117, 164
  for well near coast, 404
  for well in uniform flow, 284, 367
Water saturation, 191
Water table (*see* Phreatic surface)
Water table aquifer (*see* Phreatic aquifer)
Watershed, model of, 38
Wells:
  array of, 352
  hydraulics of, 15, 300–378
  losses in, 374–376
  near boundaries, 356–367, 480
  radius of, 302–303
  specific discharge of, 374–376
  systems of, 350–356

Well functions:
  for confined aquifer, 321
  for flowing well, 328
  for large diameter well, 329
  for leaky aquifer, 340
    with storage in semipervious layer, 344
  for phreatic aquifer, with delayed storage, 336
Well, partially penetrating (*see* Partially pene-
    trating well)

Well posed problem, 95
Wettability, 192

Yield of aquifer, 9, 12
Young's equation, 192

Zone of aeration, 22
Zone of saturation, 22